Unified Concepts in Applied Physics

EDWARD J. DIERAUF, JR.
City College of San Francisco

JAMES E. COURT
City College of San Francisco

Illustrated by Paul G. Hewitt

Prentice-Hall, Inc. Englewood Cliffs, New Jersey 07632

Library of Congress Cataloging in Publication Data

DIERAUF, EDWARD J (date)
Unified concepts in applied physics.

Includes index.
1. Force and energy. I. Court, James E.,
(date) joint author. II. Title.
QC73.D53 620.1 78-6305

ISBN 0-13-938753-6

10 9 8 7 6 5 4 3 2 1

Printed in the United States of America

PRENTICE-HALL INTERNATIONAL, INC., *London*
PRENTICE-HALL OF AUSTRALIA PTY. LIMITED, *Sydney*
PRENTICE-HALL OF CANADA, LTD., *Toronto*
PRENTICE-HALL OF INDIA PRIVATE LIMITED, *New Delhi*
PRENTICE-HALL OF JAPAN, INC., *Tokyo*
PRENTICE-HALL OF SOUTHEAST ASIA PTE. LTD., *Singapore*
WHITEHALL BOOKS LIMITED, *Wellington, New Zealand*

To Helen and Verity

Contents

3 REVIEW OF PHYSICS 27

4 FORCES 49

Preface

The idea of looking at physics in terms of concepts is not new. Any conscientious physics teacher will attempt to point out similarities between one topic and the next. But therein lies a problem. Things that turn out to have conceptual similarity and even identity are traditionally discussed only separately, in sequence, or even in separate courses.

In our treatment a concept is looked at in various physical systems to point up the similarities between mechanical, thermal, fluid, and electrical situations. Although we admit that, for example, water flowing in pipes is not in every way analogous to electric charge flowing in conductors, it turns out that remarkable and memorable similarities crop up both conceptually and mathematically from system to system. What is called resistance in electrical circuits has mechanical and fluid analogues in viscous friction and a thermal counterpart describing heat flow in conducting materials. When the right definitions are agreed upon, one unified concept of resistance applies to all systems. The student has one thing to learn in place of four. In the process he is all but forced to think of a less familiar thing such as thermal resistance in terms of the more commonly treated subject, electrical resistance. It is somewhat analogous to reading down a crossword puzzle in addition to reading across. New information is derived from the same data by looking at it differently. Such unified concepts form the basis for this beginning text in physics. Its purpose is to teach at an introductory level the physics of energy—its transfer and storage in engineering systems.

A number of engineers and scientists gathered at MIT in the 1930s to study the unified concept approach. Two of the group, Maurice Roney and Austin Fribance, worked with the Technical Education Research Center, Inc. (TERC) in Cambridge, Massachusetts, in the 1960s. They investigated the unified physics approach for students majoring in engineering technology. A number of laboratory workbooks were eventually published, but no solid text in unified physics was developed. Thus in 1972 we began work on this text to fill the vacuum.

The text, written primarily for engineering technology students, requires some knowledge of algebra and an ability to make simple technical calculations. Trigonometry is sometimes used, but there is no calculus. Some knowledge of physics is helpful but not absolutely needed.

Material in the text is adequate for two one-semester courses. We have found that formal study of the concepts can begin immediately in Chapter 4. Chapters 2 and 3 are used for reference when needed. Some apparently difficult beginning material about forces, torque, and work, which requires simple trigonometry, can be omitted without detracting from the development of the unified concept ideas. Understanding is enhanced by a concurrent introductory course in electrical circuits. A laboratory course illustrating the unified concepts is highly desirable. We have included a number of student-tested experiments in the instructor's manual.

The versatility of the unified concepts has been extended to the study of energy and its control in engineering systems. Information about this material, which covers load matching, amplification, feedback, and stability, can be requested by contacting the authors.

Several pages are needed to list the names of the many instructors and students who have taken the time to criticize and make constructive contributions over the years that we have labored. However, we want in particular to recognize Robert Angus, who helped us to get started.

San Francisco, California

EDWARD J. DIERAUF, JR.
JAMES E. COURT

1 Introduction

The objective of this textbook is to learn at an introductory level how energy is transferred and stored in engineering systems. If looked at in a special way, all these systems have things in common with respect to their energy content. These things are called unified concepts. Unified concepts stress the common relationships between systems. By studying one system, we may better understand how another system works. A unified-concept approach can make it easier to understand more complicated systems, which are mixtures of two or more systems. In accomplishing the objectives of this textbook, you will learn how to make simple technical calculations illustrating the transfer and storage of energy.

1-1 THE ENGINEERING SYSTEMS

The engineering systems that we will study are mechanical systems, fluid systems, electrical systems, and heat systems. These systems confine the energy to such things as masses, pipes, and wires. They are described very simply in Table 1-1.

TABLE 1-1 SIMPLE ENGINEERING SYSTEMS

Mechanical	
Translational systems—objects move in a straight line.	*Rotational systems*—objects move in circles about a fixed axis.
Fluid	
Hydraulic systems—movement of an incompressible fluid (a liquid).	*Pneumatic systems*—movement of a compressible fluid (a gas).
Electrical systems—movement of electrically charged particles.	*Heat systems*—movement of energy through molecular vibrations.

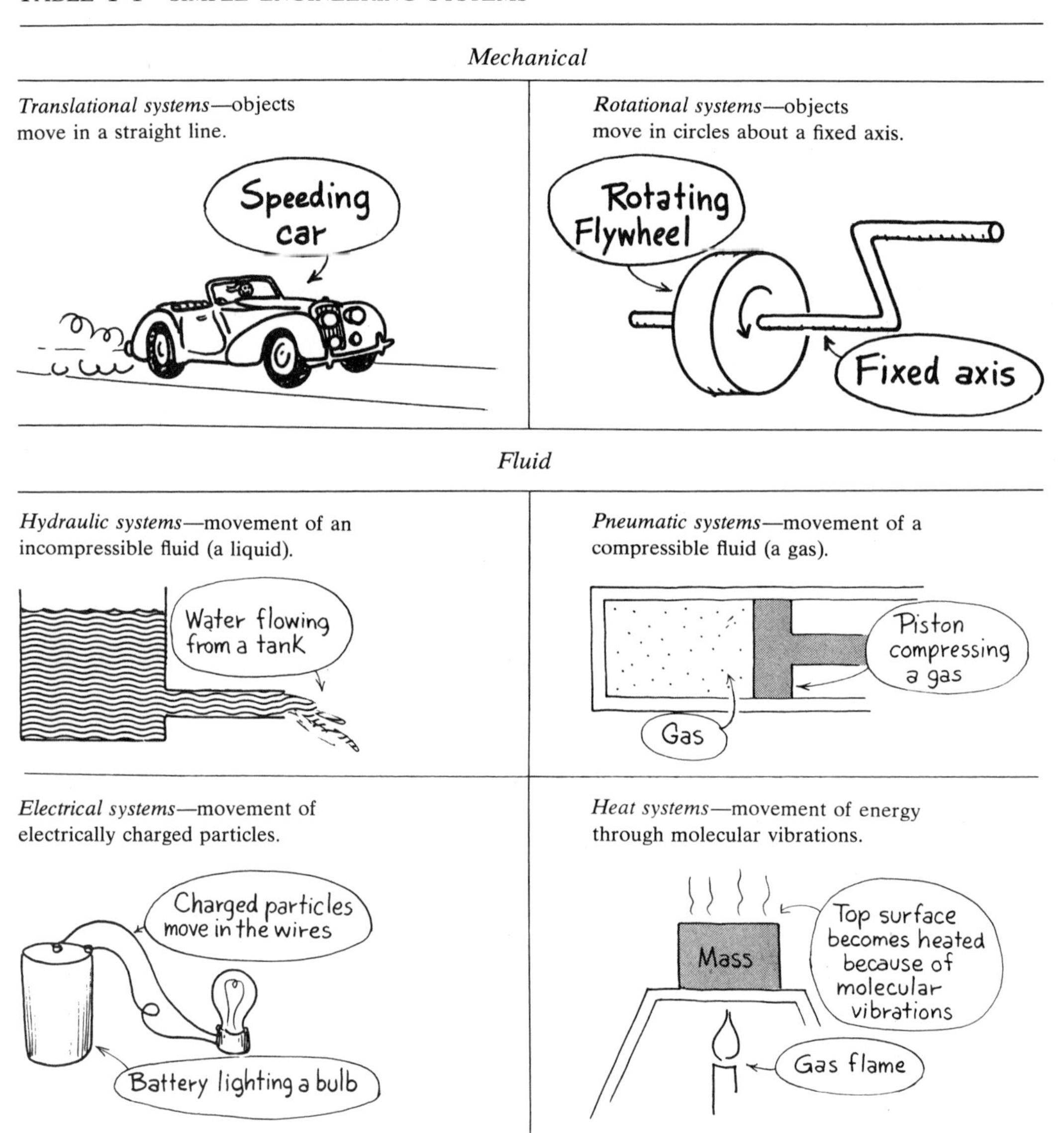

Magnetic systems (described in Chapter 8) can be compared to the other systems using unified concepts. The magnetic system is important because it interacts with the electrical system and serves as a link between electrical and mechanical systems.

Many engineering systems are combinations of the simple systems in Table 1-1. One example is a car engine. The engine has pistons and valves that move up and down in straight lines. It has gears and shafts that rotate about fixed axes. It has pumps, pipes, and hoses that direct the flow of liquids, such as water and gasoline, and gases, such as air and exhaust fumes. It has an electrical system with a battery, wires, and coils. The engine is a heat system in which heat energy is produced, transformed, and transferred.

To better understand individual systems and the interaction among systems as illustrated by the car engine, physical quantities that describe systems can be grouped into unified concepts.

1-2 THE UNIFIED CONCEPTS

The unified concepts are force, parameter, parameter rate, resistance, capacitance, inertance, time constant, impedance, and resonance. Table 1-2 shows in a very general way how the unified concepts relate to the transfer and storage of energy, and indicates the chapters in the book where the concepts are described.

TABLE 1-2 THE UNIFIED CONCEPTS

Concept	*Chapters*	*Relationship to Energy*
Force	4	Special definitions of physical quantities centered around energy
Parameter	5	
Parameter rate	6	
Resistance	7	Energy transfer
Capacitance	9	Energy storage
Inertance	10	
Time constant	11, 12, 13	Energy transfer and storage
Impedance	14, 15	Energy transfer and storage in oscillating systems
Resonance	16	

Force is responsible for energy movement in a system. Parameter and parameter rate are physical quantities that, together with force, define the energy and power in a system. Resistance relates to energy transfer. Capacitance and inertance relate to the storage of potential and kinetic energy, respectively. Time constants describe time delays in energy transfer and storage. Impedance relates to energy transfer and storage in an oscillating system. Resonance describes a situation in which a system oscillates at a natural frequency.

Many of these concepts are summarized in a table, *Unified Concepts*, on page 473 in Appendix A. Take a minute to look at this table. It is a good summary for most of the material in the text.

Our horizons are expanded when we look at unconfined energy movement, such as waves of sound and waves of electromagnetic radiation such as heat and light (Chapters 17 and 18).

You will find almost immediately that some concepts do not apply to all systems. An example of this is the definition of parameter in the heat system. In others you may have to stretch your imagination in talking about the common relationship of a concept in different systems. An example of this is the inertance of an electrical system. But the deviations are permissible if the goal is to gain an understanding of similarities between systems and to make it easier to understand the physics of these systems. Why learn several equations for power when maybe only one is necessary?

1-3 ORGANIZATION OF THE TEXT

This book is organized to show the common relationships between systems and to teach the physics of the systems centered around unified concepts. You will be amazed at how much traditional physics can be learned in this process.

Symbols used to define physical quantities appear in a table, *Physical Quantities and Their Symbols*, located in Appendix A. Another table, *Units of Measurement and Their Symbols*, is also in Appendix A. The units of measurement come primarily from two systems of units. These are the modern metric system, called the *Systeme Internationale* (SI), and the English system. Conversion between systems of units is facilitated using the *Table of Useful Conversion Factors* in Appendix A.

There are two appendices. Appendix A contains those tables already mentioned as well as a *Table of Physical Constants* and a *Table of Trigonometric Functions*. Appendix B lists technical data for a number of physical properties referred to in the text. There also is an *index* to help you find pages where a specific item is discussed.

Each chapter contains worked examples, and there are problems to be worked at the end of each chapter. The problems are graded in order of difficulty with the easier problems coming first. The answers to the odd-numbered problems appear at the end of each chapter.

Important equations are numbered and are often followed by a table describing the symbols and their units of measurement, usually in the SI and English systems of units. For example, the very important equation that work is the product of force and distance, appears as

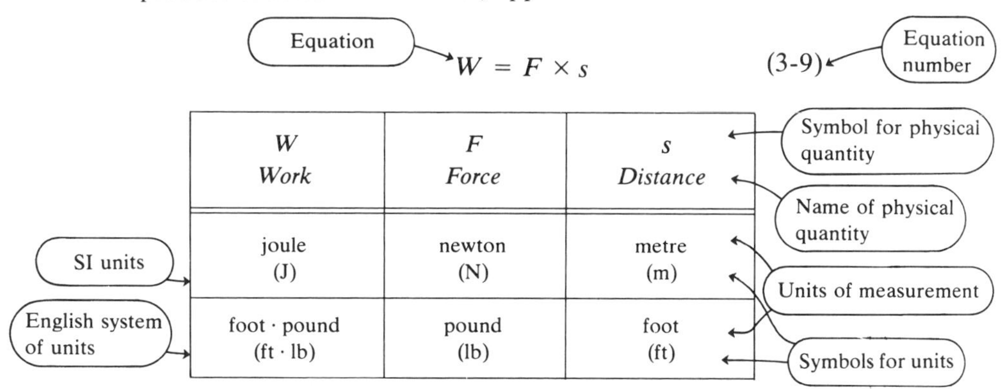

$$W = F \times s \qquad (3\text{-}9)$$

W	F	s
Work	*Force*	*Distance*
joule (J)	newton (N)	metre (m)
foot · pound (ft · lb)	pound (lb)	foot (ft)

Sections marked with an asterisk (*) may be more difficult to understand and can be passed over without losing the essential thread of the story being told. However, try reading these sections. They are not all that difficult and do contribute to a better understanding of the material. In general, the level of difficulty increases as you proceed into the text.

1-4 HOW MUCH SHOULD YOU KNOW BEFORE YOU START?

We hope you have the proper background in mathematics and physics to begin the study of the text. Our experience is that the hardest part is just plain getting started. The reader tends to become bewildered by the variety of terms, such as force, torque, pressure, degree Celsius, volt, distance, radian, cubic metre, coulomb, Btu, and joule, which appear quite quickly at the start.

You can assess how well prepared you are for all this by reading the two introductory chapters, Review of Mathematics (Chapter 2) and Review of Physics (Chapter 3). Take time to read these chapters and work the problems at each chapter's end. You don't have to know every detail in these chapters at first, but the more you know the easier you will find the material that follows. By doing this, *you* will find out where *you* stand with respect to the background in mathematics and physics needed to move ahead with confidence. And so—let's begin.

2

Review of Mathematics

The study of physics and engineering requires that you know how to make calculations. And—the calculations must be correct. One simply cannot advance the excuse that everything was okay in the calculation except the location of the decimal point when the bridge, whose strength depends on the calculation, comes crashing down.

The level of mathematics in this text is not severe. There is no calculus and very little trigonometry. You need some knowledge of algebra and geometry. You must be able to set up a simple equation, put in the numbers, and perform the calculation, getting the decimal point in the correct position. You must know how to convert from one unit to another, such as feet to inches or calories per second to watts. You must know how to draw and interpret graphs of technical data.

In addition, it is nice to know something about finding the unknown sides and angles of a right triangle using simple trigonometric relations, and to add and subtract vectors. That is what this chapter is all about. Take some time to read it over carefully. Work some of the problems at the end of the chapter. The rewards may be immense in terms of understanding the physics that follows.

2-1 CALCULATIONS USING SCIENTIFIC NOTATION

Many technical calculations involve very large and very small numbers. One problem is locating the correct position of the decimal point in the answer. For example it is not immediately obvious that $(2{,}100{,}000) \times (0.00030) = 630$. The task of locating the decimal point is made easier by putting the numbers of this calculation into scientific notation before multiplying them together.

Scientific notation form is a number (usually between 1 and 10) multiplied by some power of 10. Your first task in making a calculation is to convert the numbers involved into the scientific notation form. Some examples follow.

$$2900 = 2.9 \times 1000 = \boxed{2.9 \times 10^{3}}$$

$$380 = 3.8 \times 100 = \boxed{3.8 \times 10^{2}}$$

$$47 = 4.7 \times 10 = \boxed{4.7 \times 10^{1}}$$

$$5.6 = 5.6 \times 1 = \boxed{5.6 \times 10^{0}} \quad (10^{0} = 1 \text{ by definition})$$

$$0.65 = 6.5 \times 0.1 = \boxed{6.5 \times 10^{-1}}$$

$$0.074 = 7.4 \times 0.01 = \boxed{7.4 \times 10^{-2}}$$

$$0.0083 = 8.3 \times 0.001 = \boxed{8.3 \times 10^{-3}}$$

If two numbers written in scientific notation are *multiplied* together, the exponents (powers) *add*:

$$(2{,}100{,}000) \times (0.00030) = (2.1 \times 10^{6}) \times (3.0 \times 10^{-4})$$

$$= (2.1)(3.0) \times (10^{6} \times 10^{-4}) = 6.3 \times 10^{6+(-4)}$$

$$= \boxed{6.3 \times 10^{2}} \quad \text{or} \quad 630$$

If one number in scientific notation is *divided* by another number in scientific notation, the exponents of 10 *subtract*:

$$\frac{75{,}000}{30} = \frac{7.5 \times 10^{4}}{3.0 \times 10^{1}} = \frac{7.5}{3.0} \times 10^{4-(+1)}$$

$$= \boxed{2.5 \times 10^{3}}$$

Subtraction must take into account the sign of the exponent:

$$\frac{2100}{0.0030} = \frac{2.1 \times 10^3}{3.0 \times 10^{-3}} = \frac{2.1}{3.0} \times 10^{3-(-3)}$$

$$= 0.70 \times 10^{+6} = \boxed{7.0 \times 10^5}$$

If a number in scientific notation is *raised to a power*, the exponent of 10 is *multiplied* by the power:

$$(3000)^2 = (3.0 \times 10^3)^2 = (3.0)^2 \times (10^3)^2$$

$$= 9.0 \times 10^{3\times2} = \boxed{9.0 \times 10^6}$$

Beware of small numbers less than 1 raised to a power. They become even smaller!

$$(0.0002)^3 = (2.0 \times 10^{-4})^3 = (2.0)^3 \times (10^{-4})^3 = 8.0 \times 10^{-4\times3}$$

$$= \boxed{8.0 \times 10^{-12}}$$

The roots of numbers involve fractional powers. The square root of 9 ($\sqrt{9}$) can be written as $9^{1/2}$ since

$$9^{1/2} = (3 \times 3)^{1/2} = (3^2)^{1/2} = 3^{(2\times1/2)} = 3^1 = \boxed{3}$$

The cube root of 64 ($\sqrt[3]{64}$), which is 4, can be written as $64^{1/3}$, since

$$64^{1/3} = (4 \times 4 \times 4)^{1/3} = (4^3)^{1/3} = 4^{(3\times1/3)} = 4^1 = \boxed{4}$$

Fractional powers can help solve square root and cube root problems, such as

$$\sqrt{900} = 900^{1/2} = (9.0 \times 10^2)^{1/2} = (9.0)^{1/2} \times (10^2)^{1/2}$$

$$= 3.0 \times 10^{(2\times1/2)}$$

$$= \boxed{3.0 \times 10^1} \quad \text{or} \quad 30$$

Beware of the roots of small numbers less than 1. The roots become larger than the original number.

$$\sqrt[3]{0.000008} = (0.000008)^{1/3} = (8.0 \times 10^{-6})^{1/3} = (8.0)^{1/3} \times (10^{-6})^{1/3}$$

$$= 2.0 \times 10^{(-6\times1/3)}$$

$$= \boxed{2.0 \times 10^{-2}} \quad \text{or} \quad 0.020$$

The calculation of roots is simplified if the final exponent of 10 is an integer (whole number). The calculation of

$$\sqrt{0.0025} = (2.5 \times 10^{-3})^{1/2} = (2.5)^{1/2} \times 10^{-3/2}$$

gets us into trouble, because $10^{-3/2}$ is not readily determined. However,

$$\sqrt{0.0025} = (25 \times 10^{-4})^{1/2} = (25)^{1/2} \times (10^{-4})^{1/2} = \boxed{5.0 \times 10^{-2}} \quad \text{or} \quad 0.050$$

is easier because the exponent -2 of 10^{-2} is an integer.

Multiplication, division, powers, and roots may appear together in the same technical calculation. See if you can follow the examples below through to the final answer. Working some of the problems at the end of this chapter can give you additional practice in finding the decimal point location of an involved technical calculation.

$$\frac{(200)(12000)}{0.0030} = \frac{(2.0 \times 10^2)(1.2 \times 10^4)}{3.0 \times 10^{-3}} = \frac{2.0 \times 1.2}{3.0} \times \frac{10^2 \times 10^4}{10^{-3}}$$

$$= 0.80 \times 10^9 = \boxed{8.0 \times 10^8}$$

$$\frac{(1800)(0.0020)^3}{\sqrt{8100}} = \frac{(1.8 \times 10^3)(2.0 \times 10^{-3})^3}{(81 \times 10^2)^{1/2}} = \frac{(1.8 \times 10^3)(8.0 \times 10^{-9})}{9.0 \times 10^1}$$

$$= \frac{1.8 \times 8.0}{9.0} \times \frac{10^3 \times 10^{-9}}{10^1} = \boxed{1.6 \times 10^{-7}}$$

2-2 THE ACCURACY OF MEASUREMENTS AND CALCULATIONS

Most technical calculations involve measurements of physical quantities. The measurements are taken from instruments that are never perfectly accurate. The *accuracy* of a measurement is represented by the number of *significant figures*, which includes all the accurately known digits plus one uncertain digit.

In Fig. 2-1 the pointer and scale show two accurate digits, the 1 and the 3, plus an estimate of the pointer location between 13 and 14, say 0.4.

Figure 2-1. The reading of 13.4 has an accuracy of three significant figures.

The accuracy of the measurement in Fig. 2-2 is more difficult to determine. Suppose that we estimate the reading as 0.0017. What is the accuracy of the reading? Is it four digits? The answer is only two digits, and scientific

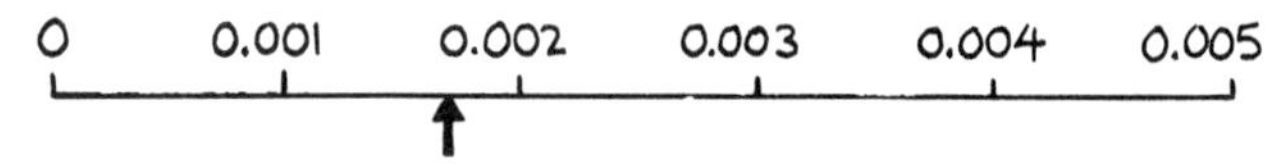

Figure 2-2. The reading of 0.0017 has an accuracy of only two significant figures.

notation helps to remove the doubt about the number of significant figures. We can imagine that the scale reads in scientific notation (Fig. 2-3). The reading is 1.7×10^{-3}, and the accuracy of two significant figures is more apparent.

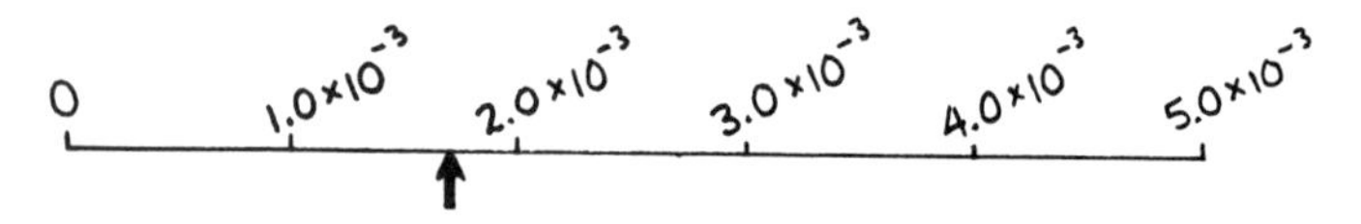

Figure 2-3. Scientific notation helps to indicate the number of significant figures in the reading.

Another scale and pointer position are shown in Fig. 2-4. The reading is somewhere between 5.74×10^4 and 5.76×10^4. A good estimate is 5.75×10^4. The accuracy of the reading is three significant figures.

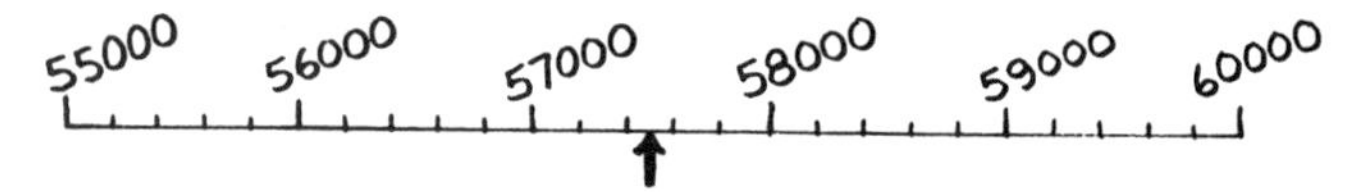

Figure 2-4. The reading 5.75×10^4 has an accuracy of three significant figures.

Calculations involving multiplication and division of measurements usually cannot be more accurate than the *least* accurate measurement in the calculation. Hence the product of the measurements 5.75×10^4 and 1.7×10^{-3} is

$$(5.75 \times 10^4)(1.7 \times 10^{-3}) = (5.75 \times 1.7)(10^4 \times 10^{-3})$$

$$= 9.775 \times 10^1 \rightarrow \boxed{98}$$

The accuracy of the answer 98 is two significant figures. It cannot be more accurate than the least accurate measurement of 1.7×10^{-3}, which has two significant figures.

In some calculations digits appear that are not measurements but are necessary to make the calculation correct. If the sides of a right triangle are 163 and 87 units, the area is

$$\tfrac{1}{2} \times 163 \times 87 = \tfrac{1}{2} \times (1.63 \times 10^2)(8.7 \times 10^1)$$

$$= (\tfrac{1}{2} \times 1.63 \times 8.7)(10^2 \times 10^1) = 7.0905 \times 10^3$$

$$\rightarrow \boxed{7.1 \times 10^3}$$

The answer's accuracy of two significant figures reflects the accuracy of the least accurate measurement, 87, and not the accuracy of the 1 or the 2 in the fraction $\frac{1}{2}$. The fraction $\frac{1}{2}$ is exactly $\frac{1}{2}$, or 0.500000 . . . , and is a part of the equation for area. It is not a measurement. A good practice is to avoid rounding off the digits of the calculation until the final answer is determined.

Calculations involving addition and subtraction of measurements do not follow the "least accurate measurement" rule. In the following examples, the arrow points to the digit that determines the cutoff point of the accuracy. Remember that the last significant figure contains some uncertainty, so it is meaningless to include digits farther to the right in the answer.

$$1.237 \times 10^2 + 8.654 \times 10^1 = \begin{array}{r} \overset{\downarrow}{123.7} \\ +\ 86.54 \\ \hline 210.24 \end{array} \rightarrow \boxed{210.2}$$

$$1.24 \times 10^3 + 3.6 \times 10^1 = \begin{array}{r} 12\overset{\downarrow}{4}0. \\ +\ 36. \\ \hline 1276. \end{array} \rightarrow 1280 = \boxed{1.28 \times 10^3}$$

$$97.6 - 95.8 = \begin{array}{r} 97.\overset{\downarrow}{6} \\ -95.8 \\ \hline \boxed{1.8} \end{array}$$

$$5.2 \times 10^3 - 1.3 = \begin{array}{r} 5\overset{\downarrow}{2}00 \\ -\quad 1.3 \\ \hline 5198.7 \end{array} \rightarrow \boxed{5.2 \times 10^3}$$

The use of electronic calculators makes it easier to locate decimal points, but many calculators do not give the accuracy of the answer. Don't show the answer from your calculator before checking its accuracy using scientific notation. If the diameter of a circle is 2.0, the circumference is *not* the answer from your calculator,

$$3.141592654 \times 2.0 = 6.283185307$$

It is 6.3.

2-3 CONVERTING THE UNITS OF A MEASUREMENT OR CALCULATION

A problem encountered in technical work is converting the units of a measurement or calculation. For example, the measurement of distance may be in inches, but the unit must be converted to feet to be used in a calculation.

Another example is converting the units of a calculation from English to SI. The power requirements of a pump may be in horsepower, but this may have to be converted into watts to find the power needed from an electric motor that drives the pump. The latter example has special implications for this text, which tries to show basic unifying concepts among various engineering systems.

Every unit has a symbol. Most of the symbols have been standardized by international agreement. All units and their symbols used in this text are in a table, *Units of Measurement and Their Symbols*, located in Appendix A. Please take a minute to look at this table and become familiar with its contents. Some examples of units and symbols from this table are as follows:

symbol	*unit*	*symbol*	*unit*	*symbol*	*unit*
ft	foot	s	second	hp	horsepower
in.	inch	min	minute	W	watt
oz	ounce	hr	hour	A	ampere
lb	pound	m	metre	mi	mile
				J	joule

These units will be used in the following examples of this section.

The conversion of the units of a measurement or calculation is made easy by treating units as if they were algebraic quantities that can be multiplied and divided. Suppose for example we wish to convert 18 inches (in.) to feet (ft). The relation between feet and inches is

$$1 \text{ ft} = 12 \text{ in.}$$

This can be written as a conversion factor in two forms: either

$$\frac{12 \text{ in.}}{1 \text{ ft}} \quad \text{or} \quad \frac{1 \text{ ft}}{12 \text{ in.}}$$

A conversion factor is multiplied by the quantity to be converted. The proper form will do two things: cancel out the unit we don't want, and replace it with the unit we do want. In this case we want to cancel out inches and replace it with feet. Which ratio is the right one?

$$18 \text{ in.} \times \frac{1 \text{ ft}}{12 \text{ in.}} \quad \text{or} \quad 18 \text{ in.} \times \frac{12 \text{ in.}}{1 \text{ ft}}$$

The first one is the clear choice:

$$18 \cancel{\text{in.}} \times \frac{1 \text{ ft}}{12 \cancel{\text{in.}}} = \boxed{1.5 \text{ ft}}$$

Here are a couple of additional examples. Convert 2.5 ounces (oz) to pounds (lb):

$$1 \text{ lb} = 16 \text{ oz}; \quad \text{conversion factor} = \frac{1 \text{ lb}}{16 \text{ oz}}$$

$$2.5 \cancel{\text{oz}} \times \frac{1 \text{ lb}}{16 \cancel{\text{oz}}} = \boxed{0.16 \text{ lb}}$$

Convert 2.4×10^2 seconds (s) to hours (hr):

$$1 \text{ hr} = 60 \text{ min}; \quad \text{conversion factor} = \frac{1 \text{ hr}}{60 \text{ min}}$$

$$1 \text{ min} = 60 \text{ s}; \quad \text{conversion factor} = \frac{1 \text{ min}}{60 \text{ s}}$$

$$(2.4 \times 10^2 \cancel{\text{s}}) \times \frac{1 \cancel{\text{min}}}{60 \cancel{\text{s}}} \times \frac{1 \text{ hr}}{60 \cancel{\text{min}}} = \boxed{6.7 \times 10^{-2} \text{ hr}}$$

Many conversion factors will not be easy to remember, especially those that relate quantities in different systems of units. Another table, *Useful Conversion Factors*, is printed in Appendix A. It should be referred to when making unit conversions such as the ones that follow.

Convert 65 feet (ft), a length, to metres (m):

$$1 \text{ m} = 3.28 \text{ ft}; \quad \text{conversion factor} = \frac{1 \text{ m}}{3.28 \text{ ft}}$$

$$65 \cancel{\text{ft}} \times \frac{1 \text{ m}}{3.28 \cancel{\text{ft}}} = \boxed{2.0 \times 10^1 \text{ m}}$$

Convert 147 newtons (N), a force, to pounds (lb):

$$1 \text{ N} = 0.225 \text{ lb}; \quad \text{conversion factor} = \frac{0.225 \text{ lb}}{1 \text{ N}}$$

$$147 \cancel{\text{N}} \times \frac{0.225 \text{ lb}}{1 \cancel{\text{N}}} = \boxed{33.1 \text{ lb}}$$

Convert 2.2 horsepower (hp) to watts (W), the SI unit of power:

$$1 \text{ hp} = 746 \text{ W}; \quad \text{conversion factor} = \frac{746 \text{ W}}{1 \text{ hp}}$$

$$2.2 \cancel{\text{hp}} \times \frac{746 \text{ W}}{1 \cancel{\text{hp}}} = \boxed{1.6 \times 10^3 \text{ W}}$$

Many measurements may require a decimal conversion before they can be used in a calculation. Examples are converting milliamperes (mA), a unit of electric current, to amperes (A), and nanoseconds (ns), a unit of time, to seconds (s). You should be familiar with the name, symbol, and meaning of the decimal prefixes. They are listed in a table, *Decimal Prefixes*, located in Appendix A. Refer to this table when making decimal conversions. Here are a couple of examples.

Convert 8.5 milliamperes (mA) to amperes (A):

$$1\ \text{mA} = 10^{-3}\ \text{A};\quad \text{conversion factor} = \frac{10^{-3}\ \text{A}}{1\ \text{mA}}$$

$$8.5\ \cancel{\text{mA}} \times \frac{10^{-3}\ \text{A}}{1\ \cancel{\text{mA}}} = \boxed{8.5 \times 10^{-3}\ \text{A}}$$

Convert 250 nanoseconds (ns) to seconds (s):

$$1\ \text{ns} = 10^{-9}\ \text{s};\quad \text{conversion factor} = \frac{10^{-9}\ \text{s}}{1\ \text{ns}}$$

$$250\ \cancel{\text{ns}} \times \frac{10^{-9}\ \text{s}}{1\ \cancel{\text{ns}}} = \boxed{250 \times 10^{-9}\ \text{s}}$$

A measurement or calculation may have more than one unit. An example is the speed of a car in miles per hour (mi/hr). Converting this measurement to feet per second (ft/s) poses no special problem. Just proceed as before except convert two units instead of one. For example, convert 60 miles per hour (mi/hr) to feet per second (ft/s):

$$1\ \text{mi} = 5280\ \text{ft};\quad \text{conversion factor} = \frac{5280\ \text{ft}}{1\ \text{mi}}$$

$$1\ \text{hr} = 3600\ \text{s};\quad \text{conversion factor} = \frac{1\ \text{hr}}{3600\ \text{s}}$$

$$60\ \cancel{\text{mi}}/\cancel{\text{hr}} \times \frac{5280\ \text{ft}}{1\ \cancel{\text{mi}}} \times \frac{1\ \cancel{\text{hr}}}{3600\ \text{s}} = \boxed{88\ \text{ft/s}}$$

Convert 300 British thermal units per minute (Btu/min) to joules per second (J/s):

$$1\ \text{J} = 9.49 \times 10^{-4}\ \text{Btu};\quad \text{conversion factor} = \frac{1\ \text{J}}{9.49 \times 10^{-4}\ \text{Btu}}$$

$$1\ \text{min} = 60\ \text{s};\quad \text{conversion factor} = \frac{1\ \text{min}}{60\ \text{s}}$$

$$300 \frac{\cancel{\text{Btu}}}{\cancel{\text{min}}} \times \frac{1\ \text{J}}{9.49 \times 10^{-4}\ \cancel{\text{Btu}}} \times \frac{1\ \cancel{\text{min}}}{60\ \text{s}} = \boxed{5.27 \times 10^{3}\ \text{J/s}}$$

The conversions are almost automatic. You don't even have to know anything about the meaning of the units as long as you use the correct conversion factors.

Sometimes the units of a measurement or calculation are squared or cubed, such as the units of area and volume. This again poses no special problem. Just remember to square or cube the conversion factor in the calculation. For example, convert 360 square inches (in.^{2}) to square feet (ft^2):

$$1 \text{ ft} = 12 \text{ in.}; \qquad 1^2 \text{ ft}^2 = 12^2 \text{ in.}^2; \qquad 1 \text{ ft}^2 = 144 \text{ in.}^2$$

$$360 \text{ in.}^2 \times \frac{1 \text{ ft}^2}{144 \text{ in.}^2} = \boxed{2.50 \text{ ft}^2}$$

Convert 1.5 cubic millimetres (mm^3) to cubic metres (m^3):

$$1 \text{ mm} = 10^{-3} \text{ m}; \qquad 1^3 \text{ mm}^3 = (10^{-3})^3 \text{ m}^3; \qquad 1 \text{ mm}^3 = 10^{-9} \text{ m}^3$$

$$1.5 \text{ mm}^3 \times \frac{10^{-9} \text{ m}^3}{1 \text{ mm}^3} = \boxed{1.5 \times 10^{-9} \text{ m}^3}$$

A number of unit conversion problems have been included at the end of this chapter. Working them will increase your skill and confidence in converting units.

2-4 CALCULATIONS INVOLVING SIMPLE GEOMETRIC OBJECTS

Many technical calculations draw upon relationships between the sides and angles of a right triangle. The sides a and b (Fig. 2-5) form the right angle and the side c, opposite the right angle, is called the hypotenuse.

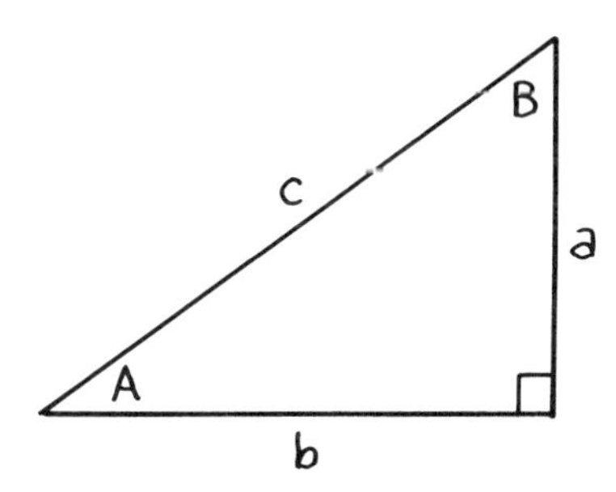

Figure 2-5. a and b are the sides of the right triangle. c is the hypotenuse.

There are a number of useful facts about the right triangle. One is that the sum of the angles A and B is equal to 90°.

$$A + B = 90°$$

Another useful fact is that the sum of the squares of the sides forming the right angle is equal to the square of the hypotenuse.

$$a^2 + b^2 = c^2$$

There are also useful trigonometric relations between the sides and angles of a right triangle. These are the sine (sin), cosine (cos), and tangent (tan).

The *sine* of an angle is the ratio of the side opposite the angle to the hypotenuse.

$$\sin A = \frac{a}{c}, \qquad \sin B = \frac{b}{c}$$

The *cosine* of an angle is the ratio of the side adjacent to the angle to the hypotenuse.

$$\cos A = \frac{b}{c}, \qquad \cos B = \frac{a}{c}$$

The *tangent* of an angle is the ratio of the side opposite the angle to the side adjacent to the angle.

$$\tan A = \frac{a}{b}, \qquad \tan B = \frac{b}{a}$$

The sines, cosines, and tangents of angles between 0 and 90° are presented in Appendix A.

Trigonometric relations are useful in solving for the unknown sides and angles of a right triangle. Two typical examples are presented below. In one example the hypotenuse and an angle are known; in the other, the two sides forming the right angle are known.

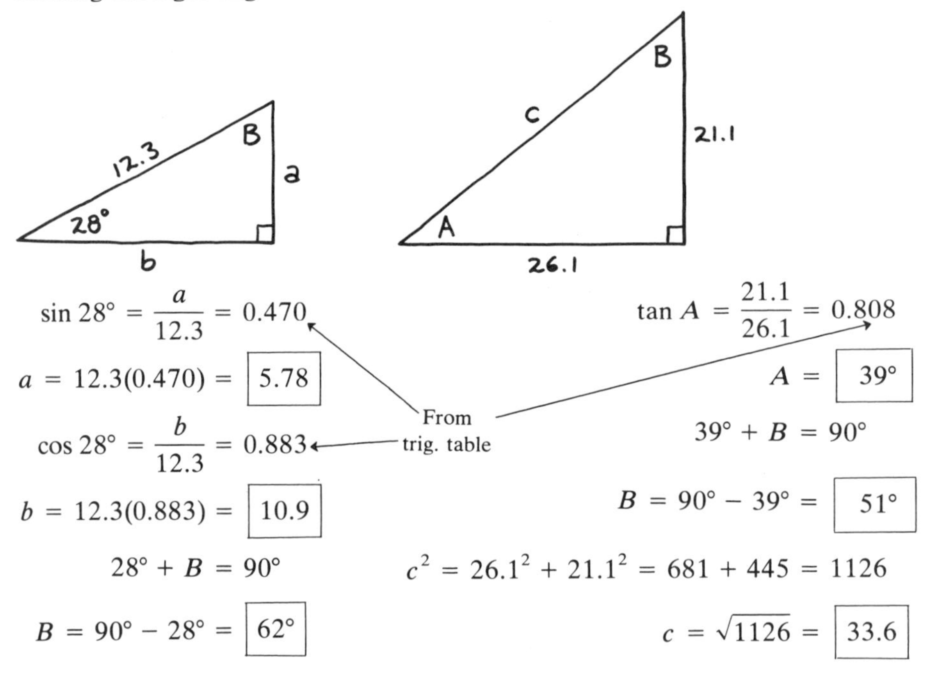

There are a number of simple geometric objects with perimeters, areas, and volumes that occur frequently in formula derivations and technical calculations. A few of the common objects and equations relating to them are shown in Table 2-1.

TABLE 2-1 COMMON OBJECTS AND RELATED EQUATIONS

Object	Equations
Circle 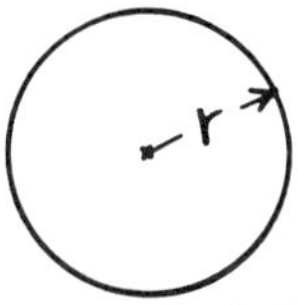	Perimeter (circumference) of the circle $s = 2\pi r$ Area of the circle $A = \pi r^2$
Rectangle 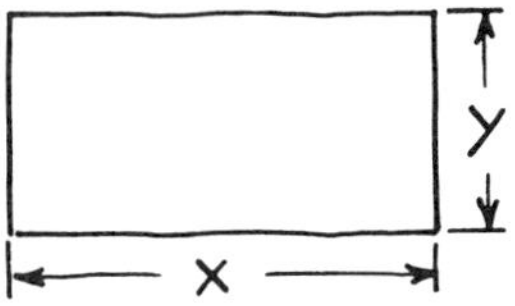	Area of the rectangle with base x and height y $A = xy$
Triangle 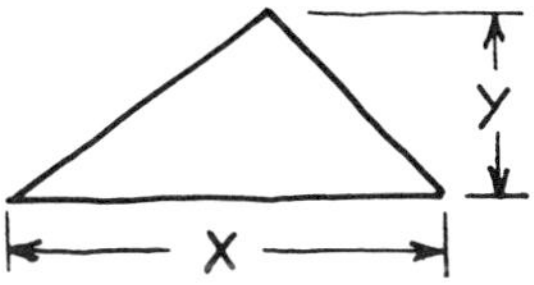	Area of the triangle $A = \frac{1}{2}xy$
Trapezoid 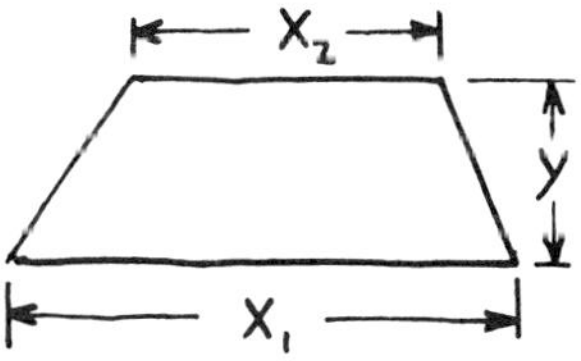	Area of the trapezoid $A = \frac{1}{2}(x_1 + x_2)y$
Box with perpendicular sides 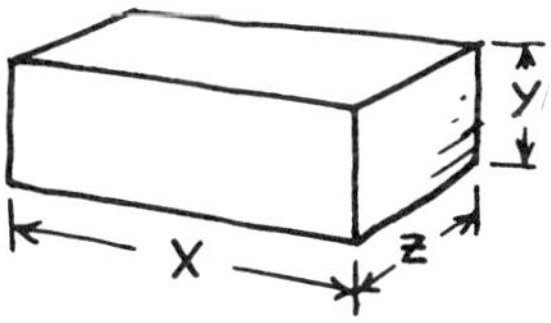	Volume of the box $V = xyz$
Cylinder 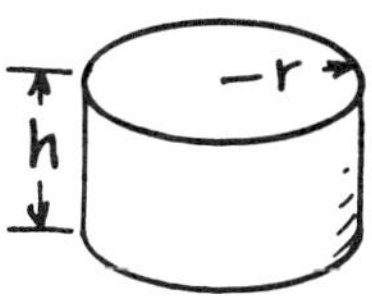	Volume of the cylinder $V = \pi r^2 h$
Sphere 	Surface area of the sphere $A = 4\pi r^2$ Volume of the sphere $V = \frac{4}{3}\pi r^3$

2-5 VECTOR ADDITION, SUBTRACTION, AND RESOLUTION

Many quantities in physics are defined completely by just specifying their magnitude (size). These quantities are called *scalars*. Examples of scalars are mass, volume, work, energy, and time. There are other quantities, called *vector* quantities, that need a direction as well as magnitude in order to be completely defined. Examples of vector quantities are displacement, force, velocity, acceleration, and momentum. Scalar quantities can be added, subtracted, multiplied, and divided using ordinary arithmetic. The same operations on vector quantities require a more elaborate treatment. This section shows how vectors are added and subtracted.

A vector is represented as an arrow. The length of the arrow is proportional to the vector's magnitude. The orientation of the arrow with respect to a reference direction and the location of the arrowhead specify the vector's direction. Two force vectors $\vec{A}$ and $\vec{B}$ are shown as arrows in Fig. 2-6. Vector

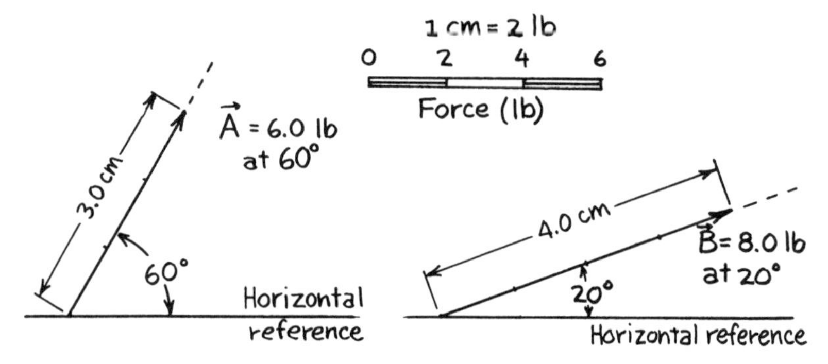

Figure 2-6. The arrows representing force vectors $\vec{A}$ and $\vec{B}$ are drawn to a scale of 1 cm = 2 lb. The arrows point in the direction of the forces.

$\vec{A}$ has a magnitude of 6.0 lb and a direction of 60° with respect to the horizontal reference. Vector $\vec{B}$ has a magnitude of 8.0 lb and a direction of 20° with respect to the same horizontal reference. A scale of 1.0 centimetre (cm) = 2.0 pound (lb) can be used in drawing the vector arrows.

Vector $\vec{B}$ can be added to $\vec{A}$ by setting up a parallelogram with vector arrows A and B as sides (Fig. 2-7). The diagonal represents the vector sum

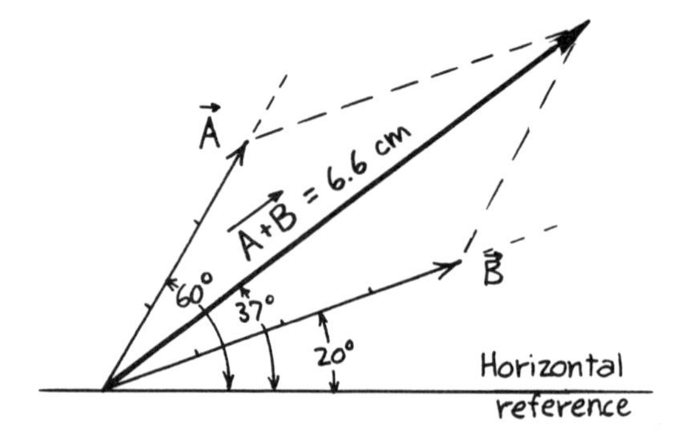

Figure 2-7. Vectors $\vec{A}$ and $\vec{B}$ are added by constructing a parallelogram. The diagonal represents the vector sum.

$\overrightarrow{A + B}$. The length of the diagonal represents the magnitude, and the orientation of the diagonal with respect to the reference direction is the direction of the vector sum. The vector sum is called the *resultant* of $\vec{A}$ and $\vec{B}$. Why don't you take a scale, protractor, and a piece of paper, lay off the two vectors, and add them together? See if you get the same answers for the resultant that we did. The length of the diagonal (6.6 cm) when multiplied by the scaling factor (1.0 cm = 2.0 lb) gives the magnitude.

$$6.6 \text{ cm} \times \frac{2.0 \text{ lb}}{1.0 \text{ cm}} = 13.2 \text{ lb} \rightarrow \boxed{13 \text{ lb}}$$

The vector sum $\overrightarrow{A + B}$ is 13 lb at an angle of 37° (using a protractor) with the reference.

Vector $\vec{B}$ can be subtracted from $\vec{A}$ by first turning $\vec{B}$ completely around so that it points in the opposite direction (Fig. 2-8). Vector B in this new

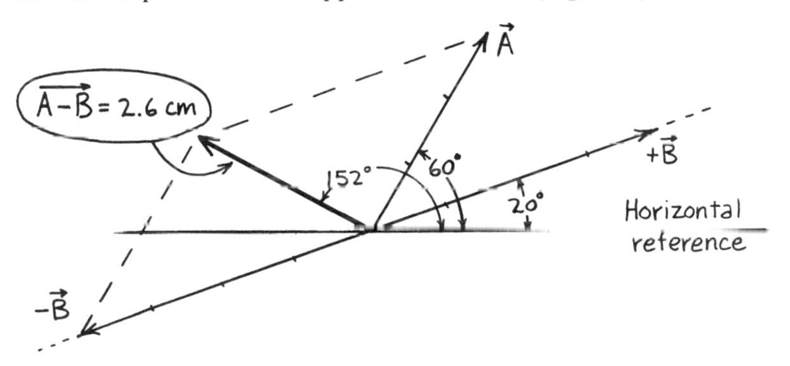

Figure 2-8. Vector $\vec{B}$ is subtracted from $\vec{A}$ by reversing $\vec{B}$ and adding to $\vec{A}$ using the parallelogram construction.

position is negative and can now be added as before to give a vector sum of $\vec{A} + (\overrightarrow{-B}) = \overrightarrow{A - B}$. The length of the diagonal (2.6 cm) when multiplied by the scaling factor gives the magnitude

$$2.6 \text{ cm} \times \frac{2.0 \text{ lb}}{1.0 \text{ cm}} = 5.2 \text{ lb}$$

The vector difference $\overrightarrow{A - B}$ is 5.2 lb at an angle of 152° (using a protractor) from the horizontal reference.

This idea of turning a vector around to subtract it is not completely new to you. Arithmetic addition and subtraction are really vector addition and subtraction along a single line. Suppose that the vectors $\vec{C} = 8$ and $\vec{D} = 6$ (Fig. 2–9) have the same direction. The vector sum $\overrightarrow{C + D} = 14$. The vector difference, $\overrightarrow{C - D} = 2$, is found by reversing the direction of $\vec{D}$ and adding.

A vector can be *resolved* into *components.* There are usually two components of interest, one horizontal and the other vertical. The vector sum of these two components equals the original vector. Another way of saying it is that the original vector can be replaced by its components. Figure 2-10 shows

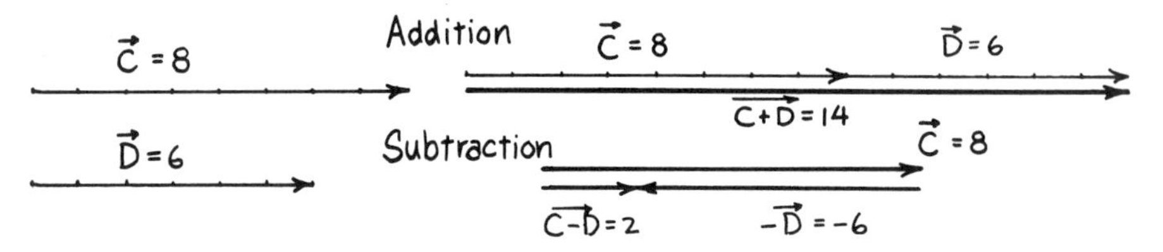

Figure 2-9. Arithmetic addition and subtraction are vector addition and subtraction along the same line.

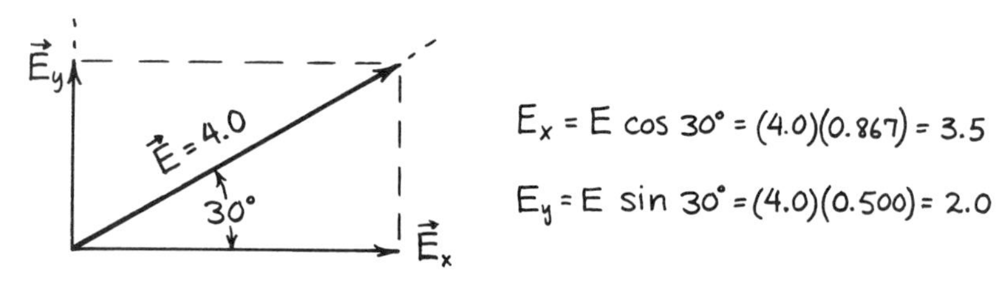

Figure 2-10. The vector $\vec{E}$ is resolved into (replaced by) horizontal component $\vec{E}_x$ and vertical component $\vec{E}_y$.

how the vector $\vec{E}$ has been resolved into a horizontal component $\vec{E}_x$ and a vertical component $\vec{E}_y$. The vector sum $\overrightarrow{E_x + E_y}$ is obviously the vector $\vec{E}$, the diagonal of the rectangle with sides $\vec{E}_x$ and $\vec{E}_y$. The magnitudes of $\vec{E}_x$ and $\vec{E}_y$ can be found graphically or by using the simple trigonometric relations indicated in Fig. 2-10.

*Vector Addition Using Trigonometry

Two vectors can be added together mathematically using trigonometry by resolving each vector into horizontal and vertical components. The horizontal components are then added together to give the horizontal component of the resultant. The same thing is done with the vertical components. Then the horizontal and vertical components of the resultant are added to get the resultant. This process is illustrated in Fig. 2-11 for vectors $\vec{A}$ and $\vec{B}$.

2-6 DRAWING AND INTERPRETING GRAPHS

Much of our technical information is presented in the form of graphs. We will continually refer to graphs in this text, looking at their shape, picking values from them, measuring tangents to them, and finding areas underneath them.

A graph visually presents the relationship between two changing quantities, for example distance and time. Suppose that the distance traveled by a moving object is recorded for different times. These data can be displayed on a graph. Scales for both axes are chosen so that they are decimal divisions that are easy to read. Each axis is labeled with the quantity being plotted, its symbol

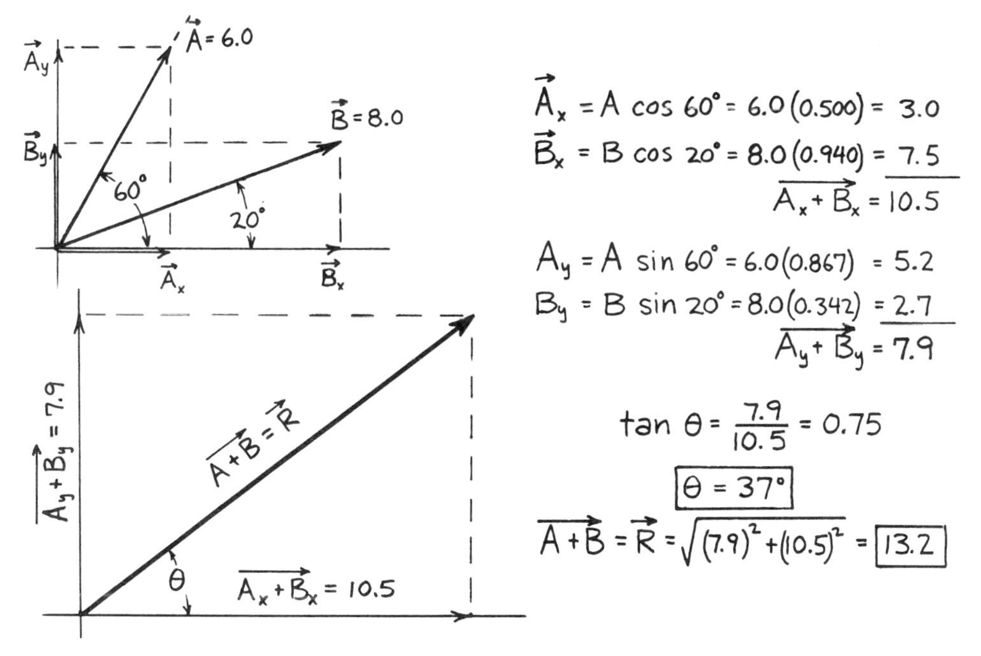

Figure 2-11. Each vector is first resolved into horizontal and vertical components. The components are added to get the components of the resultant $\vec{R}$.

and its units. The points on the curve are emphasized by placing small circles around them, and a smooth curve is drawn through the points (Fig. 2-12).

It is easy to pick off values from a well-plotted graph that has decimal divisions along each axis. For example, using Fig. 2-12, the distance is 46 m when the time is 9.6 s, and the time is 6.3 s when the distance is 20 m. It is more difficult to make such judgments when looking only at the data.

Data

Time (s)	Distance (m)
0	0.0
1	0.5
2	2.0
3	4.5
4	8.0
5	12.5
6	18.0
7	24.5
8	32.0
9	40.5
10	50.0

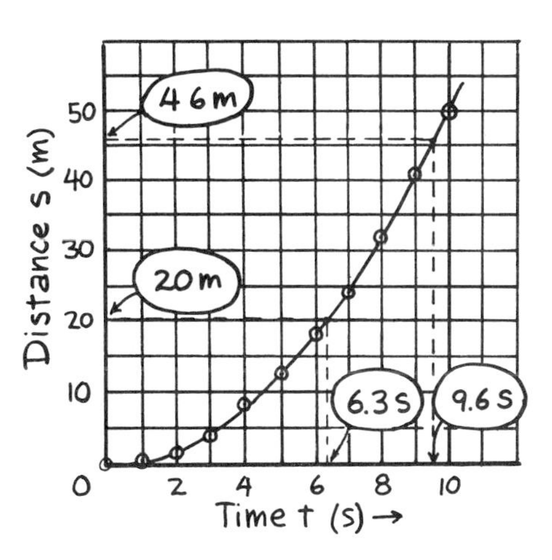

Figure 2-12. A graph of distance versus time.

The distance versus time graph is a true curve, not a straight line. Many graphs in scientific work are straight lines or approximately straight lines. When this happens, we say that there is a *linear* relation between the two plotted quantities.

*The Slope of a Curve

There are many situations where a line drawn tangent to a curve has important meaning. Figure 2-13 shows such a line, tangent to the graph of distance versus time when the time is 6.0 s. The line's position is estimated by

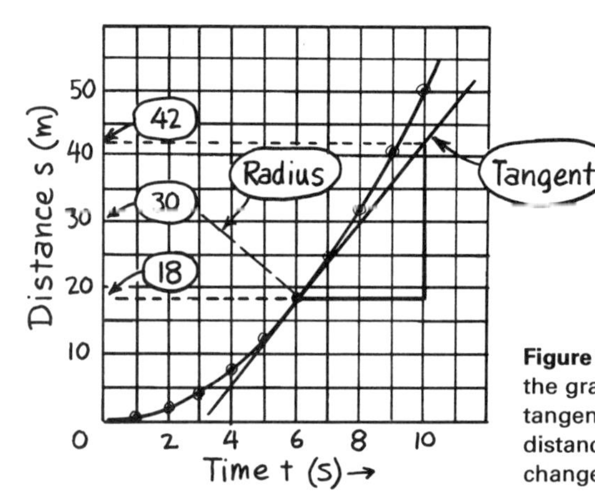

Figure 2-13. A line is drawn tangent to the graph at $t = 6.0$ s. The slope of the tangent line represents a change in distance divided by a corresponding change in time.

imagining that the curve in the immediate vicinity of the point is a portion of a large circle. The tangent is oriented to be perpendicular to the radius of this imaginary circle.

The term *slope* describes the orientation of the tangent line. The slope is defined as the ratio of the change in the quantity plotted vertically to the corresponding change in the quantity plotted horizontally. Thus the slope of the tangent at 6.0 s is

$$\frac{(30 - 18)\text{ m}}{(8.0 - 6.0)\text{ s}} = \frac{12\text{ m}}{2.0\text{ s}} = 6.0\text{ m/s} \quad \text{or} \quad \frac{(42 - 18)\text{ m}}{(10 - 6.0)\text{ s}} = \frac{24\text{ m}}{4.0\text{ s}} = 6.0\text{ m/s}$$

It makes no difference what the size of the changes are. It is the *ratio* of the changes that is important. The ratios are equal because of similar triangles. In practice, large changes are used to calculate the slope more accurately. We have already indicated that the slope of a curve may have important meaning. In this case the slope represents the velocity of the object when the time is 6.0 s.

Slopes can be either positive or negative. The best way of finding this out is by observing the orientation of the tangent. In Fig. 2-14a, the slope is positive because the tangent line extends upward to the right. In Fig. 2-14b,

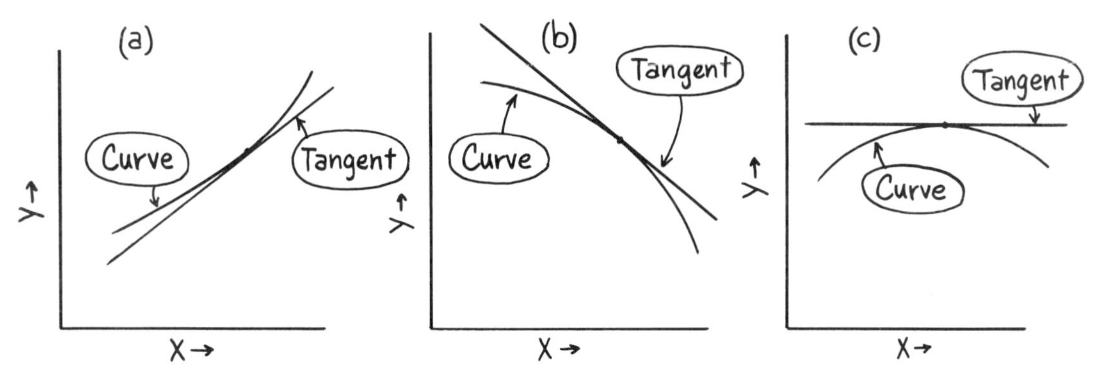

Figure 2-14. The orientation of the tangent determines the sign of the slope.

the slope is negative because the tangent extends upward to the left. Figure 2-14c shows a tangent that is horizontal. The slope is zero because a change in the horizontal quantity results in no change in the vertical quantity.

PROBLEMS

2-1. Convert the numbers from their decimal form to scientific notation.

(a) 0.00173 (b) 469,000
(c) 1080 (d) 0.0872

2-2. Convert the numbers from scientific notation to decimal form.

(a) 2.73×10^{-3} (b) 5.1×10^{4}
(c) 6.83×10^{2} (d) 4.26×10^{-6}

2-3. Perform the calculations and report your answers in scientific notation.

(a) 0.361×7200 (b) $\dfrac{1760 \times 0.0371}{834}$

(c) $\sqrt{0.176}$ (d) $(0.0082)^3$

2-4. Perform the calculations using scientific notation and report your answers in decimal form.

(a) $\dfrac{59{,}200}{0.00643}$ (b) $\dfrac{47.4 \times 6.284}{3600}$

(c) $(8000)^{1/3}$ (d) $(1640)^2$

2-5. Perform the calculations and report your answers in scientific notation to the required accuracy. Each number represents magnitude as well as accuracy.

(a) $\dfrac{471 \times 2.3}{0.000832}$ (b) $381.7 + 17.54$

2-6. Perform the calculations using scientific notation and report your answers in decimal form. Each number represents magnitude as well as accuracy.

(a) $\dfrac{0.0000162 \times 3476}{63}$

(b) $431 + 431 \times (0.00627)(2.3)$

2-7. (a) The maximum depth of a lake is 1760 ft. What is this depth in metres? (b) A person weighs 165 lb. What is the weight in newtons?

2-8. (a) The horsepower of an engine is 46. What is the engine's power rating in watts? (b) A chemical reaction releases 850 calories (cal) of heat energy. How many joules of energy does this represent?

2-9. The water pressure in a home is 75 lb/in.^2. Convert this water pressure to units of pounds/square foot and newtons/square metre.

2-10. (a) Convert 76 milligrams (mg) to kilograms (kg), 9.7 cm to metres, and 231 nanoseconds (ns) to seconds. (b) A speed limit is 88 ft/s. What is this speed limit in kilometres/hour?

2-11. (a) Convert 4.7 megohms (MΩ) to ohms (Ω), 6.3 kilocalories (kcal) to calories, and 5.2 picofarads (pF) to farads (F). (b) A litre of wine occupies how many cubic inches of volume? (A litre is the volume of a cube 10 cm on a side.)

2-12. (a) A room measures 25 ft by 40 ft. What is the area of the room in square metres? (b) A person can run a short distance at the rate of 10 yards/s. What is this rate in miles/hour?

2-13. Using Fig. 2-5 on page 15, find a, b, and B if $A = 68°$ and $c = 57.3$.

2-14. Using Fig. 2-5, find A, B, and c if $a = 7.39$ and $b = 34.7$.

2-15. What is the distance s across the river?

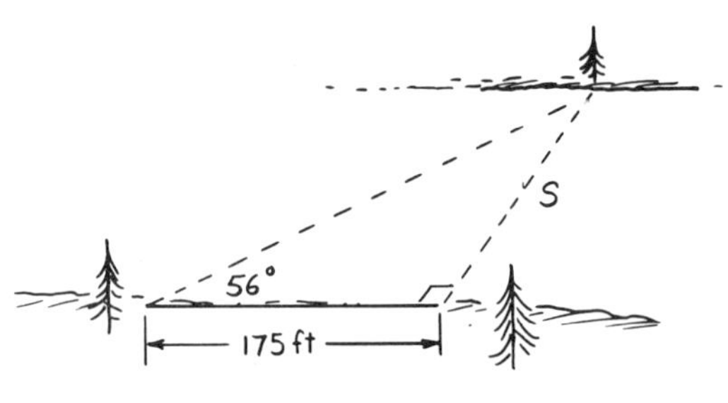

Problem 2-15

2-16. The base and height of a rectangle are 20.0 and 9.32 respectively. What is the angle formed by the rectangle's diagonal and its base?

2-17. What is the vertical height h of the roof, and what is the roof's total area?

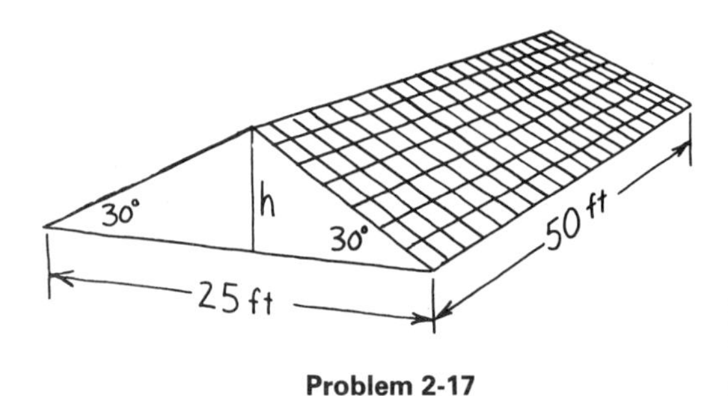

Problem 2-17

2-18. A building is 42 ft tall and casts a shadow that is 62 ft long. What is the angle within the shadow formed by the sun's rays and the ground?

2-19. What is the area under the force versus distance curve? What are the units of this area?

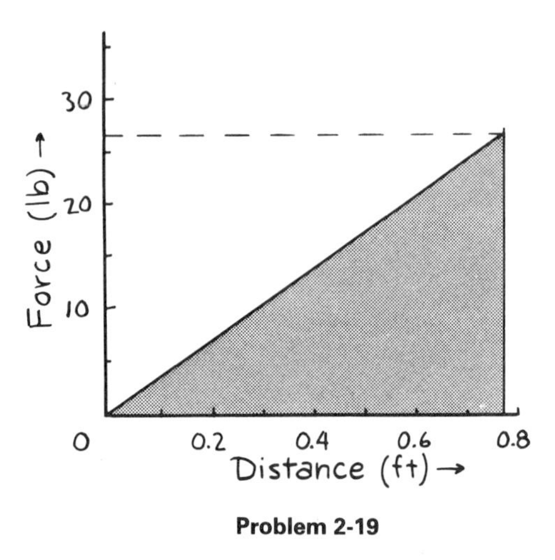

Problem 2-19

2-20. What is the area under the velocity versus time curve? What are the units of this area?

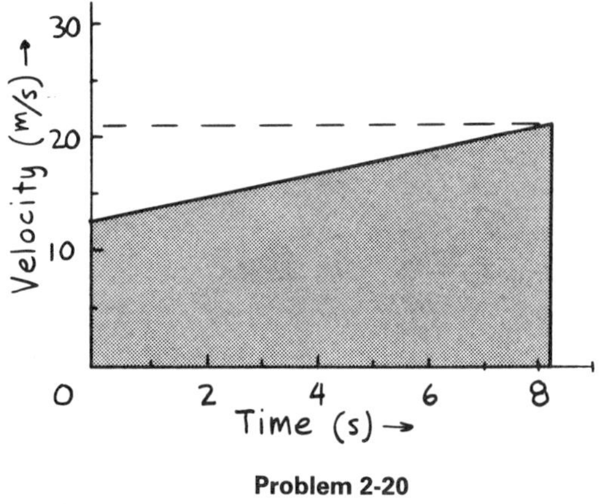

Problem 2-20

2-21. What is the volume of a small furnace in the form of a box with sides 4.87 in., 6.25 in., and 5.12 in.? What is the furnace's surface area?

2-22. How many cubic metres of water are contained within a pipe that has an inside diameter of 32 cm and a length of 5.3 m? What is the total surface area of the water? Include the area of the ends.

2-11. (a) Convert 4.7 megohms (MΩ) to ohms ings has a diameter of 1.5 ft. What is the volume of the sphere in cubic inches?

2-24. Add the force vectors graphically. Use a centimetre rule and a protractor. Then perform a vector subtraction $\overrightarrow{A - B}$ using the same technique. Be sure that your answers report both magnitude and direction.

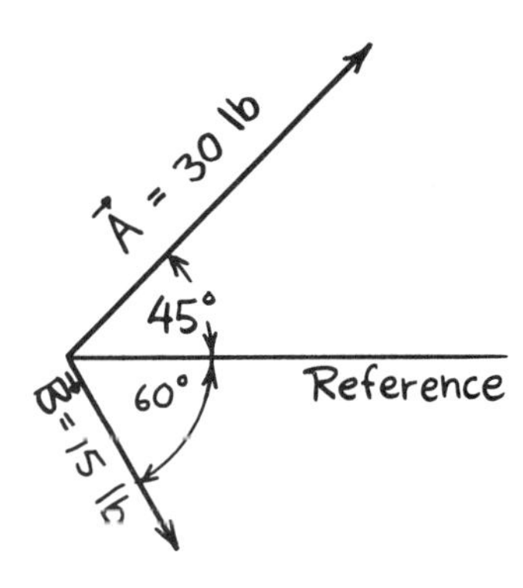

Problem 2-24

2-25. Add the velocity vectors $\vec{C}$ and $\vec{D}$ graphically. Use a centimetre rule and protractor. Then perform a vector subtraction $\overrightarrow{C - D}$ using the same technique. Be sure that your answers report both magnitude and direction.

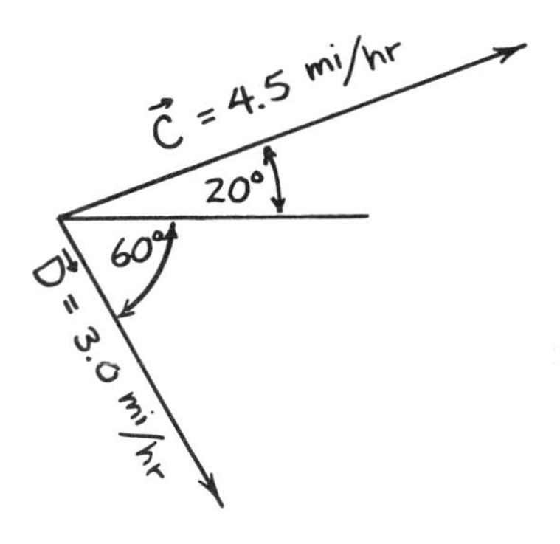

Problem 2-25

2-26. Resolve the force vector $\vec{E}$ into horizontal and vertical components. Do this using a graphical method. Verify your answers by calculating the components using trigonometry.

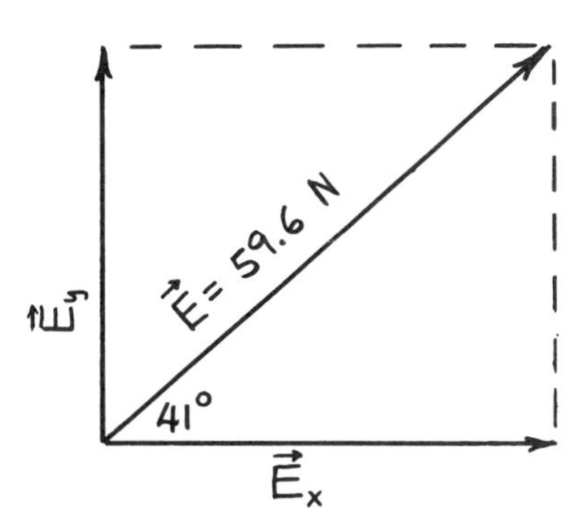

Problem 2-26

2-27. Make a plot of the volume flow rate ($\dot{V}$) versus pressure (p) data on linear graph paper. Plot the pressure on the horizontal axis. (a) What is the value of p when $\dot{V}$ is 0.70 ft^3/s? (b) What is the value of $\dot{V}$ when p is 27 lb/in.^{2}?

p (lb/in.^{2})	$\dot{V}$ (ft^3/s)
0	0
5.0	0.45
10	0.63
15	0.77
20	0.89
25	1.0
30	1.1
35	1.2

2-28. Plot the temperature T versus time t data on linear graph paper. Plot the temperature on the vertical axis. (a) What is the value of the temperature when the time is 1.7 s? (b) What is the value of the time when the temperature is 55°C?

t (s)	T (°C)
0	100.
0.5	60.6
1.0	36.7
1.5	22.2
2.0	13.5
2.5	8.2
3.0	5.0

2-29. Use a mathematical method to calculate the velocity vector sum $\overrightarrow{C + D}$ in Problem 2-25. Be sure that your answer reports both magnitude and direction.

2-30. Use a mathematical method to calculate the force vector sum $\overrightarrow{A + B}$ in Problem 2-24. Be sure that your answer reports both magnitude and direction.

2-31. What are the slopes of the pressure versus volume flow rate graph in Problem 2-27 when the pressure is 5.0 and 20 lb/in.^{2}?

2-32. What are the slopes of the temperature versus time graph in Problem 2-28 when the time is 1.0 and 2.0 s?

ANSWERS TO ODD-NUMBERED PROBLEMS

2-1. (a) 1.73×10^{-3}
(b) 4.69×10^{5}
(c) 1.08×10^{3}
(d) 8.72×10^{-2}

2-3. (a) 2.6×10^{3}
(b) 7.83×10^{-2}
(c) 4.20×10^{-1}
(d) 5.51×10^{-7}

2-5. (a) 1.3×10^{6}
(b) 3.992×10^{2}

2-7. (a) 537 m
(b) 733 N

2-9. 1.1×10^{4} lb/ft^2
5.2×10^{5} N/m^2

2-11. (a) $4.7 \times 10^{6}\ \Omega$
6.3×10^{3} cal
5.2×10^{-12} F
(b) 61 in.^{3}

2-13. $a = 53.1$
$b = 21.5$
$B = 22°$

2-15. $s = 2.6 \times 10^{2}$ ft

2-17. $h = 7.2$ ft
$A = 1.4 \times 10^{3}$ ft^2

2-19. $A = 10$ ft · lb

2-21. $V = 156$ in.^{3}
$A = 174.8$ in.^{2}

2-23. $V = 3.0 \times 10^{3}$ in.^{3}

2-25.

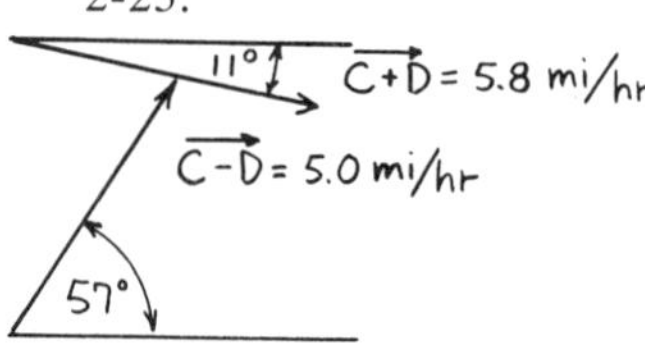

2-27. (a) $p = 12$ lb/in.^{2}
(b) $\dot{V} = 1.05$ ft^3/s

2-29.

5.73
1.06
10.5°
5.8 mi/hr

2-31. slope $= 0.045 \dfrac{\text{ft}^3/\text{s}}{\text{lb/in.}^2}$

slope $= 0.022 \dfrac{\text{ft}^3/\text{s}}{\text{lb/in.}^2}$

3

Review of Physics

This text is different from the standard beginning physics text. It involves only the portion of physics that deals with the flow and storage of energy in systems. You don't have to know a lot about these systems to start with. All we are asking is that you have some familiarity with the terms used to describe them. This will make it easier for you to study their similarities with respect to the movement of energy.

For example, have you heard of the terms force, mass, density, velocity, and acceleration used in mechanical systems? Have you ever heard of the terms pressure and viscosity used in fluid systems? Do the terms temperature, calorie, specific heat, and Btu used in heat systems ring a bell? What about the terms potential difference, resistance, current, and charge used in electrical systems? Do these terms bring back any memories? We presume that you do have some familiarity with these terms and the systems in which they are used.

The purpose of this chapter is to review those parts of a standard beginning physics course that we feel would be helpful to your understanding of what is to follow in this book. The remaining sections of this chapter should give you a base for judging just where you are with respect to the prior knowledge of physics needed to master the material in the book. Read these sections and try some of the simple problems at the chapter's end.

3-1 MOTION: DISTANCE AND TIME

Let's start with distance and time. Suppose that you are riding to work in a bus or car and you glance at the speedometer which indicates a steady speed or *velocity* of 30 miles/hour (mi/hr). If this velocity can be maintained for 1 hr, the odometer, which measures distance, will increase its reading from, say, 1234 to 1264 mi, a change of 30 mi (Fig. 3-1). This fits in with the definition of

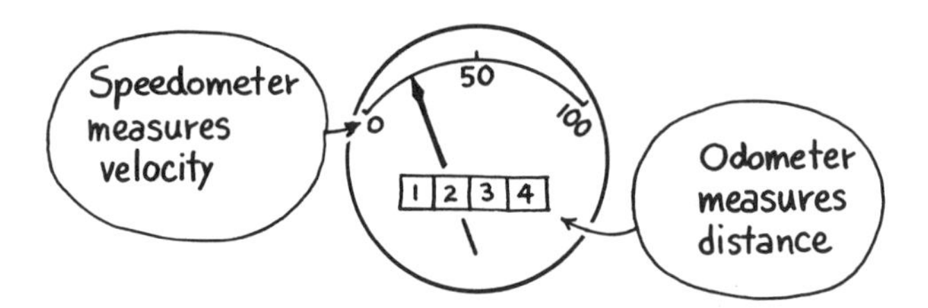

Figure 3-1. Speedometer and odometer. Instruments found on the dashboard of a car.

velocity. *Velocity* is defined as a change in distance divided by a corresponding change in time.

$$v = \frac{\Delta s}{\Delta t} \tag{3-1}$$

v *Velocity*	Δs *Change in Distance*	Δt *Change in Time*
Any combination of distance and time units	Any unit of distance	Any unit of time

The symbol Δ (Greek letter delta) denotes the change in a quantity. Read the symbol Δs as a change in distance and the symbol Δt as a change in time.

The velocity is 30 mi/hr if the car travels 30 mi in 1 hr. Velocity is a vector quantity. Its magnitude and direction (the direction of the motion) are needed to completely define it. The term *speed* is used to describe just the velocity's magnitude.

If the speedometer reading changes from one value of v to another, we say that the vehicle accelerates. *Acceleration* is the rate of change of velocity. It is defined as the change in velocity divided by the corresponding change in time.

$$a = \frac{\Delta v}{\Delta t} \tag{3-2}$$

a *Acceleration*	Δv *Change in Velocity*	Δt *Change in Time*
Any velocity unit divided by any time unit squared	Any unit of velocity	Any unit of time

Acceleration is also a vector quantity. Both the magnitude and direction (the direction of the change in velocity) are needed to completely define it.

3-2 FORCE AND MASS

Another thing that you notice when you accelerate is that a *force* is acting on you. You have to grab onto something if you are standing in the bus. If you are sitting down, you can feel the back of the seat pushing on you. *Force* is defined as a push or a pull.

Newton's Second Law

Force is always present when there is an acceleration. Rather we should say that a *net* force is present when there is an acceleration. There is no net force acting when you move at a constant velocity. Of course, there are forces present such as the force from the chair that pushes you up and the force of gravity pulling you down. These forces balance each other out and the net effect is zero. If the velocity is constant, there is no change in velocity with time, there is no acceleration, and there is no net force. If there is a change of velocity, there is an acceleration and there is a net force.

Mass is a measure of an object's opposition to change of velocity. A large mass requires more net force to accelerate it than a small mass.

The foregoing can be summarized in a famous law of physics. It is called *Newton's second law of motion* and is as follows:

$$F_{\text{net}} = ma \tag{3-3}$$

F_{net} *Net Force*	m *Mass*	a *Acceleration*
newton (N)	kilogram (kg)	$\frac{\text{metre/second}}{\text{second}}$ (m/s^2)
pound (lb)	slug (slug)	$\frac{\text{foot/second}}{\text{second}}$ (ft/s^2)

Two units in the table for Eq. (3-3) may not be familar. One is a unit of force, the newton; the other is a unit of mass, the slug. A net force of 1 newton (N) changes the velocity of a 1-kilogram (kg) mass by 1 metre (m)/second (s) during each second that the force acts. A net force of 1 pound (lb) changes the velocity of a 1-slug mass by 1 ft/s during each second that the force acts.

Acceleration of Gravity

A mass at rest has no net force acting on it. In Fig. 3-2 the person experiences two vertical forces. The earth pulls downward on the person with a force called his *weight.* This downward force of gravity is balanced by the

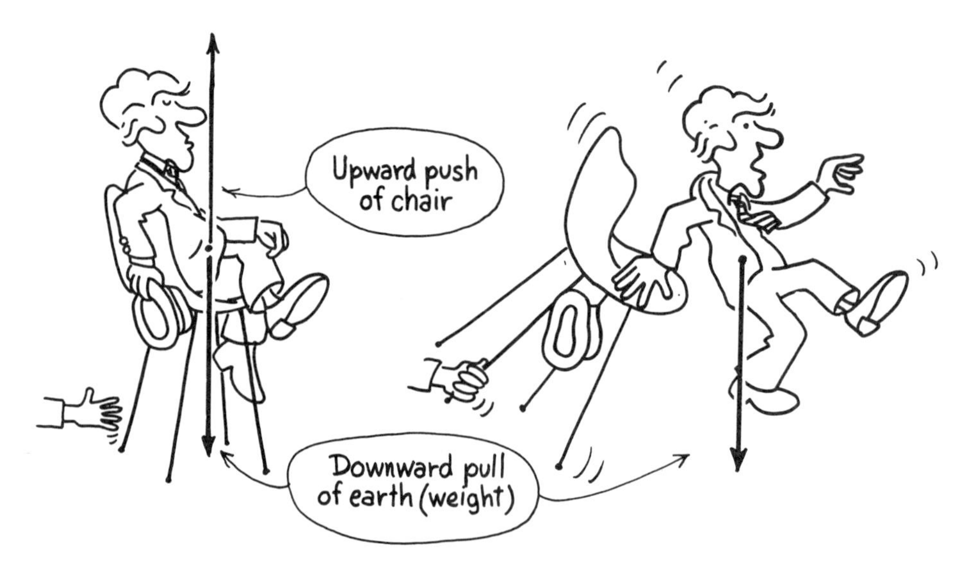

Figure 3-2. When the chair is removed, the person experiences a downward acceleration because of the unbalanced vertical force.

upward push of the chair. If the chair is suddenly removed, the person falls. His velocity changes and he accelerates. On earth at sea level this acceleration is

$$\text{acceleration of gravity } g = 32.2 \text{ ft/s}^2 = 9.81 \text{ m/s}^2 = 981 \text{ cm/s}^2$$

This is the acceleration of a freely falling body of any mass, big or small. From Newton's second law of motion, $F_{\text{net}} = ma$. But in this case the net force F_{net} is just the weight, and acceleration a is the acceleration of gravity g. Thus we get Eq. (3-4), a very important formula to calculate weight:

$$W = mg \tag{3-4}$$

W Weight	m Mass	g Acceleration of gravity
newton (N)	kilogram (kg)	$\frac{\text{metre/second}}{\text{second}}$ (m/s^2)
pound (lb)	slug (slug)	$\frac{\text{foot/second}}{\text{second}}$ (ft/s^2)

It is worth your while to try to get a "feel" for what each of these units is like. $9.81\ N = 1\ kg \times 9.81\ m/s^2$. $32.2\ lb = 1\ slug \times 32.2\ ft/s^2$. A mass of 1 slug weighs about 32 lb. A 1-kg mass weighs about 2.2 lb or 9.8 N. One newton is a little less than a quarter of a pound. Think of it as the weight of a cube of butter with one slice removed.

Grams (g) and kilograms (kg) should be familar units of mass. A kilogram is 1000 g. Water is used as a reference for mass:

One cubic centimetre of water at 4°C has a mass of 1 gram.

One cubic foot of water at 4°C weighs 62.4 pounds.

Density

The mass of 1 cm^3 of a substance can be compared to the mass of 1 cm^3 of water. For example, 11.3 cm^3 of water are needed to balance 1 cm^3 of lead (Fig. 3-3). The lead is said to be more dense. We define *density* as mass per unit volume.

$$\rho = \frac{m}{V} \tag{3-5}$$

ρ (*rho*) Density	m Mass	V Volume
kilogram/cubic metre (kg/m^3)	kilogram (kg)	cubic metre (m^3)
gram/cubic centimetre (g/cm^3)	gram (g)	cubic centimetre (cm^3)
slug/cubic foot ($slug/ft^3$)	slug (slug)	cubic foot (ft^3)

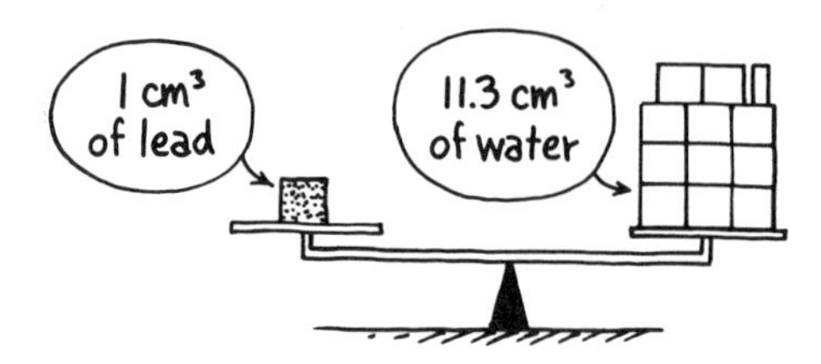

Figure 3-3. One cubic centimetre of lead balances 11.3 cubic centimetres of water.

In the above example the volume of lead is 1 cm^3 and the mass is 11.3 g, so the density of lead is 11.3 g/cm^3. Table 1 in Appendix B lists densities of various substances.

3-3 LIQUIDS, GASES, AND PRESSURE

A *solid* is a state of matter that tends to keep its shape if pulled or pushed by forces. Other common states of matter are *liquids* and *gases*, which are both classed as *fluids*. A liquid has a definite volume that takes the shape of its container, whereas a gas has neither an independent shape nor volume but tends to expand or contract to fit its container. Water exists in these three states of matter. As a solid, water is called ice; as a gas, it is called steam or water vapor. Water is a liquid at normal temperatures.

The term *pressure* is used to describe forces on fluids. Pressure p is a ratio of force F to area A. The force compresses the fluid and is always perpendicular to the area (Fig. 3-4). The pressure within a fluid is nondirectional. You

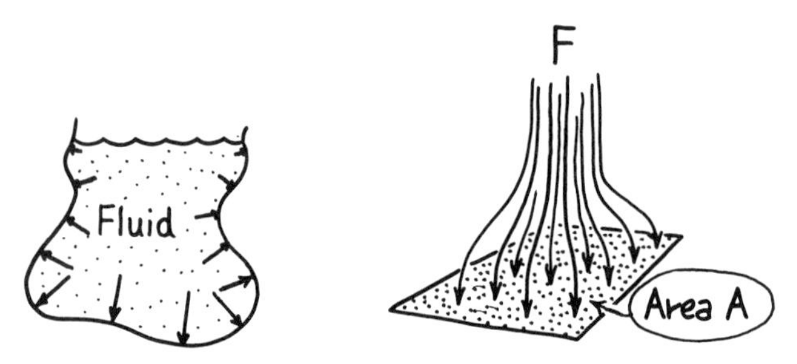

Figure 3-4. Any force exerted on or by a fluid is always perpendicular to the surface. The force is distributed over the surface—not exerted at a point.

will get the same reading if you rotate a pressure gauge to point it up, down, left, or right. The SI unit that corresponds to a pressure of 1 N/m^2 is a pascal (Pa).

$$p = \frac{F}{A} \tag{3-6}$$

p *Pressure*	F *Force*	A *Area*
pascal (Pa)	newton (N)	square metre (m^2)
pound/square foot (lb/ft^2)	pound (lb)	square foot (ft^2)

The pressure of the atmosphere at sea level is about 14.7 pounds/square inch (psi). We can think of this as being caused by a very high column of air that weighs 14.7 lb and which has a cross-sectional area of 1 $in.^2$.

The pressure on a small area at some depth in a liquid exposed to the atmosphere depends upon the weight of the column of liquid immediately above it and on the weight of the air in the atmosphere immediately above it. This pressure is called an *absolute pressure*, and is the pressure needed for most calculations involving fluids (Fig. 3-5). The pressure due to the weight of the

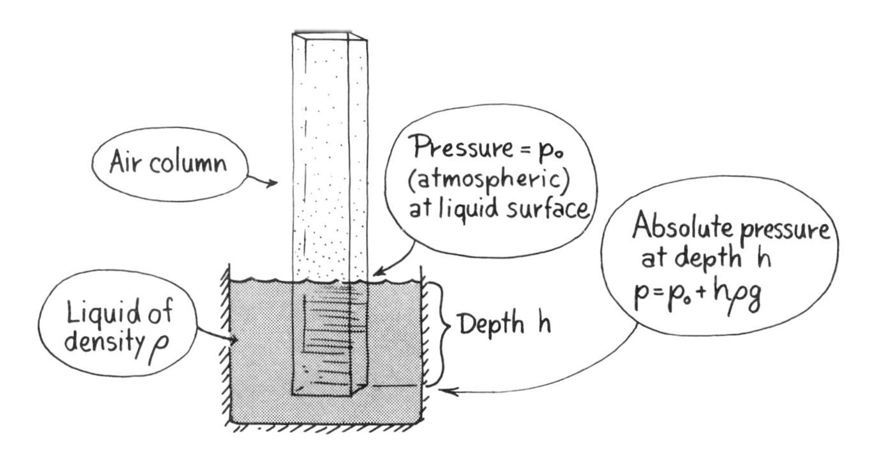

Figure 3-5. The pressure of the atmosphere must be added to the pressure from the liquid to get the absolute pressure at a point in the liquid.

liquid is proportional to the height of the liquid column or to the depth. This is easy to sense when swimming or diving. The pressure on the ears is greater the farther you get below the water surface. All this is summarized in Eq. (3-7):

$$p = p_0 + h\rho g \tag{3-7}$$

p *Absolute Pressure*	p_0 *Atmospheric Pressure*	h *Depth in Liquid*	ρ *Liquid Density*	g *Acceleration of Gravity*
pascal (Pa)		metre (m)	$\frac{\text{kilogram}}{\text{cubic metre}}$ (kg/m^3)	$\frac{\text{metre/second}}{\text{second}}$ (m/s^2)
pound/square foot (lb/ft^2)		foot (ft)	slug/cubic foot (slug/ft^3)	$\frac{\text{foot/second}}{\text{second}}$ (ft/s^2)

Pumps put pressure on liquids and gases in containers such as pipes and cylinders. Measured pressures usually do not account for atmospheric pressure. They are called *gauge pressures*. The absolute pressure p is found by adding atmospheric pressure p_0 to the gauge pressure p_g.

$$p = p_0 + p_g \tag{3-8}$$

p *Absolute Pressure*	p_0 *Atmospheric Pressure*	p_g *Gauge Pressure*
pascal (Pa)		
pound/square foot (lb/ft^2)		

The gauge pressure is the difference between the absolute pressure inside the pipe and the atmospheric pressure outside of the pipe (Fig. 3-6).

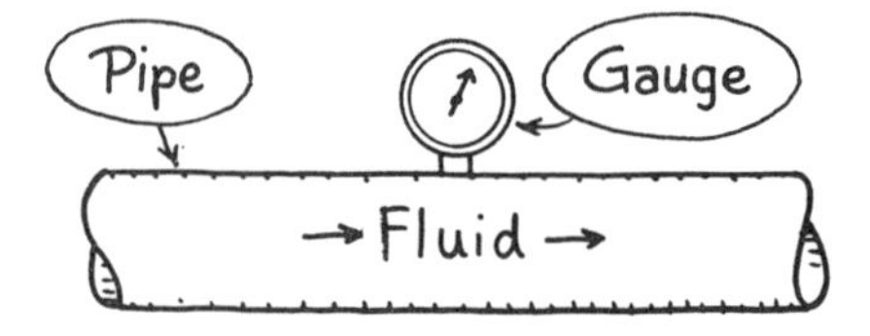

Figure 3-6. Gauge pressure is measured by a pressure gauge.

Boyle's Law

Pressures have a dramatic effect on changing the volume of a gas. In a very slowly performed experiment for which there is no temperature change, doubling the absolute pressure cuts the volume by one half (Fig. 3-7). Experiments show that, as long as the temperature of the gas remains constant and the volume changes are not too severe, the volume is inversely proportional to

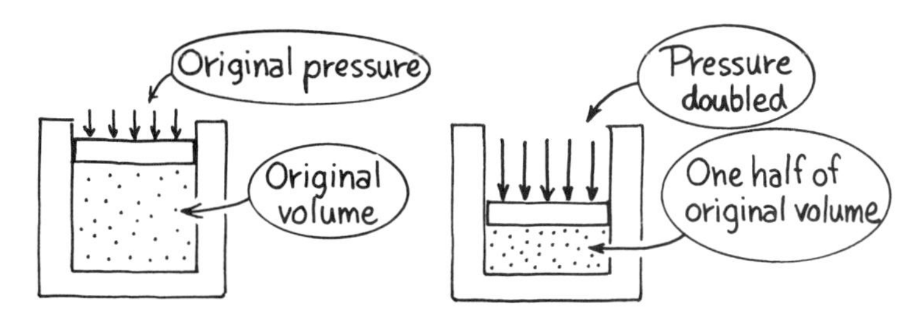

Figure 3-7. The volume is reduced by half when the absolute pressure doubles *if* the temperature of the gas does not change during compression.

absolute pressure. Another way of saying this is that the product of absolute pressure and volume remains constant.

$$\text{absolute pressure} \times \text{volume} = \text{constant}$$

This is called Boyle's law.

Hooke's Law

Pressures do not have much effect on compressing liquids and solids. But the effect, although small, can be measured. For many materials the amount of deformation (bend, twist, stretch, or change in volume) depends directly on the amount of applied stress, that is, on how hard the material is bent, twisted, stretched, or squeezed. When the stress is tensional, this relation is called Hooke's law. Small deformations of material are used to detect forces and pressures.

3-4 WORK AND ENERGY

Let us now go back and take another look at forces and distances. Suppose that the force in Fig. 3-8 moves the block from A to B. We say that the force does

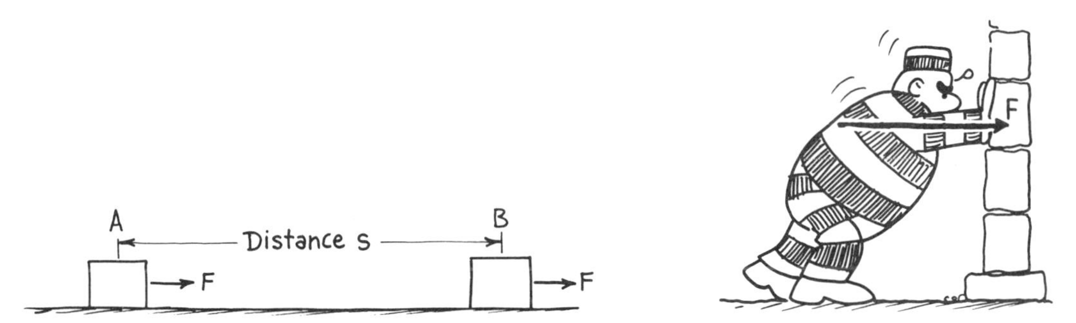

Figure 3-8. The force F does work in moving the block from A to B. This work W is defined as the product Fs. The man pushing against the wall does not do any work because the wall doesn't move.

work on the block, and arbitrarily define this work as

$$W = Fs \tag{3-9}$$

W Work	F Force	s Distance
newton · metre (N · m) = joule (J)	newton (N)	metre (m)
foot · pound (ft · lb)	pound (lb)	foot (ft)

The SI unit for work is the joule (J). One joule (J) is equivalent to 1 newton-metre (N · m) of work.

Power

Another interesting thing about the definition of work is that it does not involve time. The movement from A to B can occur in a second or a day or a million years. Regardless of how long it takes, the same work is done. *Power* is a term used to describe the rate of doing work. More power is needed to move the block in 1 s than in 1 day because the rate of doing work is greater when the time interval is smaller. The SI unit for power is the watt (W). One watt (W) is equivalent to 1 joule/second (J/s) of power.

$$P = \frac{W}{t} \tag{3-10}$$

P Power	W Work Done	t Elapsed Time
$\frac{\text{joule}}{\text{second}}$ (J/s) = watt (W)	joule (J)	second (s)
$\frac{\text{foot} \cdot \text{pound}}{\text{second}}$ (ft · lb/s)*	foot · pound (ft · lb)	

*550 ft · lb/s is 1 horsepower (hp).

Energy

We now look at three situations in which work is done by a force. In the first situation (Fig. 3-9) the force (equal to the weight of the block) moves the block upward. The work done by the lifting force is force × distance = weight × height, in this case.

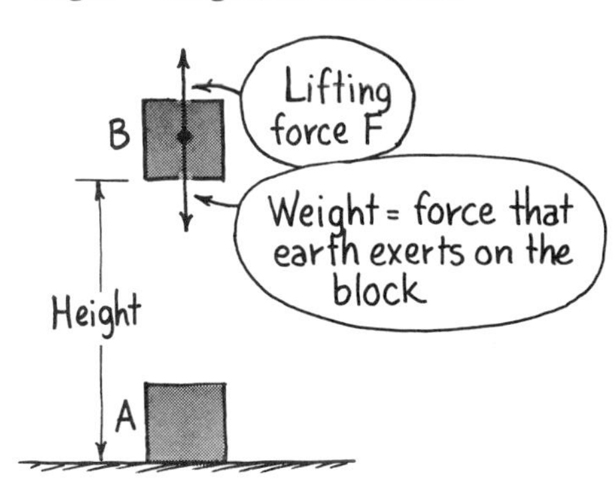

Figure 3-9. Work is done by the force in lifting the block from *A* to *B*. The work done is the product of the weight and the height.

Here we introduce the concept of *energy*. Energy is defined as the *ability to do work*. The transfer and storage of energy is what this book is all about. Energy exists in batteries as chemical energy, which can be converted into electrical energy, which in turn can be converted into heat energy or energy in a beam of light or energy in the sound waves of a radio or the mechanical energy of a motor.

The block at the higher position *B* has a greater ability to do work than in the lower position *A*. We intuitively feel this if the block is heavy and we are standing underneath it. The block at *B* in Fig. 3-10 has more energy because of the work done by the lifting force. The *change* in energy from *A* to *B* is equal to the work done by the force.

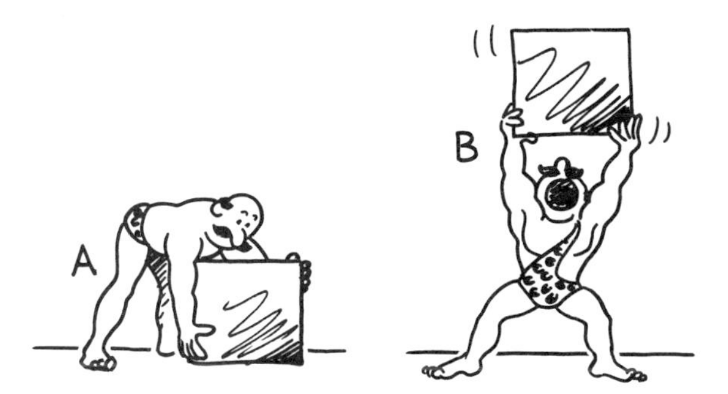

Figure 3-10. The block at *B* has more energy than at *A*. This is because work was done in lifting the block.

This does not mean that the energy at A is zero. Remember, it is possible to be standing under position A, too. This illustrates that a position or reference is needed for measuring energy changes. Energy stored by virtue of position is called *potential energy*. In this particular case,

$$E_P = Wh \tag{3-11}$$

E_P *Potential Energy*	W *Weight*	h *Height*
joule (J)	newton (N)	metre (m)
foot · pound (ft · lb)	pound (lb)	foot (ft)

Now consider two other situations where the block moves horizontally rather than vertically. First, the block, starting from rest, is pulled over a very smooth surface (Fig. 3-11). This time the force accelerates the block. The block

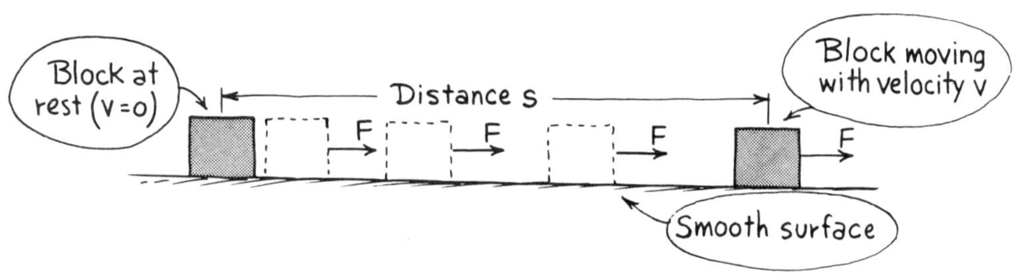

Figure 3-11. The force F accelerates the block starting from rest at A to a velocity v at B. In the process the force does work (force × distance).

gains velocity. The force does work (force × distance) in changing the velocity of the block. The block with a higher velocity has greater ability to do work. We intuitively know this if a car approaches while we are crossing the street (Fig. 3-12). The block with its velocity has more energy because of the work

Figure 3-12. It is easy to believe that a moving car has more energy than a car at rest.

done by the force. Physicists call the energy stored by virtue of the motion of the block *kinetic energy*. Using Newton's second law (Eq. 3-3), it can be shown that the kinetic energy depends on the mass of the block and the square of its velocity.

$$E_K = \tfrac{1}{2}mv^2 \tag{3-12}$$

E_K Kinetic Energy	m Mass	v Velocity
joule (J)	kilogram (kg)	metre/second (m/s)
foot · pound (ft · lb)	slug (slug)	foot/second (ft/s)

In the third situation (Fig. 3-13), the block moves horizontally over a rough surface. Friction between the block and the surface prevents any acceleration, and the block just bumps along at constant velocity. The force

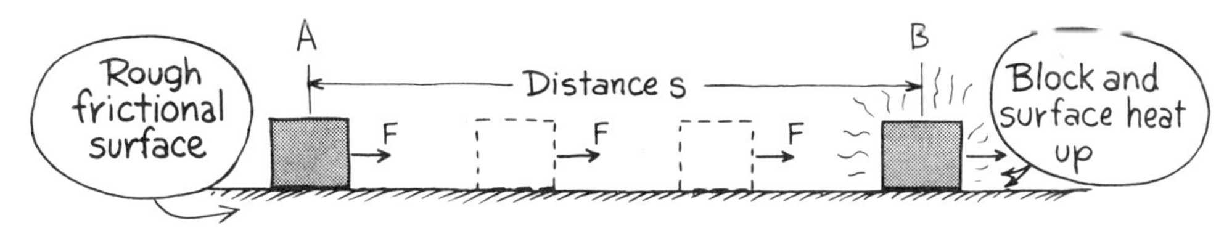

Figure 3-13. The force does work in moving the block from *A* to *B* at constant speed. The work done appears as heat energy.

still does work on the block. This work does not appear as potential energy or kinetic energy. The block and the surface do, however, heat up. They get warm. There is a *temperature* change. *Heat* is another form of energy. In this case, the heat is energy converted from mechanical work.

Physicists have discovered that the concept of energy helps to explain a lot of physical observations in our world if the assumption is made that work and energy in all their forms must be accounted for. This accounting is called the *law of conservation of energy*. Any work done on an object, whether it be a block or a molecule or an electron, must appear as energy in some form. Potential energy, kinetic energy, and heat energy are a few of energy's forms. The work done on the block of Fig. 3–13 did not show up as a change of potential or kinetic energy. It appeared as heat. But what is heat?

3-5 HEAT AND TEMPERATURE

Heat is considered as a different form of energy, although it is actually the kinetic energies of vibration of all the molecules in an object plus some energy that binds the molecules together. In a solid, each molecule vibrates back and forth; the more violent the vibration, the greater the heat energy. In a gas, the molecules are free to move. They travel in straight lines, only changing direction if they collide with another molecule or hit the sides of their container. The collisions of molecules against the sides of their container account for the pressure of the gas.

Temperature

Heat energy is detected by temperature, a measure of relative hotness or coldness. Temperature, however, is actually a measure of the average molecular kinetic energy of an object. Two common temperature scales are the Fahrenheit and Celsius scales. The temperature of melting ice is zero degrees Celsius (°C), and the temperature of boiling water at normal atmospheric pressure is 100°C. The corresponding temperatures on the Fahrenheit scale are 32°F and 212°F. The name *centigrade*, which has been widely used, is the same as Celsius. Celsius is now the accepted term. You might also need to know that one Celsius degree is 1.8 or 9/5 times the size of a Fahrenheit degree.

Suppose that we have a gas inside a cylinder and carefully measure its volume at the ice-bath temperature of 0°C or 32°F. If the cylinder is heated and the *same* pressure is maintained on the piston, the gas expands. The gas occupies a larger volume. This expansion of matter with a temperature increase is common to most materials. The "cold" volume and the "hot" volume with their corresponding temperatures are plotted in Fig. 3-14 on a volume versus temperature graph. A line connecting the two points is a very good indicator of what volume to expect for intermediate temperatures between 0 and 100°C. If we extend (extrapolate) this line backward, we find that it intersects the temperature axis at −273°C. This represents the temperature of the gas when it has no volume at all. Physicists reason that this would be the temperature at which there is no kinetic energy of motion of the molecules. They are all at rest, occupying very little space. Such a temperature is called *absolute zero*.

Two temperature scales, the *Kelvin* and the *Rankine*, use absolute zero as the starting point for measuring temperature. They are called absolute temperature scales. One kelvin (K) represents the same temperature change as a Celsius degree. One Rankine degree (°R) represents the same temperature change as a Fahrenheit degree. The absolute temperature scales are used in making calculations on gases.

Equivalence of Mass and Energy

Matter at zero kelvin (0 K) is characterized as having no kinetic energy of motion of its molecules. It has no heat energy. Actually, there is still

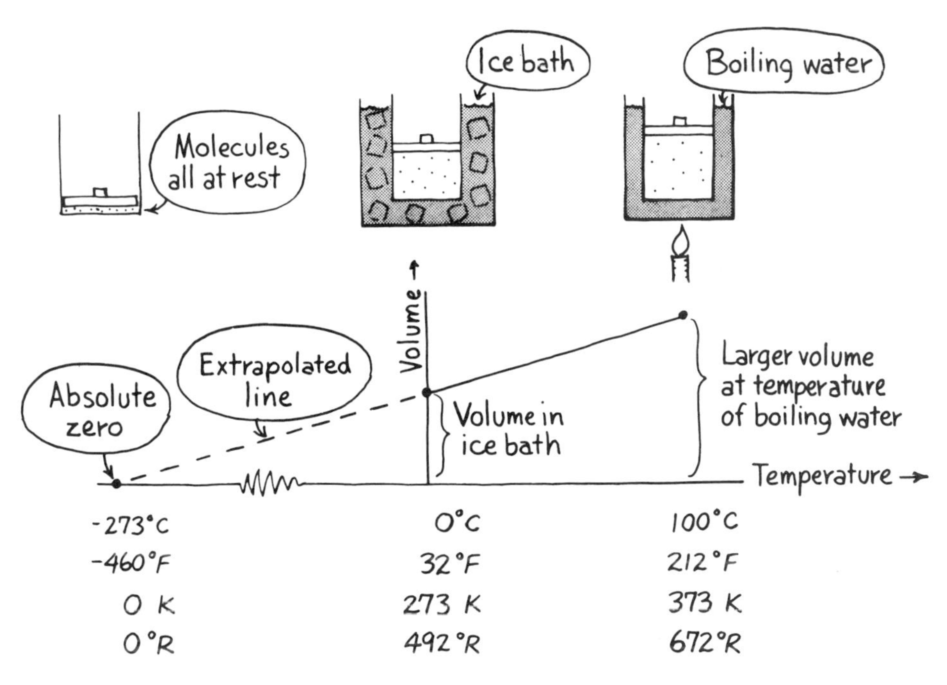

Figure 3-14. Volume of the gas at constant pressure varies uniformly with temperature. Extrapolation of graph shows lowest possible temperature at −273°C. Absolute zero is the starting point for Kelvin and Rankine temperature scales.

a lot of energy locked up in the mass of the electrons, protons, and neutrons that make up matter. Matter is another form of energy. This energy, predicted by Albert Einstein, is the product of the mass and the square of the velocity of light.

$$E = mc^2 \tag{3-13}$$

E *Energy*	m *Mass*	c *Velocity of Light**
joule (J)	kilogram (kg)	metre/second (m/s)
foot · pound (ft · lb)	slug (slug)	foot/second (ft/s)

**Note*: Velocity of light = 3.0×10^8 m/s.

Heat Units

The terms calorie and British thermal unit (Btu) are used to describe the quantity of heat energy stored in matter above absolute zero. A *calorie* is defined as the quantity of heat needed to change the temperature of 1 g of water by 1°C. A *Btu* is defined as the quantity of heat needed to change the temperature of 1 lb of water by 1°F (Fig. 3-15).

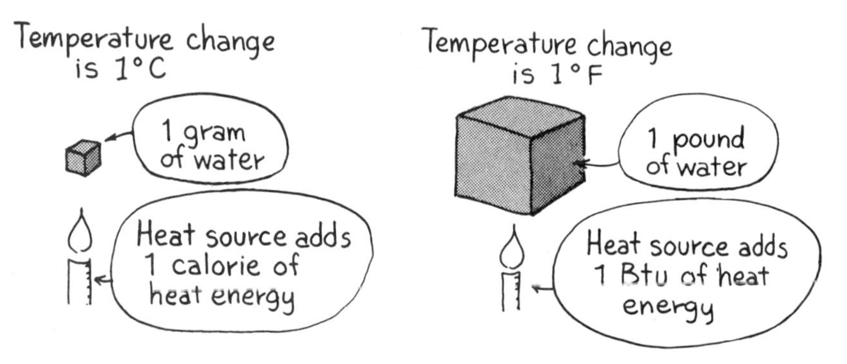

Figure 3-15. The calorie and Btu are defined as the amount of heat needed to change the temperature of a given amount of water by 1 degree.

Specific Heat

Other materials, such as metals and insulators, do not require as much heat to change their temperature. The *specific heat* of a material is a measure of the heat energy needed to change its temperature. The equation to calculate quantity of heat involved in a temperature change is

$$Q = mc(T_f - T_o) \tag{3-14}$$

Q *Quantity of Heat*	m *Mass or Weight*	c *Specific Heat*	T_f *Final Temperautre*	T_o *Original Temperature*
calorie (cal)	gram (g)	$\frac{\text{calorie}}{\text{gram} \cdot \text{degree Celsius}}$ (cal/g · °C)	Celsius degree (°C)	
British thermal unit (Btu)	pound (lb)	$\frac{\text{Btu}}{\text{pound} \cdot \text{degree Fahrenheit}}$ (Btu/lb · °F)	Fahrenheit degree (°F)	

Table 3, Appendix B, lists specific heats of various materials. For example, the specific heat of iron is 0.113. This means that only 0.113 cal of heat is needed to change the temperature of 1 g of iron by 1°C, or that 0.113 Btu of heat is needed to change the temperature of 1 lb of iron by 1°F.

Phase Changes

Heat is needed to change matter from one *phase* to another (i.e., solid to liquid and liquid to gas). The heat needed to turn a solid to liquid is called the *latent heat of fusion*. It is associated with the processes of melting or freezing. The heat needed to turn liquid to gas is called the *latent heat of vaporization*. Heat of vaporization is associated with the processes of vaporization or condensation. Boiling is vaporization at atmospheric pressure. Both phase changes occur at constant temperature. The most common example is again water. Water in its solid state is ice, which melts at 0°C. About 80 cal of heat are needed to change 1 g of ice into water at 0°C. About 540 cal of heat are needed to change 1 g of water into steam at 100°C under a pressure of 1 atmosphere (atm). The equations used to calculate the quantity of heat needed for a phase change are as follows:

$$Q = mL_f \tag{3-15}$$

solid to liquid (melting) or liquid to solid (freezing)

$$Q = mL_v \tag{3-16}$$

liquid to gas (vaporization) or gas to liquid (condensation)

Q *Quantity of Heat*	m *Mass or Weight*	L_f *Latent Heat of Fusion*	L_v *Latent Heat of Vaporization*
calorie (cal)	gram (g)	calorie/gram (cal/g)	
British thermal unit (Btu)	pound (lb)	British thermal unit/pound (Btu/lb)	

Values of latent heats for some other substances are shown in Table 2, Appendix B. Much more extensive lists are available in physics and engineering handbooks.

3-6 ELECTRICITY

Matter is made up of atoms. The most common atomic model, shown in Fig. 3-16, is of a nucleus containing protons and neutrons surrounded by electrons

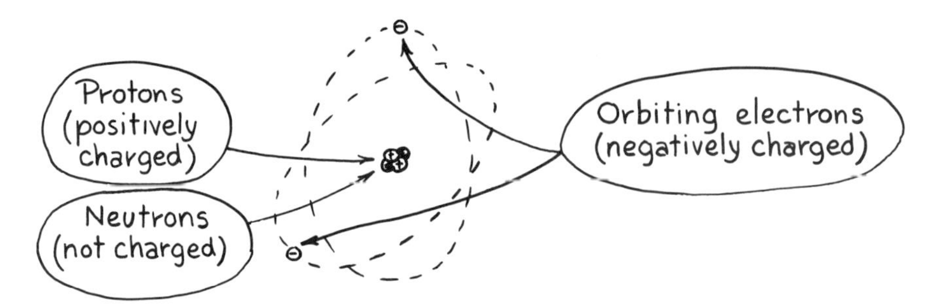

Figure 3-16. Diagrammatic sketch of a helium atom. Nucleus composed of two protons and two neutrons is surrounded by two orbiting electrons.

that orbit the nucleus. Electrons repel electrons and protons repel protons; but protons and electrons are attracted to each other. This mysterious behavior is attributed to a property called *charge*. By convention, electrons are negatively charged and protons positively charged. Unlike charges attract, and like charges repel. Neutrons have no charge. They are neither attracted nor repelled by charged particles. Metallic elements like copper easily lose some of their orbiting electrons. These electrons move through the metal or a conducting liquid or gas under the influence of electric forces. The forces can come from a cluster of like-charged particles as is found at the terminals of a battery or on the plates of a charged capacitor.

Electrical Energy

Energy considerations are important in explaining some of the things that we are able to observe in an electric circuit. A very simple circuit consisting of a battery, a switch, and a light bulb, all connected by wires, is shown in Fig. 3-17. Also shown is a mechanical system that includes an engine, a rope tow, a hill with snow on it, and a skier. In the electrical circuit, positive charge is concentrated at the terminal of the battery marked *B*. When the switch is closed, this charge moves through the wires. It passes through the light bulb and back to the negative post of the battery at *A*. In this text we will talk about movement of positive charge, because many definitions in electricity are based upon this convention. Actually, it is the negatively charged electron that moves in this circuit, and in the opposite direction, too.

Work must be done by the heat engine to lift the skier to the top of the hill using the ski tow in the mechanical system. The skier is the "particle" that gains mechanical potential energy. He gets it from the engine, which converts heat energy into mechanical potential energy. In the electrical system, a positive charge moves from *A* to *B* through the battery. The positive charge is repelled by the positive charges at *B* and attracted by the negative charges at *A*. Work is required to force the positive charge to occupy a position at *B*.

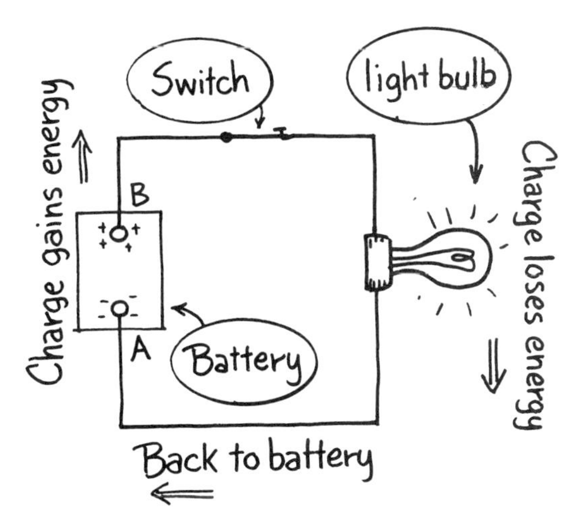

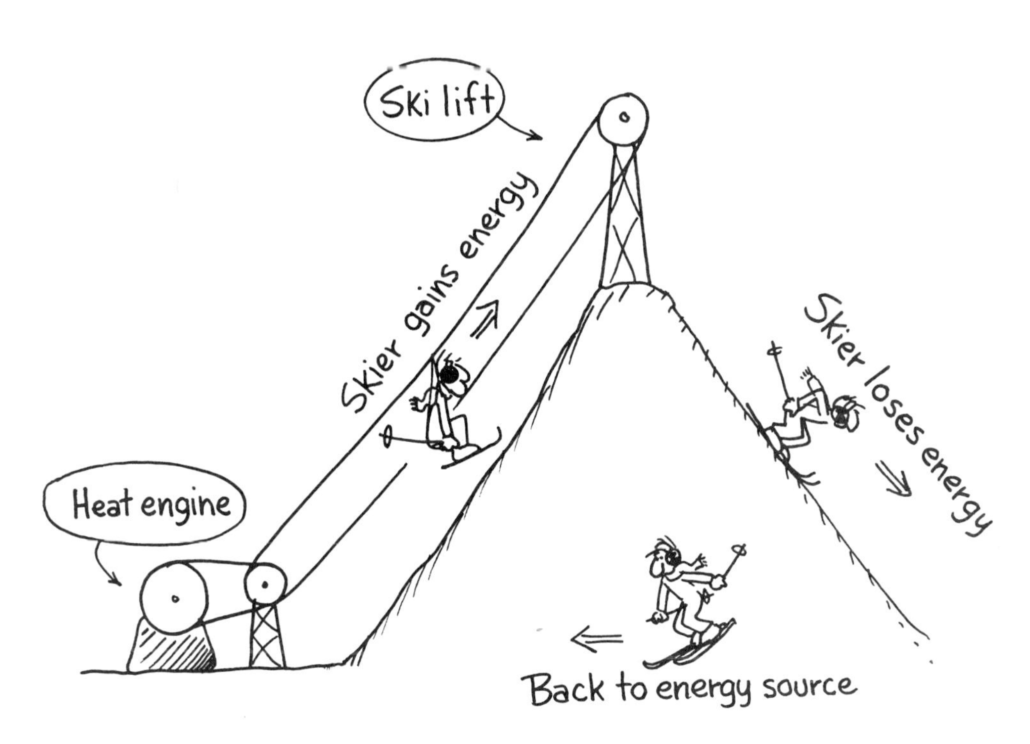

Figure 3-17. The movement of energy is similar in the mechanical ski tow system and in the simple electrical circuit.

Here the positive charge has a higher *electrical potential energy*, just as the skier at the top of the hill has a higher mechanical potential energy. A measure of the electrical potential energy is the *voltage* or the *electromotive force* (emf) or the *potential difference* of the battery. In this particular case, the electromotive force is defined as

$$E = \frac{W}{Q} \tag{3-17}$$

E *Electromotive Force*	*Q* *Quantity of Charge*	*W* *Work Done on (Energy Added to) the Charge*
volt (V)	coulomb (C)*	joule (J)

* One coulomb is equivalent to the charge on 6.25×10^{18} electrons.

The work for all this comes from the conversion of chemical energy to electrical potential energy.

Electromotive force in the electrical system is analogous to the height of the skier at the top of the hill in the mechanical system. The skier starts down the hill. Some of the potential energy is converted into kinetic energy as he quickly attains a terminal velocity. He is held back from further acceleration by friction forces. From then on, the potential energy lost as the skier descends appears as heat energy. The potential energy lost and the heat energy gained are equal, according to the energy conservation law. At the bottom of the hill, the skier has just enough energy left to barely make it to the ski tow before stopping. In this cycle the energy from the heat engine eventually appears as energy used to overcome friction forces producing the heat. The skier is the vehicle for this energy exchange.

Likewise, when the switch is closed, the positive charge moves through the wires and through the light bulb, where the potential energy of the charge is converted into the energy of heat and light. The loss in potential energy by a given amount of charge is measured by a potential difference across the light bulb. The charge limps around to the negative pole of the battery and completes the cycle. In this cycle, chemical energy from the battery is converted into heat and light energy at the light bulb. The charged particle is the vehicle for this energy exchange.

3-7 SUMMARY

This brief review of physics should give you an idea of the background material needed for this text. It should also permit you to assess how well you understand the background material. If you understood only parts of this chapter, you may want to refer to a general beginning physics text so that you can read up on those topics in which you are weak.

Most of what follows is carefully explained. We do assume, however, that you have some of the background knowledge covered in this review chapter. Try your hand at some of the review problems at the end of this chapter before moving on.

PROBLEMS

3-1. A satellite has a velocity of 40,000 mi/hr. How many miles will it travel in one day?

3-2. An object moves a distance of 1.4 m in 7.0 s. What is its velocity in metres/second?

3-3. A car, starting from rest, attains a velocity of 30 m/s in 8.0 s. What is the acceleration of the car?

3-4. A block of wood slides down an inclined plane and changes its velocity from 1.0 to 7.0 ft/s in 2.0 s. What is the block's acceleration?

3-5. What is the mass in slugs of 1 ft^3 of water?

3-6. What is the weight in newtons of a 3.0-kg mass?

3-7. What is the mass of 25 cm^3 of lead?

3-8. If 2.5 ft^3 of a substance weighs 235 lb, what is the "weight density" of the material? ("Weight density" is *weight* per unit volume.)

3-9. How much force is exerted by a pressure of 60 psi on an area of 64 $in.^2$?

3-10. What gauge pressure is placed on a gas enclosed by a piston and a cylinder if the cylinder's diameter is 0.10 m and the force on the piston is 250 N?

3-11. If the gauge pressure is 25 psi, what is the absolute pressure?

3-12. What is the absolute pressure in pounds/square foot at a depth of 100 ft in water? (*Hint*: Convert the atmospheric pressure of 14.7 psi to lb/ft^2 before calculating the pressure due to the depth of water.)

3-13. A horizontal force of 15.0 N moves a block 3.0 m along a flat surface. How much work is done by the force?

3-14. How far would the block in Problem 3-13 have to move if the force does 500 J of work?

3-15. What power is needed to lift a 2000-lb object a vertical distance of 8.0 ft in 10 s?

3-16. How many joules of energy are expended in a 500-W heating element during a 60-s time interval?

3-17. How much potential energy is stored in a 150-lb weight lifted 1000 ft above the ground?

3-18. How many joules of energy are lost when a 25-kg mass falls a distance of 6.0 m?

3-19. What velocity does a 1.6-kg mass need, to have a kinetic energy of 180 J?

3-20. What is the kinetic energy of a 2.0-oz bullet fired from a gun with a velocity of 1600 ft/s?

3-21. The volume of a gas is 1.0 m^3 at 0°C. What volume will the gas occupy at 100°C if the pressure remains constant while the gas is heated? (*Hint*: Use Fig. 3-14.)

3-22. The mass of a person is 70 kg. (This represents a weight of 154 lb.) How much energy is this mass equivalent to?

3-23. How many calories of energy are removed from a 50-g mass of water when it is cooled from 20 to 0°C?

3-24. How many Btu are needed to heat 250 lb of water from 60 to 180°F?

3-25. How many Btu are needed to change the temperature of 15 lb of aluminum from 68 to 212°F?

3-26. In a heat experiment, 88 cal of heat energy are added to a 40-g mass. The temperature of the mass changes from 20 to 40°C. What is the specific heat of the mass?

3-27. How many calories are removed from 200 g of water during the freezing process at 0°C?

3-28. How many Btu of heat energy are needed to melt 300 lb of zinc? The zinc is already at its melting temperature.

3-29. How many Btu of heat energy are needed to boil away 5.0 lb of water at 212°F?

3-30. How many calories are removed from 50 g of ethyl alcohol vapor when it condenses to a liquid?

3-31. An electric charge of 400 C gains 600 J of energy from a battery. What is the battery's emf?

3-32. An electric charge of 4.0 C passes through a 12-V battery. How many joules of energy are added to the charge?

ANSWERS TO ODD-NUMBERED PROBLEMS

3-1. 9.6×10^5 mi
3-3. 3.8 m/s^2
3-5. 1.94 slug
3-7. 2.8×10^2 g
3-9. 3.8×10^3 lb
3-11. 40 psi
3-13. 45 J
3-15. 1.6×10^3 ft · lb/s
3-17. 1.5×10^5 ft · lb
3-19. 15 m/s
3-21. 1.4 m^3
3-23. 1.0×10^3 cal
3-25. 4.7×10^2 Btu
3-27. 1.6×10^4 cal
3-29. 4.9×10^3 Btu
3-31. 1.5 V

4

Forces

Our formal study of unified concepts in applied physics begins by looking at something common and tangible in the systems we will study. It is called ***force****, and it is something that we can identify with in a practical way. Force in mechanical systems is a push or pull. This is perhaps the most familiar force we know about in the real world. It is the thing that moves or tends to move objects, and this leads directly into the idea of work and energy. More correctly, it is a net force or force difference that moves or tends to move objects.*

Force-like quantities in the other systems are carefully chosen to relate to work and energy in the same way that they do in the mechanical translational system. The unified concept idea of "force" becomes ***torque*** *in the mechanical rotational system,* ***pressure*** *in fluid systems,* ***temperature*** *in the heat system, and* ***potential difference*** *or voltage in the electrical system. Force moves objects in straight lines, torque twists objects about an axis of rotation, pressure pushes fluids through pipes, a temperature difference is necessary for heat to flow, and potential difference is needed for electric charge to flow in wires.*

This chapter will not look at work and energy relations, but only at the force relations. So let's take a close look at force, torque, pressure, temperature and voltage. These are not easy concepts, so you should study carefully what follows.

4-1 FORCE IN MECHANICAL TRANSLATIONAL SYSTEMS

A mechanical translational system is one in which an object moves in a straight line. The direction and the kind of motion are determined by the forces acting on the object. The definition of force can be complicated, but we will define it simply as a push or a pull.

The units of force are the newton (N) in SI units and the pound (lb) in the English system of units. Forces can also be measured in ounces (oz) and dynes (dyn). The dyne is the force unit in the CGS system of units. Conversion factors between these units are listed in the table of conversion factors in Appendix A. For our present use, a few conversion factors are given in Table 4-1.

TABLE 4-1 FORCE UNITS AND CONVERSION FACTORS

1 newton (N) = 0.225 pound (lb)
1 pound (lb) = 16 ounces (oz)
1 newton (N) = 10^5 dyne (dyn)

Spring scales and strain gauges are examples of instruments used to measure force. Balances are used to weigh objects. Weight is the downward force that the earth exerts on an object. Pounds, kilograms, and grams are popular units of weight. Of the three, pounds is the only correct unit of weight. A 1-kilogram (kg) mass weighs 9.81 N at sea level. Hence multiply the mass in kilograms by 9.81 to find the weight in newtons. A 1-gram (g) mass weighs 981 dyn at sea level. Hence multiply the mass in grams by 981 to get the weight in dynes. The relation between mass and weight is reviewed for you in Sec. 3-2.

Forces with the Same Line of Action

A force has a magnitude and a *line of action.* A line of action is in the same direction as the force and passes through the point where the force is applied. Figure 4-1 shows that force, a vector quantity, can be represented by an arrow. The length of the arrow is the force's magnitude, and the direction of the arrow is the force's direction.

More than one force may act along the same line. In Fig. 4-2a, the weight and the rope tension have the same line of action. The weight's line of action passes through the center of gravity of an object and is always directed downward. In Fig. 4-2b, the car experiences equal but opposite forces.

Figure 4-1. The force's line of action is in the same direction as the force and passes through the point where the force is applied. Arrows are used to represent forces.

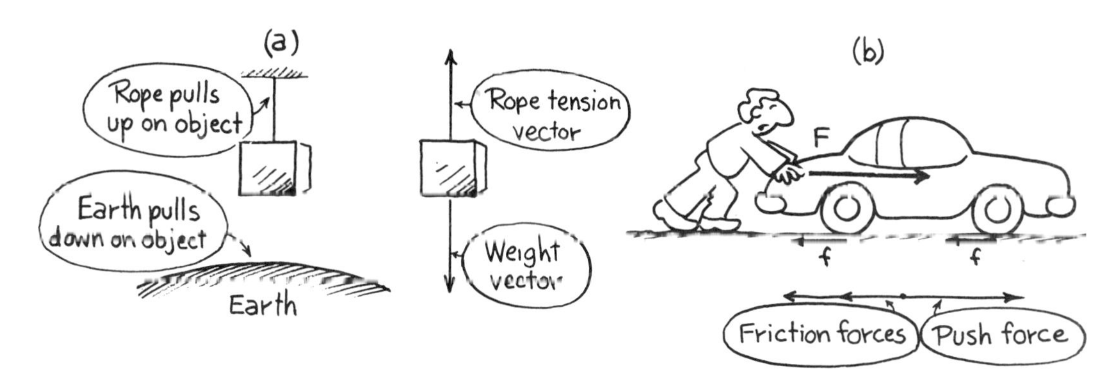

Figure 4-2. (a) The weight and rope tension are equal and have the same line of action. The net force on the object is zero. (b) The pushing force F is balanced by the car's friction forces f. The net horizontal force is zero. Vertical forces due to the car's weight are not shown on the figure.

In both of these examples the vector sum of the forces is zero. There is no net force. In situations like this the velocity is constant. There may be no motion (zero velocity) or uniform motion with constant velocity but no acceleration.

Forces may have the same line of action but be unbalanced. We can cut the rope holding the object (Fig. 4-3a) and it will fall, or the man can push harder on the car, causing it to accelerate to the right (Fig. 4-3b). In these examples the net force is not zero but simply the *difference* between the opposing forces. The direction of the net force is the direction of the larger force. The difference between the forces also happens to be the result of a vector addition in which the two forces have opposite directions. As you would expect, the objects will accelerate (velocity changes with time) when the net force is not zero.

Figure 4-3. The net force is the difference between the opposing forces when the opposing forces are unequal.

*Forces with Different Lines of Action

Sometimes the forces' lines of action are in different directions. Figure 4-4 shows a skier sliding down a steep frictionless hill. The weight of the skier (vertical) and the surface force (perpendicular to the surface) have lines of action in different directions. The net force on the skier is the vector sum of the

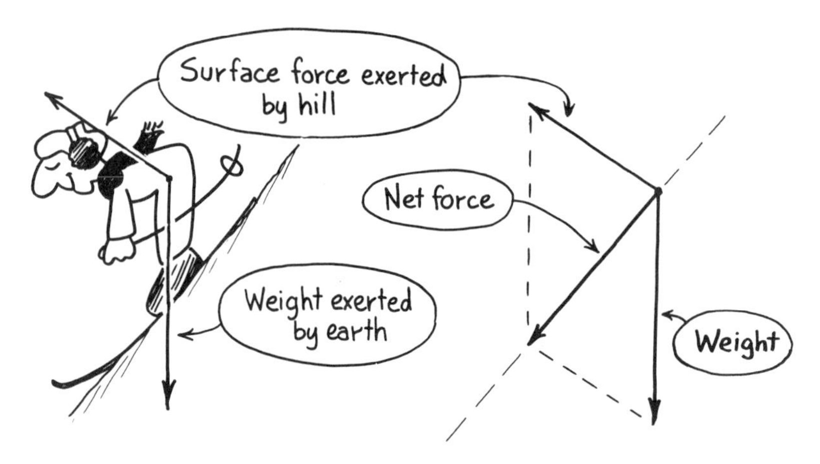

Figure 4-4. When the lines of action are in different directions, the net force is found by constructing a force parallelogram.

two forces and can be found by constructing a force parallelogram. The weight and surface forces are the sides of the parallelogram, and the net force is the diagonal. The net force points in the same direction as the hill slope. The net force can be calculated using the graphical or mathematical methods described in Sec. 2-5. Example 4-1 shows an easier problem.

Example 4-1 An object weighing 4.0 N falls to the ground on a windy day. The horizontal wind force is 3.0 N to the right. What is the net force on the object?

Solution

Draw the weight and wind forces as vectors. The force parallelogram is a rectangle. The diagonal F_{net} is the net force and the angle α is its direction.

$$F_{net} = \sqrt{(3.0\text{ N})^2 + (4.0\text{ N})^2} = \boxed{5.0\text{ N}}$$

$$\tan\alpha = \frac{4.0\text{ N}}{3.0\text{ N}} = 1.33 \Rightarrow \alpha = \boxed{53°}$$

4-2 TORQUE IN MECHANICAL ROTATIONAL SYSTEMS

A mechanical rotational system is one in which an object rotates about a fixed axis. The direction and kind of rotational motion are determined by the torques acting on the object. Torque is the product of a force and its moment arm about the axis of rotation. The force F, moment arm ℓ, and the axis of rotation are shown in Fig. 4-5. The axis of rotation is perpendicular to the

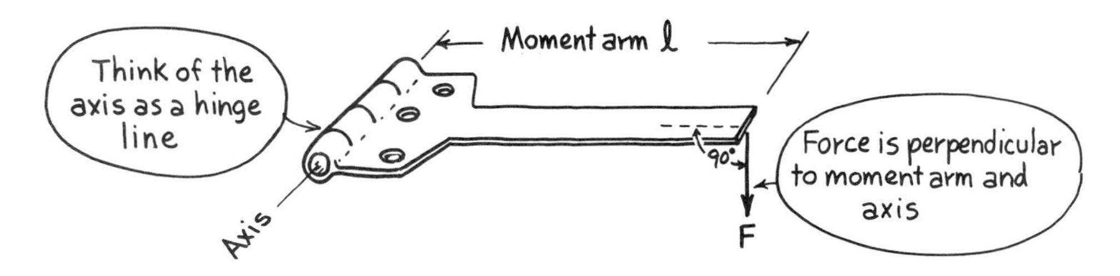

Figure 4-5. Torque exerted by force F about the axis (hinge line) is equal to $F\ell$. Axis, moment arm, and force are mutually perpendicular.

paper and perpendicular to the horizontal moment arm, which in turn is perpendicular to the force's vertical line of action. The equation for torque τ, when everything is as shown in Fig. 4-5, is

$$\tau = F\ell \tag{4-1}$$

τ Torque	F Force	ℓ Moment Arm
newton · metre (N · m)	newton (N)	metre (m)
pound · foot (lb · ft)	pound (lb)	foot (ft)

The direction of the torque is the direction that the shaft tends to rotate if acted upon by the force. In Fig. 4-5, this direction is the same as the direction in which the hands of a clock move, so we say the torque's direction is clockwise. In this text we shall use the symbol (⟳) to indicate the clockwise direction. The torque's direction would have been counterclockwise (⟲) if the force had pointed upward instead of downward.

Torques are more complicated than forces because torque is the product of force and moment arm. The two torques in Fig. 4-6 are equal even though

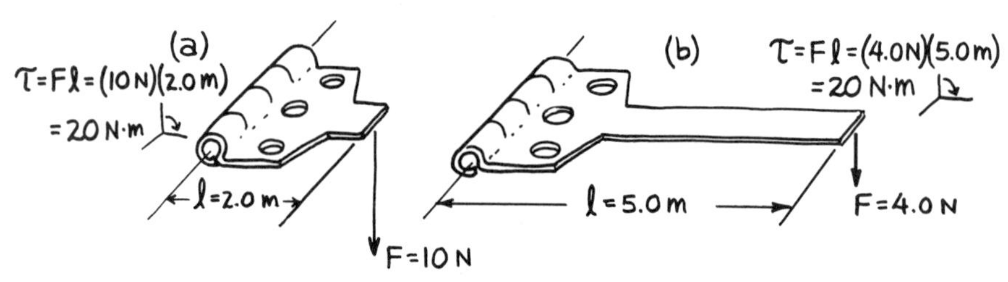

Figure 4-6. The torque of the 10-N force has the same effect as the torque of the 4.0-N force because the moment arm of the 10-N force is smaller.

the forces are different. The force in Fig. 4-6b is smaller, but it is being applied to a longer moment arm. Its turning effect is the same as that of the larger force in Fig. 4-6a.

Torque wrenches and dynamometers are examples of instruments that measure torque. These instruments are actually force meters. The force is located at a fixed distance from its axis of rotation (Fig. 4-7).

Net Torque

More than one torque can act about an axis of rotation. Figure 4-8 shows three situations. In Fig. 4-8a, forces F_A and F_B exert equal but opposite torques on the stick. The *net* torque is the difference, which is zero. The stick will be either at rest or it will rotate with a constant angular velocity. In Fig. 4-8b, the net torque is the difference between the two torques and is in the

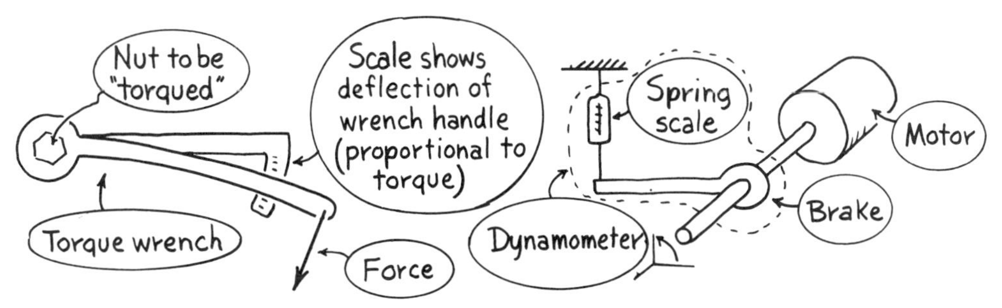

Figure 4-7. Instruments for measuring torques, such as the torque wrench and the dynamometer, are actually force meters operating with a fixed moment arm.

direction of the larger torque. The stick will rotate about the axis with increasing angular velocity in a clockwise direction. Figure 4-8c shows unbalanced torques causing the stick to accelerate in the counterclockwise direction.

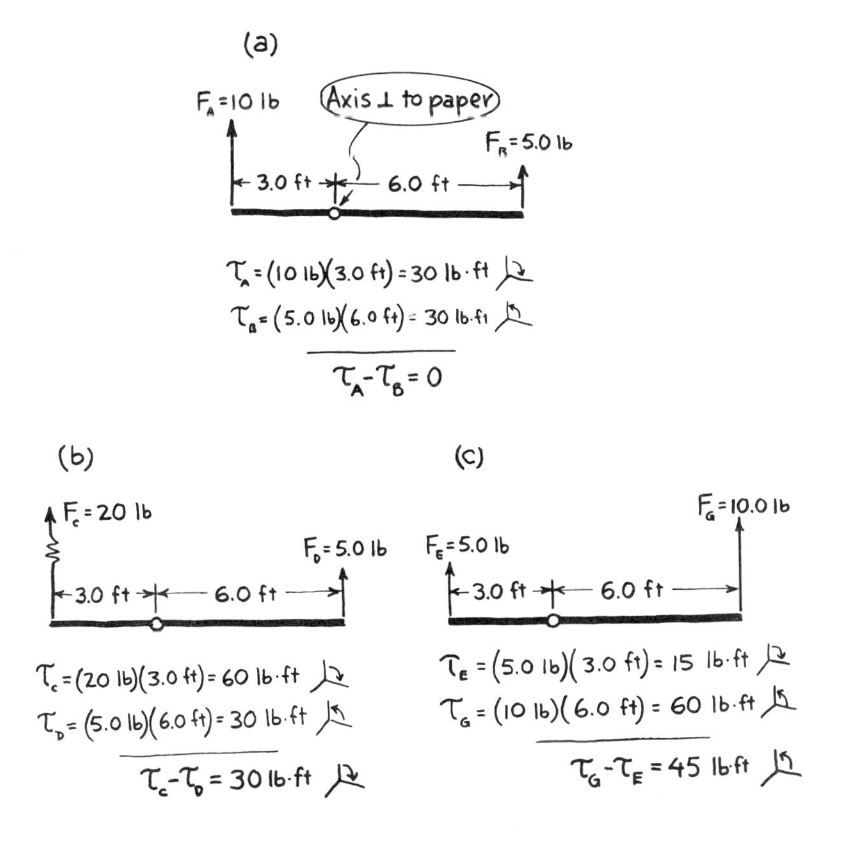

Figure 4-8. Net torques are found by calculating the difference between the clockwise and counterclockwise torques.

*More Complicated Torques

The force's line of action and its moment arm may not be perpendicular to each other. Figure 4-9 shows such a situation, in which a force of

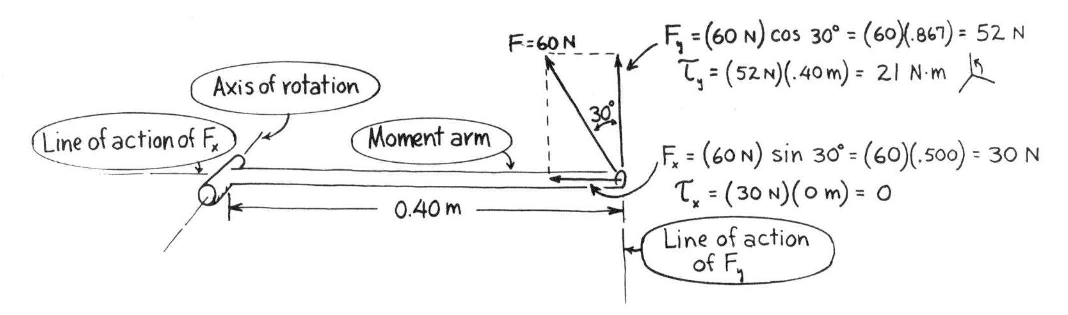

Figure 4-9. The torque due to a force whose line of action is not perpendicular to the moment arm can be found by resolving the force into components.

60 N acts at an angle of 30° from the vertical. The torque can be found by resolving the force into two components. One component F_y is perpendicular to the moment arm, and the other F_x is parallel to the moment arm. The perpendicular component exerts a counterclockwise torque τ_y of 21 N · m. The horizontal component F_x has a line of action that intersects the axis of rotation. The moment arm of this component is zero; thus F_x exerts no twisting effect or torque about the axis. The torque of the original 60-N force is only that of its vertical component, 21 N · m. Here is another example.

Example 4-2 A motor exerts a clockwise 9.0-ounce · inch (oz · in.) torque on the gear. What force F must be exerted against a tooth to keep the gear from moving?

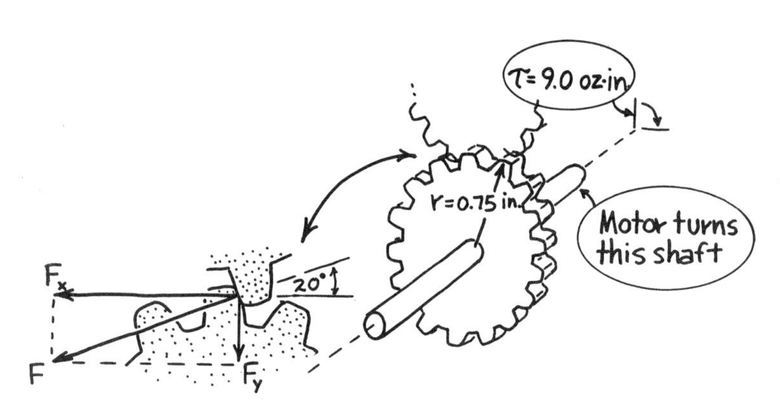

Example 4.2

Solution

The net torque is zero if the force exerts a counterclockwise torque of 9.0 oz · in. This torque is supplied by the force's horizontal component.

$$\left\{\begin{matrix} \tau = 9.0 \text{ oz} \cdot \text{in.} \\ \ell = 0.75 \text{ in.} \end{matrix}\right\}$$

$$\tau = F_x \ell$$

$$F_x = \frac{\tau}{\ell} = \frac{9.0 \text{ oz} \cdot \text{in.}}{0.75 \text{ in.}} = 12 \text{ oz}$$

$$F = \frac{F_x}{\cos 20°} = \frac{12 \text{ oz}}{0.940} = 12.77 \rightarrow \boxed{13 \text{ oz}}$$

4-3 PRESSURE IN FLUID SYSTEMS

A fluid is a liquid or a gas, and pressure is responsible for its movement. Pressure has already been defined in Sec. 3-3, which you should carefully read again. The main points of this review are as follows:

1. Pressure p is the ratio of the force F exerted on a surface to the area A of the surface. $p = F/A$ (Eq. 3-6)

2. The force F is always perpendicular to area A and always tends to compress the fluid.

3. The pressure at a point in a fluid is the same in all directions.

4. The pressure of the atmosphere is on the average about 14.7 pounds per square inch (psi).

5. Measured pressures usually do not account for atmospheric pressure. Such measured pressures are called *gauge* pressures. The absolute pressure p used in most calculations is found by adding atmospheric pressure p_0 to the gauge pressure p_g. $p = p_0 + p_g$ (Eq. 3-8)

6. Gauge pressure p_g at a depth h in a uniform fluid of density ρ is given by $p_g = h\rho g$. (g is acceleration of gravity.)

7. Pressure can dramatically change the volume of a gas, but has very little effect on the volume of a liquid.

Most of these points are illustrated in the following problem.

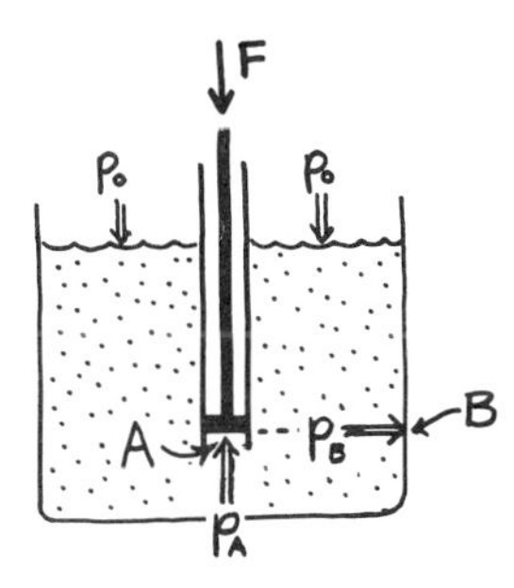

Example 4.3

Example 4-3 A piston has a cross-sectional area of 0.015 square metres (m^2). It is pushed by a force F to a depth of 3.0 m into a tank of gasoline.

(a) What is the absolute pressure at point A directly under the piston?

(b) What is the horizontal pressure at point B, which is at the same depth as point A?

(c) What downward force is exerted on the piston to hold it in position?

Solution to (a)

Appendix B, Table 1

$$\begin{Bmatrix} h = 3.0\,\text{m} \\ g = 9.81\,\text{m/s}^2 \end{Bmatrix}$$

$$p_A = p_0 + h\rho g = 1.013 \times 10^5 \frac{\text{N}}{\text{m}_2} + 3.0\,\text{m}\left(680\frac{\text{kg}}{\text{m}^3}\right)\left(9.81\frac{\text{m}}{\text{s}^2}\right)$$

$$= (1.013 \times 10^5 + 0.20 \times 10^5)\frac{\text{N}}{\text{m}^2} = \boxed{1.21 \times 10^5 \text{ pascals (Pa)}}$$

Solution to (b)

Since pressure is the same in all directions at a given depth,

$$p_B = p_A = \boxed{1.21 \times 10^5 \text{ Pa}}$$

Solution to (c)

$$\begin{Bmatrix} A = 0.015\,\text{m}^2 \\ = 1.5 \times 10^{-2}\,\text{m}^2 \end{Bmatrix}$$

$$F = (p_A - p_0)A = (1.21 \times 10^5 - 1.01 \times 10^5)\frac{\text{N}}{\text{m}^2}(1.5 \times 10^{-2}\,\text{m}^2)$$

$$= \boxed{3.0 \times 10^2 \text{ N}}$$

Pressure Units

The units of pressure come in all shapes and sizes. A popular unit of pressure is the pound per square inch (psi). The pound per square foot (lb/ft^2) and the newton per square metre (N/m^2) are important units of pressure because they are used for calculations in English and SI units. The SI unit for a pressure of $1\ \text{N/m}^2$ is called a pascal (Pa). Other units have special names. An atmosphere (atm) is the average pressure at sea level and is used to measure large pressures. The unit inches of water is used to measure pressures related to the flow of gases. The torr is the pressure exerted by a column of mercury 1 millimetre (mm) high and is used to measure very low pressures. You should know how to convert back and forth between units. Table 4-2 shows a number of pressure conversion factors. These factors are also included in the table of conversion factors in Appendix A.

TABLE 4-2 UNITS OF PRESSURE

1 atmosphere (atm)	= 1.013×10^5 newtons/metre^2 (N/m^2) or pascal (Pa)
	= 14.7 pounds/inch^2 (psi)
	= 2.12×10^3 pounds/foot^2 (lb/ft^2)
	= 407 inches of water
	= 760 torr (torr)

Pressure Instruments

The manometer and the Bourdon tube pressure gauge are examples of instruments that measure pressure. The manometer is a U-shaped tube partially filled with a liquid, usually water or mercury. The difference in heights of the liquid in the arms of the tube is a measure of the difference between atmospheric pressure p_0 and the pressure p being measured. Figure 4-10a shows a manometer measuring a pressure larger than atmospheric pressure. Figure 4-10b shows a manometer measuring a pressure of less than 1 atm.

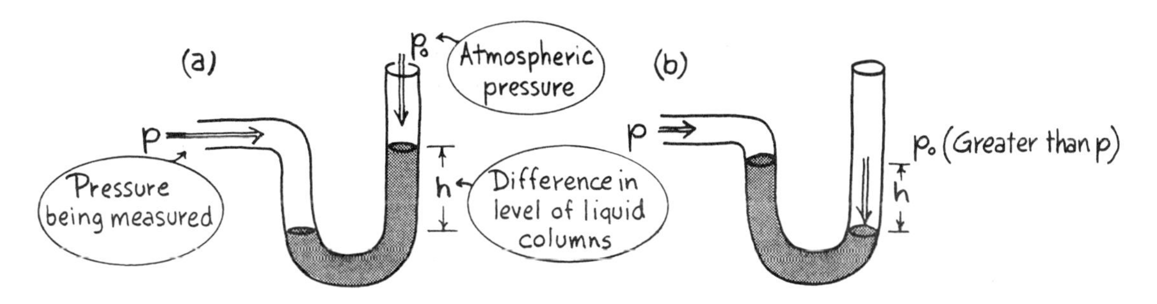

Figure 4-10 The difference in liquid levels indicates the difference between the unknown pressure in one arm and the atmospheric pressure in the other arm.

The Bourdon gauge has a hollow metallic tube that tends to straighten out when the pressure increases inside the tube. The tube deflection is a measure of the pressure difference between the atmosphere and the pressure inside the tube (Fig. 4-11). Both the manometer and the Bourdon tube measure gauge pressure.

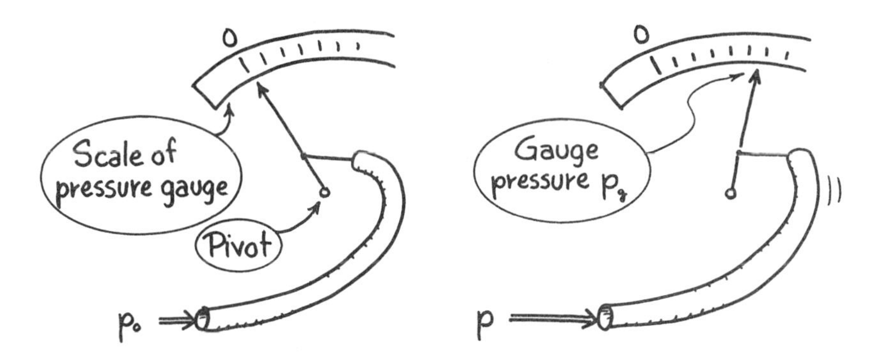

Figure 4-11. The Bourdon tube tends to flatten out when pressure increases inside the tube. The pointer indicates the gauge pressure.

4-4 TEMPERATURE IN HEAT SYSTEMS

Force in a heat system involves the notion of temperature, not too precisely defined as a measure of relative hotness or coldness. A more formal definition relates temperature of an object to the average kinetic energy of its molecules. Kinetic energy (Sec. 3-4) is energy of motion.

The molecules of metal making up the hot plate at *A* (Fig. 4-12) jiggle back and forth rapidly owing to the flame. These molecules have gained kinetic

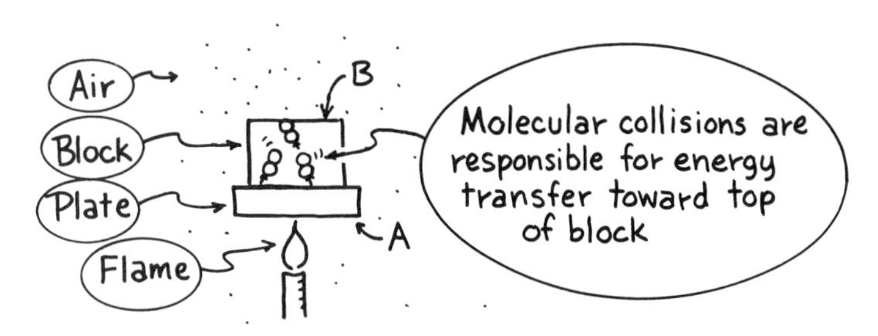

Figure 4-12. Heat flows from the hot surface *A* to the cooler surface *B* through the block. A temperature difference is needed for the movement of heat energy.

energy from the flame (i.e., they are at a high temperature). Fast-moving plate molecules hit block molecules at the bottom surface. The energized block molecules strike neighbors immediately above. The domino effect ultimately causes all block molecules to vibrate faster, gaining energy by collisions from below and transmitting energy through collisions upward. If tiny thermometers were installed in the block, they would show a steady decrease in the temperature of the block from the hot lower surface to the warm upper surface. There the block transmits heat to the air. The warmed air molecules are replaced through natural circulation with other, colder air molecules, which in turn are warmed by the block.

This rather simple explanation involving the motion of molecules illustrates two important ideas:

1. There is a flow of energy or heat from the hot plate to the air through the block. We know that there is a heat flow because warmed air molecules continually leave the top surface.
2. The direction of heat flow is from the hot to the cold object.

The condition that drives the heat is the *temperature difference*, and this is the force-like quantity in a heat system.

The Celsius (°C) and Fahrenheit (°F) scales are used to measure temperature. Celsius and centigrade are two names for the same temperature scale. Celsius is now preferred over centigrade. The freezing point of water is 0°C or

32°F. The normal boiling point of water is 100°C or 212°F. This establishes a relation between the two temperature scales. Equation (4-2) permits you to convert from one temperature scale to the other.

$$T_C = \tfrac{5}{9}(T_F - 32) \tag{4-2}$$

T_C Celsius Temperature	T_F Fahrenheit Temperature
degrees Celsius (°C)	degrees Fahrenheit (°F)

In many situations a temperature difference is all that is required in a calculation. Conversion of temperature difference is easy since

$$1°\text{C} = 1.8°\text{F}$$

Two other temperature scales are important for making calculations on gases. These absolute temperature scales use absolute zero as the starting point for measuring temperature. Take another look at Sec. 3-5 to refresh your memory about the Kelvin (K) and the Rankine (°R) scales and about absolute zero, which is −273°C or −460°F. The Celsius degree is used in the Kelvin scale and the Fahrenheit degree is used in the Rankine scale. Kelvin and Rankine temperatures are related to Celsius and Fahrenheit temperatures using

$$T_K = T_C + 273 \tag{4-3}$$

$$T_R = T_F + 460 \tag{4-4}$$

T_K Kelvin Temperature	T_C Celsius Temperature	T_R Rankine Temperature	T_F Fahrenheit Temperature
kelvin (K)	degrees Celsius (°C)	degrees Rankine (°R)	degrees Fahrenheit (°F)

The kelvin is the official SI unit for temperature. Note that the symbol for its unit is (K), not (°K).

Temperature calculations are illustrated by Example 4-4.

Example 4-4 Heat flows from a hot surface of 125°C to a cooler surface of 43°C.

(a) What is the temperature difference between the hot and cold surfaces in degrees Celsius and degrees Fahrenheit?

(b) What is the temperature of the hot surface in degrees Fahrenheit?
(c) What is the absolute temperature of the cold surface in kelvins?

Solution to (a)

$$\Delta T_C = 125 - 43 = \boxed{82°\text{C}}$$

$$\Delta T_F = (82°\text{C})\left(\frac{1.8°\text{F}}{1°\text{C}}\right) = \boxed{148°\text{F}}$$

Solution to (b)

$$T_C = \tfrac{5}{9}(T_F - 32) = \frac{5T_F - 160}{9} \Rightarrow T_F = \frac{9T_C + 160}{5} = \frac{(9)(125) + 160}{5} = \boxed{257°\text{F}}$$

Solution to (c)

$$T_K = T_C + 273 = 43 + 273 = \boxed{316\ \text{K}}$$

Examples of temperature-measuring instruments are the common thermometer and the thermocouple. The thermocouple generates a very small electric voltage, which is related to the temperature being measured.

4-5 ELECTROMOTIVE FORCE IN ELECTRICAL SYSTEMS

The battery in an electrical system is a device that forces electrically charged particles to leave one of its terminals and travel through an external circuit to the other terminal. The battery is referred to as a "seat" of *electromotive force* (emf), and this force-like quantity is measured as a *potential difference* in volts. Potential difference E is the force for the electrical system. Its SI unit is the *volt* (V). Fortunately, the volt is the only unit of potential difference with which we need to be concerned.

Section 3-6 gives a definition for potential difference. The definition involves the notions of work, energy, and charge. A more formal treatment of

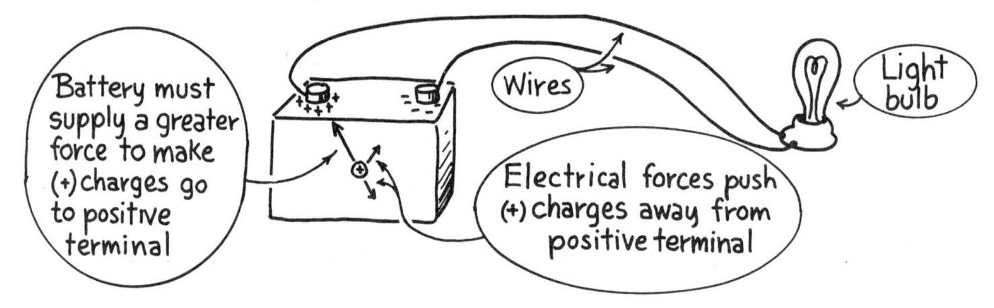

Figure 4-13. The battery forces charged particles to move through the wires and the light bulb of the external circuit. The battery must exert a force on the charged particles.

potential difference will be delayed until the next chapter, where charge, work, and energy are discussed in more detail. A force must be exerted on a positive charge to move it toward the positive terminal of a battery (Fig. 4-13). Many people do indeed therefore think of voltage as a force or electrical pressure that causes charge to move in electrical circuits.

4-6 FOUR DIFFERENT KINDS OF FORCE

All through the rest of this book we will be talking about forces that arise in different ways and do different things. These can be classified into four categories that you should start to become familiar with. The setup illustrated in Fig. 4-14 includes a dashpot (shock absorber), an unstretched coil spring, and a block of material (a mass). All three are connected to a lightweight yoke that can be acted on by an outside force, labeled F_E. If there is going to be any change in the system (i.e., if it is to be "excited"), F_E will have to do it. Thus F_E will be called the *excitation force*. F_E can come from anywhere. It can be applied by tying a rope to the yoke, by grabbing with your hand, by pulling it with a magnet, or whatever. All that we know is that it exists. What causes it does not matter for our purposes.

excitation force F_E = outside force that puts energy into a system

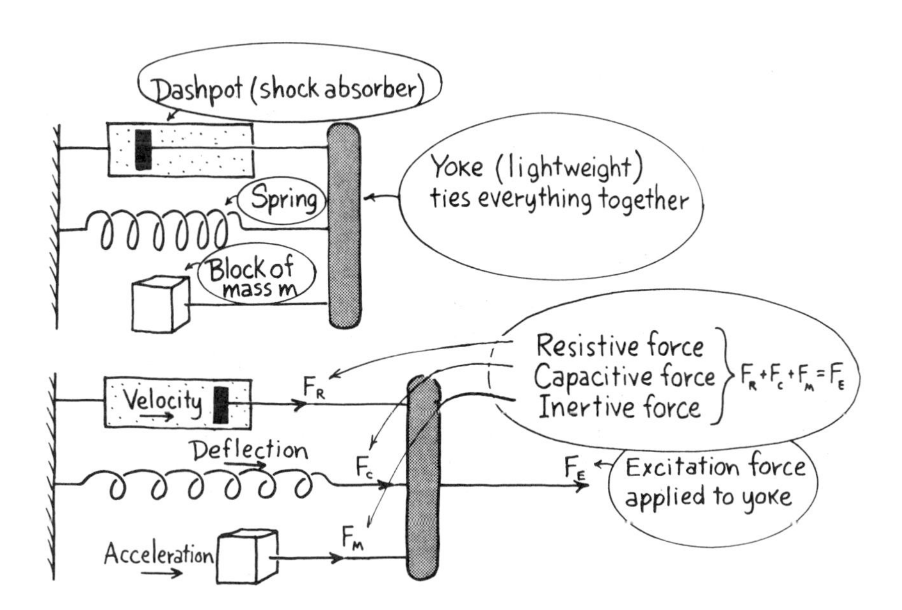

Figure 4-14. An excitation force can do three kinds of jobs, analogous to overcoming friction, stretching a spring, and accelerating an object.

Suppose now that F_E is applied to the yoke so that it begins to move. The dashpot expands, the spring stretches, and the block accelerates. We will look at the three elements one at a time to see what effects they produce.

Resistive Force

The dashpot is a device that produces a frictional force—something that shows up only when there is motion or at least a tendency to move. The dashpot and analogous devices are called *resistors*, and they have a couple of things in common in the various systems. (Resistors will be discussed at length in Chapter 7.)

1. Whatever work is done on a resistor (i.e., whatever energy is put into a resistor) is changed to heat and is effectively lost from the system.
2. The frictional force that appears in the resistor is either constant or depends somehow on how fast the system is moving.

The force generated by a resistor always opposes the motion. Thus part of the excitation force always must be used up in the resistor if there is to be any movement. It is this part of the excitation force that we call *resistive force.*

resistive force F_R = that part of the excitation force used up in the resistor

Resistive forces in other systems include the pressure to overcome the frictional pressure loss (back pressure) when fluid flows in a pipe, and the potential difference required to force electric charge through an electrical resistor.

Capacitive Force

If the spring in Fig. 4-14 is unstretched to begin with, it exerts no force on the yoke. But when the yoke moves to the right, the spring stretches and pulls back. If the yoke moves to the left, the spring compresses a little and pushes forward. Again we are looking at a device in which a force arises to oppose whatever we do to it. There are some definite differences, however, between the spring and the dashpot. The spring and analogous things in other systems are called *capacitors* and are characterized by the following (all to be discussed in Chapter 9):

1. Work that is done on (energy put into) a capacitor is not lost. It is only stored and will remain so as long as we hold the spring stretched out or compressed.

2. The amount of force that it takes to stretch or compress (deflect) the spring depends directly on how much deflection there is.

So another part of the excitation force can be used up in the capacitor, and it is this that we call *capacitive force.*

capacitive force F_C = that part of the excitation force used up in the capacitor

Capacitive forces in other systems include the torque needed to wind up a spring, the pressure needed to force water into the bottom of a tank, and the potential difference necessary to force charge onto the plates of an electrical capacitor.

Inertive Force

As the excitation force F_E begins to move the system of Fig. 4-14, one other sort of force appears. The block has mass, and according to Newton's second law, it will accelerate only if acted upon by a net force. Mass and similar things in other systems describe a property called *inertia*, which is discussed at length in Chapter 10. Devices that contain inertia are characterized as follows:

1. Whatever work is done (i.e., the energy that is put into an inertial device) goes toward making it "move." The energy is not lost, but remains "stored" as long as the motion continues.
2. The force applied to make a mass move is directly related to the rate at which the motion changes—the acceleration of the block in the system of Fig. 4-14.

Thus whatever is left of the excitation force after F_R and F_C are taken out is used in overcoming inertia (i.e., to accelerate the system). This leftover force is commonly referred to as the *net* force or the *resultant* force acting on the system. Because of its relation to the inertial part of the system, we may also call it the *inertive force* and label it F_M.

net force = F_{net} = inertive force F_M = that part of the excitation force acting to accelerate the sytem

In Chapter 10 we will give the name *inertance* to such things as mass and its analogs in other systems. Inertive forces in other systems include the torque needed to accelerate a circular saw blade, the pressure exerted on water to make it flow when a valve is suddenly opened, and the potential difference involved in building a magnetic field about a coil in an electrical circuit.

Table 4-3 summarizes some of what has just been discussed.

TABLE 4-3 SUMMARY OF FOUR DIFFERENT KINDS OF FORCE

Name	*Symbol*	*Source*	*Characteristics*
Excitation force	F_E	Any outside source	No restriction
Resistive force	F_R	Overcomes "friction" within system	Work done is lost as heat May be related to how "fast" system is moving
Capacitive force	F_C	Overcomes opposition of springs and the like	Work done is stored Related to amount of deflection from equilibrium
Net force or Inertive force	F_{net} F_M	Acts to "accelerate" the system	Work done is stored Related to amount of "acceleration"

Figure 4-15 summarizes many of the important elements of Table 4-3 and of the unified concepts in general. Excitation force F_E is responsible for transfer of energy into a system. Resistive force F_R transfers energy out of the system via a resistor. The other two forces account for the energy that is stored within the system. Study this carefully. This book is all about how that energy enters, how some leaves the system, and how the rest is stored.

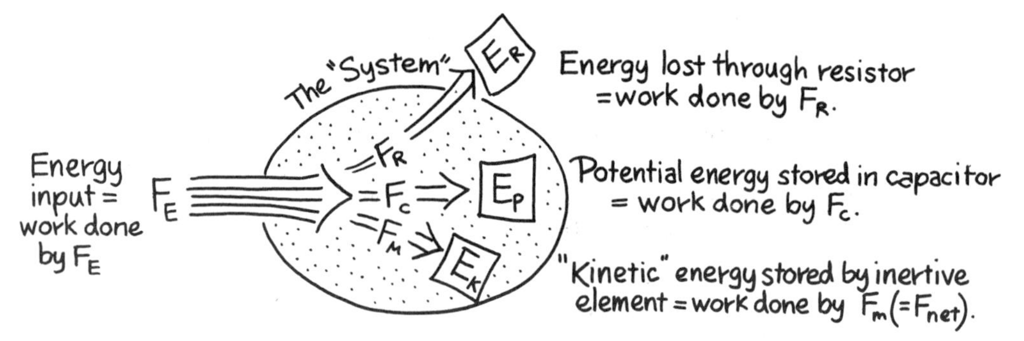

Figure 4-15. Schematic look at a system containing a resistor, a capacitor, and an inertive element.

4-7 SUMMARY

The unified concept "force" is force in the mechanical translational system, torque in the mechanical rotational system, pressure in the fluid system, temperature in the heat system, and potential difference in the electrical system. Net force or force difference moves or tends to move things in systems. The "things" that help to measure movement are called *parameters*, which is the title of the next chapter.

Table 4-4 summarizes the various force-like quantities and their defining equations, symbols, and units.

TABLE 4-4 SUMMARY OF FORCE-LIKE QUANTITIES

System	*Force and symbol*	*Equation*	*SI Unit (Symbol)*	*Other Units*
Mechanical translational	Force F		newton (N)	lb, oz, dyn
Mechanical rotational	Torque τ	$\tau = Fl$	newton·metre (N · m)	lb · ft, oz · in.
Fluid	Pressure p	$p = F/A$	pascal (Pa) = (N/m^2)	psi, atm, lb/ft^2, inches of water, torr
Heat	Temperature difference T		kelvin (K)	°F, °C, °R
Electrical	Potential difference E or V		volt (V)	None

PROBLEMS

4-1. What is the weight of a 350-g mass in (a) newtons (N), (b) dynes (dyn), and (c) pounds (lb)?

4-2. A spring scale reads 145 lb. What is the scale reading in newtons?

4-3. What mass in kilograms (kg) weighs 15 lb?

4-4. What is the net horizontal force on the 800-g mass in newtons? What is the direction of the net force? The surface and the pulleys are frictionless.

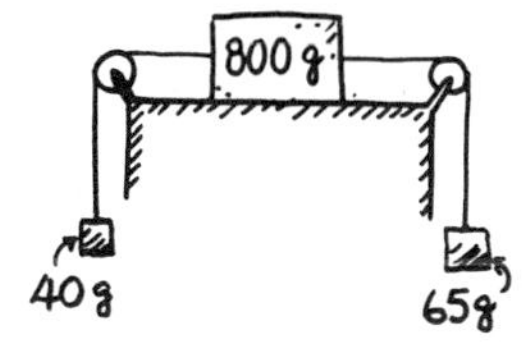

Problem 4-4

4-5. The spring scale reads 35 lb. What torque is exerted by it around the axis of rotation? What is the direction of the torque?

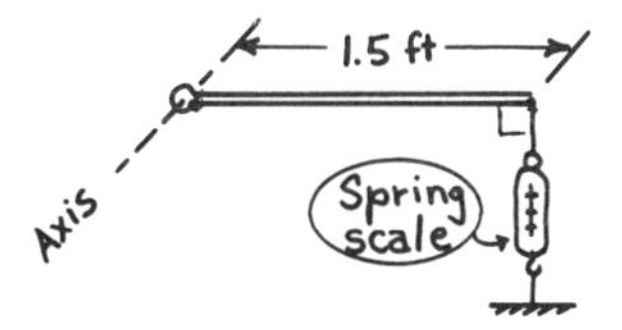

Problem 4-5

4-6. What torque is exerted by the 750-g mass about the pulley's axis of rotation? Give your answer in newton metres (N · m). What is the direction of the net torque?

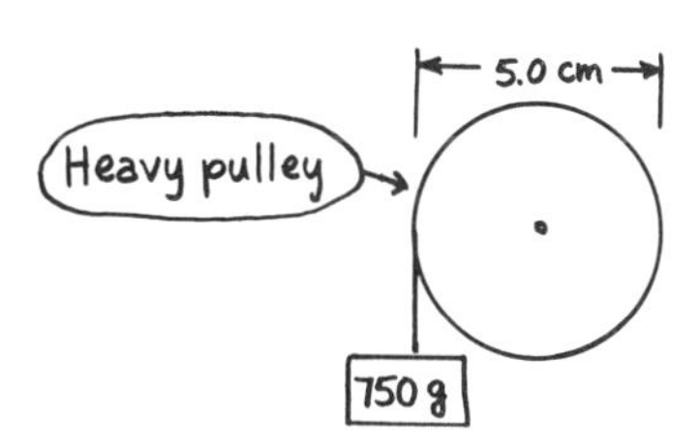

Problem 4-6

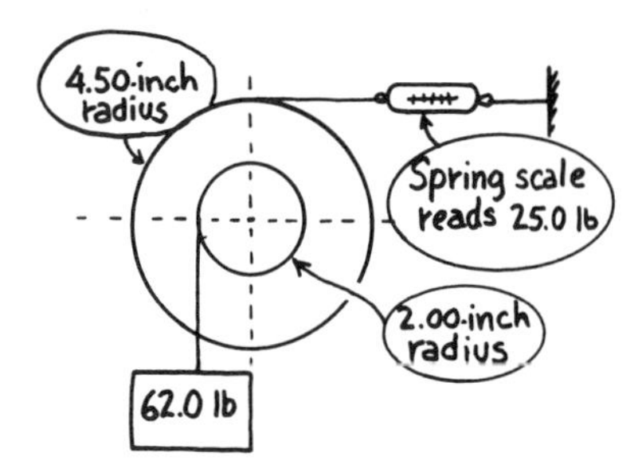

Problem 4-7

4-7. What is the net torque exerted on the flywheel about its axis of rotation? Give your answer in pound · inches (lb · in.). What is the direction of the net torque?

4-8. What is the net torque on the center pulley? Give your answer in newton · metres. What is the direction of the net torque? The two outer pulleys are frictionless.

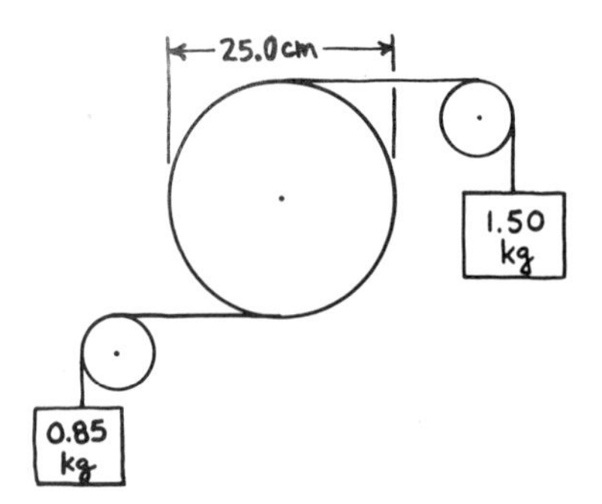

Problem 4-8

4-9. A force of 150 lb acts against a piston of diameter 2.00 in. that encloses a gas inside a cylinder. (a) What pressure in pounds per square inch (psi) is exerted by the piston on the trapped gas? (b) convert this pressure into pascals (Pa).

4-10. A 13.0-kg mass rests on top of a piston that encloses a gas within a cylinder. The cross-sectional area of the piston is 0.25 square metres (m^2). What pressure in pascals is exerted on the gas by the piston? Neglect the piston's weight and atmospheric pressure in making the calculation.

4-11. If a gauge pressure is 250 lb/ft^2, what is the absolute pressure in pounds/square inch (psi)?

4-12. The pressure at one point along a straight, horizontal pipe containing a moving fluid is 60 psi and the pressure at another point further downstream is 2.0 atm. What is the pressure difference between the two points? Express your answer in pounds per square foot.

4-13. Convert a pressure of 30 psi into (a) pascals; (b) pounds/square foot.

4-14. Convert a pressure of 0.00125 atm into (a) torr (mm of mercury); (b) inches of water.

4-15. A research submersible (submarine) is being considered for use in a large freshwater lake to get a look at the bottom. The vessel has a circular window of 8.00-in. diameter. What is the total force on the window at the maximum depth of 1640 ft? Neglect atmospheric pressure in making this calculation.

4-16. A Celsius thermometer measures the temperature of a sick person as 38.2°C. What is this temperature in degrees Fahrenheit?

4-17. The melting point of aluminum is 1220°F. What is this in degrees Celsius?

4-18. The temperature inside a furnace is 524°C and the temperature of its surface is 162°C.

(a) What is the temperature difference across the walls in degrees Fahrenheit?

(b) What is the outside (surface) temperature in degrees Fahrenheit, degrees Rankine, and kelvins?

4-19. Convert the following temperatures as indicated:

(a) "Room temperature," 72°F, to degrees Celsius.

(b) The highest temperature I've ever worked in, 116°F, to degrees Celsius.

(c) The freezing point of mercury, −40°C, to degrees Fahrenheit.

4-20. The steepest drivable street in San Francisco has a slope of 17.5° from the horizontal. A 2000-kg car is parked parallel to the street and its tires are not against the curb.

(a) What force in newtons must the brakes exert to keep the car from rolling down the hill?

(b) How many pounds is this?

4-21. A heavy object is acted upon by a force of 200 N whose line of action is slanted upward at an angle of 26° with respect to the horizontal surface.

(a) How much of this force tends to move the object horizontally along the surface?
(b) How much of this force tends to lift the object vertically away from the surface?
(c) What is the magnitude of the 200-N force in pounds?

4-22. Spring scale A reads 75 N. What does spring scale B read if the net horizontal force on the block is zero? The surface is frictionless.

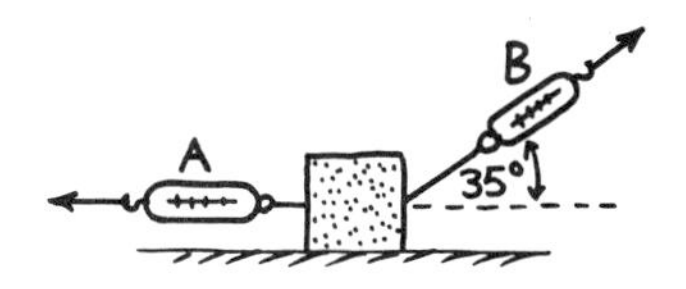

Problem 4-22

4-23. A 500-N weight is supported by two ropes as shown. What is the force in each rope? Their resultant just balances the 500-N weight.

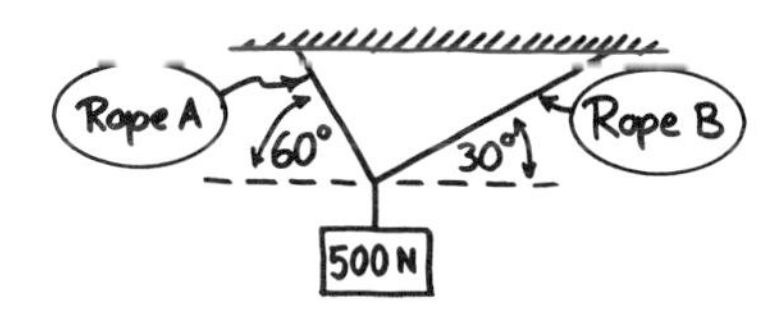

Problem 4-23

4-24. A construction crane lifts a 1600-lb load. The load is then pulled aside with a rope so that the crane cable makes a 15° angle with the vertical.
(a) What is the tension in the cable?
(b) What is the tension in the rope?

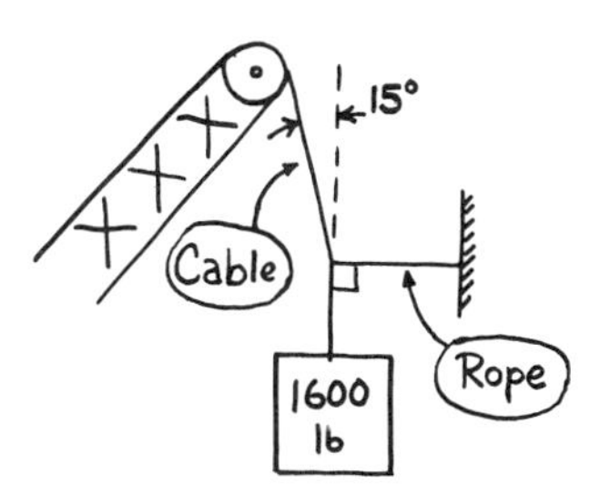

Problem 4-24

4-25. How large must force F be such that the net torque about point C is zero?

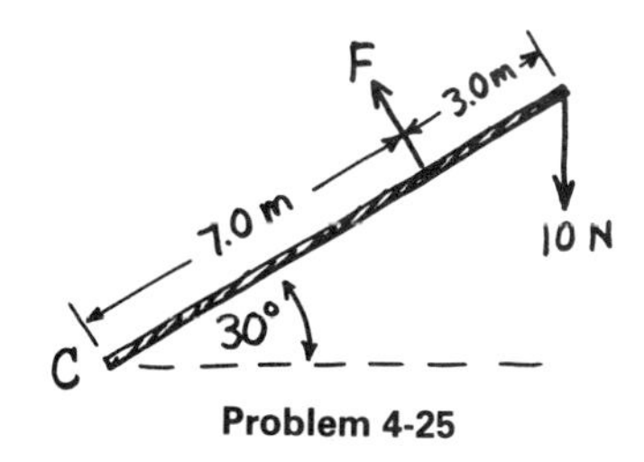

Problem 4-25

4-26. The 5.0-g mass rotates the pulley of a viscosimeter (an instrument used to measure frictional forces in fluids).
(a) What is the weight of the suspended mass in dynes?
(b) What torque in dyne · centimetres is applied to the pulley?

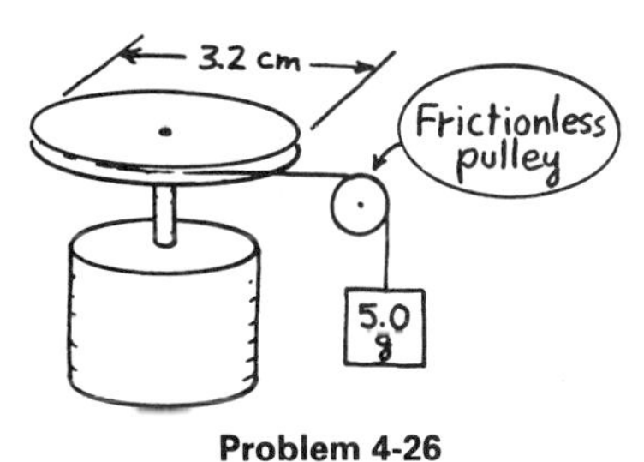

Problem 4-26

4-27. What is the tension force in the guy wire if the net torque about point A on the mast is zero?

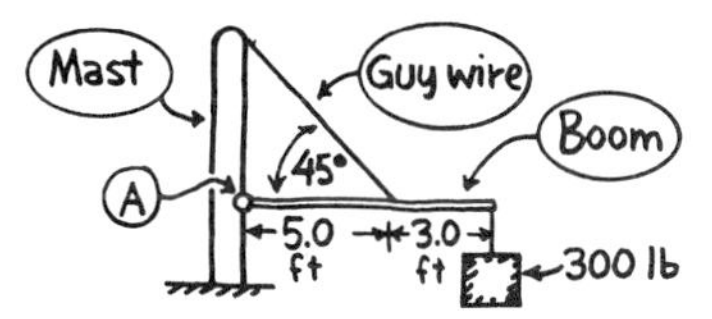

Problem 4-27

4-28. What is the resultant (net) torque on the shaft in pound · inches? What is the direction of this torque? Convert your answer to newton · metres.

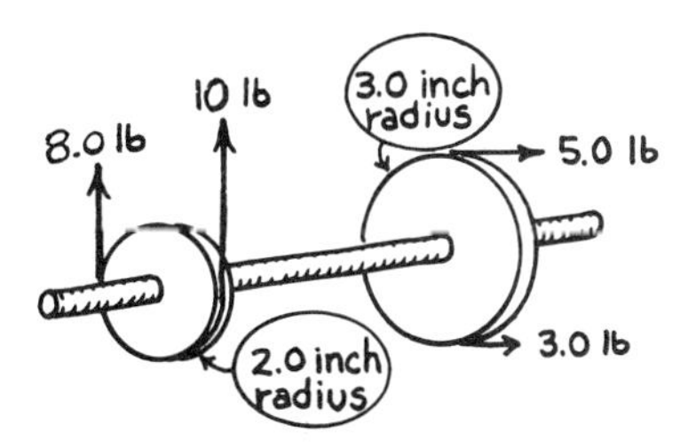

Problem 4-28

4-29. What absolute pressure does each manometer measure? The liquid is mercury. Give your answer in pascals.

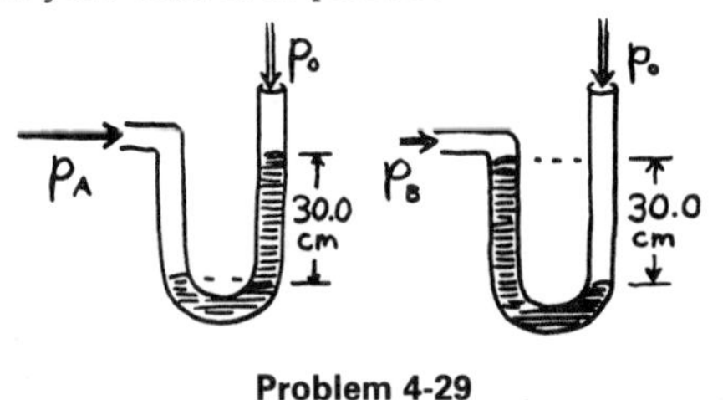

Problem 4-29

4-30. A rectangular swimming float is 12.0 ft long and 8.0 ft wide. The float itself weighs 600 lb. What must be the total weight of swimmers on the float so that the float's bottom surface sinks exactly 0.50 ft into the water? Assume that it is in fresh water, not in the ocean. (*Hint*: First calculate the force from the pressure on the bottom of the float.)

4-31. In the system of Fig. 4-14, if F_E = 20 lb, F_C = 4 lb, and F_M = 7 lb, what force is needed to overcome the friction of the dashpot?

4-32. If it takes 35 N to stretch the spring as much as shown in Fig. 4-14, and the dashpot is producing 20 N of frictional force, how much of a 100-N excitation force goes toward making the system move faster?

ANSWERS TO ODD-NUMBERED PROBLEMS

4-1. (a) 3.43 N
(b) 3.43×10^5 dyn
(c) 0.772 lb
4-3. 6.8 kg
4-5. 52 lb · ft
4-7. 11 lb · in.
4-9. (a) 47.7 psi
(b) 3.29×10^5 Pa
4-11. 16.4 psi
4-13. (a) 2.1×10^5 Pa
(b) 4.3×10^3 lb/ft^2
4-15. 3.57×10^4 lb
4-17. 660°C
4-19. (a) 22°C
(b) 47°C
(c) −40°F
4-21. (a) 180 N
(b) 87.7 N
(c) 45.0 lb
4-23. 433 N; 250 N
4-25. 12 N
4-27. 680 lb
4-29. 1.41×10^5 Pa
0.613×10^5 Pa
4-31. 9 lb

5 Parameters

The selection of the force-like quantities in Chapter 4 is based upon their contribution to movement within the systems. Forces become a unified concept because they play the same role within the various systems. Nothing changes its motion unless a force, torque, pressure, electromotive force, or temperature difference is present.

The next unified concept to be introduced may not seem so easy to identify in a given system. A ***parameter*** *is a physical quantity whose measured value helps to define the behavior of a system. Each system has many parameters in the normal sense. Force, weight, density, mass, acceleration, potential energy, kinetic energy, momentum, and impulse are examples of physical quantities—parameters—whose measured value helps to define the behavior of a mechanical system. Yet in this text we will select only one physical quantity for each system and call it parameter. Except for heat systems, we will define parameter such that the product of force and parameter is work.*

$$\boxed{\textit{work} = \textit{force} \times \textit{parameter}}$$

Once the force is chosen for a given system, the selection of the parameter is narrowed to a single choice. For mechanical, fluid, and electrical systems these choices are distance, angle, volume, and charge. Fortunately, the choices are all familiar physical quantities. The heat system, we shall discover, does not fit this simple pattern and so an exception must be made.

5-1 DISTANCE IN MECHANICAL TRANSLATIONAL SYSTEMS

Section 3-4 discusses the idea of work and the relationship between work and energy. Take some time to go back and review this very important section. A force must move an object from one place to another in order to do work on the object. If the force and the displacement (distance moved) are in the same direction, then work is the product of force and distance. If the force's line of action is in a direction different from the displacement, then only the component of the force in the direction of motion enters into the equation for work. Figure 5-1 shows these possibilities.

$$W = Fs \qquad (3\text{-}9) \qquad\qquad W = (F \cos \theta)s \qquad (5\text{-}1)$$

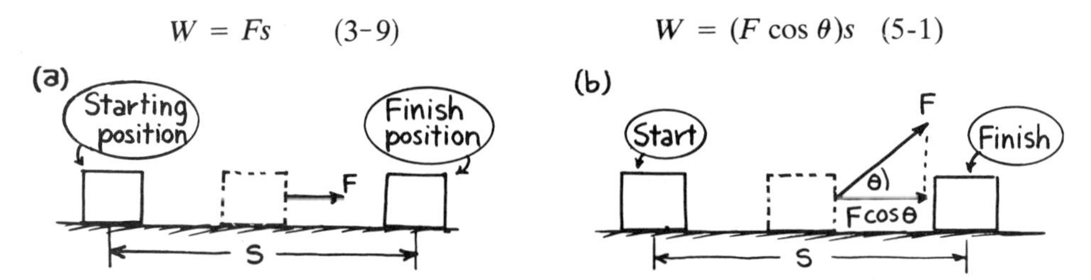

Figure 5-1. (a) The force and distance moved are in the same direction. (b) The force's line of action is in a direction different from the motion.

F *Force*	s *Distance*	θ *Angle*	W *Work*
newton (N)	metre (m)	degree(°) or radian (rad)	joule (J)
pound (lb)	foot (ft)		foot · pound (ft · lb)

Distance is the parameter in this system because force times distance is work. A few instruments that measure distance are metre sticks, vernier calipers, and micrometers. The odometer of Sec. 3-1 measures the distance a car moves.

The unit of distance or displacement in the English system of units is the *foot* (ft). The SI unit of distance is the *metre* (m).

A number of useful distance conversion factors are listed below. They also appear in the table, *Useful Conversion Factors*, printed in Appendix A.

$$1 \text{ metre (m)} = 3.28 \text{ feet (ft)}$$

$$1 \text{ metre (m)} = 39.4 \text{ inches (in.)}$$

$$1 \text{ inch (in.)} = 2.54 \text{ centimetres (cm)}$$

$$1 \text{ mile (mi)} = 5280 \text{ feet (ft)}$$

$$1 \text{ kilometre (km)} = 0.621 \text{ mile (mi)}$$

A useful conversion factor for work or energy is

$$1 \text{ joule (J)} = 0.738 \text{ foot} \cdot \text{pound (ft} \cdot \text{lb)}$$

Remember that 1 J of work is equivalent to 1 newton · metre (N · m) of work.

Example 5-1 A horizontal 55-lb force moves a heavy object a horizontal distance of 16 ft.

(a) How much work is done by the force in foot · pounds?
(b) How much work is done by the force in joules?
(c) How many metres does the object move?

Solution to (a)

Since the force and distance are in the same direction:

$\begin{Bmatrix} F = 55 \text{ lb} \\ s = 16 \text{ ft} \end{Bmatrix}$

$$W = Fs = 55 \text{ lb}(16 \text{ ft}) = \boxed{8.8 \times 10^2 \text{ ft} \cdot \text{lb}}$$

Solution to (b)

$$8.8 \times 10^2 \text{ ft} \cdot \text{lb}\left(\frac{1 \text{ J}}{0.738 \text{ ft} \cdot \text{lb}}\right) = \boxed{1.2 \times 10^3 \text{ J}}$$

Solution to (c)

$$16 \text{ ft}\left(\frac{1 \text{ m}}{3.28 \text{ ft}}\right) = \boxed{4.9 \text{ m}}$$

5-2 ANGLE IN MECHANICAL ROTATIONAL SYSTEMS

Angle of rotation is the parameter for this system. In Fig. 5-2, point A at the end of the hinged stick rotates in a counterclockwise direction. The angle through which point A rotates is θ. This is the same angle through which point B rotates, even though point B is closer to the rotational axis. The distance moved by point A is the arc length s. The arc radius is r. A common measure of the angle θ is *degrees* (°), there being 360° in a full circle. Another measure

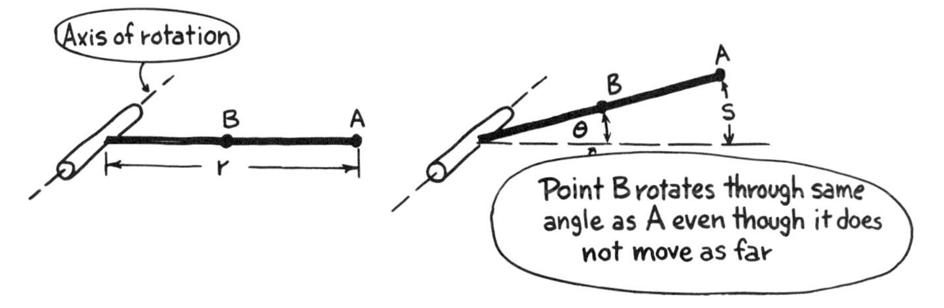

Figure 5-2. Point A rotates through an angle θ traveling a distance s. The distance from A to the axis of rotation is r, and the angle θ in radians is the ratio of s to r.

of angle is the revolution, one revolution being equivalent to 360°. The measure of angle that we will use most is the *radian* (rad). An angle measured in radians is the ratio of the arc length s to the radius r.

$$\theta = \frac{s}{r} \tag{5-2}$$

θ *Angle*	s *Length*	r *Radius*
radian (rad)	metre (m)	
	foot (ft)	

The radian is dimensionless; that is, it has no units at all since angle is a ratio of two lengths. It is easy to convert back and forth between degrees and radians. In one full rotation of 360° the arc length s is the circle's circumference $2\pi r$, and the ratio s/r is $2\pi r/r$ or 2π rad. Hence 2π rad equals 360°, or 1 rad equals $360/2\pi = 57.3°$.

$$1 \text{ radian (rad)} = 57.3 \text{ degree } (°)$$

$$1 \text{ revolution (rev)} = 6.28 \text{ radian (rad)}$$

Appendix A gives the radian equivalent for whole angles between zero and 90°. A couple of instruments that measure angle are the protractor and a revolution counter.

Let's see if the radian measure of angle satisfies the definition of parameter in the mechanical rotational system. Remember that torque is the "force" in this system. Hence force times parameter or torque times angle should equal work. In Fig. 5-3, a force F rotates point A at the end of the stick along an arc length s. The work done by this force is

$$W = F \times s \tag{3-9}$$

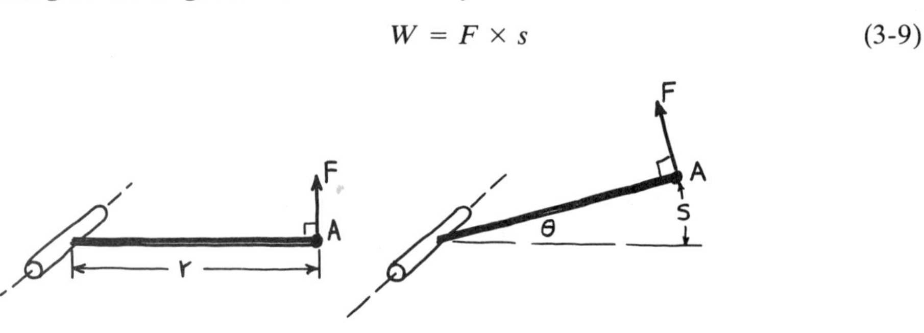

Figure 5-3. The force F moves a distance s and does work $W = Fs$ on the rotating stick. The torque τ is the product $F \times r$. The angle θ in radians is the ratio s/r.

since the force and distance moved are in the same direction. (Both force and distance change direction, but they change direction together, being at all times perpendicular to the radius r.) Multiply the right side of Eq. (3-9) by r/r. The force becomes torque ($\tau = Fr$), and the length becomes angle measured in radians ($\theta = s/r$).

$$W = Fs \times \frac{r}{r}$$

$$= Fr \times \frac{s}{r}$$

$$W = \tau\theta \tag{5-3}$$

W *Work*	θ *Angle*	τ *Torque*
joule (J)	radian (rad)	newton · metre (N · m)
foot · pound (ft · lb)		pound · foot (lb · ft)

So, indeed, angle θ *in radians* is the parameter for this system, since the product of torque and angle is work.

Example 5-2 A force of 25 N does 15 J of work on the hinged stick of Fig. 5-3. Point A is 0.20 m from the axis of rotation.

(a) What torque is exerted on the stick?
(b) Through what angle in radians does the hinged stick rotate?
(c) How many degrees is the angle of rotation?

Solution to (a)

$\left\{ \begin{array}{l} F = 25\text{N} \\ r = 0.20\text{ m} \end{array} \right\}$

$$\tau = Fr = 25\text{ N}(0.20\text{ m}) = \boxed{5.0\text{ N}\cdot\text{m}}$$

Solution to (b)

$\{W = 15\text{J}\}$

$$W = \tau\theta$$

$$\theta = \frac{W}{\tau} = \frac{15\text{ J}}{5.0\text{ N}\cdot\text{m}} = \boxed{3.0\text{ rad}}$$

Solution to (c)

$$3.0\text{ rad}\left(\frac{57.3^\circ}{1\text{ rad}}\right) = \boxed{170^\circ}$$

5-3 VOLUME IN FLUID SYSTEMS

The unified concept of parameter in fluid systems is *volume V*. Figure 5-4 illustrates several popular ways of measuring liquid volume. In the first method, we draw upon the fact that the volume in this particular case is in the form of a cylinder. If the radius r and the height h can be measured, the volume can be calculated. Table 2-1 lists formulas for simple areas and volumes. Referring to this table, we find that the volume V of a cylinder is $V = \pi r^2 h$. In the second method, either the liquid's mass or weight is

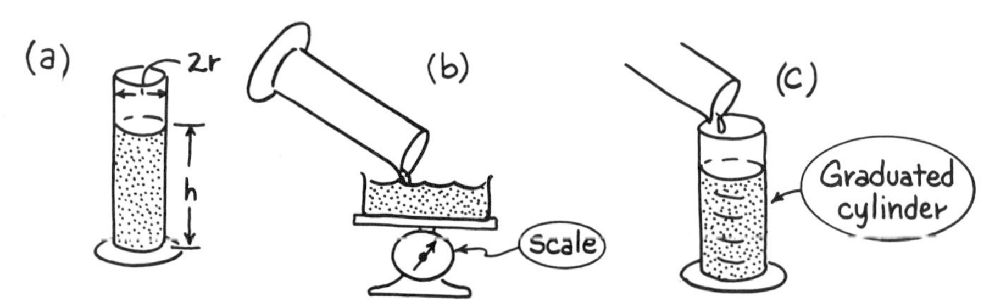

Figure 5-4. The volume of a liquid can be measured in several ways. In (a) the height h and radius r are measured. The volume V is calculated using $V = \pi r^2 h$. In (b) the liquid is weighed and the volume calculated using the density equation $V = m/\rho$. In (c) the volume is measured directly in a graduated cylinder.

measured. The volume is calculated using the definition of density $\rho = m/V$, as discussed in Sec. 3-2. A third method is to simply pour the liquid into a graduated cylinder and measure the volume directly. Example 5-3 illustrates the first two methods.

Example 5-3 Two volumes of gasoline are to be compared. The first volume is in a cylinder with radius 6.00 in. and height 18.0 in. The second volume of gasoline weighs 52.0 lb. Which volume is greater?

Solution

$$r = 6.00 \text{ in.}\left(\frac{1 \text{ ft}}{12 \text{ in.}}\right) = 0.500 \text{ ft}$$

$$h = 18.0 \text{ in.}\left(\frac{1 \text{ ft}}{12 \text{ in.}}\right) = 1.50 \text{ ft}$$

Let V_1 be the first volume and V_2 the second volume.

$$V_1 = \pi r^2 h = \pi(0.500 \text{ ft})^2(1.50 \text{ ft}) = \boxed{1.18 \text{ ft}^3}$$

$$V_2 = \frac{m}{\rho} = \frac{mg}{\rho g} = \frac{52.0 \text{ lb}}{42.0 \text{ lb/ft}^3} = \boxed{1.24 \text{ ft}^3}$$

↖ App. B, Table 1

The second volume is greater.

Measurement of a volume of gas may be more difficult. A liquid cannot be easily compressed, but a gas can. The density of a gas depends upon its temperature and pressure. Gas volume can be calculated by measuring the dimensions of the container, but calculations of volume based upon weight must reflect the gas temperature and pressure. Appendix B, Table 1, lists several gas densities at a pressure of 1 atm and a temperature of 0°C.

Let's see if volume satisfies the definition for parameter in fluid systems. Remember that pressure is the "force" for this system. Hence force times distance or pressure times volume should equal work. In Fig. 5-5, the piston's

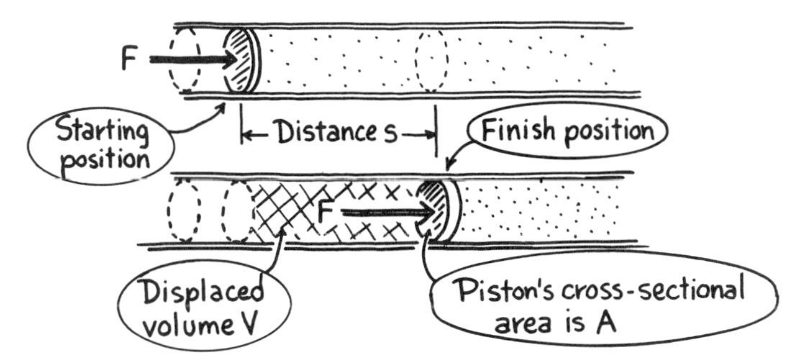

Figure 5-5. The steady force F pushes the piston a distance s to the right. The moving piston displaces the volume $V = As$. The piston pressure is $p = F/A$.

cross-sectional area is A. A steady force F moves the piston to the right a distance s. The force does work on the fluid. Using Eq. (3-9), this work is,

$$W = F \times s \tag{3-9}$$

Multiply the right side of this equation by A/A. The force becomes pressure ($p = F/A$), and the distance becomes volume ($V = As$):

$$W = Fs \times \frac{A}{A}$$

$$\underset{\text{Work}}{W} = \underset{\text{Pressure}}{\frac{F}{A}} \times \underset{\text{Volume}}{(As)}$$

$$W = pV \tag{5-4}$$

W *Work*	p *Pressure*	V *Volume*
joule (J)	pascal (Pa)	cubic metre (m^3)
foot · pound (ft · lb)	pound/square foot (lb/ft^2)	cubic foot (ft^3)

So volume V is indeed the parameter for the fluid system, because the product of pressure and volume is work. The unit of volume is the cubic metre (m^3) in SI units and the cubic foot (ft^3) in the English system of units.

Example 5-4 A steady force of 75 lb, acting on a piston with a cross-sectional area of 4.0 in.^{2}, displaces 30 lb of water through a small opening at the base of a cylinder.

(a) What volume of water is displaced by the piston?
(b) What pressure in pounds/square foot does the piston exert?
(c) How much work is done by the force F?

Solution to (a)

$\{W = 30\text{ lb}\}$

This W means "weight"

$$V = \frac{\mathrm{W}}{\rho g} = \frac{30\text{ lb}}{62.4\text{ lb/ft}^3} = \boxed{0.48\text{ ft}^3}$$

Appendix B, Table 1, for water

Solution to (b)

$\begin{Bmatrix} A = 4.0\text{ in.}^2 \\ F = 75\text{ lb} \end{Bmatrix}$

$$A = 4.0\text{ in.}^2\left(\frac{1\text{ ft}^2}{144\text{ in.}^2}\right) = 2.78 \times 10^{-2}\text{ ft}^2$$

$$p = \frac{F}{A} = \frac{75\text{ lb}}{2.78 \times 10^{-2}\text{ ft}^2} = \boxed{2.7 \times 10^3\text{ lb/ft}^2}$$

Solution to (c)

$$\mathrm{W} = pV = (2.7 \times 10^3\text{ lb/ft}^2)(0.48\text{ ft}^3) = \boxed{1.3 \times 10^3\text{ ft} \cdot \text{lb}}$$

This W means "work"

In the examples discussed so far, force and displacement are in the same direction. Sometimes the displacement may be opposite to the force direction. In Fig. 5-6a, the force and displacement are in the same direction. The piston

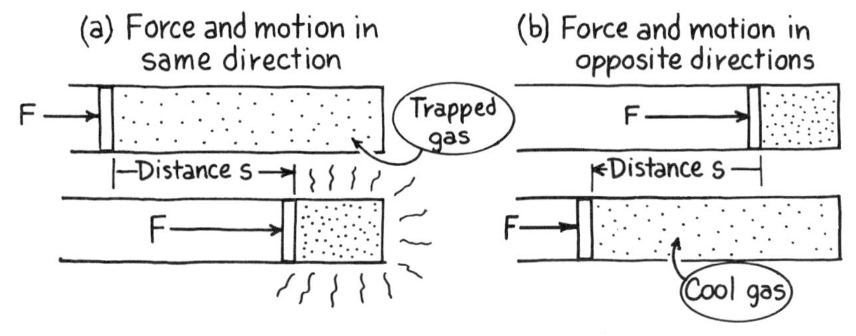

Figure 5-6. In (a) the force F and distance s are in the same direction. The force does work on the gas, that is, the gas compresses and heats up. In (b) the force and distance are in opposite directions. The gas does work by pushing the piston backward. The gas expands, loses some of its energy, and cools down. Both situations are complicated by the fact that the force is continually changing.

does work on the gas in this case, and the gas gains energy. It is compressed and heats up. In Fig. 5-6b, the force and displacement are in opposite directions. The gas does work on the piston and loses some energy by expanding and cooling down. These situations are further complicated by the fact that the pressure continually changes during the displacement. Equation (5-3) still applies, however, for very small displacements. We will discuss changing forces in more detail in later chapters.

5-4 CHARGE IN ELECTRICAL SYSTEMS

Section 3-6 discusses the atomic nature of matter. It talks about electrons, protons, and neutrons. Electrons and protons are charged particles that attract or repel each other according to the following rule: *like charges repel*; *unlike charges attract.*

The amount of *charge* Q is parameter in the electrical system. It is a measure of the excess electrons or protons in a given location. Any bit of matter that you can see includes incredibly large numbers of charged particles, but normally there are as many positively charged ones as negatively charged ones. Only if there is more of one than the other do we call a body charged.

The SI unit of charge is the coulomb (C). A coulomb represents a grouping of 6.25×10^{18} excess electrons or protons. Why does the coulomb represent such a large number of electrons or protons? The answer relates to the almost incomprehensible number of atoms that make up matter. Suppose that we have a mass of copper that is 63.5 g, the atomic weight of copper. This is a volume of about 7 cm^3, a bit less than 1 in.^{3}. The number of free electrons moving, if the cube were part of an electric circuit, is about 6×10^{23} electrons. Hence the charge available in this small mass of copper is

$$6 \times 10^{23} \text{ free electrons} \times \frac{1\text{ C}}{6.25 \times 10^{18} \text{ electrons}}$$

$$= \text{(approximately) } 10^5 \quad \text{or} \quad 100{,}000 \text{ C}$$

So maybe the coulomb is not so big after all! The charge of an electron or proton is sometimes useful to know. The charge of one electron is 1.6×10^{-19} C.

Section 3-6 defines electromotive force E as

$$E = \frac{W}{Q} \tag{3-17}$$

where W is the work done or energy added to an amount of charge Q. A battery whose emf is 1 volt (V) adds 1 J of energy to each coulomb of charge that passes through it. A volt is a joule per coulomb. Fortunately, the coulomb and volt are the only practical units of charge and emf that we need to be concerned with. Equation (3-16) can be rearranged to give

$$W = EQ \tag{5-5}$$

W *Work*	E *Electromotive Force*	Q *Charge*
Joule (J)	volt (V)	coulomb (C)

Equation (5-5) is in agreement with the unifying idea that work is the product of force and parameter. The force in an electrical system is emf. If you have trouble thinking of a volt as a joule per coulomb, you may want to think of emf as an electrical pressure analogous to pressure in a fluid system. Many people find this a useful way of looking at emf.

Example 5-5 A 1.5-V battery adds 12 J of energy to some charge.
(a) What amount of charge moved through the battery?
(b) How many electrons does this represent?

Solution to (a)

$$\left\{\begin{matrix} E = 1.5\text{ V} \\ W = 12\text{ J} \end{matrix}\right\}$$

$$E = \frac{W}{Q}$$

$$Q = \frac{W}{E} = \frac{12\ J}{1.5\ V} = \boxed{8.0\ C}$$

Solution to (b)

$$8.0\ \text{C}\left(\frac{6.25 \times 10^{18}\ \text{electrons}}{1\ \text{C}}\right) = \boxed{5.0 \times 10^{19}\ \text{electrons}}$$

5-5 QUANTITY OF HEAT IN HEAT SYSTEMS

The parameter for the heat system is *quantity of heat.* Section 3-5 discusses quantity of heat and gives definitions for two of its units, the calorie (cal) and the British thermal unit (Btu). These definitions are as follows:

1. A *calorie* is the quantity of heat needed to raise the temperature of 1 gram of water by 1 degree Celsius.
2. A *Btu* is the quantity of heat needed to raise the temperature of 1 pound of water by 1 degree Fahrenheit.

The specific heat c of a material is a measure of its ability to absorb heat, and the quantity of heat Q needed to change the temperature of a material can be found using

$$Q = mc(T_f - T_0) \tag{3-14}$$

where m is the mass of material, and T_f and T_o are the final and original temperatures, respectively. Values of specific heat for various materials can be found in Appendix B, Table 3.

Heat is a form of energy, and conversion between heat units and energy units may sometimes be necessary. A few useful conversion factors shown here are taken from the table, *Useful Conversion Factors*, located in Appendix A.

$$1 \text{ British thermal unit (Btu)} = 252 \text{ calories (cal)}$$

$$1 \text{ joule (J)} = 0.239 \text{ calorie (cal)}$$

$$1 \text{ joule (J)} = 9.49 \times 10^{-4} \text{ British thermal unit (Btu)}$$

$$1 \text{ joule (J)} = 0.738 \text{ foot} \cdot \text{pound (ft} \cdot \text{lb)}$$

Up to this point, the selection of the "force" and the parameter has worked out very nicely. The force-like quantity is something that gets things moving, the parameter is a useful and familiar physical quantity, and the product of force and parameter is work or energy. Unfortunately, as you have been warned, the heat system does not fit so nicely into this pattern. The product of force and parameter is temperature times energy, which is obviously not energy. Another parameter could have been chosen (such as energy per degree of temperature), but quantity of heat as a parameter fits in with other ideas, such as heat transfer and heat storage, that we will deal with later. We want to maintain analogies to fluid and electrical systems because such analogies are used in studying complicated heat problems. Quantity of heat is the best selection for the heat system parameter in order to make these analogies. So we will stick with quantity of heat as parameter even though it fails the basic test that the product of force and parameter is energy.

Example 5-6 (a) How much heat in Btu's must be added to 45 lb of copper to change its temperature from 68 to 212°F?

(b) How many joules of energy does this represent?

Solution to (a)

$$\begin{Bmatrix} m = 45 \text{ lb} \\ T_f = 212°\text{F} \\ T_o = 68°\text{F} \end{Bmatrix}$$

Appendix B, Table 3

$$Q = mc(T_f - T_o) = 45 \text{ lb}\left(0.093 \frac{\text{Btu}}{\text{lb} \cdot °\text{F}}\right)[(212 - 68)°\text{F}]$$

$$= 602 \rightarrow \boxed{6.0 \times 10^2 \text{ Btu}}$$

Solution to (b)

$$602 \text{ Btu}\left(\frac{1 \text{ J}}{9.49 \times 10^{-4} \text{ Btu}}\right) = \boxed{6.3 \times 10^5 \text{ J}}$$

Note that m stands for weight when $Q = mc(T_f - T_o)$ is used with English units.

TABLE 5-1 FORCE AND PARAMETER QUANTITIES

System	Force-like Quantity			Parameter			Work = Force × Parameter (5-1)	
	Name	*Symbol*	*Units*	*Name*	*Symbol*	*Units*		
Mechanical translational	Force	F	newton (N) pound (lb)	Distance	s	metre (m) foot (ft)	$W = Fs$	(3-9)
Mechanical rotational	Torque	τ	newton · metre (N · m) pound · foot (lb · ft)	Angle	θ	radian (rad) radian (rad)	$W = \tau\theta$	(5-3)
Fluid	Pressure	p	pascal (Pa) pound/square foot (lb/ft^2)	Volume	V	cubic metre (m^3) cubic foot (ft^3)	$W = pV$	(5-4)
Electrical	Electromotive force (emf)	E	volt (V)	Charge	Q	coulomb (C)	$W = EQ$	(5-5)
Heat	Temperature	T	Celsius degree (°C) Fahrenheit degree (°F)	Quantity of heat	Q	calorie (cal) British thermal unit (Btu)	Analogy fails	

5-6 SUMMARY

Force and parameter are the cornerstones upon which the unified concepts are built. They have been selected so that analogies can be made when we study how energy is transferred and stored in systems. The criterion for their selection is that the product of force and parameter is equal to work or energy. The force-like quantity was picked first as something that gets things moving. Parameter was then chosen to fit the basic equation that says force times parameter equals energy. Fortunately, this gives us parameters that are useful and familiar physical quantities. The whole picture fits together nicely except for the heat system, in which the force (temperature) and parameter (heat energy) product does not equal energy. In spite of this, many useful analogies can be made between the heat system and other systems. Table 5-1 summarizes all this.

Our task now is to study how energy is transferred and stored in the various systems. Before doing this we will look at how parameter changes with time. This is the subject of the next chapter.

PROBLEMS

5-1. A horizontal force of 150 N pushes a block 4.0 m along a horizontal surface. How much work is done by the force?

5-2. A force pulls a block 15 ft along a horizontal surface. How much work is done by the force if its magnitude is 45 lb and its line of action is inclined upward at an angle of 20° with the horizontal?

5-3. A shaft has a radius of 4.0 in. Through what angle in radians does the shaft rotate if:
(a) the angle of rotation is 30°?
(b) the shaft turns 1.5 revolutions?
(c) a point on the surface moves an arc length of 6.0 in.?

5-4. A 30-N weight falls 0.10 m and unwinds a string wrapped around a pulley of diameter 0.18 m.
(a) What torque is exerted on the pulley?
(b) Through what angle does the pulley rotate? (Express your answer in radians.)
(c) How much work is done by the torque?

5-5. The cross-sectional area of a pipe is 0.010 m^2 and the length is 1.5 m. Find the volume of water in cubic metres inside the pipe and find the mass of the water in kilograms.

5-6. A steady pressure of 2.00 atm forces 300 ft^3 of propane gas through a hose. How much energy in foot · pounds is needed to do this?

5-7. (a) How much charge in coulombs is represented by an accumulation of 15,000 excess electrons?
(b) How much charge is represented by a deficiency of 80,000 electrons?

5-8. The charge on one plate of a capacitor is a negative 5.0×10^{-6} C. What kind of atomic particle is in excess on this plate and how many of them are present?

5-9. How much heat energy in calories is needed to change the temperature of 120 g of silver from 22 to 50°C?

5-10. In an experiment, 500 cal of heat energy changes the temperature of 40 g of a material from 20 to 78°C. What kind of material might this be?

5-11. A 2.0-ft^3 container filled with loose earth is lifted 50 ft vertically upward.
(a) What force in pounds is needed to lift the container?
(b) How much energy is expended in doing this?

5-12. The small pulley makes two complete revolutions in a counterclockwise direction.
(a) How far does a dot painted on the belt move?
(b) How many revolutions does the larger pulley make? Assume that there is no slippage between the belt and the pulleys.

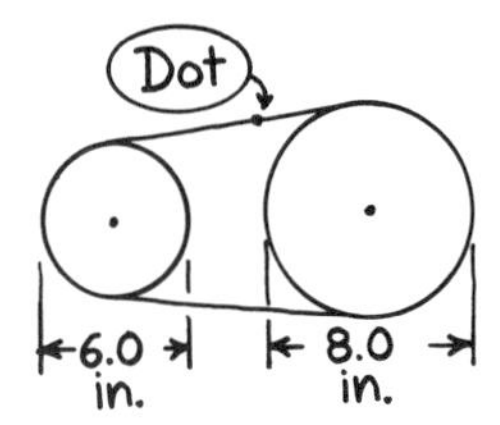

Problem 5-12

5-13. Approximate the volume in cubic centimetres of water in the U-tube. The inside diameter of the tube is 1.00 cm.

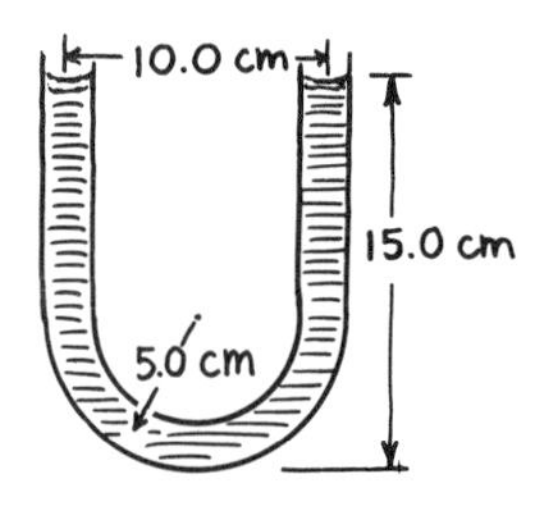

Problem 5-13

5-14. A pressure of 4.0×10^3 Pa (a pressure equivalent to a water column approximately 40 cm high) forces 50 cm^3 of water through a small-diameter capillary tube. How much energy in joules is needed to do this? (*Remember*: a pascal (Pa) is 1 N/m^2.)
5-15. If 240 J of energy is added to 10 C of charge in a battery, how much energy would be added to 25 C of charge?
5-16. A typical 12-V car battery stores 1.5×10^5 C of charge. How much energy in joules is needed to charge it?
5-17. How much heat energy in calories must be removed from 50 g of mercury to change it from a liquid to a solid at −40°C? (*Hint*: Refer to Sec. 3-5 and Appendix B, Table 2.)
5-18. How much heat energy in Btu's is needed to boil away 60 lb of water at 212°F and 1 atm of pressure? (*Hint*: Refer to Sec. 3-5 and Appendix B, Table 2.)
5-19. Which force does more work, a 60-N force that moves a distance of 0.50 cm or a 13-lb force that moves a distance of 0.62 in.?
5-20. (a) How large a force is needed to move a 50-lb block slowly up a frictionless surface inclined at an angle of 30° with the horizontal?
(b) How much work is done by this force in moving the block 12 ft along the surface?
5-21. A steady torque of 10.0 lb · ft acts against a rotating flywheel. The flywheel comes to rest after making 50 complete revolutions. How much energy in foot · pounds was needed to stop the flywheel?
5-22. The 250-N force moves 0.10 m downward along a portion of a circular arc.
(a) Through what angle does the lever rotate?
(b) How much work is done by the force?
(c) How far is the end of the heavy object lifted?

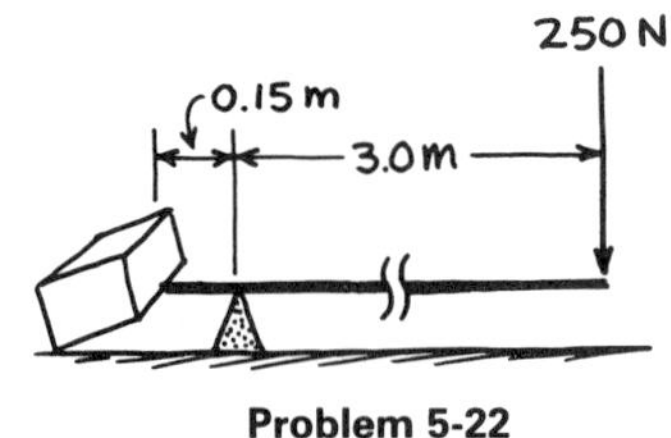

Problem 5-22

5-23. A small pump expends 20,000 J of energy in displacing a liquid volume of 4.0 m^3. What steady pressure in pascals acted on the liquid?
5-24. A fan exerts a steady pressure difference of 0.300 psi when displacing the air in a room that measures 15.0 by 12.0 by 10.0 ft.
(a) How much work in foot · pounds is done by the fan?
(b) How many pounds of air are displaced? The room temperature is 0°C.
5-25. A proton is in the space between two oppositely charged parallel plates. The plates are separated by a distance of 0.010 cm and there is a potential difference of 24 V between the plates.
(a) How much work is required to move the proton from the negative plate to the positive plate?

(b) What force in newtons acts on the proton? Assume that this force is constant.

5-26. An electron in the space between two oppositely charged parallel plates picks up 4.8×10^{-17} J of energy as it moves from the positive to the negative plate. The plates are separated by a distance of 2.0 cm.

(a) What (constant) force acts on the electron?

(b) What is the potential difference across the plates?

5-27. The upper 20 m of water in a large lake warms up from about 5°C in the winter to an average of about 17°C by the end of the summer. The lake's area is 5.0×10^8 m^2. How much heat has to be absorbed in order to produce this change? Give your answer in joules.

5-28. About 7.8×10^6 cal of heat energy is released when 1 kg of coal burns completely. How much coal is ideally needed to supply the heat to warm the lake of Problem 5-27?

ANSWERS TO ODD-NUMBERED PROBLEMS

5-1. 6.0×10^2 J

5-3. (a) 0.52 rad
(b) 9.4 rad
(c) 1.5 rad

5-5. 0.015 m^3; 15 kg

5-7. (a) -2.4×10^{-15} C
(b) $+13 \times 10^{-15}$ C

5-9. 1.9×10^2 cal

5-11. (a) 1.5×10^2 lb
(b) 7.6×10^3 ft · lb

5-13. 28.0 cm^3

5-15. 6.0×10^2 J

5-17. 1.4×10^2 cal

5-19. 60-N force does 0.30 J
13-lb force does 0.91 J

5-21. 3.14×10^3 ft · lb

5-23. 5.0×10^3 Pa

5-25. (a) 3.8×10^{-18} J
(b) 3.8×10^{-14} N

5-27. 5.0×10^{17} J

6

Parameter Rate

*In this chapter you encounter the idea of a **rate**. A rate is a **ratio** of something divided by something else. An example is a baseball player's batting average. The average is computed by dividing the number of hits by the number of times at bat. This can be called a rate, the rate of hits per try. A player with 80 hits in 320 tries has a batting average or a batting rate of 80 hits/320 tries = 0.250 hit/try. Rates do not necessarily remain constant. Our player may have a good month getting 32 hits in 80 tries for a batting rate of 32 hits/80 tries = 0.400 hit/try. If our hero has a perfect day at the plate, getting 4 out of 4, his batting rate for the game is 4 hits/4 tries = 1.000 hit/try.*

Clearly, the batting rate under discussion is subject to change with the passage of time. Although baseball statisticians don't compute such things, it is perfectly possible to compute a rate at which the batting average changes. If the average rises from 0.230 hit/try to 0.270 hit/try in two weeks, we could say that there has been a change in the batting rate of

$$\frac{0.270-0.230}{2}\,\frac{hit/try}{week}=0.020\,\frac{hit/try}{week}$$

The last calculation is an example of a time rate of change. Chapter 5 studied a number of parameters, such as distance, angle, volume, charge, and quantity of heat. This chapter studies the rate at which these parameters change with time.

6-1 DISTANCE–TIME GRAPHS: VELOCITY

As an introduction to the concept of parameter rate, let us look at how a parameter (distance) changes with time for a car moving along a highway (Fig. 6-1). Suppose that we hover overhead in a traffic-spotting helicopter and make a video tape of the scene below. The car is first noticed at point A driving eastward at a moderate speed. At point B, 100 metres (m) from A, the road sign says "End of Speed Zone." The driver hits the throttle and steadily speeds up until he gets to point C, 114 m from B.

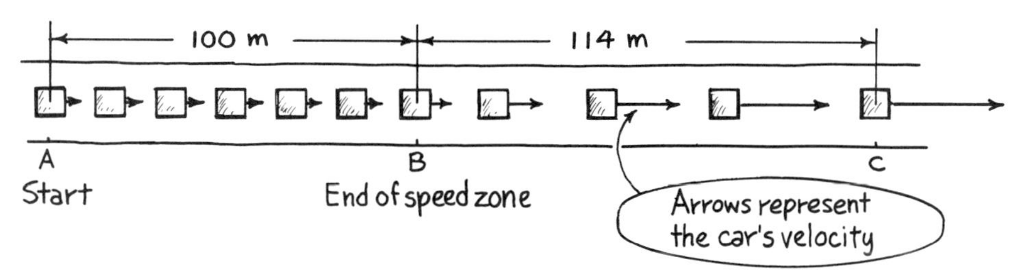

Figure 6-1. The traffic scene.

Later, viewing the results, we are able to pick values of position and time from the pictures. Data from between points A and B are listed and plotted in Fig. 6-2. Before getting too deeply into this, you may want to go back and read about velocity (Sec. 3-1) and slopes of graphs (Sec. 2-6).

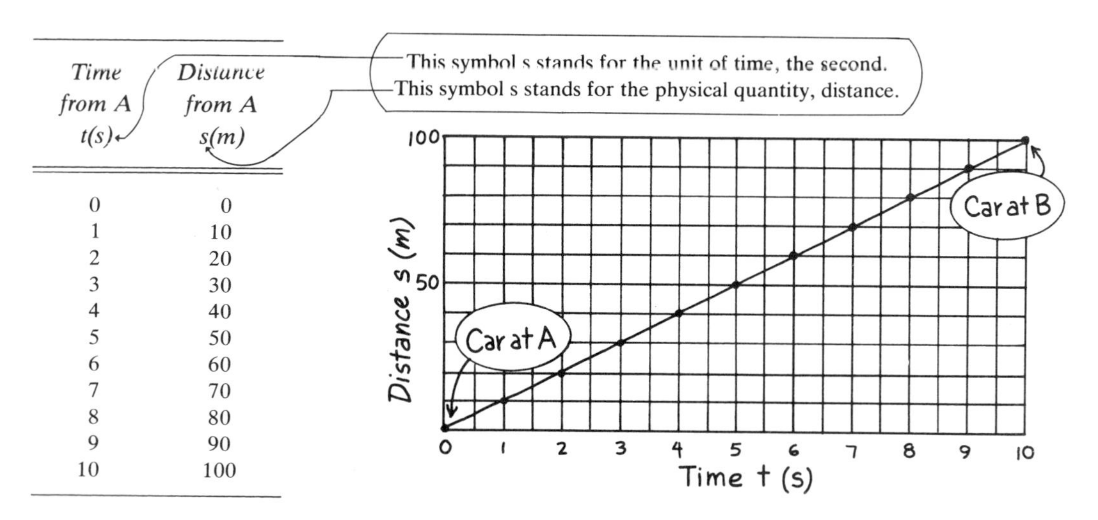

Time from A t(s)	*Distance from A* s(m)
0	0
1	10
2	20
3	30
4	40
5	50
6	60
7	70
8	80
9	90
10	100

Figure 6-2. Distance–time graph for car between points A and B.

Velocity is defined as the time rate of change of distance. It is clear from the data and the curve that the distance changes by 10 metres for every 1.0-second (s) change in time. The velocity of the car is constant at 10 m/1.0 s or 10 m/s. The straight line distance–time graph tells us that the car's speed did not change between points A and B.

Data from between points B and C are tabulated and plotted in Fig. 6-3.

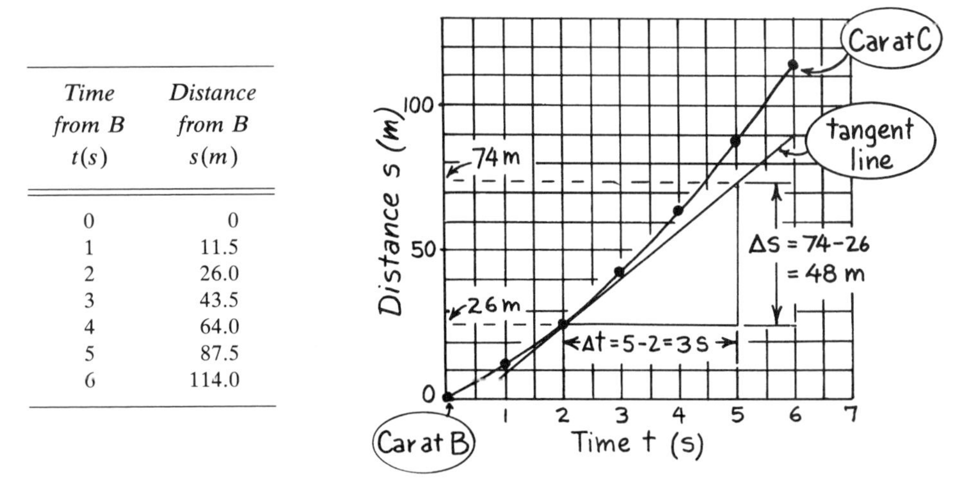

Time from B $t(s)$	Distance from B $s(m)$
0	0
1	11.5
2	26.0
3	43.5
4	64.0
5	87.5
6	114.0

Figure 6-3. Distance–time graph for car between points B and C.

This graph is not a straight line. The velocity is continually changing, and we must go to a more formal definition of velocity to explain what is happening. Velocity is defined as the slope of a distance versus time graph. Its value at a particular time is found by drawing a tangent line and calculating its slope ($\Delta s/\Delta t$):

$$v = \frac{\Delta s}{\Delta t} \qquad (3\text{-}1)$$

The symbol Δs stands for an incremental change in distance, and the symbol Δt stands for a corresponding incremental change in time. Both Δs and Δt are measured with respect to the tangent line.

Let's go back to Fig. 6-3 and calculate some velocities. What is the velocity when $t = 2$ s? Eyeball in a tangent line at $t = 2$ s. Use a plastic rule to do this. Pick off distances along the tangent line corresponding to $\Delta t = 5.0 - 2.0 = 3.0$ s. The distances are 74 and 26 m, respectively. This makes

$\Delta s = 74 - 26 = 48$ m. Hence at $t = 2.0$ s

$$v = \frac{\Delta s}{\Delta t} = \frac{48\text{ m}}{3.0\text{ s}} = 16\text{ m/s}$$

The increments Δs and Δt picked to measure velocity can be any size. In a mathematical treatment using calculus, the increments are very small; but it is better to use larger increments when finding the slope graphically, because it is more accurate. The ratio $\Delta s/\Delta t$ will be the same no matter what size increments are picked.

Try and find the velocity when the time is 5.0 s. Eyeball in a tangent line, pick some increments, and make a simple division to find $\Delta s/\Delta t$. Your answer should be close to 25 m/s.

6-2 VELOCITY–TIME GRAPHS: ACCELERATION

Velocity and time for each second as the car travels between points A and B are listed and plotted in Fig. 6-4.

Time from A $t(s)$	*Velocity* $v(m/s)$
0	10
1	10
2	10
3	10
4	10
5	10
6	10
7	10
8	10
9	10
10	10

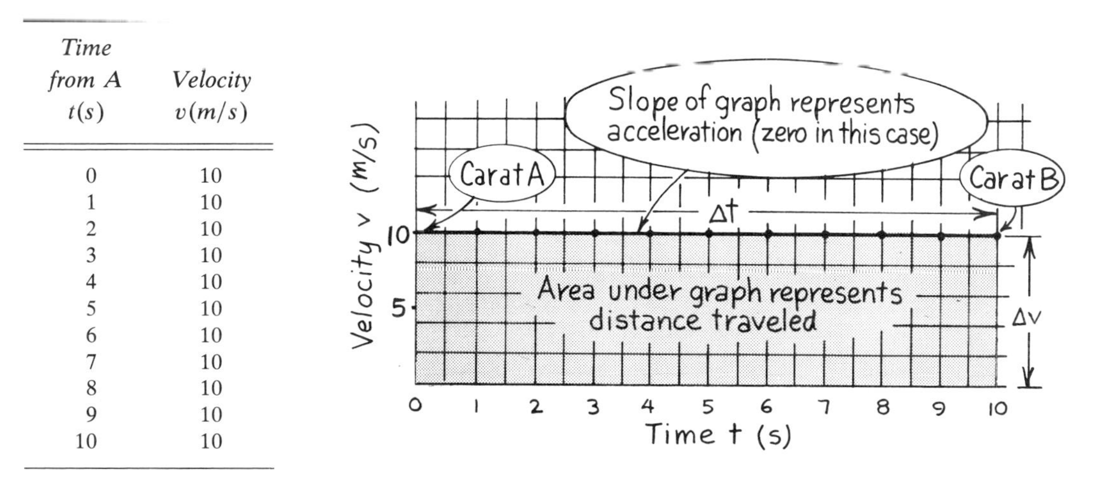

Figure 6-4. Velocity–time graph for car between points A and B. The horizontal graph represents zero acceleration.

As the car moved from A to B, its velocity did not change. The car's speedometer reading was a constant 10 m/s. Thus a table of velocity and time between A and B is quite simple, and a graph of velocity versus time is a straight horizontal line. Such a system with a constant speed is said to be in a *steady-state* condition.

The graph along with the velocity and time axes in Fig. 6-4 defines a rectangle of base Δt and height v. According to the definition of velocity,

$$v = \frac{\Delta s}{\Delta t} \tag{3-1}$$

Solving for Δs,

$$\underset{\text{Area of rectangle}}{\Delta s} = \overset{\text{Height of rectangle}}{v}\ \underset{\text{Length of rectangle}}{\Delta t}$$

The area under the velocity versus time graph is a graphical representation of the distance traveled, Δs. In the car's case, $v = 10\ \text{m/s}$, $\Delta t = 10\ \text{s}$, and $\Delta s = v\ \Delta t = (10\ \text{m/s})(10\ \text{s}) = 100\ \text{m}$, the distance between points A and B. This idea of area representing distance turns out to be valid not only in this simple constant velocity case but even when the velocity changes.

We have already found that at $t = 2.0\ \text{s}$ the car's velocity is $16\ \text{m/s}$. Let's go back to Fig. 6-3 and draw tangents accurately, computing velocities at 1-s intervals as the car moves from B to C. The results are listed and plotted in Fig. 6-5.

Time from B t *(s)*	*Velocity* v *(m/s)*
0	10
1	13
2	16
3	19
4	22
5	25
6	28

Figure 6-5. Velocity–time graph of car between points *B* and *C*. Straight line graph represents *constant acceleration.*

The rising graph tells us that the car's velocity is increasing. The car accelerates between points B and C. It is no longer in a steady-state condition.

Acceleration is defined as the time rate of change of velocity. A more formal definition of acceleration is the slope of a velocity versus time graph. Acceleration is found by drawing a tangent line and calculating its slope.

$$\underset{\text{Acceleration}}{a} = \frac{\Delta v}{\Delta t} \qquad (3\text{-}2)$$

The symbol Δv stands for an incremental change in velocity, and Δt is the corresponding time interval.

The car's acceleration is constant from B to C because the velocity versus time graph in Fig. 6-5 plots as a straight line. The constant acceleration can be computed from the slope of the graph. For example, between 2 and 6 s

$$a = \frac{\Delta v}{\Delta t} = \frac{(28 - 16)\text{m/s}}{(6.0 - 2.0)\text{s}} = 3.0\ \text{m/s}^2$$

or between 0 and 5 s

$$a = \frac{\Delta v}{\Delta t} = \frac{(25 - 10)\text{m/s}}{(5.0 - 0)\text{s}} = 3.0\ \text{m/s}^2$$

Again, any increment will do for finding the slope, and the bigger the increments the better. The units of acceleration are (velocity units/time units) and can be expressed as such, for example, (m/s)/s. More commonly the time units are combined and the whole thing written as distance/time^2, for example, m/s^2.

Example 6-1 Calculate the distance between points B and C by finding the total area under the velocity–time graph of Fig. 6-5.

Solution

The area that represents distance traveled is in the form of a trapezoid with two parallel sides of 10 m/s and 28 m/s and a base that represents 6.0 s. The area of a trapezoid from Sec. 2-4 is

$$\text{area} = \tfrac{1}{2}(\text{sum of sides}) \times \text{base}$$

$$= \tfrac{1}{2}(10 + 28)\text{m/s} \times (6.0\ \text{s}) = \boxed{114\ \text{m}}$$

6-3 KINEMATIC EQUATIONS

The study of motion (position and time) goes by the name of *kinematics*. So far we have been looking at kinematics from a graphical standpoint. Although it works and leads to better understanding, it takes time to draw tangents and find slopes as we have been doing. Also, as you probably have found, it is hard to get consistent answers. Results depend on such things as how well the tangent is eyeballed and how well the increments are measured. It would be faster and more accurate to have a set of mathematical statements (equations) that could be relied on. We will look at three equations that will solve any kinematics problem that you are likely to encounter. For the first time now, be warned that *the equations coming up work only when acceleration is constant.* If acceleration changes with time (i.e., if the velocity–time graph is not a straight line), these equations are not used.

Let us look again at a velocity–time graph like Fig. 6-5 (see Fig. 6-6).

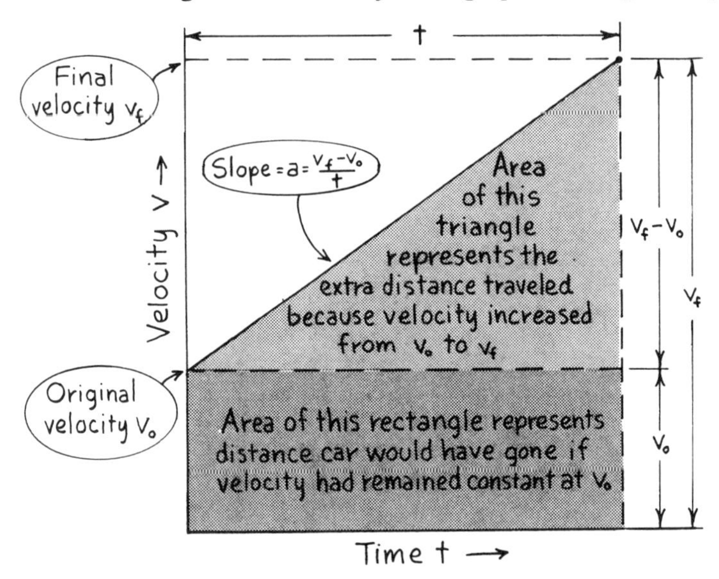

Figure 6-6. Velocity–time graph showing constant acceleration for t seconds.

Acceleration a, the slope of the graph, is a velocity increment Δv divided by the corresponding time increment Δt:

$$\text{slope} = \text{acceleration } a = \frac{\Delta v}{\Delta t} = \frac{v_f - v_o}{t}$$

Rearranging, $v_f - v_o = at$ and

$$v_f = v_o + at \qquad (6\text{-}1)$$

<table>
<tr><th>v_f
Final Velocity</th><th>v_o
Original Velocity</th><th>a
Constant Acceleration</th><th>t
Time</th></tr>
<tr><td colspan="2">metre/second
(m/s)</td><td>metre/second / second
(m/s^2)</td><td rowspan="2">second
(s)</td></tr>
<tr><td colspan="2">foot/second
(ft/s)</td><td>foot/second / second
(ft/s^2)</td></tr>
</table>

An equation for distance traveled is easy to come up with if we remember that area under the curve represents distance. The total distance can be thought of as made of two parts represented by the rectangle and thc triangle as illustrated in Fig. 6-6. Study this figure before you continue toward Eq. (6-2)

area of rectangle (distance at constant v_o)
$= \text{altitude} \times \text{base} = v_o t$

area of triangle (extra distance due to acceleration)
$= \frac{1}{2} \text{altitude} \times \text{base} = \frac{1}{2}(v_f - v_o)t$

Since $v_f - v_o = at$ from Eq. (6-1)

$$\text{area of triangle} = \tfrac{1}{2}(v_f - v_o)t = \tfrac{1}{2}(at)t = \tfrac{1}{2}at^2$$

and

$$\text{total area} = v_o t + \tfrac{1}{2}at^2$$

But the total area is the total distance traveled s. Hence

$$s = v_o t + \tfrac{1}{2}at^2 \qquad (6\text{-}2)$$

s *Distance Traveled*	v_o *Original Velocity*	t *Elapsed Time*	a *Constant Acceleration*
metre (m)	metre/second (m/s)	second (s)	$\frac{\text{metre/second}}{\text{second}}$ (m/s²)
foot (ft)	foot/second (ft/s)		$\frac{\text{foot/second}}{\text{second}}$ (ft/s²)

Equations (6-1) and (6-2) both include time. It is convenient (though not absolutely necessary) to have an equation that does not have t in it. This can be done by solving Eq. (6-1) for t and substituting the t into Eq. (6-2). Without going through the motions, the result is

$$v_f^2 = v_o^2 + 2as \qquad (6\text{-}3)$$

v_o *Original Velocity*	v_f *Final Velocity*	a *Constant Acceleration*	s *Distance Traveled*
metre/second (m/s)		$\frac{\text{metre/second}}{\text{second}}$ (m/s²)	metre (m)
foot/second (ft/s)		$\frac{\text{foot/second}}{\text{second}}$ (ft/s²)	foot (ft)

Equations (6-1), (6-2), and (6-3) are the mechanical translational forms of all you need to know to solve kinematic problems involving *constant* acceleration. Remember (second warning) that *if acceleration changes these equations do not apply*. Now for some examples.

Example 6-2 The car that has been discussed moves 100 m from A to B at constant speed of 10 m/s. Use a kinematic equation to find how long it takes.

Solution

In all kinematics problems (and any problem for that matter), it is best to start by writing down what is known and what is to be found. Then choose one of the equations in which everything is known except what you are looking for. Equation (6-2) is the right one for this problem.

$$\left\{\begin{array}{l} v_0 = 10\text{ m/s} \\ v_f = 10\text{ m/s} \\ a = 0 \\ s = 100\text{ m} \\ t = ? \end{array}\right\}$$

$$s = v_o t + \tfrac{1}{2}at^2 = v_o t \qquad \text{since } a \text{ is zero}$$

$$t = \frac{s}{v_o} = \frac{100\text{ m}}{10\text{ m/s}} = \boxed{10\text{ s}}$$

Example 6-3 Between B and C the car's velocity increases at a constant rate of 3.0 m/s^2.

(a) What is the velocity 4.5 s after passing point B?
(b) How far does it travel in that time?

Solution to (a)

Equation (6-1) will solve this problem.

$$\left\{\begin{array}{l} v_0 = 10\text{ m/s} \\ a = 3.0\text{ m/s}^2 \\ t = 4.5\text{ s} \\ v_f = ? \end{array}\right\}$$

$$v_f = v_o + at = \underbrace{10\frac{\text{m}}{\text{s}}}_{\text{Velocity at start (point } B)} + \underbrace{\left(3.0\frac{\text{m}}{\text{s}^2}\right)(4.5\text{ s})}_{\text{Velocity gained from acceleration}} = 23.5 \rightarrow \boxed{24\frac{\text{m}}{\text{s}}}$$

Solution to (b)

Either Eq. (6-2) or Eq. (6-3) will solve this. Using Eq. (6-2),

$$\left\{\begin{array}{l} v_0 = 10\text{ m/s} \\ v_f = 23.5\text{ m/s} \\ a = 3.0\text{ m/s}^2 \\ t = 4.5\text{ s} \\ s = ? \end{array}\right\}$$

$$s = v_o t + \tfrac{1}{2}at^2$$

$$= \left(10\frac{\text{m}}{\text{s}}\right)(4.5\text{ s}) + \frac{1}{2}\left(3.0\frac{\text{m}}{\text{s}^2}\right)(4.5\text{ s})^2 = 75.4 \rightarrow \boxed{75\text{ m}}$$

Using Eq. (6-3),

$$v_f^2 = v_o^2 + 2as \Rightarrow s = \frac{v_f^2 - v_o^2}{2a} = \frac{(23.5\text{ m/s})^2 - (10\text{ m/s})^2}{2(3.0\text{ m/s}^2)} = 75.4 \rightarrow \boxed{75\text{ m}}$$

You will find kinematics problems very simple if you keep these equations in mind and if you apply them reasonably. Do *not* get caught in the trap of memorizing special cases, such as, for example, when $v_o = 0$ reducing Eq. (6-1) to $v_f = at$. This is not anything new and will lead you astray if v_o is not zero.

*6-4 UNIFIED KINEMATICS

Our task in this book is to take information gained from one system and use it to help understand other systems. All the discussion so far has been about mechanical translational things. Parameter in that system is taken as distance. *Parameter rate* is the time rate of change of parameter. In the mechanical translational system, parameter rate is velocity.

We will be dealing with time rates from now on. For the sake of simplicity we will adopt the commonly used "dot" notation. For example instead of writing $\Delta s/\Delta t$, we will write the s with a dot over it ($\dot{s}$), and call it "s-dot" when we talk about it.

$v = \dot{s}$ *Velocity* = "*s-dot*"	Δs *Change of Distance*	Δt *Change of Time*
metre/second (m/s)	metre (m)	second (s)
foot/second (ft/s)	foot (ft)	

$$v = \dot{s} = \frac{\Delta s}{\Delta t} \quad \text{(parameter rate in a mechanical translational system)} \tag{6-4}$$

We have gone one step beyond parameter rate in the examples discussed. Acceleration has been defined as the time rate of change of velocity. We can extend the "dot" notation as follows:

$a = \dot{v} = \ddot{s}$ *Acceleration* = "*v-dot*" = "*s-double dot*"	Δv *Change of Velocity*	Δt *Change of Time*
$\frac{\text{metre/second}}{\text{second}}$ (m/s^2)	metre/second (m/s)	second (s)
$\frac{\text{foot/second}}{\text{second}}$ (ft/s^2)	foot/second (ft/s)	

$$a = \dot{v} = \ddot{s} = \frac{\Delta v}{\Delta t} \quad \text{(rate of change of parameter rate in a mechanical translational system)} \tag{6-5}$$

A letter with two dots above it is read as "double dot." It means the rate of change of the rate of change of that quantity.

When we want to talk in general terms, the symbol Q is used for parameter. Thus in unified concept terms, we have Equations (6-6) through (6-10):

Q	Parameter
$\dot{Q}$	Parameter rate o = original f → final
$\ddot{Q}$	Rate of change of parameter rate (constant)
ΔQ	Change of parameter
$\Delta \dot{Q}$	Change of parameter rate
t	Time
Δt	Change of time

Definitions

$$\dot{Q} = \frac{\Delta Q}{\Delta t} \tag{6-6}$$

$$\ddot{Q} = \frac{\Delta \dot{Q}}{\Delta t} \tag{6-7}$$

Kinematics equations for constant $\ddot{Q}$

$$\dot{Q}_f = \dot{Q}_o + \ddot{Q}t \tag{6-8}$$

$$Q = \dot{Q}_o t + \tfrac{1}{2}\ddot{Q}t^2 \tag{6-9}$$

$$\dot{Q}_f^2 = \dot{Q}_o^2 + 2\ddot{Q}Q \tag{6-10}$$

Equations (6-6) through (6-10) are general unified concept forms of Eqs. (6-1) through (6-3). $\dot{Q}$ is velocity in the mechanical translational system. Rotational $\dot{Q}$ is *angular velocity*, fluid $\dot{Q}$ is *volume flow rate*, thermal $\dot{Q}$ is *heat rate*, and electrical $\dot{Q}$ is charge flow rate or *current*.

6-5 MECHANICAL ROTATIONAL KINEMATICS

Parameter in a mechanical rotational system is angle, represented by the symbol θ (theta). Parameter rate is called *angular velocity* ω (omega), and the rate of change of angular velocity is *angular acceleration* α (alpha). Except for the symbols, the definitions of ω and α are just like those for v and a:

Definitions

$$\omega = \dot{\theta} = \frac{\Delta \theta}{\Delta t} \tag{6-11}$$

$$\alpha = \dot{\omega} = \ddot{\theta} = \frac{\Delta \omega}{\Delta t} \tag{6-12}$$

θ *Angle*	$\omega = \dot{\theta}$ *Angular Velocity*	$\alpha = \dot{\omega} = \ddot{\theta}$ *Angular Acceleration*	t *Time*
radian (rad)	radian/second (rad/s)	$\frac{\text{radian/second}}{\text{second}}$ (rad/s^2)	second (s)

Another unit of angular velocity is revolutions per minute.

$$1 \text{ revolution/minute (rpm)} = 0.1047 \text{ radian/second (rad/s)}$$

The radian/second is the unit you must use in most calculations.

The three kinematic equations (6-1) through (6-3) can be used just as well for mechanical rotational problems as for mechanical translational problems. Let's take a look at some examples.

Example 6-4 The tires on the car of the introductory example have a radius of 35 cm (0.35 m).

(a) Through what angle in radians does each wheel turn as the car moves the 100 m from A to B?

(b) What is the angular velocity of a wheel between points A and B?

Solution to (a)

$\left\{ \begin{matrix} s = 100 \text{ m} \\ r = 0.35 \text{ m} \\ \theta = ? \end{matrix} \right\}$

The angle θ in radians is defined by Eq. (5-2) as

$$\theta = \frac{s}{r}$$

$$= \frac{100 \text{ m}}{0.35 \text{ m}} = 286 \rightarrow \boxed{2.9 \times 10^2 \text{ rad}}$$

Solution to (b)

Since v is constant, ω must also be constant.

$$\omega = \frac{\Delta\theta}{\Delta t} = \frac{286 \text{ rad}}{10 \text{ s}} = 28.6 \rightarrow \boxed{29 \text{ rad/s}}$$

Example 6-5 As the car moves from B to C, it accelerates and so do the wheels. The distance of 114 m is covered in 6.0 s.

(a) Through what angle did a wheel turn?
(b) What was the angular acceleration of a wheel?
(c) What is the wheel's angular velocity at C?

Solution to (a)

$\left\{ \begin{matrix} s = 114 \text{ m} \\ r = 0.35 \text{ m} \end{matrix} \right\}$

$$\theta = \frac{s}{r} = \frac{114 \text{ m}}{0.35 \text{ m}} = 326 \rightarrow \boxed{3.3 \times 10^2 \text{ rad}}$$

Solution to (b)

$$\left\{\begin{array}{l} \theta = 326 \text{ rad} \\ \omega_o = 28.6 \text{ rad/s} \\ t = 6.0 \text{ s} \\ \alpha = ? \end{array}\right\}$$

Equation (6-2) will do the job if written in rotational terms.

$$\theta = \omega_o t + \tfrac{1}{2}\alpha t^2 \Rightarrow \alpha = \frac{2(\theta - \omega_o t)}{t^2}$$

$$\alpha = \frac{2[326 \text{ rad} - (28.6 \text{ rad/s})(6.0 \text{ s})]}{(6.0 \text{ s})^2} = 8.58 \rightarrow \boxed{8.6 \text{ rad/s}^2}$$

Solution to (c)

$$\left\{\begin{array}{l} \theta = 326 \text{ rad} \\ \omega_o = 28.6 \text{ rad/s} \\ t = 6.0 \text{ s} \\ \alpha = 8.58 \text{ rad/s}^2 \\ \omega_f = ? \end{array}\right\}$$

Either Eq. (6-1) or (6-3) will do if written in rotational terms. Use Eq. (6-1):

$$\omega_f = \omega_o + \alpha t = \underset{\substack{\uparrow \\ \text{Original value} \\ \text{of angular} \\ \text{velocity}}}{28.6 \text{ rad/s}} + \underbrace{(8.58 \text{ rad/s}^2)(6.0 \text{ s})}_{\substack{\text{Increase in angular} \\ \text{velocity due to} \\ \text{acceleration}}} = 80.1 \rightarrow \boxed{80 \text{ rad/s}}$$

Now is a good time to point out some simple relations between distance moved and angle turned for a point moving on a circle. Figure 6-7 shows a

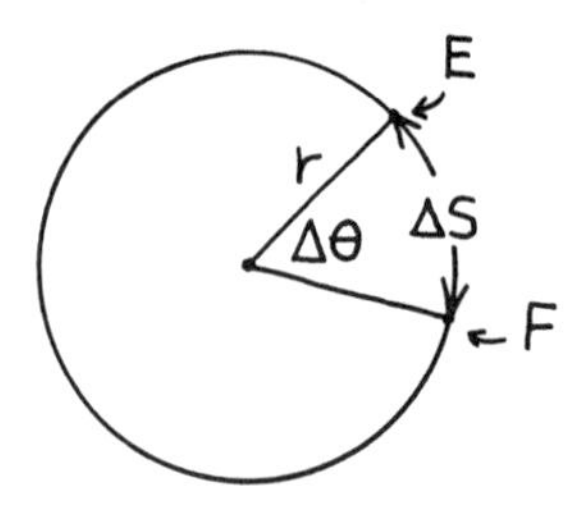

Figure 6-7. A point moves from E to F along an arc of length Δs in Δt seconds.

point that moves an arc length Δs from E to F in Δt seconds. The angle turned is $\Delta\theta$. From the definition of a radian,

$$\Delta\theta = \frac{\Delta s}{r} \tag{5-2}$$

and

$$\Delta s = r\Delta\theta$$

Divide both sides by the time interval Δt:

$$\frac{\Delta s}{\Delta t} = r\frac{\Delta\theta}{\Delta t}$$

or

$$v = r\omega \tag{6-13}$$

The velocity v is called a *tangential velocity* because the point's motion is tangent to the circle. If the angular velocity changes by $\Delta\omega$ in Δt seconds,

$$\Delta v = r\Delta\omega$$

Again divide both sides by the time interval Δt:

$$\frac{\Delta v}{\Delta t} = r\frac{\Delta \omega}{\Delta t}$$

$$a = r\alpha \tag{6-14}$$

The acceleration a is called a *tangential acceleration.*

Some problems become easier if you know these relations. For instance, look at part (b) in Example 6-4. We know that the translational velocity between points A and B is 10 m/s. Thus $v = r\omega$,

$$\omega = \frac{v}{r} = \frac{10 \text{ m/s}}{0.35 \text{ m}} = 28.6 \rightarrow \boxed{29 \text{ rad/s}}$$

In Example 6-5, part (b), we know that the translational acceleration is 3.0 m/s^2, and so

$$\alpha = \frac{a}{r} = \frac{3.0 \text{ m/s}^2}{0.35 \text{ m}} = 8.57 \rightarrow \boxed{8.6 \text{ rad/s}^2}$$

In Example 6-5, part (c), the velocity is 28 m/s when the time is 6.0 s. This information comes from the table associated with Fig. 6-5. Hence

$$\omega = \frac{v}{r} = \frac{28 \text{ m/s}}{0.35 \text{ m}} = \boxed{80 \text{ rad/s}}$$

6-6 FLUID KINEMATICS

The parameter in fluid systems is *volume* V, and parameter rate is called *volume flow rate* $\dot{V}$. Rate of change of volume flow rate—analogous to acceleration—has no special name. We will use the symbol $\ddot{V}$ (V double dot) for it.

Definitions

$$\dot{V} = \frac{\Delta V}{\Delta t} \tag{6-15}$$

$$\ddot{V} = \frac{\Delta \dot{V}}{\Delta t} \tag{6-16}$$

V *Volume*	$\dot{V}$ *Volume Flow Rate*	$\ddot{V}$ *Rate of Change of Volume Flow Rate*	t *Time*
cubic metre (m^3)	cubic metre/second (m^3/s)	$\frac{\text{cubic metre/second}}{\text{second}}$ (m^3/s^2)	second (s)
cubic foot (ft^3)	cubic foot/second (ft^3/s)	$\frac{\text{cubic foot/second}}{\text{second}}$ (ft^3/s^2)	

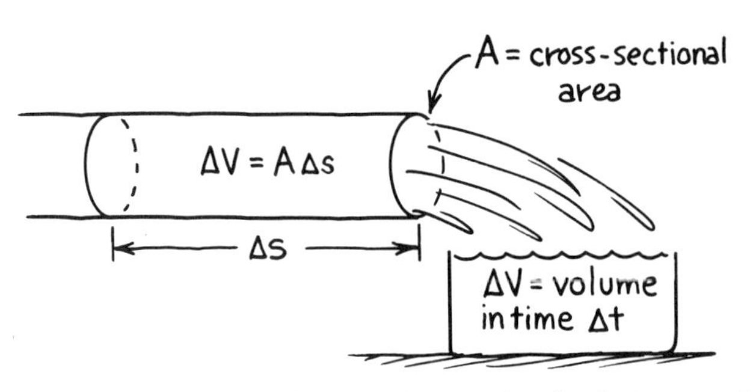

Figure 6-8. A volume of fluid ΔV leaves the pipe in Δt seconds.

Again the three kinematic equations (6-1) through (6-3) can be utilized for fluid kinematic problems, this time with V, $\dot{V}$, and $\ddot{V}$ in place of distance s, velocity v, and acceleration a, respectively.

Perhaps the easiest way to visualize volume flow rate $\dot{V}$ is to consider it as a volume of liquid ΔV collected in a graduated bucket during a certain time interval Δt.

There is a simple and useful relation between the volume flow rate $\dot{V}$ in a pipe and the average speed v at which the fluid moves. Think of the fluid coming out of the pipe in Δt seconds as though it were a cylinder of length Δs, cross-sectional area A, and volume ΔV (Fig. 6-8).

By definition

$$\dot{V} = \frac{\Delta V}{\Delta t}$$

But

$$\Delta V = A\,\Delta s$$

Therefore,

$$\dot{V} = \frac{A\,\Delta s}{\Delta t} = Av$$

$$\dot{V} = Av \tag{6-17}$$

$\dot{V}$ *Volume Flow Rate*	A *Cross-sectional Area of Pipe*	v *Speed of Fluid in Pipe*
$\frac{\text{cubic metre}}{\text{second}}$ (m^3/s)	square metre (m^2)	metre/second (m/s)
$\frac{\text{cubic foot}}{\text{second}}$ (ft^3/s)	square foot (ft^2)	foot/second (ft/s)

It is important that you recognize the difference between volume flow rate $\dot{V}$ and velocity v. Both are considered to be parameter rates, and as you have just seen, velocity is involved in the idea of volume flow rate. Which one is used depends strictly upon what system you are dealing with. If parameter is distance, then parameter rate is v. If parameter is volume, the parameter rate is $\dot{V}$. Don't mix them up. Let's look at a few example problems about volume flow rate.

Example 6-6 Water comes out of a hose of cross-sectional area 1.0 in.^{2} at a volume flow rate of 0.050 ft^3/s. At what speed is the water moving?

Solution

$$\left\{\begin{matrix} A = 1.0 \text{ in.}^2 \\ \dot{V} = 0.050 \text{ ft}^3/\text{s} \end{matrix}\right\}$$

$$A = 1.0 \text{ in.}^2\left(\frac{1 \text{ ft}^2}{12^2 \text{ in.}^2}\right) = 6.94 \times 10^{-3} \text{ ft}^2$$

$$\dot{V} = Av \Rightarrow v = \frac{\dot{V}}{A} = \frac{0.050 \text{ ft}^3/\text{s}}{6.94 \times 10^{-3} \text{ ft}^2} = \boxed{7.2 \text{ ft/s}}$$

Example 6-7 Water pours out of a pipe 4.0 cm^2 in cross section. If the water travels at a velocity of 2.0 m/s, what is the volume flow rate?

Solution

$$\left\{\begin{matrix} A = 4.0 \text{ cm}^2 \\ v = 2.0 \text{ m/s} \end{matrix}\right\}$$

$$A = 4.0 \text{ cm}^2\left(\frac{1 \text{ m}^2}{100^2 \text{ cm}^2}\right) = 4.0 \times 10^{-4} \text{ m}^2$$

$$\dot{V} = Av = (4.0 \times 10^{-4} \text{ m}^2)(2.0 \text{ m/s}) = \boxed{8.0 \times 10^{-4} \text{ m}^3/\text{s}}$$

Example 6-8 A valve is set to allow a constant volume flow rate of 0.20 m^3/s. What volume of water flows through the valve in 17 s?

Solution

$$\left\{\begin{matrix} \dot{V}_o = \dot{V}_f = 0.20 \text{ m}^3/\text{s} \\ \ddot{V} = 0 \\ t = 17 \text{ s} \\ V = ? \end{matrix}\right\}$$

The kinematic equation

$$V = \dot{V}_o t + \tfrac{1}{2}\ddot{V}t^2$$

similar to $s = v_o t + \frac{1}{2}at^2$ [Eq. (6-2)] will work for this problem.

$$V = \dot{V}_o t + \tfrac{1}{2}\ddot{V}t^2 = \underbrace{(0.20 \text{ m}^3/\text{s})(17 \text{ s})}_{\text{Volume at constant original flow rate}} + \underbrace{\tfrac{1}{2}(0 \text{ m}^3/\text{s}^2)(17 \text{ s})^2}_{\text{No extra volume because flow rate does not change}} = \boxed{3.4 \text{ m}^3}$$

6-7 THERMAL KINEMATICS

Parameter in the heat system is quantity of heat energy Q. Parameter rate, denoted by $\dot{Q}$, will be called *heat rate.* Notice that since energy is taken as parameter in this system, parameter rate is an *energy rate* or *power.* Section 3-4 reviews the idea of energy rate or power. The SI unit of power is the joule/second (J/s), which is the same as a watt (W). There are other more popular heat rate units such as Btu/hour and calorie/second. Rate of change of heat rate $\ddot{Q}$ does not have any special name, although it might well be thought of as thermal acceleration. Equations (6-8) through (6-10) apply to thermal problems without even a change of symbols.

$$\text{Definitions} \begin{cases} \dot{Q} = \dfrac{\Delta Q}{\Delta t} & \text{(6-18)} \\ \ddot{Q} = \dfrac{\Delta \dot{Q}}{\Delta t} & \text{(6-19)} \end{cases}$$

Q *Quantity of Heat Energy*	$\dot{Q}$ *Heat rate = Power*	$\ddot{Q}$ *Rate of Change of Heat Rate*	t *Time*
calorie (cal)	calorie/second (cal/s)	$\frac{\text{calorie/second}}{\text{second}}$ (cal/s^2)	second (s)
British thermal unit (Btu)	$\frac{\text{British thermal unit}}{\text{second}}$ (Btu/s)	$\frac{\text{British thermal unit/second}}{\text{second}}$ (Btu/s^2)	

Heat can be measured in a variety of ways. Several methods are indicated in the following examples.

Example 6-9 The heat exchanger shown transfers heat efficiently from a source to water that flows through it. Suppose that 10 kg of water flows through the heat exchanger in 2.0 min. The inlet and outlet water temperatures are 60 and 80°C, respectively. Find the heat rate in watts.

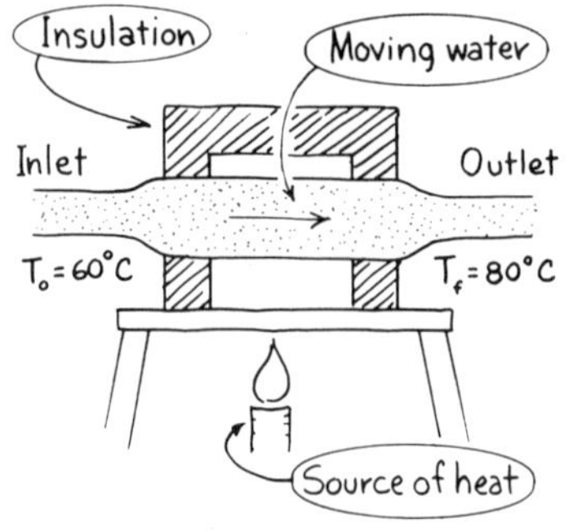

Solution

$$\left\{\begin{aligned} m &= 10\text{ kg}\left(\frac{1000\text{ g}}{1\text{ kg}}\right) \\ &= 10\times10^3\text{ g} \\ T_f &= 80°\text{C} \\ T_o &= 60°\text{C} \\ t &= 2.0\text{ min}\left(\frac{60\text{ s}}{1\text{ min}}\right) \\ &= 120\text{ s} \\ Q &= ? \\ \dot{Q} &= ? \end{aligned}\right\}$$

The specific heat c of a substance is defined by Eq. (3-14) as the amount of heat needed to change the temperature of a unit mass by 1°. We must first find the quantity of heat transferred using Eq. (3-14) and then divide by time to get the heat rate.

$$Q = mc(T_f - T_o)$$

(Appendix B, Table 3)

$$= (10 \times 10^3\text{ g})(1.00\text{ cal/(g.°C)}[(80 - 60)°\text{C}] = 2.0 \times 10^5\text{ cal}$$

$$\dot{Q} = \frac{\Delta Q}{\Delta t} = \frac{2.0 \times 10^5\text{ cal}}{120\text{ s}} = 1.67 \times 10^3\text{ cal/s}$$

$$= 1.67 \times 10^3\text{ cal/s}\left(\frac{1\text{ W}}{0.239\text{ cal/s}}\right) = \boxed{7.0 \times 10^3\text{ W}}$$

Example 6-10 Suppose that the heat transferred through a metal block boils water at its top surface. In 10 min, 2.0 kg of water are boiled away or evaporated. Find the average heat rate in watts.

Solution

$$\left\{\begin{aligned} m &= 2.0\text{ kg}\left(\frac{1000\text{ g}}{1\text{ kg}}\right) \\ &= 2000\text{ g} \\ L_v &= 540\text{ cal/g} \\ t &= 10\text{ min}\left(\frac{60\text{ s}}{1\text{ min}}\right) = 600\text{ s} \\ Q &= ? \\ \dot{Q} &= ? \end{aligned}\right\}$$

The heat of vaporization L_v of a substance is the amount of heat that must be added to change a unit mass from the liquid to the gaseous state at the boiling temperature. This is discussed in Sec. 3-5. Appendix B, Table 2, includes some values of L_v for various substances. For water, L_v = 540 cal/g.

$$Q = mL_v \qquad (3\text{-}16)$$

$$= 2000\text{ g}(540\text{ cal/g}) = 1.08 \times 10^6\text{ cal}$$

$$\dot{Q} = \frac{\Delta Q}{\Delta t} = \frac{1.08 \times 10^6\text{ cal}}{600\text{ s}} = 1.80 \times 10^3\text{ cal/s}$$

$$= 1.80 \times 10^3\text{ cal/s} \times \left(\frac{1\text{ W}}{0.239\text{ cal/s}}\right) = \boxed{7.5 \times 10^3\text{ W}}$$

Example 6-11 The power supplied to an electric heater can be measured with a wattmeter. All the energy is converted to heat and the meter effectively becomes a heat rate meter. Suppose that the wattmeter reads a constant 5.0×10^3 W. How much heat (in Btu) is produced in 15 min?

$$\left\{\begin{aligned} \dot{Q}_o &= \dot{Q}_f = 5.0\times10^3\text{ W} \\ &= 5.0\times10^3\,\frac{\text{J}}{\text{s}} \times \frac{9.49\times10^{-4}\text{ Btu}}{1\text{ J}} \\ &= 4.745\,\frac{\text{Btu}}{\text{s}} \\ \ddot{Q} &= 0 \quad (\dot{Q}\text{ is constant}) \\ t &= 15\text{ min}\left(\frac{60\text{ s}}{1\text{ min}}\right) = 900\text{ s} \\ Q &= ? \end{aligned}\right\}$$

Solution

Of the three kinematic equations, (6-9) will work:

$$Q = \dot{Q}_o t + \tfrac{1}{2}\ddot{Q}t^2$$

$$= \underbrace{(4.745\text{ Btu/s})(900\text{ s})}_{\text{Heat at constant original rate}} + \underbrace{\tfrac{1}{2}(0\text{ Btu/s}^2)(900\text{ s})^2}_{\text{No extra heat because heat rate doesn't change}} = \boxed{4.3 \times 10^3\text{ Btu}}$$

6-8 ELECTRICAL KINEMATICS

Parameter in the electrical system is quantity of charge measured in coulombs. The parameter rate or charge flow rate is called *electric current I*. As with fluid systems, there is a danger of confusing parameter rate with velocity. Charge is carried by electrons that drift from one atom to another in a conductor (Sec. 3-6). Although electrical impulses or signals travel through a wire at the speed of light, the charge-carrying electrons themselves only drift at very small velocities such as a few millimetres per second. As in a fluid system, parameter rate is not the speed but the amount of parameter per second that flows past a given point in the circuit. An instrument that measures charge flow rate or current is the ammeter. It can be thought of as a device that counts passing electrons in a 1-s time interval and converts the count to quantity of charge. Remember that 1 coulomb (C) of charge is approximately the charge of 6.25×10^{18} electrons.

The SI unit for current is coulomb/second (C/s), which is given the familiar name of ampere (A). The electrical equivalent of acceleration $\ddot{Q}$, or $\dot{I}$, has no special name. It can be referred to as the rate of change of current or simply "I dot." Its units are ampere/second (A/s). The three kinematic equations (6-1) through (6-3) work very nicely for electrical situations, as long as $\dot{I}$ is constant.

$$\text{Definitions} \begin{cases} I = \dot{Q} = \dfrac{\Delta Q}{\Delta t} & \text{(6-20)} \\[2ex] \dot{I} = \ddot{Q} = \dfrac{\Delta I}{\Delta t} & \text{(6-21)} \end{cases}$$

Q *Electric Charge*	$\dot{Q} = I$ *Electric Current*	$\ddot{Q} = \dot{I}$ *Rate of Change of Current (I dot)*	t *Time*
coulomb (C)	$\dfrac{\text{coulomb}}{\text{second}}$ = ampere (C/s) (A)	$\dfrac{\text{coulomb/second}}{\text{second}} = \dfrac{\text{ampere}}{\text{second}}$ (C/s^2) (A/s)	second (s)

Unified concept ideas help us to distinguish between current and voltage. Voltage is analogous to pressure or push in a fluid system, whereas current is analogous to the volume flow rate.

One thing that creates some confusion is the direction of electric current. We will adopt the convention illustrated in Fig. 6-9 that considers current to be a flow of positive charge. Thus our current flows from the (+) side of a battery through the external circuit to the (−) side. Actually, the particles that drift in this case and in most cases are electrons moving in the opposite direction.

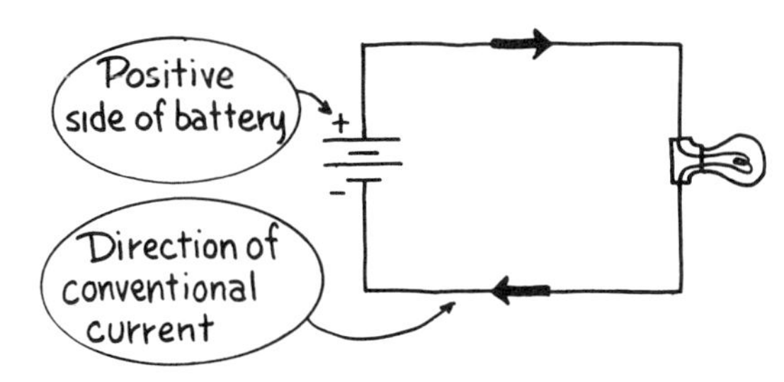

Figure 6-9. Conventional current is thought of as positive charge flowing out of the (+) terminal through the external circuit and back into the (−) terminal of the battery.

Here are a couple of examples that illustrate the relation between current, charge, and time.

Example 6-12 An ammeter measures a constant 0.50 A when placed in an electric circuit.

(a) How many coulombs of charge pass through the circuit in a 1-min time interval?

(b) How many electrons does this represent?

Solution to (a)

$$\left\{\begin{array}{l} \dot{I}=0 \\ I_o=I_f=0.50\text{ A} \\ t=1.0\text{ min}\left(\frac{60\text{ s}}{1\text{ min}}\right)=60\text{ s} \\ Q=? \end{array}\right\}$$

Use Eq. (6-2) written in electrical terms:

$$Q = I_o t + \tfrac{1}{2}\dot{I}t^2$$

$$= \underbrace{0.50\text{ A}(60\text{ s})}_{\text{Charge that flows at original constant rate}} + \underbrace{\tfrac{1}{2}(0\text{ A/s})(60\text{ s})^2}_{\text{No extra charge because current does not change}} = \boxed{30\text{ C}}$$

Solution to (b)

$$30\text{ C} \times \frac{6.25 \times 10^{18}\text{ electrons}}{1\text{ C}} = \boxed{1.9 \times 10^{20}\text{ electrons}}$$

Example 6-13 An automobile storage battery is rated at 100 ampere-hours. This means that the battery can maintain a current of 1.0 A for 100 hr. How much charge does the battery transfer as it discharges?

Solution

$$\left\{\begin{array}{l} I_o=I_f=1.0\text{ A} \\ \dot{I}=0 \quad (I_o\text{ is assumed not to change}) \\ t=100\text{ hr}\left(\frac{3600\text{ s}}{1\text{ hr}}\right) \\ \quad =3.6\times10^5\text{ s} \\ Q=? \end{array}\right\}$$

Equation (6-2) should do the job:

$$Q = I_o t + \tfrac{1}{2}\dot{I}t^2$$

$$= \overbrace{1.0\text{ A}(3.6 \times 10^5\text{ s})}^{\text{Charge at constant original current}} + \underbrace{\tfrac{1}{2}(0\text{ A/s})(3.6 \times 10^5\text{ s})^2}_{\text{No extra charge because current does not change}}$$

$$= \boxed{3.6 \times 10^5\text{ C}}$$

6-9 SUMMARY—AND A WORD OF ENCOURAGEMENT

This chapter so far includes 21 numbered equations, and you may be thinking, "Good grief, how can anyone remember all that?" The answer is that really there are only three equations to worry about, plus two definitions that you may already have been familiar with. The two definitions and the three kinematic equations stated in general form are

Definitions

$$\dot{Q} = \frac{\Delta Q}{\Delta t} \tag{6-6}$$

$$\ddot{Q} = \frac{\Delta \dot{Q}}{\Delta t} \tag{6-7}$$

Kinematic equations

$$\dot{Q}_f = \dot{Q}_o + \ddot{Q}t \tag{6-8}$$

$$Q = \dot{Q}_o t + \tfrac{1}{2}\ddot{Q}t^2 \tag{6-9}$$

$$\dot{Q}_f^2 = \dot{Q}_o^2 + 2\ddot{Q}Q \tag{6-10}$$

These equations will solve the parameter–time problems that you come across as long as $\ddot{Q}$ is constant. It matters little in what form you learn these equations. If you think in terms of mechanics, you may wish to remember them in terms of s, v, and a. If you are electrically oriented, Q, I, and $\dot{I}$ might be more memorable. Whatever works best for you is the way you should do it. As long as you remember which quantities are analogous in the various systems, you will have no trouble. Please don't forget (final warning!) that *the kinematic equations work only when $\ddot{Q}$ does not change.* Later, starting in Chapter 11, we will deal with situations in which $\ddot{Q}$ does change. Table 6-1 at the end of the chapter is a handy reference for keeping track of what $\dot{Q}$ and $\ddot{Q}$ mean in the various systems.

6-10 ENERGY RATE: POWER

Of the five types of systems we are considering, the thermal or heat system is the odd ball. Parameter in the heat system is energy, and therefore the heat system parameter rate is an energy rate or *power.* Energy rate is a very important idea for other systems too, and the equations describing it fit in nicely with the unified concepts developed so far.

Energy rate or power is defined in Sec. 3-4 as

$$P = \frac{W}{t} \tag{3-10}$$

The units of power are watts or foot · pounds/second, but the unit of horsepower is often used. You should have in mind what it is:

$$1 \text{ horsepower (hp)} = 550 \text{ foot} \cdot \text{pounds/second (ft} \cdot \text{lb/s)}$$

P Power	W Amount of Work Done or Energy Transferred	t Elapsed Time
joule/second = watt (J/s) (W)	joule (J)	second (s)
$\frac{\text{foot} \cdot \text{pound}}{\text{second}}$ (ft · lb/s)	foot · pound (ft · lb)	

Note that the same symbol W is used both for the unit watt and the physical quantity work. Don't confuse them.

Power-conversion factors are included in a *Table of Useful Conversion Factors* in Appendix A.

An equation for energy rate or power in all but the heat system is very easy to derive. For example, in the mechanical translational system

$$W = Fs \tag{3-9}$$

Divide both sides of the equation by t:

$$\frac{W}{t} = \frac{Fs}{t} = F\frac{s}{t}$$

But W/t is power and s/t is velocity. Therefore,

$$P = Fv \quad \text{(mechanical translational)} \tag{6-22}$$

P Power	F Force	v Velocity
joule/second = watt (J/s) (W)	newton (N)	metre/second (m/s)
foot · pound/second (ft · lb/s)	pound (lb)	foot/second (ft/s)

In unified concepts terms, Eq. (6-22) becomes

$$P = F\dot{Q} \quad \text{(general unified concepts form)} \tag{6-23}$$

P Power	F Force	$\dot{Q}$ Parameter Rate
appropriate units of power	appropriate units of force	appropriate units of parameter rate

and for the other systems

$$P = \tau\omega \quad \text{(mechanical rotational)} \tag{6-24}$$

P *Power*	τ *Torque*	ω *Angular Velocity*
joule/second = watt (J/s) (W)	newton · metre (N · m)	radian/second (rad/s)
foot · pound/second (ft · lb/s)	pound · foot (lb · ft)	

$$P = p\dot{V} \quad \text{(fluid)} \tag{6-25}$$

P *Power*	p *Pressure*	$\dot{V}$ *Volume Flow Rate*
joule/second = watt (J/s) (W)	pascal (Pa)	$\frac{\text{cubic metre}}{\text{second}}$ (m^3/s)
foot · pound/second (ft · lb/s)	$\frac{\text{pound}}{\text{square foot}}$ (lb/ft^2)	$\frac{\text{cubic foot}}{\text{second}}$ (ft^3/s)

$$P = EI \quad \text{(electrical)} \tag{6-26}$$

P *Power*	E *Electromotive Force*	I *Electric Current*
joule/second = watt (J/s) (W)	volt (V)	ampere (A)

These power equations hold provided both the force and the parameter rate remain constant during the time interval in which the power is measured. If either changes, we can still fall back and use the basic equation for power

$$P = \frac{W}{t} \tag{3-10}$$

provided we know how much energy was added during the time interval. This is called an average power. Or we can reduce the time interval and use

$$P = F\dot{Q} \tag{6-23}$$

to calculate the power using the force and parameter rate for a particular time instant. Such ideas are useful in studying systems in which the force and parameter rate change with time. Let's look at a couple of steady-state power examples.

Example 6-14 A tow truck pulling a small car provides a steady force of 3000 N. The velocity is constant at 10 m/s. Find the power provided by the tow truck to the car in horsepower.

Solution

$$\begin{Bmatrix} F = 3000\text{ N} \\ v = 10\text{ m/s} \end{Bmatrix}$$

$$P = Fv = 3000\text{ N}\,(10\text{ m/s}) = 30{,}000\text{ N}\cdot\text{m/s or W}$$

$$30{,}000\text{ W}\left(\frac{1\text{ hp}}{746\text{ W}}\right) = \boxed{40\text{ hp}}$$

TABLE 6-1 PARAMETER RATES

System	*Parameter Rate*	*Symbol*	*Equation*	*SI Unit*	*Symbol*	*Power Equation*
Mechanical translational	Velocity	v or $\dot{s}$	$v = \Delta s/\Delta t$	metre/second	m/s	$P = Fv$
Mechanical rotational	Angular velocity	ω or $\dot{\theta}$	$\omega = \Delta\theta/\Delta t$	radian/second	rad/s	$P = \tau\omega$
Fluid	Volume flow rate	$\dot{V}$	$\dot{V} = \Delta V/\Delta t$	$\frac{\text{cubic metre}}{\text{second}}$	m^3/s	$P = p\dot{V}$
Electrical	Electric current	I or $\dot{Q}$	$I = \Delta Q/\Delta t$	ampere	A	$P = EI$
Heat	Heat rate	$\dot{Q}$	$\dot{Q} = \Delta Q/\Delta t$	joule/second or watt	J/s = W	—

Unified Concept Equations

Definitions $\begin{cases} \dot{Q} = \Delta Q/\Delta t \\ \ddot{Q} = \Delta \dot{Q}/\Delta t \end{cases}$

Power $P = F\dot{Q}$

Kinematic equations $\begin{cases} \dot{Q}_f = \dot{Q}_o + \ddot{Q}t \\ Q = \dot{Q}_o t + \frac{1}{2}\ddot{Q}t^2 \\ \dot{Q}_f^2 = \dot{Q}_o^2 + 2\ddot{Q}Q \end{cases}$

Example 6-15 A 3.0-V flashlight battery maintains a steady current of 0.25 A through the bulb. How much power is the battery producing?

Solution

$$\left\{\begin{matrix} E = 3.0\ \text{V} \\ I = 0.25\ \text{A} \end{matrix}\right\} \qquad P = EI = 3.0\ \text{V}(0.25\ \text{A}) = \boxed{0.75\ \text{W}}$$

So ends the chapter specifically devoted to velocity, acceleration, and analogous quantities. It is not by any means the last you will hear of such things. An understanding of the basic ideas presented in this chapter is important if you are to comprehend what follows. Parameter rates and changing parameter rates will appear continually in the definitions and discussions that comprise the rest of this book. For example, the basic definition of resistance, the subject of the next chapter, includes parameter rate. Before you continue it will be well worth your while to do as many of the following problems as you can. In working these problems use Table 6-1, which summarizes most of what has been discussed in the chapter.

PROBLEMS

6-1. A car travels at a constant speed of 88 ft/s for 5.0 min. How far does the car go?

6-2. An airplane takes 5.0 hr to fly 3000 miles across the country. What is its speed (assumed constant) in feet/second?

6-3. The speed of light is 3.00×10^8 m/s. If the distance from the sun to earth is 1.5×10^8 km, how long does it take for light to travel from sun to earth?

6-4. A subway train passes through an 8.0-km tunnel in 4.5 min.
(a) What is the train's speed in metres/second?
(b) How much power in watts is supplied by the train's engine if the excitation force needed to move the train is 5.0×10^4 N?

6-5. A grindstone turns at a steady rate of 2000 rpm (revolution/minute). A motor supplies a steady excitation torque of 1.5 lb · ft to do this.
(a) Through how many radians does the grindstone turn in 20 s?
(b) How much power is needed to rotate the grindstone? Give your answer in foot · pounds/second.

6-6. How many radians does a 28-in. diameter bicycle wheel turn through on a 10-mi trip?

6-7. A long-playing (12-in. diameter) phonograph record turns at 33.3 rpm. How far does a point on the outside edge of the record travel during the $\frac{1}{2}$ hr that it plays?

6-8. A circular saw blade of radius 15 cm rotates at 1800 rpm.
(a) What is the angular velocity in radians/second?
(b) What is the speed in metres/second of a point on the rim?

6-9. A pipe of inside diameter 5.0 cm delivers 1.0 m^3 of water in 1.0 hr. At what speed in metres/second does the water flow in the pipe?

6-10. The oceans of the world contain $1.32 \times 10^9\ \text{km}^3$ (cubic kilometres) of water. All the rivers of the world have a combined flow rate estimated at $1.03 \times 10^6\ \text{m}^3/\text{s}$. How long would it take the rivers to fill the ocean basins at this rate?

6-11. A pump can deliver 0.10 ft^3/s of water continuously.
(a) How much water can be pumped in a 24-hr day?
(b) What power is required if the water's pressure increases by 1100 lb/ft^2 as it passes through the pump? Give your answer in horsepower.

6-12. Air conditioners are commonly rated in terms of the "refrigeration ton," which is equivalent to 12,000 Btu/hr. How much heat can be extracted from a room in 30 s by a "12-ton" air conditioner?

6-13. A 10-kg block of ice takes just 8.0 days to melt completely in an insulated chest. What is the heat rate (assumed constant) through the walls of the chest? Give the answer in calories/second.

6-14. Heat flows through a conductor at a rate of 20 watt. (One watt is 1 joule/second.) Calculate the amount of heat in calories that flows in 120 s.

6-15. A water heater with a tank that holds $0.12\ m^3$ has a recovery time of 8.0 min. This means that it will heat a tank full of water from the input temperature of 15°C to the output temperature of 55°C in 8.0 min. At what average rate is heat transferred to the water?

6-16. If 7.3×10^{12} electrons pass a point in a circuit during 40 ms of time, what is the current? (Assume that the current remains constant.)

6-17. A steady current of 0.83 A flows through a light bulb filament for 3.0 hr. The current comes from a 6.0-V battery.

(a) How much charge passes through the filament?

(b) How many electrons does this represent?

(c) What power is supplied by the battery?

6-18. The current in a detection circuit is 15×10^{-6} A. How many electrons pass a given point in the circuit in 10 ms?

6-19. A car traveling at 60 mi/hr is slowed down at a constant rate of $4.4\ ft/s^2$. How far will the car go before it slows to half its original speed?

6-20. A box of hundred dollar bills falls out of an airplane and is clocked at the following speeds and times as it falls. Plot a graph of v versus t, with v on the vertical axis.

v(ft/s)	0	30	48	64.5	72	73.5	75	75
t(s)	0	0.5	1.0	2.0	3.0	4.0	5.0	6.0

6-21. Using the graph of Problem 6-20, estimate the acceleration when $t = 2.0$ s.

6-22. If you threw a baseball straight down from the top of a 150-ft tower at an initial velocity of 40 ft/s, how long would it take the baseball to reach the ground?

6-23. A car rolls down a hill with constant acceleration of $3.0\ m/s^2$.

(a) What is its velocity after 5.0 s if it starts from rest ($v_o = 0$)?

(b) How far does it travel in 5.0 s?

(c) What would its velocity be after rolling that far *if* it had been given an initial velocity of 4.0 m/s?

6-24. A cylinder of radius 15 cm rotates at 1800 rpm. When the power is cut off, the cylinder decelerates at a constant rate to a complete stop in 20 s. What angle in radians does it turn through while stopping?

6-25. A water valve initially allows a flow of 8.0 gal/min. The valve is then opened at a constant rate so that 25 s later the flow rate is 27.0 gal/min.

(a) Find the rate of change of volume flow rate (assumed to be constant).

(b) How many gallons of water flow through the valve in those 25 s?

6-26. Water from a fire hose can reach a maximum height of 20 m when the hose is pointed straight up.

(a) How fast is the water moving when it comes out of the hose?

(*Note*: This is a simple kinematics problem in which $s = 20$ m, $v_f = 0$, and $a = g = 9.81\ m/s^2$. Be careful about the sign of a.)

(b) The fire hose has a nozzle of diameter 2.0 cm. What is the volume flow rate in cubic metres/second?

6-27. Heat initially flows at the rate of 20 W. (One watt is 1 joule/second.) How much heat in calories flows if the heat rate is steadily reduced at the rate of 0.40 W/s to 5.0 W?

6-28. An ammeter reading has been decreasing at a constant rate of 0.20 A/s for 17 s.

(a) What was the original current if the ammeter now reads 7.1 A?

(b) How much charge passed through the ammeter during those 17 s?

6-29. An ammeter reading increases at a constant rate from 2.0 to 13.0 A. During that time 825 C of charge passes through the ammeter.

(a) What is the rate of change of current ($\ddot{Q}$)?

(b) How long does it take for the 11.0-A change to occur?

6-30. An electrical device called a capacitor collects charge on its plates according to the picture as recorded on an oscilloscope.
(a) What is the current at $t = 0$ s?
(b) What is the current at $t = 10$ ms?

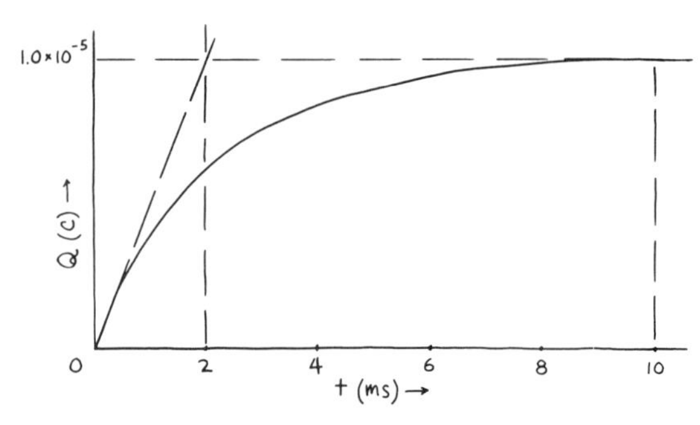

Problem 6-30

6-31. A voltage control is turned up steadily so that current to a heater increases from zero at a constant rate of 0.10 A/s.
(a) What will the current be after 30 s?
(b) How much charge flows during that 30-s time interval?
(c) What will the current be when a total of 100 C passes a given point in the circuit?

ANSWERS TO ODD-NUMBERED PROBLEMS

6-1. 2.6×10^4 ft
6-3. 5.0×10^2 s
6-5. (a) 4.2×10^3 rad
(b) 3.1×10^2 ft · lb/s
6-7. 3.1×10^3 ft
6-9. 0.14 m/s
6-11. (a) 8.6×10^3 ft^3
(b) 0.20 hp
6-13. 1.2 cal/s
6-15. 1.0×10^4 cal/s
6-17. (a) 9.0×10^3 C
(b) 5.6×10^{22} electron
(c) 5.0 W
6-19. 6.6×10^2 ft
6-21. 11 ft/s^2
6-23. (a) 15 m/s
(b) 38 m
(c) 16 m/s
6-25. (a) 1.3×10^{-2} gal/s^2
(b) 7.3 gal
6-27. 1.1×10^2 cal
6-29. (a) 1.00×10^{-1} A/s
(b) 1.10×10^2 s
6-31. (a) 3.0 A
(b) 45 C
(c) 4.5 A

7

Resistance and Energy Loss

In any physical system in which parameter can "flow," there must be a force from outside of the system to get things going (i.e., to initiate the flow). A car begins to move because the road surface pushes on its tires. Water starts to flow as a result of the pressure difference created by a pump. Electric charge flows under the influence of a potential difference (emf) maintained by a battery.

A force has to be there to keep things going—to maintain the flow— for in the act of flowing, energy is invariably lost from the system because of "friction" of one kind or another. If you switch off a car's engine the car eventually comes to a stop. If you shut down the pump, water stops flowing. Unplug the TV and the picture disappears. Turn off the stove and the pot cools.

In each of these systems, there is a built-in tendency to resist the flow. Whenever a force is applied, there appears an internal opposition to whatever the applied force is trying to do. The purpose of this chapter is to study the effects of a system's internal opposition to the movement of energy from one place to another.

We first study the resistive force F_R and its relation to the excitation force F_E, the outside force that gets things going. The next sections discuss resistance, the relation between resistive force and parameter rate. Electrical resistance is defined by Ohm's law. Resistance in other systems is analogous, but some systems may not follow Ohm's law. One example is Coulomb or "dry" friction in mechanical systems.

Section 7-4 studies the work done by the resistive force. This work is lost from the system. Section 7-5 describes how such things as physical dimensions and type of material affect a system's resistance. Some situations are discussed where resistance departs from its Ohm's law behavior, becoming nonlinear. For example, in fluids a quantity called the Reynolds number predicts the transition from laminar flow to turbulent flow. Finally, in Sec. 7-6 resistors are put together in series and parallel combinations.

7-1 RESISTIVE FORCE

Section 4-6 identified four kinds of forces that may act on or within a system. Excitation force F_E is the outside force that gets things going. It is responsible for any flow that is initiated. It is the force that puts energy into a system. The excitation force may be distributed among three tasks. Capacitive force F_C and inertive (net) force F_M are responsible for the input of potential and kinetic energy, respectively. Resistive force F_R is that part of the excitation force needed to overcome the friction of mechanical and fluid systems and its analogues in thermal and electrical systems. The work done by the resistive force is always lost from the system, most commonly in the form of heat. The energy is lost in that it cannot be reclaimed by the system that produces it. For example the heat produced by an electric stove may be put to good use such as heating soup. That energy is lost, however, in that it cannot be put back into the stove's electrical system.

Mechanical and Fluid Systems

In any system in which particles of matter move past one another, collisions between particles are bound to occur. Such collisions are wasteful of energy to some degree, because the particles do not rebound with the same energy with which they collide. If, for example, one solid object is dragged across another (Fig. 7-1), the surface irregularities bang into one another. Some of the energy of collision is changed to heat and effectively lost from the system. The force required to compensate for these collisions is the resistive force F_R.

If the surfaces are separated by a layer of fluid (Fig. 7-2), the collisions are largely avoided, and heat is produced only as fluid molecules slide around one another. Less energy is lost, less heat is produced, and less frictional force is exerted when the fluid is present. Thus less of the excitation force is needed to overcome friction, and the resistive force F_R is decreased.

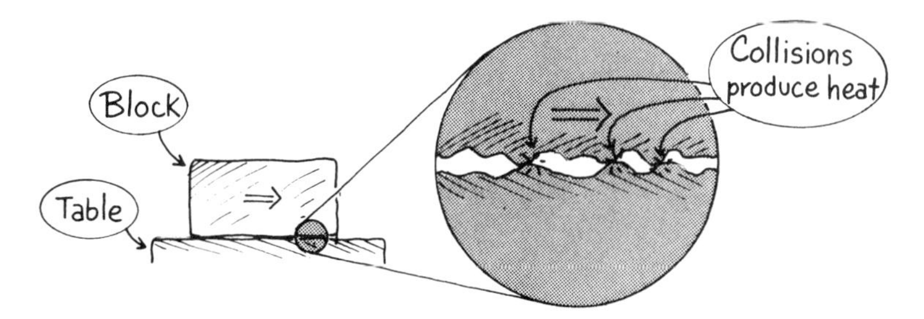

Figure 7-1. Even the smoothest surface is not truly smooth. Irregularities on one surface bump into those on the other surface as the block slides across the table. The collisions produce heat, which is energy lost from the system.

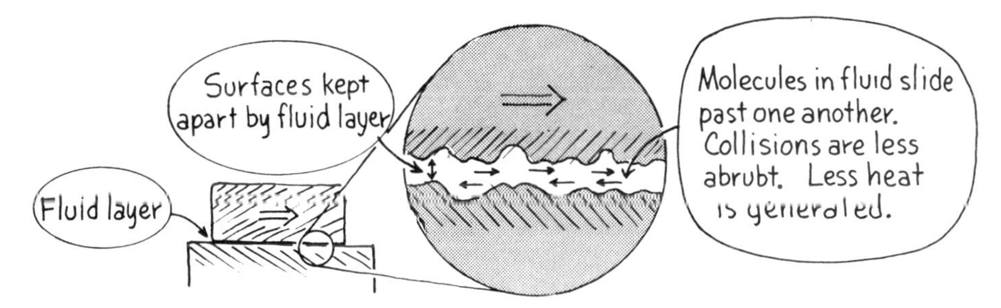

Figure 7-2. A fluid film between solid surfaces keeps the irregularities from interlocking.

In the case of a fluid moving through a pipe (Fig. 7-3), the molecules at the outer edge are in contact with the inside walls of the pipe. A frictional force is exerted by the pipe on the water. That outer layer of molecules tends to slow up the next layer, and so on to the center. It's as though the fluid were a cluster of well-greased pipes, one inside the other. The innermost layer attains the greatest velocity and the outer layer the least. As the fluid moves along, friction

Figure 7-3. Orderly fluid flow in a pipe. At low velocities concentric layers of fluid slide past one another in an orderly way. Inside layers travel faster than outside layers.

saps energy from the system, and the pressure decreases downstream (Fig. 7-4a). The part of the pump's pressure used to overcome friction is designated as the resistive force.

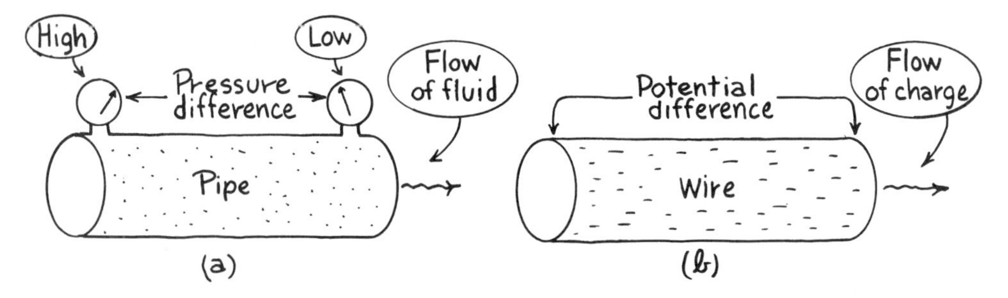

Figure 7-4. The analogy is good between resistive forces in fluid and electrical systems.

Electrical and Heat Systems

An enlarged sketch of a wire with its free electrons (Fig. 7-4b) looks like a sketch one would make of a pipe containing fluid. As in the fluid system, the electrical "pressure" decreases downstream. In this case, however, the pressure drop is a potential difference.

Electric charge (Secs. 3-6 and 5-4) is carried in a conductor by electrons that drift from atom to atom under the influence of electric forces. These "free" electrons can be thought of as a cloud of unattached particles moving like a swarm of bees through a crowd of people. A given bee is not associated with a particular person in the crowd. The rate of progress of the swarm is limited as random collisions result in transfer of energy to the crowd. In the process the bees suffer, too, losing some of their energy. Electrical resistive force (potential difference) is a measure of energy lost by an average electron "bee" as it interacts with the "crowd" of atoms.

As explained in Sec. 5-5, heat systems do not fit quite as neatly into the unified-concepts scheme. The problem is that energy itself has been chosen as parameter. In many situations it is helpful to think of heat as a sort of fluid that can be shoved around from place to place, if one can muster the proper "pressure" or temperature difference. When visualized this way, the behavior of heat in passing through a conductor is analogous to water in a pipe or to charge in a wire. In fact, electrons are primarily responsible for the conduction of heat in metals.

To make the analogy easier, consider a device called a heat pump that sucks in our heat fluid at one end and squirts it out the other. The heat pump, like any other seat of excitation force, requires an outside source of energy to do its job. The excitation force manufactured by the heat pump is measured in terms of temperature difference (Fig. 7-5).

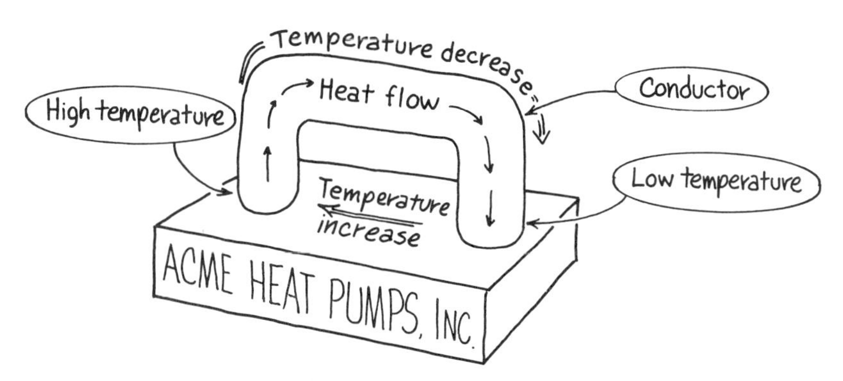

Figure 7-5. The heat pump, like a water pump or battery, must maintain an excitation force (temperature difference) in order to keep parameter flowing in the system.

A metallic conductor is connected between the high- and low-temperature ports. Heat begins to flow into the conductor at the high-temperature end and out at the low-temperature end in a manner reminiscent of the fluid and electrical systems. The heat pump's entire effort is to maintain the temperature difference and flow of heat once constant steady-state temperatures are established in different parts of the system.

7-2 FRICTION FORCES IN MECHANICAL SYSTEMS

Frictional force and its analogues in other systems arise only in response to actual or impending movement in a system. Such forces are always in the direction opposite to the direction of flow. The size or magnitude of a frictional force (and thus the resistive force F_R) is independent of the parameter rate $\dot{Q}$ in some systems, but in most situations F_R changes when $\dot{Q}$ changes.

Dry Friction

In the case of a solid object sitting on a horizontal surface (Fig. 7-6a), the frictional force f between block and table cannot be measured or even recognized as existing until force F is applied. If F is not very large, the block will not move because of an equal but opposite force f_s of *static* (standing still) friction. The block sticks to the table largely owing to the interlocking of surface irregularities (Fig. 7-6b). Force F must be increased to overcome this original stickiness before any motion can occur. The friction will increase along with F until f_s reaches its maximum possible value. At that point any additional increase in F will make the block move. Curiously, as soon as motion begins, the friction force changes. Surface bumps have been pulled out of interlocking holes, and as long as motion is maintained the irregularities skim from one high point to another. Thus the force f_k of *kinetic* (moving) friction is usually

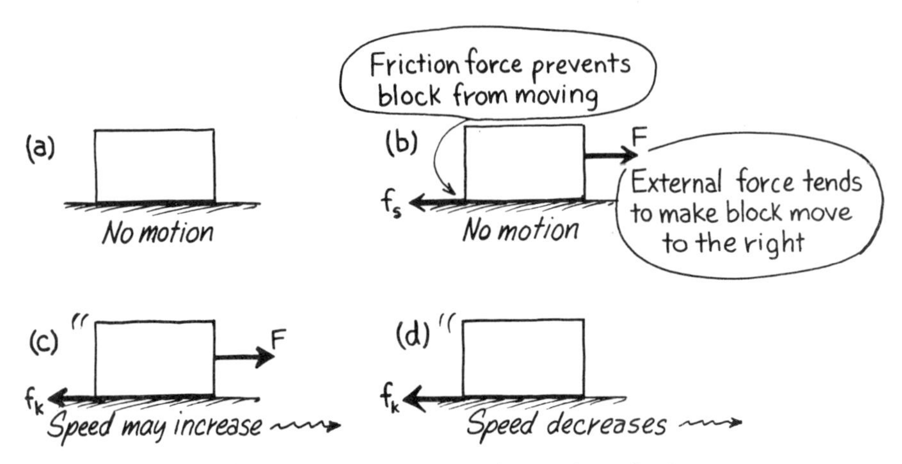

Figure 7-6. The friction force f takes on one of a number of values depending upon the conditions of external force and motion.

less than force f_s for static friction (Fig. 7-6c). The kinetic friction force acts as long as the block moves, even if F is removed (Fig. 7-6d).

Experiments show that once such a system is moving the frictional force is essentially constant. Changing the speed (parameter rate) seems to have little or no effect on the frictional force between two solid surfaces. A graph of frictional force versus velocity (Fig. 7-7) is simply a straight line parallel to the velocity axis.

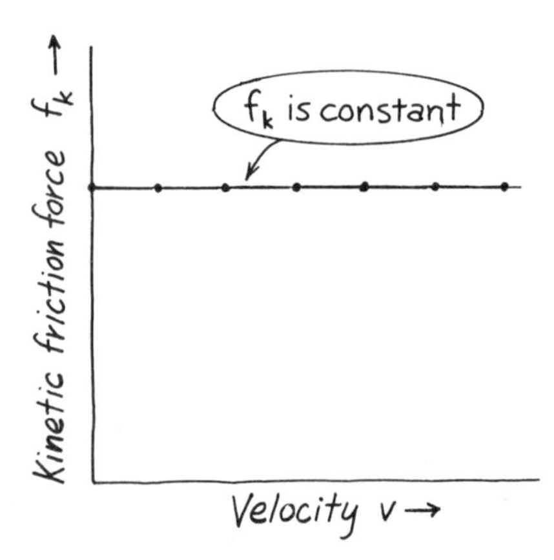

Figure 7-7. Resistive force (equal to friction force f_k) is independent of parameter rate (velocity) for solid surfaces.

The area of contact between solid surfaces does not matter either (Fig. 7-8a). What does matter is the force with which the two surfaces are held together. This force, perpendicular to the surfaces, is called the *normal force N*. In Fig. 7-8, the normal force is just the weight of the bricks. If another brick is piled on top of the first, the normal force becomes twice as much, and twice the frictional force results (Fig. 7-8b). Frictional force f is directly proportional to normal force N for most solid surfaces.

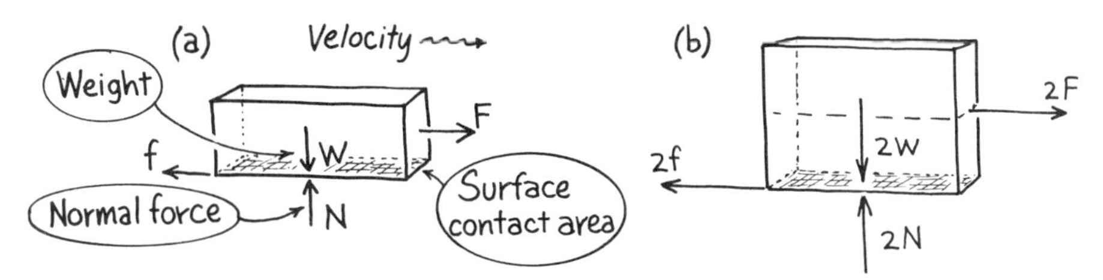

Figure 7-8. The force of friction between dry solid surfaces depends mainly on the kind of surfaces and the force holding them together.

The other important factor that affects the value of f is the nature of the surfaces involved. For a given set of surfaces, the constant ratio of frictional force f to normal force N is a measure of the frictional properties of the surfaces. This ratio is called the *coefficient of friction* μ (Greek letter mu). It has been determined experimentally for various combinations of materials, surface conditions, and starting conditions. Such values serve only as a guide to

$$\mu = \frac{f}{N} \tag{7-1}$$

μ *Coefficient of Friction*	f *Frictional Force*	N *Normal Force (perpendicular to surface)*
Dimensionless ratio (units cancel out)	newton (N)	
	pound (lb)	

expected frictional forces because of the possible wide variation in condition (roughness, temperature, lubrication, etc.) of the surfaces. Values of μ given by various sources vary widely even for the same materials. Such values must be used with restraint. It is reasonable, however, to use the idea of coefficient of friction if conditions are constant. The following examples illustrate its use.

Example 7-1 An increasing horizontal force acts on an empty 15-kg box at rest on a horizontal floor. The box does not move until the force becomes 45 N.

(a) What is the coefficient of static friction μ_s between box and floor?

(b) What force would be needed to start the box if 80 kg of rock are added to the box?

Solution to (a)

$\begin{Bmatrix} m = 15 \text{ kg} \\ F = 45 \text{ N} = f_s \end{Bmatrix}$

W=mg, F, f_s, N

$$N = W = mg = 15 \text{ kg}(9.81 \text{ m}/s^2) = 147 \text{ N}$$

$$\mu_s = \frac{f_s}{N} = \frac{45 \text{ N}}{147 \text{ N}} = 0.306 \rightarrow \boxed{0.31}$$

Solution to (b)

$\{m = 15 + 80 = 95 \text{ kg}\}$

$$N = W = mg = 95 \text{ kg}(9.81 \; m/s^2) = 932 \text{ N}$$

$$f_s = \mu_s N = 0.306(932 \text{ N}) = 285 \rightarrow \boxed{2.8 \times 10^2 \text{ N}}$$

Example 7-2 The loaded box of Example 7-1 can be kept moving at constant speed by a force of 160 N.

(a) What is the coefficient of kinetic friction μ_k?
(b) What force would be needed to keep the empty box moving?

Solution to (a)

$\begin{Bmatrix} N = 932 \text{ N} \\ \text{(Example 7-1b)} \\ F = 160 \text{ N} = f_k \end{Bmatrix}$

$$\mu_k = \frac{f_k}{N} = \frac{160 \text{ N}}{932 \text{ N}} = 0.172 \rightarrow \boxed{0.17}$$

Solution to (b)

$\begin{Bmatrix} N = 147 \text{ N} \\ \text{(Example 7-1a)} \end{Bmatrix}$

$$f_k = \mu_k N = 0.172(147 \text{ N}) = 25.2 \rightarrow \boxed{25 \text{ N}}$$

Viscous Friction

The frictional force produced by the molecules of a fluid sliding past one another differs from "dry" friction in many respects. For one thing, a fluid (liquid or gas) cannot produce static friction. Fluids are defined as substances that have no shear strength. The molecules of a fluid are not rigidly joined, so that any stress at all will start flow. Opposition to flow is produced only when the molecules are moving and colliding with one another and the walls of their container.

With no more information than this, we can see that a graph of friction versus velocity or rate of flow is different than for solid objects. The graphs of Figs. 7-9 and 7-10 pass through the origin and frictional force can only increase as rate of flow increases. The graph's shape depends upon the characteristics of the flow. If the fluid flow is very orderly, with one layer slipping smoothly past another, the flow is said to be *laminar.* Laminar flow is most nearly achieved when thin layers of viscous (thick) fluid flow at low speeds over smooth surfaces.

A dashpot (shock absorber) is an example of a device whose operation depends upon the viscous properties of fluids. It consists of a piston that moves in a cylinder enclosing a fluid (Fig. 7-9). The friction force due to the piston movement in the viscous fluid is proportional to the piston velocity for moderate piston speeds. Thus a graph of frictional opposition versus velocity (resistive force versus parameter rate) turns out to be like the straight line shown in Fig. 7-9.

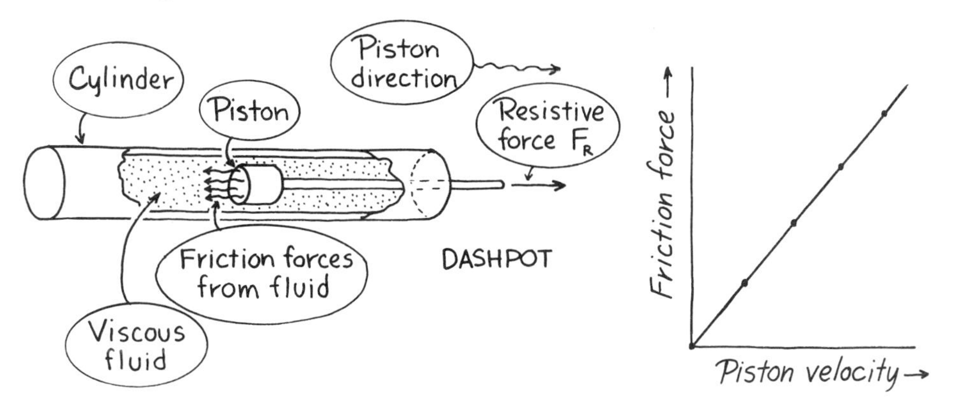

Figure 7-9. The friction force due to piston movement in the viscous fluid is proportional to the piston velocity.

The behavior of fluids in most situations is much more complex than purely laminar flow. When flow rates reach a critical value for a system, the flow becomes *turbulent*. The fluid circulates in complex patterns, wasting energy and increasing frictional opposition. Turbulence invariably produces a force versus parameter rate graph of increasing slope as sketched in Fig. 7-10. The exact relation between resistive force and parameter rate depends on the system. Force may be proportional to the square (second power) or cube (third power) of parameter rate or, in fact, to any whole or fractional power greater than 1. The job of automobile designers should be to make that graph as nearly

Figure 7-10. A vehicle like a pickup truck encounters a large frictional force because of complex turbulent flow of air through which it moves.

straight as possible to avoid unnecessary friction and the waste of energy that it causes. Many vehicles encounter air friction roughly proportional to the cube of velocity. Small wonder that fuel economy goes down rapidly as speed increases.

7-3 RESISTANCE

The relation between resistive force and parameter rate in a given system can best be shown by graphs such as those in Figs. 7-9 and 7-10. The relation can also be described by giving ratios of resistive force to parameter rate. This is particularly useful if the ratio is constant, as it would be for the straight line graph of Fig. 7-9. Such ratios are commonly used in electrical and thermal systems and in fluid systems that involve laminar flow. The ratio of resistive force F_R to parameter rate $\dot{Q}$ is called *resistance R*.

$$\text{Definition of resistance:} \quad \underset{\text{Resistance}}{R} = \frac{\overset{\text{Resistive force}}{F_R}}{\underset{\text{Parameter rate}}{\dot{Q}}} \tag{7-2}$$

We define resistance in this way because it is convenient to do so. The concept of resistance has not been handed down to us from on high nor discovered hiding some place. It is just a ratio of force to parameter rate. Such a definition is called *Ohm's law* in the electrical system.

The value of R for a given system is not necessarily constant. It is constant only where the force–parameter rate graph is a straight line (Fig. 7-11). A device that requires resistive force is a *resistor.* If R for a resistor is constant, the resistor is said to be *linear* in reference to the straight line graph.

We will describe resistance in the various physical systems by stating its Ohm's law definition, Eq. (7-2), and working an example problem illustrating its application.

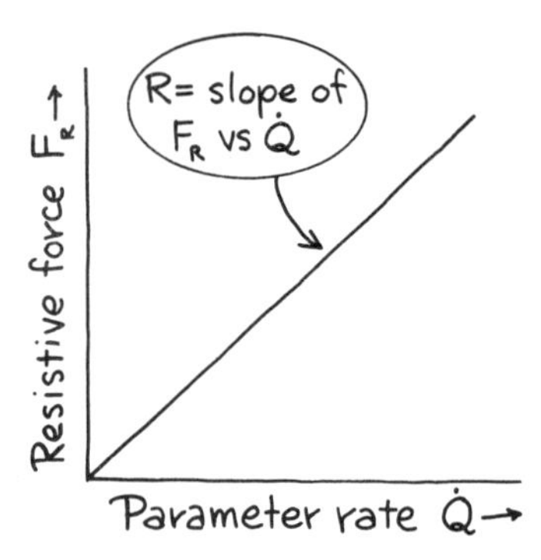

Figure 7-11. A linear resistor has an F_R versus $\dot{Q}$ curve that is a straight line that passes through the origin. Resistance, the ratio $F_R/\dot{Q}$, is the slope of the line.

Resistance in an electrical system is the ratio of potential difference V_R across a resistor to the current I through the resistor. This is Ohm's law.

$$R = \frac{V_R}{I} \tag{7-3}$$

R Resistance	V_R Resistor Potential Difference	I Current
$\frac{\text{volt}}{\text{ampere}}$ = ohm (Ω)	volt (V)	ampere (A)

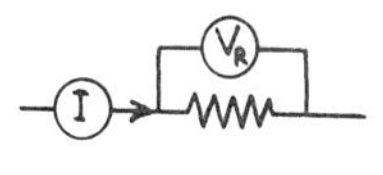

Example 7-3 The battery emf E is 6.0 V. When the switch is closed the ammeter reads 0.30 A.

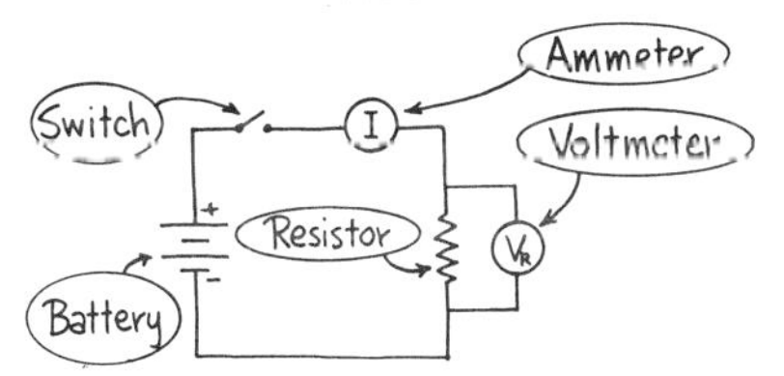

(a) What is the potential difference V_R across the resistor?
(b) What is the resistance R of the resistor?
(c) What will the current be if a 24-V battery is used in place of the 6.0-V battery? Assume that the resistor is linear.

Solution to (a)

When steady state is reached, the entire 6.0 V of the excitation force is used to produce heat in the resistor. Resistive force V_R becomes equal to the excitation force E.

$$\boxed{V_R = 6.0\ \text{V}}$$

Solution to (b)

$\left\{\begin{matrix} V_R = 6.0\ \text{V} \\ I = 0.30\ \text{A} \end{matrix}\right\}$

$$R = \frac{V_R}{I} = \frac{6.0\ \text{V}}{0.30\ \text{A}} = \boxed{20\ \Omega}$$

Solution to (c)

$\left\{\begin{matrix} V_R = 24\ \text{V} \\ R = 20\ \Omega \end{matrix}\right\}$

$$I = \frac{V_R}{R} = \frac{24\ \text{V}}{20\ \Omega} = \boxed{1.2\ \text{A}}$$

Thermal resistance R in a heat system is the ratio of temperature difference between hot T_2 and cold T_1 surfaces to the heat rate $\dot{Q}$. Normally, the system is in the steady state (constant temperatures and heat rate) when these measurements are made. Heat always flows from the hot to cold surface.

T_2 (Hot) T_1 (Cold) $\dot{Q}$

$$R = \frac{T_2 - T_1}{\dot{Q}} \qquad (7\text{-}4)$$

R *Thermal Resistance*	T_2 *Hot Temperature*	T_1 *Cold Temperature*	$\dot{Q}$ *Heat Rate*
$\frac{\text{Celsius degree}}{\text{calorie/second}}$ $\left(\frac{°C}{cal/s}\right)$	Celsius degree (°C)		calorie/second (cal/s)
$\frac{\text{Fahrenheit degree}}{\text{calorie/second}}$ $\left(\frac{°F}{Btu/hr}\right)$	Fahrenheit degree (°F)		British thermal unit/hour (Btu/hr)

Other units of temperature (kelvin and degrees Rankine) and heat rate (watt, Btu/s) can be used to define thermal resistance.

Example 7-4 An electric heater is contained inside a box of insulating material. The heater draws 200 W of electrical power. The system in the steady state has a constant outside surface temperature of 275°C, while the constant inside temperature is 550°C. Assume that all the electrical energy is converted to heat energy.

(a) Find the thermal resistance of the insulating box.

(b) What temperature difference would one expect if the power input were increased to 300 W? Assume that the thermal resistance remains constant.

Solution to (a)

$$\left\{\begin{array}{l} T_2 = 550°C \\ T_1 = 275°C \\ \dot{Q} = 200\ W \end{array}\right\}$$

$$R = \frac{T_2 - T_1}{\dot{Q}} = \frac{(550 - 275)°C}{200\ W} = \boxed{1.38\frac{°C}{W}}$$

Solution to (b)

$$\left\{\begin{array}{l} R = 1.38°C/W \\ \dot{Q} = 300\ W \end{array}\right\}$$

$$T_2 - T_1 = R\dot{Q} = \left(1.38\frac{°C}{W}\right)(300\ W) = \boxed{412°C}$$

Resistance R in the fluid system is the ratio of resistive pressure p_R to the volume flow rate $\dot{V}$.

High pressure p_R Low pressure $\dot{V}$

$$R = \frac{p_R}{\dot{V}}$$

R *Fluid Resistance*	p_R *Resistive Pressure*	$\dot{V}$ *Volume Flow Rate*
$\frac{\text{pascal}}{\text{cubic metre/second}}$ $\left(\frac{\text{Pa}}{\text{m}^3/\text{s}}\right)$	pascal (Pa)	cubic metre/second (m^3/s)
$\frac{\text{pound/square foot}}{\text{cubic foot/second}}$ $\frac{\text{lb/ft}^2}{\text{ft}^3/\text{s}}$	pound/square foot (lb/ft^2)	cubic foot/second (ft^3/s)

Other units of pressure (psi, inches of water, torr, etc.) and volume flow rate (gal/min, cm^3/s, etc.) can be used to define fluid resistance. Just make sure in working problems that the units of resistance are consistent with the units of your problem.

Example 7-5 The volume flow rate of water through a small glass tube was measured for varying pressures supplied by a water column of height h. For convenience, pressure is measured in centimetres of water and thus is equal to the height h.

Pressure h (cm of water)	*Volume Flow Rate* $\dot{V}$ *(cm³/s)*
0	0.0
35	1.3
42	1.6
55	2.1
65	2.4

(a) Show that the tube is a linear resistor and find its resistance.
(b) What value of h produces a volume flow rate $\dot{V}$ of 0.90 cm^3/s?

Solution to (a)

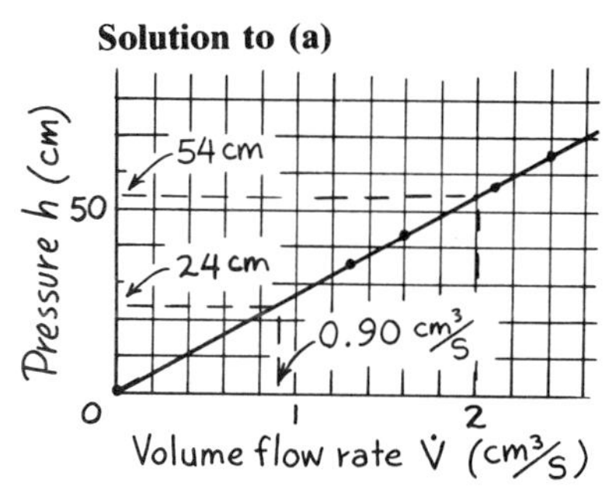

Perhaps the best way to test for linearity is to plot a graph of h versus $\dot{V}$. The points plot in a line that is very nearly straight and that passes through the origin. Resistance is constant and the resistor is linear. Use its slope to calculate R.

$$R = \frac{p_R}{\dot{V}} = \frac{h}{\dot{V}} = \frac{54 \text{ cm}}{2.0 \text{ cm}^3/\text{s}} = \boxed{27 \frac{\text{cm}}{\text{cm}^3/\text{s}}}$$

Solution to (b)

$$\left\{\begin{array}{l} R = 27\ \text{cm}/(\text{cm}^3/\text{s}) \\ \dot{V} = 0.90\ \text{cm}^3/\text{s} \end{array}\right\} \qquad h = R\dot{V} = \left(27\,\frac{\text{cm}}{\text{cm}^3/\text{s}}\right)\left(0.90\,\frac{\text{cm}^3}{\text{s}}\right) = \boxed{24\ \text{cm}}$$

The answer can also be picked directly from the graph.

Resistance R in a mechanical translational system is the ratio of resistive force F_R to velocity v.

$$R = \frac{F_R}{v} \qquad (7\text{-}6)$$

R *Resistance*	F_R *Resistive Force*	v *Velocity*
newton / metre/second $\left(\frac{\text{N}}{\text{m/s}}\right)$	newton (N)	metre/second (m/s)
pound / foot/second $\left(\frac{\text{lb}}{\text{ft/s}}\right)$	pound (lb)	foot/second (ft/s)

Equation (7-6) may be applied to systems with viscous friction in which force changes with velocity. In systems with dry friction, force tends to be independent of velocity and Eq. (7-6) is of little use.

Example 7-6 A force of 95.0 N causes the piston of a bicycle pump to move at a constant velocity of 0.300 m/s.
(a) What is the resistance of the bicycle pump?
(b) What force is needed to move the piston at a velocity of 0.120 m/s?

Solution to (a)

$$\left\{\begin{array}{l} F_R = 95.0\ \text{N} \\ v = 0.300\ \text{m/s} \end{array}\right\} \qquad R = \frac{F_R}{v} = \frac{95.0\ \text{N}}{0.300\ \text{m/s}} = \boxed{317\,\frac{\text{N}}{\text{m/s}}}$$

Solution to (b)

$$\left\{\begin{array}{l} R = 317\,\frac{\text{N}}{\text{m/s}} \\ v = 0.120\ \text{m/s} \end{array}\right\} \qquad F_R = Rv = \left(317\,\frac{\text{N}}{\text{m/s}}\right)(0.120\ \text{m/s}) = \boxed{38.0\ \text{N}}$$

Resistance in a mechanical rotational system is the ratio of resistive torque τ_R to angular velocity ω.

$$R = \frac{\tau_R}{\omega} \tag{7-7}$$

R *Resistance*	τ_R *Resistive Torque*	ω *Angular Velocity*
$\frac{\text{newton} \cdot \text{metre}}{\text{radian/second}}$ $\frac{\text{N} \cdot \text{m}}{\text{rad/s}}$	newton · metre (N · m)	radian/second (rad/s)
$\frac{\text{pound} \cdot \text{foot}}{\text{radian/second}}$ $\frac{\text{lb} \cdot \text{ft}}{\text{rad/s}}$	pound · foot (lb · ft)	

Example 7-7 What torque must be supplied so that the shaft of a motor can rotate at 200 rad/s against a resistive load of 1.25×10^{-2} N · m/(rad/s)?

Solution

$$\left\{ \begin{aligned} R &= 1.25 \times 10^{-2} \frac{\text{N} \cdot \text{m}}{\text{rad/s}} \\ \omega &= 200 \text{ rad/s} \end{aligned} \right\}$$

$$\tau_R = R\omega = \left(1.25 \times 10^{-2} \frac{\text{N} \cdot \text{m}}{\text{rad/s}}\right)\left(200 \frac{\text{rad}}{\text{s}}\right) = \boxed{2.50 \text{ N} \cdot \text{m}}$$

Definitions of resistance, presented in this section for the various systems, are summarized in Table 7-2 at the end of the chapter.

7-4 ENERGY LOSS FROM RESISTORS

The ideas of work and energy (Sec. 3-4 and Chapter 5) make it possible to do things in physics that are difficult or not possible otherwise. In Chapter 5, parameter was defined so that the product of force and parameter is work:

$$\text{work} = \text{force} \times \text{parameter}$$

The one exception to this is the heat system in which the parameter itself is energy.

The work W done by a resistive force F_R is equivalent to the energy E_R lost from a system through a resistor:

$$W = E_R = F_R Q \tag{7-8}$$

(W: Work done by resistive force; E_R: Energy lost from system)

Another way to look at the same thing is in the form of a graph. We have seen (Sec. 7-2) that resistive force can be constant with dry friction or it can change as parameter rate changes. In Sec. 6-2, it was shown that area under a velocity–time graph represents distance. Using the same sort of argument, the area under a force–parameter graph represents the work done by the force as the parameter changes (Fig. 7-12).

Figure 7-12. Work done by a resistive force is represented by the area under a graph of F_R versus Q, regardless of how F_R changes.

Power, the rate of doing work, was discussed in Sec. 6-10 and is defined as $P = W/t$ [Eq. (3-10)], where W is the work done in a time interval t. Power using unified-concept terms is the product of force and parameter rate, $P = F\dot{Q}$ [Eq. (6-23)] if the force is kept constant. Power lost from systems owing to a constant resistive force is

$$P = F_R\dot{Q} \tag{7-9}$$

The units of power are watts (W) if F and $\dot{Q}$ are in terms of newtons (N), metres (m), and seconds (s). In the English system the unit of power is foot · pounds/second if F and $\dot{Q}$ are in terms of pounds (lb), feet (ft), and seconds (s). Equations (7-8) and (7-9) do *not* apply to heat systems since the parameter rate is already by definition power.

The various forms that $W = F_RQ$ and $P = F_R\dot{Q}$ may take are summarized in Table 7-2 at the end of the chapter. Refer to this table in working the following examples in which resistive force is kept constant.

Example 7-8 In Example 7-2 a box of rocks was dragged along at constant speed by a force of 160 N.

(a) Draw a graph of resistive force against distance as the box is dragged 12.0 m.
(b) Calculate the energy lost and the power consumed if the speed is 2.00 m/s.

Solution to (a)

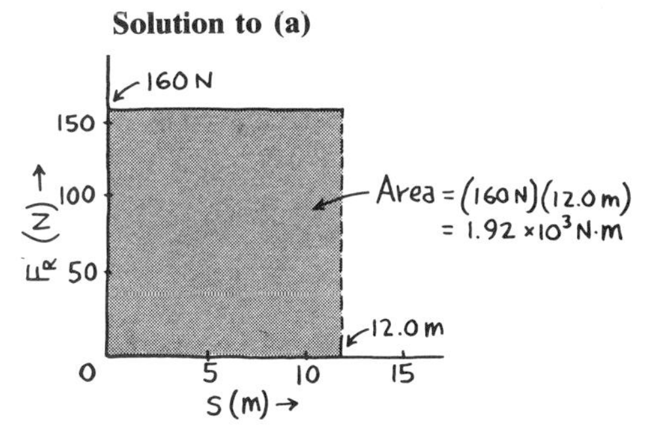

Note that the energy lost is the area of the rectangle.

Solution to (b)

$$\left\{\begin{aligned} F_R &= 160\ \text{N} \\ Q = s &= 12.0\ \text{m} \\ \dot{Q} = v &= 2.00\ \text{m/s} \end{aligned}\right\}$$

$$E_R = F_R Q = F_R s = 160\ \text{N}(12.0\ \text{m}) = \boxed{1.92 \times 10^3\ \text{J}}$$

$$P = F_R \dot{Q} = F_R v = 160\ \text{N}(2.00\ \text{m/s}) = \boxed{3.20 \times 10^2\ \text{W}}$$

Example 7-9 The voltmeter reading is constant at 12.0 V across a 60.0-Ω resistor.
(a) How much power is dissipated as heat?
(b) How much energy is dissipated in 5.00 min?

Solution to (a)

$$\left\{\begin{aligned} V_R &= 12.0\ \text{V} \\ R &= 60.0\ \Omega \end{aligned}\right\}$$

$$I = \frac{V_R}{R} = \frac{12.0\ \text{V}}{60.0\ \Omega} = 0.200\ \text{A}$$

$$P = F_R \dot{Q} = V_R I = 12.0\ \text{V}(0.200\ \text{A}) = \boxed{2.40\ \text{W}}$$

Solution to (b)

$$\left\{\begin{aligned} t &= 5.00\ \text{min} \\ &= 5.00\ \text{min} \times \left(\frac{60\ \text{s}}{1\ \text{min}}\right) \\ &= 300\ \text{s} \end{aligned}\right\}$$

$$E_R = Pt = (2.40\ \text{J/s})(300\ \text{s}) = \boxed{720\ \text{J}}$$

(1 W = 1 J/s)

Example 7-10 A torque of 0.075 N · m is just enough to overcome the friction in the bearings of a circular saw. If the saw runs for 2.0 min at 3300 rpm, how much heat is generated in the bearings?

Solution

$$\left\{\begin{aligned} \tau_R &= 0.075\ \text{N}\cdot\text{m} \\ \omega &= 3300\ \text{rpm} \\ &= 3300\ \text{rpm} \times \left(\frac{0.1047\ \text{rad/s}}{1\ \text{rpm}}\right) \\ &= 346\ \text{rad/s} \\ t &= 2.0\ \text{min} \\ &= 2.0\ \text{min} \times \left(\frac{60\ \text{s}}{1\ \text{min}}\right) \\ &= 120\ \text{s} \end{aligned}\right\}$$

$$P = \tau_R \omega = (0.075\ \text{N}\cdot\text{m})(346\ \text{rad/s}) = 26\ \text{J/s}$$

$$E_R = Pt = (26\ \text{J/s})120\ \text{s} = \boxed{3.1 \times 10^3\ \text{J}}$$

7-5 RESISTIVITY AND CONDUCTIVITY

As we have seen in Sec. 7-2, it is possible to establish a quantity μ (coefficient of friction) that describes the frictional properties of materials. The value of μ allows one to roughly predict how much friction results under certain conditions. Other quantities are commonly used to describe system behavior in which resistive force changes, especially those in which F_R is proportional to $\dot{Q}$.

Electrical System

Experiments show that the resistance of a conductor depends upon at least four different things, including its temperature. One useful quantity called *resistivity* ρ (Greek letter rho) describes the resistive properties of a given material, taking into account its size and shape. All else being equal, the resistance of a wire increases with length and decreases as its cross-sectional area becomes larger. Mathematically, this can be written

Resistance "Proportional to"

$$R \propto \frac{\ell}{A}$$

ℓ ← Length

A ← Cross-sectional area

The proportionality sign can be replaced by an "equals" sign (=) and a proportionality constant.

$$R = \text{proportionality constant} \times \frac{\ell}{A}$$

The constant is the resistivity ρ, and the relation is usually written

$$R = \rho \frac{\ell}{A} \tag{7-10}$$

R *Electrical Resistance*	ρ *Resistivity*	ℓ *Length*	A *Cross-sectional Area*
ohm (Ω)	ohm · square metre / metre ($\Omega \cdot m^2/m$)	metre (m)	square metre (m^2)
	ohm · circular mil / foot (Ω · circular mil/ft)	foot (ft)	circular mil (circular mil)

Resistivity describes how material opposes the flow of electric charge. If resistivity is large, the material is a good *insulator.* If resistivity is small, the material is a good *conductor.* Table 5, Appendix B gives values of ρ for a number of materials. Note that the metallic materials, such as copper, with small values of ρ head the list of conductors, whereas the insulators such as glass have very large values of ρ. Semiconductor materials, such as silicon and germanium, are in between. The units of resistivity include ohm · circular mil/foot. This is a practical unit whose use is illustrated in the following example.

Example 7-11 A resistor is to be made from 100 ft of AWG 30 aluminum wire. (In the American wire gauge scale [AWG] a size 30 wire has a diameter of 0.010 in.) Find the resistance.

Solution

$$\left\{\begin{array}{l} d = 0.010 \text{ in.} = 10 \text{ mil} \\ \ell = 100 \text{ ft} \end{array}\right\}$$

The hardest task is converting diameter to area. This is made easier because the area of 1 circular mil is defined as the area of a circle whose diameter is 0.001 in. or 1 mil. The area of any circle in circular mils becomes just the square of the diameter measured in mils.

$$A = d^2 = (10)^2 = 100 \text{ circular mils}$$

$$R = \rho\frac{\ell}{A} = \left(16\frac{\Omega \cdot \text{circular mil}}{\text{ft}}\right)\frac{100 \text{ ft}}{100 \text{ circular mils}}$$

(16: Appendix B, Table 5)

$$= \boxed{16\ \Omega}$$

Although resistivity is the more commonly used form of this information, it is perfectly reasonable to use its reciprocal, called *conductivity* σ (Greek letter sigma).

$$\sigma = \frac{1}{\rho} \tag{7-11}$$

(σ: Electrical conductivity; ρ: Electrical resistivity)

Most materials (not all) show an increase in electrical resistance with temperature. The predictable variation of R with temperature is put to good use in the temperature-measuring device called a thermistor, in which the electrical resistance is measured and interpreted as temperature.

Heat System

Everything that has just been said about electrical systems applies to heat systems, except, of course, the units are different. The heat-conducting behavior of a material is traditionally described as its *thermal conductivity k* instead of resistivity. As in the electrical system, thermal resistance R varies with the length ℓ (or thickness, which better describes most heat systems) and inversely with the area A of a heat conductor. The relation is written

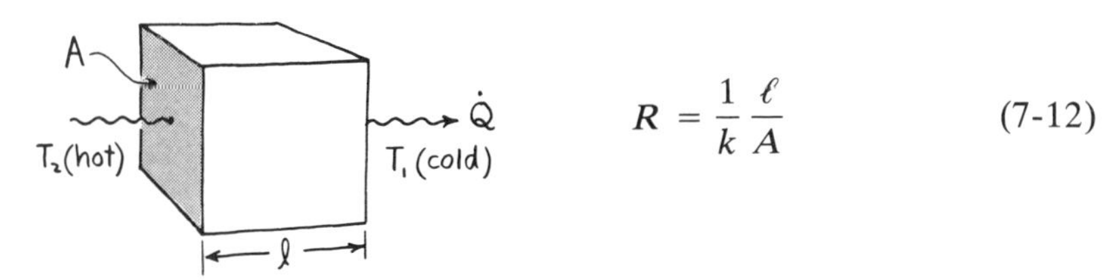

$$R = \frac{1}{k}\frac{\ell}{A} \qquad (7\text{-}12)$$

R *Thermal Resistance*	k *Thermal Conductivity*	ℓ *Thickness*	A *Cross-sectional Area*
$\frac{\text{Celsius degree}}{\text{calorie/second}}$ $\frac{°\text{C}}{\text{cal/s}}$	$\frac{(\text{cal/s}) \cdot \text{cm}}{\text{cm}^2 \cdot °\text{C}}$	centimetre (cm)	square centimetre (cm^2)
$\frac{\text{Fahrenheit degree}}{\text{Btu/hour}}$ $\frac{°\text{F}}{\text{Btu/hr}}$	$\frac{(\text{Btu/hr}) \cdot \text{in.}}{\text{ft}^2 \cdot °\text{F}}$	inch (in.)	square foot (ft^2)

Thermal conductivity, the inverse of resistivity, describes how well a material conducts heat. If k is small, the resistance is large and the material is a good thermal insulator. If k is large, the material is a good heat conductor. Table 4, Appendix B, gives the value of k for a number of materials. The metals that head the list have large values of k. They are good heat conductors. The insulators, such as rock wool, have small values of k. Example 7-12 illustrates the use of thermal conductivity.

Example 7-12 A refrigerator door, 2.5 ft wide and 5.0 ft tall, is insulated with rock wool 2.0 in. thick.

(a) Find the door's thermal resistance.

(b) What is the heat rate if the temperatures are 68°F outside and 25°F inside?

Solution to (a)

$$\left\{\begin{aligned} A &= 5.0\ \text{ft}(2.5\ \text{ft}) \\ &= 12.5\ \text{ft}^2 \\ k &= 0.29\frac{(\text{Btu/hr}) \cdot \text{in.}}{\text{ft}^2 \cdot °\text{F}} \end{aligned}\right\} \text{App. B, Table 4}$$

$$R = \frac{1}{k}\frac{\ell}{A} = \left(\frac{1}{0.29\dfrac{(\text{Btu/hr}) \cdot \text{in.}}{\text{ft}^2 \cdot °\text{F}}}\right)\left(\frac{2.0\ \text{in.}}{12.5\ \text{ft}^2}\right) = \boxed{0.55\frac{°\text{F}}{\text{Btu/hr}}}$$

Solution to (b)

$\begin{Bmatrix} T_2 = 68°F \\ T_1 = 25°F \end{Bmatrix}$

$$\dot{Q} = \frac{T_2 - T_1}{R} = \frac{(68 - 25)°F}{0.55°F/(Btu/hr)} = \boxed{78\ Btu/hr}$$

It is interesting to know that the commonly quoted "R value" for building insulation is a measure of its resistance. For example, a material rated R-19 has a resistance of 19°F/(Btu/hr) per square foot of area.

Fluid System

Unfortunately, the relation between length, area, and material starts to break down for fluid systems. The opposition of a pipe to fluid flow depends upon the pipe's length and area. It also depends on a lot of other things, such as roughness of the pipe's inner surface, the frictional properties of the fluid itself, and how fast the fluid moves inside the pipe.

Experiments show that, if a fluid moves through a pipe with laminar flow, the system's resistance becomes greater in direct proportion to the length of the pipe, just as in the electrical and thermal systems. In such a fluid system, however, resistance R varies inversely with the *square* of the pipe's cross-sectional area. This is a departure from the equivalent expressions for electrical and heat systems [Eqs. (7-10) and (7-12)]. Mathematically,

Resistance ↘ Length ↙

$$R \propto \frac{\ell}{A^2}$$

↗ "Proportional to" ↖ Cross-sectional area *squared*

Another difference is in the proportionality constant, called a *friction factor.* In the electrical and thermal cases there is only one "fluid" (electric charge, heat energy) common to all conductors in the system. The electrical or thermal proportionality constant (resistivity, 1/conductivity) describes the behavior of the conductor. In fluid systems the friction depends on both the conductor and the great variety of fluids that can be conducted. If we consider only laminar flow in long, straight, smooth pipes (a very special case), the proportionality constant need only describe the properties of the fluid and not the pipe. In this special case

(See table on p. 134 for units.)

Laminar flow friction factor ↙

$$R = (8\pi\eta)\frac{\ell}{A^2} \tag{7-13}$$

The friction factor $8\pi\eta$ depends upon *viscosity* η (Greek letter eta). The viscosity of a fluid describes its "thickness" or tendency to oppose flow. For example, motor oil is more viscous than gasoline, and the viscosity of molasses

R *Fluid Resistance*	η *Fluid Viscosity*	ℓ *Length*	A *Cross-sectional Area*
$\frac{\text{pascal}}{\text{cubic metre/second}}$ $\left(\frac{\text{Pa}}{\text{m}^3/\text{s}}\right)$	$\frac{\text{kilogram}}{\text{metre} \cdot \text{second}}$ $\left(\frac{\text{kg}}{\text{m} \cdot \text{s}}\right)$	metre (m)	square metre (m^2)
$\frac{\text{pound/square foot}}{\text{cubic foot/second}}$ $\left(\frac{\text{lb/ft}^2}{\text{ft}^3/\text{s}}\right)$	$\frac{\text{slug}}{\text{foot} \cdot \text{second}}$ $\left(\frac{\text{slug}}{\text{ft} \cdot \text{s}}\right)$	foot (ft)	square foot (ft^2)

is greater than that of water. The common phrase "slower then molasses in January" points up the fact that viscosity itself changes with temperature. The value of η for water at 20°C is about 1.0×10^{-3} kg/(m · s). For glycerin, which runs less freely than water, the value is about 2.0×10^{-3} kg/(m · s).

****Reynolds number.*** In many practical situations, fluid does not flow under such ideal conditions. Roughness, velocity, low viscosity, and change of direction all promote turbulent flow that is complicated almost beyond belief. It has been found empirically (i.e., strictly by experiment) that the type of flow (laminar or turbulent) depends upon a combination of conditions that can be put together in a quantity called the *Reynolds number* N_R:

$$N_R = \frac{\rho v d}{\eta} \tag{7-14}$$

N_R *Reynolds Number*	ρ *Fluid Density*	v *Fluid Velocity*	η *Fluid Viscosity*	d *Pipe Diameter*
Dimensionless (units cancel out)	$\frac{\text{kilogram}}{\text{cubic metre}}$ (kg/m^3)	$\frac{\text{metre}}{\text{second}}$ (m/s)	$\frac{\text{kilogram}}{\text{metre} \cdot \text{second}}$ $\left(\frac{\text{kg}}{\text{m} \cdot \text{s}}\right)$	metre (m)
	$\frac{\text{slug}}{\text{cubic foot}}$ (slug/ft^3)	$\frac{\text{foot}}{\text{second}}$ (ft/s)	$\frac{\text{slug}}{\text{foot} \cdot \text{second}}$ $\left(\frac{\text{slug}}{\text{ft} \cdot \text{s}}\right)$	foot (ft)

The Reynolds number makes it possible to predict the type of flow in a given system. The flow is laminar for any fluid moving in cylindrical pipes if N_R is less than 2000. If N_R is greater than 4000, the flow is turbulent. For values between 2000 and 4000, the flow is in a state of transition between laminar and turbulent. These values of N_R apply regardless of the system of units employed. Again, it should be stressed that this is all based purely on observation.

Example 7-13 Water with viscosity $\eta = 1.0 \times 10^{-3}$ kg/(m · s) flows through a small glass tube of inside diameter 2.0 mm. What is the greatest velocity for which laminar flow can be expected?

Solution

$$\left\{\begin{aligned} N_R &= 2000 \\ d &= 2.0\text{ mm} \times \left(\frac{1\text{ m}}{10^3\text{ mm}}\right) \\ &= 2.0 \times 10^{-3}\text{ m} \\ \rho &= 1.0 \times 10^3\text{ kg/m}^3 \end{aligned}\right\}$$

App. B, Table 1

Laminar flow occurs until $N_R = 2000$.

$$N_R = \frac{\rho v d}{\eta} \Rightarrow v = \frac{N_R \eta}{\rho d}$$

$$v = \frac{2000[1.0 \times 10^{-3}\text{ kg/(m} \cdot \text{s)}]}{(1.0 \times 10^3\text{ kg/m}^3)(2.0 \times 10^{-3}\text{ m})} = \boxed{1.0\text{ m/s}}$$

Note: This corresponds to a volume flow rate $\dot{V}$ of

$$\dot{V} = Av = \frac{\pi}{4}d^2 v = \frac{\pi}{4}(2.0 \times 10^{-3}\text{ m})^2\left(1.0\frac{\text{m}}{\text{s}}\right) = 3.1 \times 10^{-6}\text{ m}^3/\text{s}$$

$$3.1 \times 10^{-6}\frac{\text{m}^3}{\text{s}} \times \frac{(10^2)^3\text{ cm}^3}{1\text{ m}^3} = 3.1\text{ cm}^3/\text{s}$$

Example 7-14 (a) What is the fluid resistance of the tube in Example 7-13 if its length is 0.15 m?

(b) What pressure difference across the tube is needed to produce the flow rate calculated in Example 7-13?

Solution to (a)

$$\left\{\begin{aligned} \eta &= 1.0 \times 10^{-3}\text{ kg/(m} \cdot \text{s)} \\ \ell &= 0.15\text{ m} \\ A &= \frac{\pi}{4}d^2 \\ &= \frac{\pi}{4}(2.0 \times 10^{-3})^2\text{ m}^2 \\ &= 3.1 \times 10^{-6}\text{ m}^2 \end{aligned}\right\}$$

$$R = (8\pi\eta)\frac{\ell}{A^2} = \frac{8\pi[1.0 \times 10^{-3}\text{ kg/(m} \cdot \text{s)}]0.15\text{ m}}{(3.1 \times 10^{-6}\text{ m}^2)^2}$$

$$= \boxed{3.8 \times 10^8\frac{\text{Pa}}{\text{m}^3/\text{s}}}$$

Solution to (b)

$\{\dot{V} = 3.1 \times 10^{-6}\text{ m}^3/\text{s}\}$

$$p_R = R\dot{V} = \left(3.8 \times 10^8\frac{\text{N/m}^2}{\text{m}^3/\text{s}}\right)\left(3.1 \times 10^{-6}\frac{\text{m}^3}{\text{s}}\right) = \boxed{1.2 \times 10^3\text{ Pa}}$$

Note: This is equivalent to the pressure at the base of a column of water that is

$$1.2 \times 10^3\text{ Pa} \times \frac{407\text{ in. water}}{1.013 \times 10^5\text{ Pa}} \times \frac{2.54\text{ cm}}{1\text{ in.}} = 12\text{ cm high.}$$

Table 7-1 at the end of the chapter summarizes the resistivity, conductivity, and friction factor equations for electrical, thermal, and fluid systems.

7-6 RESISTORS IN SERIES AND PARALLEL

A resistor is rarely all by itself in a system. They can be joined together in two basic combinations called series and parallel arrangements. We start with the electrical system and use it as a model for extending the discussion to other systems.

Electrical System as a Model

In the series arrangement (Fig. 7-13a), current from the battery must pass through all three resistors on its way around the circuit. There is no other path. Current through each resistor is the same as the current I from the battery. The potential at the battery's positive terminal is E volts above that at the negative terminal. Thus a charge moving around the circuit encounters three drops of potential V_1, V_2, and V_3, which must add up to the value of the potential rise E. Current and voltage equations associated with the series circuit are as follows:

Series connected electrical system:

$$I = I_1 = I_2 = I_3 \quad \text{(currents are equal)}$$

$$E = V_1 + V_2 + V_3 \quad \text{(voltages are additive)}$$

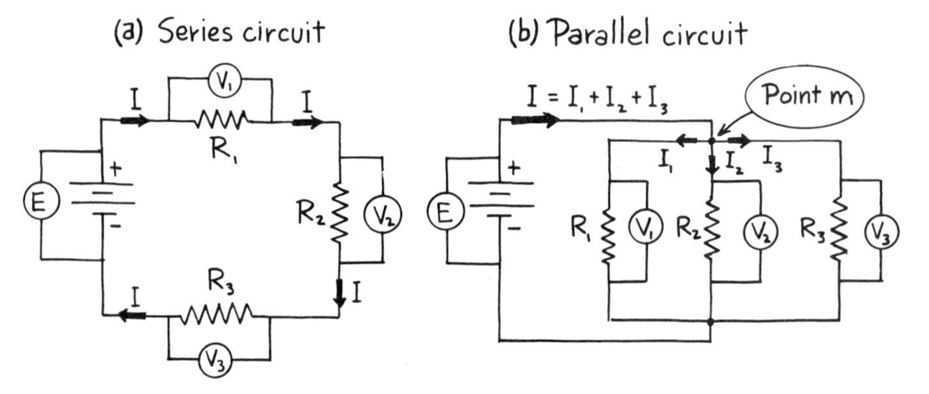

Figure 7-13. Electrical resistors in series and parallel arrangements.

In the parallel circuit of Fig. 7-13b, current leaving the battery has a choice of three routes. Any current I entering a junction such as at point m must equal the sum of the currents ($I_1 + I_2 + I_3$) leaving the junction. Charge can lose energy in any one of the three resistors. Whichever resistor chosen, the charge ends up on the other side at the potential of the battery's negative terminal. Thus the potential drops across all three resistors are equal. Current and voltage equations associated with the parallel circuit are as follows:

Parallel connected electrical system:

$$I = I_1 + I_2 + I_3 \quad \text{(currents are additive)}$$

$$E = V_1 = V_2 = V_3 \quad \text{(voltages are equal)}$$

The electrical system serves as a model to define series and parallel for other physical systems. Using the general symbols F for force and $\dot{Q}$ for parameter rate, we have the following:

Definition of series system:

$$F = F_1 + F_2 + F_3 + \cdots \quad \text{(forces additive)} \tag{7-15}$$

$$\dot{Q} = \dot{Q}_1 = \dot{Q}_2 = \dot{Q}_3 = \cdots \quad \text{(parameter rates equal)} \tag{7-16}$$

Definition of parallel system:

$$F = F_1 = F_2 = F_3 = \cdots \quad \text{(forces equal)} \tag{7-17}$$

$$\dot{Q} = \dot{Q}_1 + \dot{Q}_2 + \dot{Q}_3 + \cdots \quad \text{(parameter rates additive)} \tag{7-18}$$

These equations should be used as definitions to determine whether a physical system is series or parallel connected. The definitions apply not only to systems involving resistive forces, but to any combination of resistive, capacitive, and inertive forces. This becomes useful later when we look at systems with all three types of forces in series and parallel combinations.

The distinction between series and parallel probably looks quite simple at this stage, but, as you will see, some thermal and many mechanical systems are not as easy to spot as electrical and fluid systems. It is useful to be able to classify a given system arrangement in order to know how to predict the net effect of the components acting together.

Other Systems

Figure 7-14 illustrates series and parallel arrangements of resistors in other systems. The fluid arrangements are analogous to their electrical counterparts. The series heat system also fits with no trouble. In the parallel heat system, there are multiple paths (two in this simple sketch) for the heat to take as it moves from the warm source through the conducting material to the cooler side. The problem with the analogy is that, to be strictly correct, the cool surfaces of both thermal resistors must be at exactly the same thermal "potential" (temperature) as are the low-potential ends of the electrical resistors of Fig. 7-13b. This uniformity of temperature may not actually be maintained in practice.

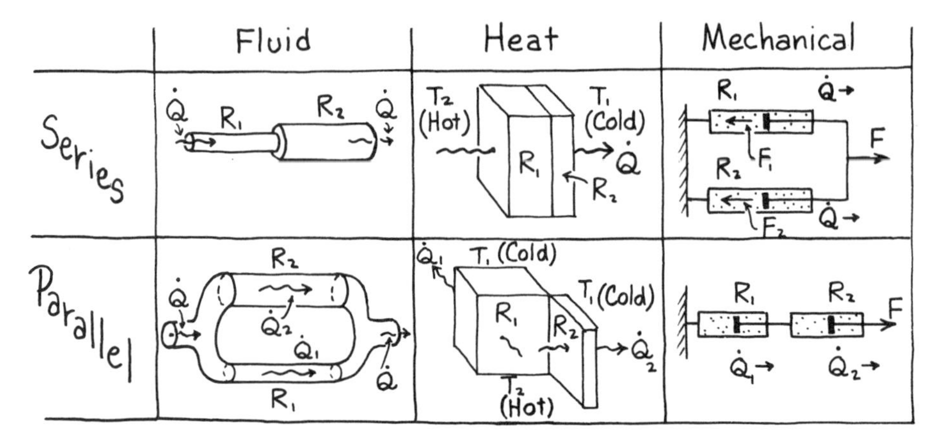

Figure 7-14. Series and parallel arrangements of fluid and thermal resistors look like the equivalent electrical systems. The mechanical series and parallel arrangements look wrong at first, but fit the definitions of Eqs. (7-15) through (7-18).

The mechanical systems really look bad at first. If series in the other systems means joining the resistors end to end and parallel means side by side, why is it not so for the mechanical resistors? The answer lies in the definitions of Eqs. (7-15) through (7-18). When a force is applied to the side-by-side arrangement, the dashpots must expand together at the same speed ($\dot{Q} = \dot{Q}_1 = \dot{Q}_2$) while they produce frictional forces proportional to their resistances ($F = F_1 + F_2$). In the end-to-end situation the applied force is transmitted as a tension equally along the string of dashpots ($F = F_1 = F_2$). Since the distance that force F moves during expansion is the sum of the extensions of the two dashpots, $\dot{Q} = \dot{Q}_1 + \dot{Q}_2$.

It is an interesting exercise to consider various mechanical devices rigged in different ways and to categorize them as series or parallel or combinations of the two. We will come back to the topic in Chapters 15 and 16.

Equivalent Resistance

It is often necessary to know the combined effect or *equivalent resistance* of a number of resistors in combination. For the series system, Eq. (7-15) says that

$$F = F_1 + F_2 + F_3 + \cdots \tag{7-15}$$

But $R = F_R/\dot{Q}$ [Eq. (7-2)] or $F_R = R\dot{Q}$. Substitute in Eq. (7-15) to get

$$R\dot{Q} = R_1\dot{Q}_1 + R_2\dot{Q}_2 + R_3\dot{Q}_3 + \cdots$$

Equation (7-16) says that all the $\dot{Q}$'s are the same, so we can divide through by $\dot{Q}$ to get the following:

Equivalent resistance in series connected systems:

$$R = R_1 + R_2 + R_3 + \cdots \tag{7-19}$$

For resistors in parallel, Eq. (7-18) says that

$$\dot{Q} = \dot{Q}_1 + \dot{Q}_2 + \dot{Q}_3 + \cdots \tag{7-18}$$

But $\dot{Q} = F_R/R$. Substitute to get

$$\frac{F_R}{R} = \frac{F_{R1}}{R_1} + \frac{F_{R2}}{R_2} + \frac{F_{R3}}{R_3} + \cdots$$

And, since all the F's are equal in a parallel system, we can divide through by F_R to obtain the following:

Equivalent resistance in parallel-connected systems:

$$\frac{1}{R} = \frac{1}{R_1} + \frac{1}{R_2} + \frac{1}{R_3} + \cdots \tag{7-20}$$

If only two resistors make up the parallel arrangement,

$$\frac{1}{R} = \frac{1}{R_1} + \frac{1}{R_2}$$

and

$$R = \frac{R_1 \times R_2}{R_1 + R_2} \tag{7-21}$$

Figure 7-15 shows generalized series and parallel systems.

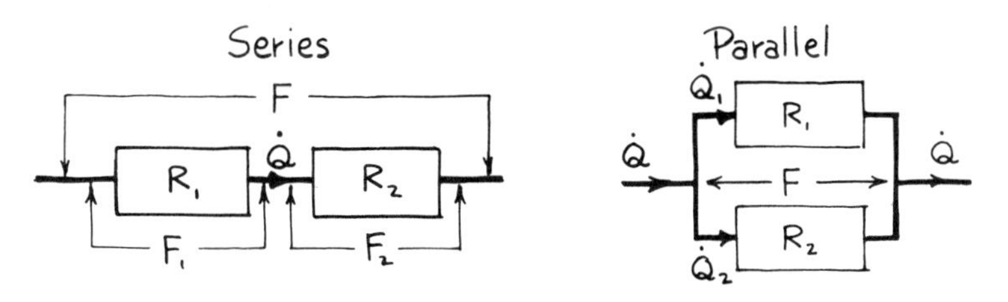

Figure 7-15. Forces and parameter rates in series- and parallel-connected systems.

The following four examples illustrate the use of these equations in series and parallel arrangements of different systems.

Example 7-15 Two small-bore glass capillary tubes allowed flow rates of 1.0 and 3.0 cm^3/s when attached individually to the source supplying a constant 40 cm of water pressure. What flow rates would be expected when the tubes are put together in parallel and series arrangements?

Solution

$\left\{\begin{aligned} \dot{V}_1 &= 1.0\ \text{cm}^3/\text{s} \\ \dot{V}_2 &= 3.0\ \text{cm}^3/\text{s} \end{aligned}\right\}$ Flow rates are additive in parallel systems.

$$\dot{V} = \dot{V}_1 + \dot{V}_2 = (1.0 + 3.0)\text{cm}^3/\text{s} = \boxed{4.0\ \text{cm}^3/\text{s}}$$

$\left\{\begin{aligned} p_R &= h = 40\ \text{cm of water} \\ R_1 &= \frac{p_R}{\dot{V}_1} = \frac{40\ \text{cm of water}}{1.0\ \text{cm}^3/\text{s}} = 40\,\frac{\text{cm of water}}{\text{cm}^3/\text{s}} \\ R_2 &= \frac{p_R}{\dot{V}_2} = \frac{40\ \text{cm of water}}{3.0\ \text{cm}^3/\text{s}} = 13\,\frac{\text{cm of water}}{\text{cm}^3/\text{s}} \end{aligned}\right\}$ Find the equivalent series resistance and use it to get the flow rate.

$$R = R_1 + R_2 = 40 + 13 = 53\,\frac{\text{cm of water}}{\text{cm}^3/\text{s}}$$

$$\dot{V} = \frac{p_R}{R} = \frac{40\ \text{cm of water}}{53\ \text{cm of water}/(\text{cm}^3/\text{s})} = \boxed{0.75\ \text{cm}^3/\text{s}}$$

Example 7-16 Suppose that the refrigerator door of Example 7-12 is covered with a 2.0-in. sheet of ice on its inside surface.

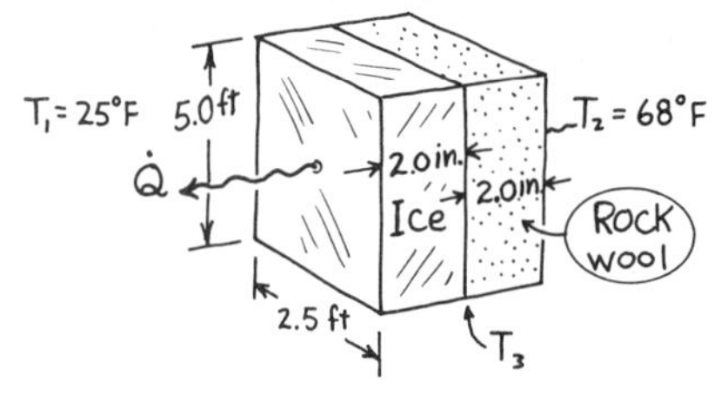

(a) What is the heat rate with the ice layer present?

(b) What is the temperature at the interface between the ice and the rock wool?

Solution to (a)

$\left\{\begin{aligned} A &= 5.0\ \text{ft}(2.5\ \text{ft}) = 12.5\ \text{ft}^2 \\ k &= 12\,\frac{(\text{Btu/hr})\cdot\text{in.}}{\text{ft}^2\cdot{}^\circ\text{F}} \\ R_1 &= 0.55\,\frac{{}^\circ\text{F}}{\text{Btu/hr}} \end{aligned}\right\}$ ← App. B, Table 4; ← Example 7-12

This is a series arrangement because the same heat flows through both materials. The rock wool's resistance has been calculated at 0.55°F/(Btu/hr). The resistance R_2 of the ice is

$$R_2 = \frac{1}{k}\frac{\ell}{A} = \left(\frac{1}{12\,\frac{(\text{Btu/hr})\cdot\text{in.}}{\text{ft}^2\cdot{}^\circ\text{F}}}\right) \times \left(\frac{2.0\ \text{in.}}{12.5\ \text{ft}^2}\right) = 0.013\,\frac{{}^\circ\text{F}}{\text{Btu/hr}} \quad (\text{Ice})$$

The resistances are additive.

$$R = R_1 + R_2 = 0.55 + 0.013 = 0.563\,\frac{{}^\circ\text{F}}{\text{Btu/hr}}$$

and the new heat rate is

$$\dot{Q} = \frac{T_2 - T_1}{R} = \frac{(68 - 25)^\circ\text{F}}{0.563^\circ\text{F}/(\text{Btu/hr})} = \boxed{76\ \text{Btu/hr}}$$

The ice layer reduces the heat rate from 78 to 76 Btu/hr.

Solution to (b)

The temperature difference across the rock wool layer is $T_2 - T_3$.

$$R_1 = \frac{T_2 - T_3}{\dot{Q}} \Rightarrow T_2 - T_3 = R_1\dot{Q} = \left(0.55\,\frac{°\text{F}}{\text{Btu/hr}}\right)76\text{ Btu/hr} = 42°\text{F}$$

and the temperature at the interface is

$$T_3 = T_2 - 42 = 68 - 42 = \boxed{26°\text{F}}$$

The results of this calculation do not imply that you should never defrost your refrigerator. The ice layer impedes the flow of heat from inside the box through the coils to the circulating cold refrigerant, too. The resistance of this path should be kept small by periodically removing the ice that forms on the coils.

Example 7-17 Two dashpots (shock absorbers) are acted upon individually by a 12-N force. R_1 expands at a speed of 0.10 m/s and R_2 at a speed of 0.25 m/s. At what speed does the 12-N force move if the dashpots are series connected and parallel connected?

Solution

$$\left\{\begin{array}{l} F_R = 12\text{ N} \\ v_1 = 0.10\text{ m/s} \\ v_2 = 0.25\text{ m/s} \end{array}\right\}$$

First calculate the values of R_1 and R_2.

$$R_1 = \frac{F_R}{v_1} = \frac{12\text{ N}}{0.10\text{ m/s}} = 120\,\frac{\text{N}}{\text{m/s}}$$

$$R_2 = \frac{F_R}{v_2} = \frac{12\text{ N}}{0.25\text{ m/s}} = 48\,\frac{\text{N}}{\text{m/s}}$$

The resistances add when in series.

$$R = R_1 + R_2 = 120 + 48 = 168\text{ N/(m/s)}$$

and

$$v = \frac{F_R}{R} = \frac{12\text{ N}}{168\text{ N/(m/s)}} = \boxed{0.071\text{ m/s}}$$

The reciprocal resistances add when in parallel:

$$\frac{1}{R} = \frac{1}{R_1} + \frac{1}{R_2} \Rightarrow R = \frac{R_1 R_2}{R_1 + R_2} = \frac{(120)(48)}{120 + 48} = 34.3\,\frac{\text{N}}{\text{m/s}}$$

and

$$v = \frac{F_R}{R} = \frac{12\text{ N}}{34.3\text{ N/(m/s)}} = \boxed{0.35\text{ m/s}}$$

Often systems are neither completely series nor parallel connected, but a combination of both. Example 7-18 shows one method of solving these more complicated problems.

Example 7-18 Electrical resistors of 10, 20, and 30 Ω are arranged as shown. What is the equivalent resistance of the entire circuit?

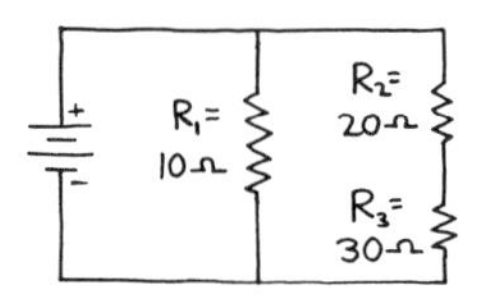

Solution

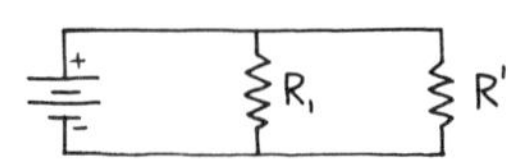

R_2 and R_3 are in series with each other, so that their equivalent resistance R' is

$$R' = R_2 + R_3 = 20 + 30 = 50\ \Omega$$

R_1 and R' are a parallel combination. Their equivalent resistance R is

$$R = \frac{R_1 R'}{R_1 + R'} = \frac{10(50)}{10 + 50} = \boxed{8.3\ \Omega}$$

7-7 SUMMARY

In this chapter we have looked at the first of three basic forces, *resistive force*, at the concept of *resistance*, and at the properties of *resistors* in various systems and arrangements.

1. Resistive force F_R is the force needed to overcome friction in mechanical and fluid systems and its analogues in thermal and electrical systems.

2. In "dry" friction situations, the resistive force depends only upon the normal force holding two surfaces together and the nature of the surfaces.

3. In viscous friction situations, the friction force changes with velocity. This behavior is found in other systems, too. The Ohm's law ratio of resistive force to parameter rate is called resistance.

$$R = \frac{F_R}{\dot{Q}} \tag{7-2}$$

The form of Eq. (7-2) in the various systems is shown in Table 7-1.

4. The work done by resistive force F_R turns into energy E_R that is lost from the system, usually in the form of heat. This energy can be represented as the area under a graph of resistive force versus parameter. The one exception is the heat system, in which the parameter by definition is energy. Energy and power loss equations in the various systems are shown in Table 7-1.

TABLE 7-1 RESISTANCE DEFINITIONS; ENERGY AND POWER LOSS EQUATIONS

System	*Resistive Force* F_R	*Parameter* Q	*Parameter Rate* $\dot{Q}$	*Resistance Equation* $R = F_R/\dot{Q}$	*Energy Loss Equation* $E_R = F_R Q$	*Power Loss Equation* $P = F_R\dot{Q}$
Mechanical translational	Resistive force F_R	Distance s	Velocity v	$R = F_R/v$	$E_R = F_R s$	$P = F_R v$
Mechanical rotational	Resistive torque τ_R	Angle θ	Angular velocity ω	$R = \tau_R/\omega$	$E_R = \tau_R\theta$	$P = \tau_R\omega$
Fluid	Resistive pressure p_R	Volume V	Volume flow rate $\dot{V}$	$R = p_R/\dot{V}$	$E_R = p_R V$	$p = p_R\dot{V}$
Electrical	Potential difference V_R	Charge Q	Current I	$R = V_R/I$	$E_R = V_R Q$	$P = V_R I$
Heat	Temperature difference $T_2 - T_1$	Quantity of heat Q	Heat rate $\dot{Q}$	$R = \dfrac{T_2 - T_1}{\dot{Q}}$	Equations do not apply	

5. Resistance in electrical, thermal, and fluid systems depends upon physical dimensions and other factors described by a proportionality constant. Table 7-2 summarizes this information.

TABLE 7-2 RESISTANCE EQUATIONS IN THE ELECTRICAL, THERMAL, AND FLUID SYSTEMS

Electrical	Thermal	Fluid (laminar flow only)
$R = \rho\dfrac{\ell}{A}$ (7-10), ρ = Resistivity	$R = \dfrac{1}{k}\dfrac{\ell}{A}$ (7-12), k = Thermal conductivity	$R = (8\pi\eta)\dfrac{\ell}{A^2}$ (7-13), η = Viscosity

6. Resistors can be arranged in various ways, but any arrangement can be analyzed in terms of series or parallel combinations. Figure 7-15 summarizes this information.

In Chapter 8 we will take some time out to look at magnetism, a system that is different yet similar in some ways to those we have studied so far. Chapters 9 and 10 introduce the other two basic forces, capacitive force and inertive force. These are forces whose work is not lost from a system but stored within for later use. Quantities called capacitance and inertance will be defined that, like resistance, describe force and energy relations in the systems.

PROBLEMS

7-1. A force of 9.0 lb is needed to start a 20-lb block moving. After it starts, a force of 6.0 lb will keep it moving. Find the coefficients of static friction μ_s and kinetic friction μ_k.

7-2. What force is needed to keep a 40-N block moving on a horizontal surface if the coefficient of kinetic friction is 0.35?

7-3. A box of apples weighing 50 lb falls from an airplane. An R-meter (new invention—patent pending) attached to the box tells us that the resistance due to air friction is 0.40 lb/(ft/s). Assume that the resistance remains constant.
(a) Find the frictional force on the box when its velocity is 60 mi/hr.
(b) Find the terminal velocity of the box in feet/second. (*Hint*: The terminal or highest velocity occurs when the frictional force equals the weight.)

7-4. A streamlined object falling through the air experiences 50 N of frictional force at a speed of 20 m/s.
(a) If the resistance is constant, what will the frictional force be at a speed of 55 m/s?
(b) If the terminal velocity is 75 m/s, what is the mass of the object?
(*Hint*: The terminal or highest velocity occurs when the frictional force equals the weight.)

7-5. A small motor develops 12 oz · in. of torque while rotating at 4000 rpm against a viscous resistive load.
(a) What is the load's resistance in English units of lb · ft/(rad/s)?
(b) What power is needed to overcome the friction?

7-6. In Example 7-5, find the pressure drop in centimetres of water when the flow rate is 1.2 cm^3/s.

7-7. A fan draws 900 ft^3/min of air through a pipe. The pressure difference between the ends of the pipe is 6.00 in. of water.
(a) What is the pipe's resistance in English units of $(lb/ft^2)/(ft^3/s)$?
(b) What power is consumed in overcoming the friction?

7-8. The temperature inside a room is 70°F and outside it is 30°F. 4.0×10^4 Btu of heat is lost each day through the window glass and 15×10^4 Btu of heat is lost through the walls, floor, and ceiling. What is the thermal resistance of the room as a whole? Give your answer in °F/(Btu/hr).

7-9. A refrigerator consumes power at an average rate of 500 W and keeps the inside temperature at 5.0°C when the outside temperature is 24.0°C.
(a) What is the thermal resistance of the refrigerator box in °C/W?
(b) Convert the result to English units of °F/(Btu/hr).

7-10. An insulated picnic chest contains melting ice that keeps the inside temperature at 0°C. If the chest remains unopened, 10 kg of ice lasts 3 days (72 hr) when the outside temperature is 35°C.
(a) Find the heat rate through the chest in watts.
(b) What is the thermal resistance of the chest?
(*Hint*: According to Appendix B, Table 2, it takes 80 cal to melt 1 g of ice.)

7-11. A 100-W electric light bulb is connected to a 120-V source. What is the resistance of the bulb's filament?

7-12. A certain resistor allows a current of 2.5 mA when the potential difference across its terminals is 9.0 V. The current changes to

3.9 mA when the potential difference increases to 14.0 V. Is this a linear resistor? Explain your answer.

7-13. A simple electrical circuit consists of a battery connected to a resistor. 20.0 C of charge passes through the resistor giving up 480 J of energy, which is converted to heat in 10.0 s.

(a) What is the current?

(b) What is the potential difference across the resistor?

(c) What is the resistance of the resistor?

(d) What power is dissipated in the resistor?

7-14. A 35.0-Ω resistor is to be wound using wire of diameter 4.00×10^{-3} in.

(a) What length of copper wire is needed?

(b) What length of iron wire would be needed?

7-15. Compute the resistance in ohms of a copper transmission wire whose cross-sectional area is 2.00 cm^2 and whose length is 500 km.

7-16. Suppose that the dimensions and potential difference associated with a wire resistor can be changed in various ways. What sort of change in the current I would take place in each of the following cases?

(a) The length ℓ is doubled.

(b) The area A is halved.

(c) Potential difference V_R is doubled.

(d) The material is changed from aluminum to iron.

7-17. Which offers more resistance to the flow of heat: a piece of steel whose area is 20.0 ft^2 and whose thickness is 25.0 in., or a common brick wall of area 100 ft^2 and thickness 2.50 in.? Show how you get your answer.

7-18. Suppose that the dimensions and temperatures associated with a firebrick wall can be changed in various ways. What sort of change in the heat rate ($\dot{Q}$) would take place in each of the following cases?

(a) The wall thickness is doubled.

(b) The hot side temperature increases.

(c) The area is doubled.

(d) Concrete is substituted for firebrick.

7-19. The viscosity of water is 1.10×10^{-5} slug/(ft · s) at room temperature. What is the resistance of 120 ft of tubing whose cross-sectional area is 0.015 ft^2? What assumption is needed in making this calculation?

7-20. Suppose that pipe dimensions, pressures, and fluid can be changed in various ways. What sort of change in flow rate ($\dot{V}$) would take place in each of the following cases?

(a) Length ℓ is doubled.

(b) Glycerin is substituted for water.

(c) The pressures at the points separated by the length ℓ are increased by equal amounts.

(d) Area A is doubled.

7-21. Resistor R_1 draws 0.10 A from a 3.0-V source. Resistor R_2 draws 0.20 A from a 12.0-V source. What currents are drawn from a 24.0-V source if both resistors are (a) connected in series? (b) connected in parallel?

7-22. Calculate the equivalent resistance of the following combinations.

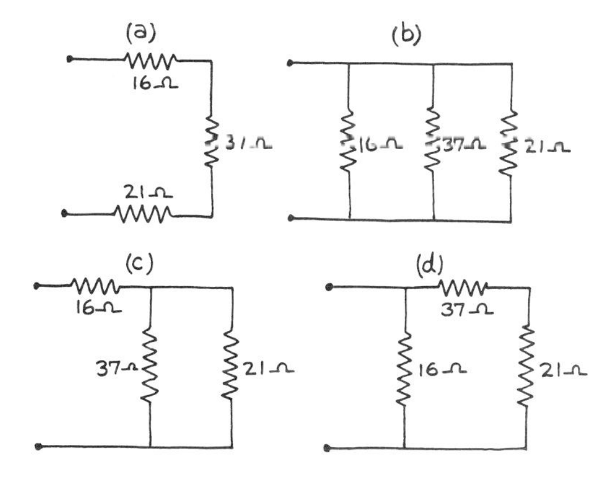

Problem 7-22

7-23. Two small glass tubes are components of a fluidic circuit. Their resistances are 27 and 56 cm of water/(cm^3/s) when measured individually. A flow rate of 1.4 cm^3/s is to be maintained through the system. Calculate the pressure needed in centimetres of water if the tubes are (a) series connected, and (b) parallel connected.

7-24. A box-shaped ice chest insulated with rock wool has a lid 24 in. long and 15 in. wide. The sides of the box are 12 in. high. The sides and bottom of the chest are 2.0 in. thick and the lid is 1.5 in. thick.

(a) Calculate the thermal resistance of the box excluding the lid.

(b) Calculate the thermal resistance of the lid by itself.

(c) Calculate the thermal resistance of the entire chest (with lid on).

7-25. Each wall of a square box furnace made of firebrick is 1.00 in. thick and 0.80 ft^2 in area.

(a) What is the furnace's thermal resistance?

(b) What is the heat rate in Btu/hour if the inside and surface temperatures are maintained at 350 and 170°F?

7-26. A 100-lb block rests on an inclined plane. The block does not start to move by itself until the angle of incline with respect to the horizontal is increased to 30°.

(a) Find the coefficient of static friction μ_s.

(b) Find the friction force on the block after it begins to slide. μ_k is 0.32.

7-27. The force on the end of a pickup in the groove of a long-playing record is equivalent to the weight of 2.0 g. The coefficient of sliding friction between needle and groove is 0.50.

(a) What is the weight of the pickup in newtons?

(b) What is the heat rate in Btu/hour if the pickup and the groove?

7-28. A 50-lb block can be pulled along at constant speed by a 10-lb force acting at an angle of 20° above the horizontal. What is the coefficient of kinetic friction?

7-29. A glass sphere (a marble) of mass 12 g is dropped into a vertical tube of water. The marble attains terminal velocity almost instantly, and thereafter travels the 1.0-m length of the tube in 6.0 s.

(a) Find the system's resistance.

(b) If a 43-g steel sphere of the same size is used in place of the glass marble, how long will it take to reach the bottom of the tube? Assume that the resistance is the same for both spheres.

7-30. An electric oven maintains a temperature of 450°F inside while the outside surface stays at 140°F. The heating coil cycles on and off, averaging 14 s on and 42 s off each cycle. When on, the coil draws 7.5 A from a 220-V source. What is the thermal resistance of the oven? Express your answer in degrees Celsius/watt.

7-31. The frictional drag on a particular automobile varies with the square of the velocity. A resistive force of 50 lb is needed to keep it moving at 20 mi/hr.

(a) What is the frictional force at a constant speed of 60 mi/hr?

(b) What power is needed to keep the car moving at 20 mi/hr and 60 mi/hr?

ANSWERS TO ODD-NUMBERED PROBLEMS

7-1. 0.45, 0.30

7-3. (a) 35 lb
(b) 1.2×10^2 ft/s

7-5. (a) 1.5×10^{-4} lb · ft/(rad/s)
(b) 26 ft · lb/s

7-7. (a) 2.08 (lb/ft^2)/(ft^3/s)
(b) 468 ft · lb/s

7-9. (a) 3.8×10^{-2} °C/W
(b) 2.0×10^{-2} °F/(Btu/hr)

7-11. 144Ω

7-13. (a) 2.00 A
(b) 24.0 V
(c) 12.0 Ω
(d) 48.0 W

7-15. 42.5 Ω

7-17. Brick, $R = 5.8 \times 10^{-3}$ °F/(Btu/hr), offers more resistance than steel, $R = 3.6 \times 10^{-3}$ °F/(Btu/hr).

7-19. 1.5×10^2(lb/ft^2)/(ft^3/s) laminar flow

7-21. (a) 0.27 A
(b) 1.2 A

7-23. (a) 1.2×10^2 cm of water
(b) 26 cm of water

7-25. (a) 0.029°F/(Btu/hr)
(b) 6.3×10^3 Btu/hr

7-27. (a) 2.0×10^{-2} N
(b) 9.8×10^{-3} N

7-29. (a) 0.71 N/(m/s)
(b) 1.7 s

7-31. (a) 4.5×10^2 lb
(b) 1.5×10^3 and 40×10^3 ft · lb/s

8

Another System—Magnetism

Magnetism is important because of its interaction with electricity. Magnetic forces arise whenever a charged particle moves. It is convenient to distinguish between electrical force on a charge at rest and magnetic force on a charge in motion, but in reality the forces cannot be separated. Because of their interaction, it is impossible to go deeper into the study of electricity without some knowledge of magnetism.

Magnetism can be used to control the movement of liquids and gases, to accurately regulate temperature and velocities, to actuate mechanical and hydraulic linkages, and to convert mechanical energy into electrical energy and electrical energy into mechanical energy. Some knowledge of magnetism is needed to understand electromagnetic waves (Chapter 18). These are just a few of the examples of its importance.

It is always hard to learn about a system for the first time. There are new terms and new ideas that must be assimilated in order to understand the new material. We will try to ease this burden by using unified concepts to relate magnetism to other systems you already know something about. If the concepts are applied rigorously, theoretical differences show up right away. Nevertheless, unified concepts can help.

To start with we will examine how the unified concepts of force, parameter rate, and resistance apply to magnetic circuits. The comparison is made in Section 8-3. The latter part of the chapter discusses induced electromotive forces. Comparisons can be made with other systems using unified concepts, and again these comparisons can be helpful in understanding the fairly difficult idea of an induced electromotive force.

8-1 MAGNETIC FIELDS AND MAGNETIC FLUX

A permanent magnet, bent into the form of a horseshoe, has a north pole and a south pole. The air gap between poles is said to be occupied by a *magnetic field*, which can be detected by placing a conducting wire in the air gap as shown in Fig. 8-1. The wire is connected through a switch to a battery, and any current in the wire is detected by the ammeter. No force acts on the wire if no

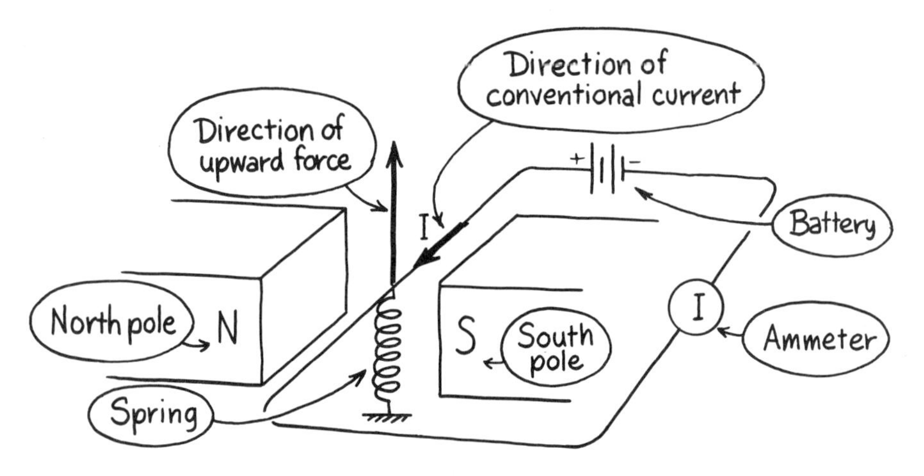

Figure 8-1. The magnetic field between the poles of a magnet can be detected by observing an upward force on a wire that carries electrical current.

current flows (switch open). When the switch is closed, a current I flows and the wire is forced upward. The magnetic field is detected by the force that can be measured, for example, by a spring deflection. The force is reversed (down instead of up) if either the current direction is reversed or the magnet's poles are reversed. Nobody knows why magnetic forces exist. Nature just behaves that way.

The magnetic field can be visualized by drawing a bunch of lines or arrows that leave the north pole and enter the south pole of the magnet (Fig. 8-2). These lines are called *magnetic flux* ϕ (Greek letter phi). The unit of magnetic flux in SI units is the *weber* (Wb). Again in our imagination, the concentration of these fictitious flux lines represents the *magnetic field strength* B. The magnetic field strength can be thought of as the number of lines, measured in

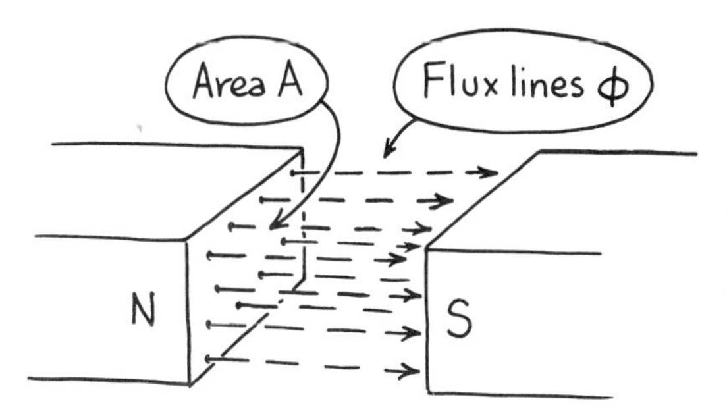

Figure 8-2. The magnetic field is visualized by imagining that a bunch of arrows called magnetic flux leave the north pole, pass through cross-sectional area A, and enter the south pole of the magnet. The concentration of the lines (ϕ/A) represents the strength B of the magnetic field.

webers, passing perpendicularly through a unit of area. The unit of magnetic field strength in SI units is the weber/metre^2 (Wb/m^2), which is called a *tesla* (T). The magnetic field strength B is sometimes referred to as *magnetic flux density*.

$$B = \frac{\phi}{A} \tag{8-1}$$

B Magnetic Field Strength	ϕ Magnetic Flux	A Area
$\frac{\text{weber}}{\text{square metre}}$ = tesla (Wb/m^2) = (T)	weber (Wb)	square metre (m^2)

The strength of the magnetic field is measured in terms of the current I flowing in the wire perpendicular to the magnetic flux lines (Fig. 8-3) and the force F exerted on the moving charges in the conductor by the magnetic field. The magnetic field strength is calculated and basically defined by the equation

$$B = \frac{F}{\ell I} \tag{8-2}$$

B Magnetic Field Strength	F Force	ℓ Length	I Current
tesla (T)	newton (N)	metre (m)	ampere (A)

The length ℓ is that portion of the wire lying within the magnetic field.

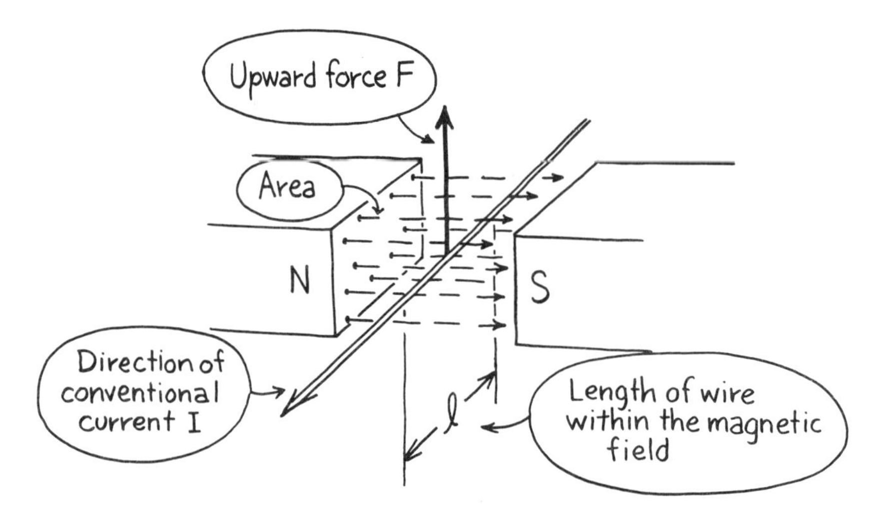

Figure 8-3. The strength of the magnetic field B is measured in terms of the current I moving perpendicular to the field and the upward force F exerted on the length of wire ℓ within the field.

Notice that a tesla has the units of N/(m · A). These units are individually familiar to us and suggest that the magnetic field strength can be found by measuring a force, a length, and a current. This is indeed true, although fields are not usually measured that way. The magnetic field of the earth has a strength of about 3×10^{-5} T, and the magnetic field of a strong permanent magnet can be as large as 1 T.

Example 8-1 In Fig. 8-1 the wire length within the field is 0.020 m, the pole cross-sectional area is $4.0 \times 10^{-4}\ \text{m}^2$, and the magnetic field is 0.10 T.

(a) Calculate the magnetic flux in the space between the poles.
(b) Calculate the upward force if the current is 1.5 A.

Solution to (a)

$$\left\{\begin{array}{l} \ell = 2.0 \times 10^{-2}\ \text{m} \\ A = 4.0 \times 10^{-4}\ \text{m}^2 \\ B = 0.10\ \text{T} \end{array}\right\}$$

$$B = \frac{\phi}{A} \Rightarrow \phi = BA = 0.10\ \text{T}(4.0 \times 10^{-4}\ \text{m}^2)$$

$$= \boxed{4.0 \times 10^{-5}\ \text{Wb}}$$

Solution to (b)

$$\{I = 1.5\ \text{A}\}$$

$$B = \frac{F}{\ell I} \Rightarrow F = B\ell I = 0.10\ \text{T}(2.0 \times 10^{-2}\ \text{m})(1.5\ \text{A})$$

$$= \boxed{3.0 \times 10^{-3}\ \text{N}}$$

The force resulting from the interaction between magnetic fields and electric currents has important engineering applications in electric motors and measuring instruments.

8-2 MAGNETIC FIELD OF A CURRENT

Another strange thing that happens in nature is that a moving charge or current generates its own magnetic field. The flux lines around a long straight wire form concentric circles in planes perpendicular to the wire (Fig. 8-4). Note

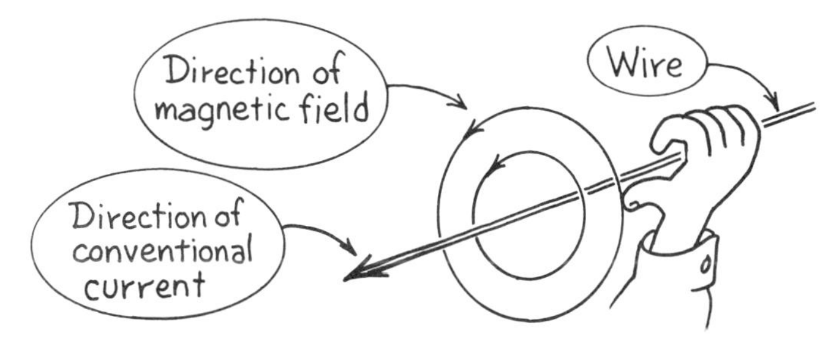

Figure 8-4. A moving charge or current generates its own magnetic field in the form of concentric circles if the wire is long and straight. The plane of the concentric circles is perpendicular to the wire.

the direction of the magnetic field with respect to the direction of the conventional current in the wire. The magnetic field is in the same direction as the curled fingers of your right hand if your thumb points in the direction of the positively charged conventional current. This way of remembering things is called the *right-hand rule*.

More often than not the wire is wound in the shape of a *coil* or *solenoid*. This concentrates the magnetic field inside the windings, as shown in Fig. 8-5.

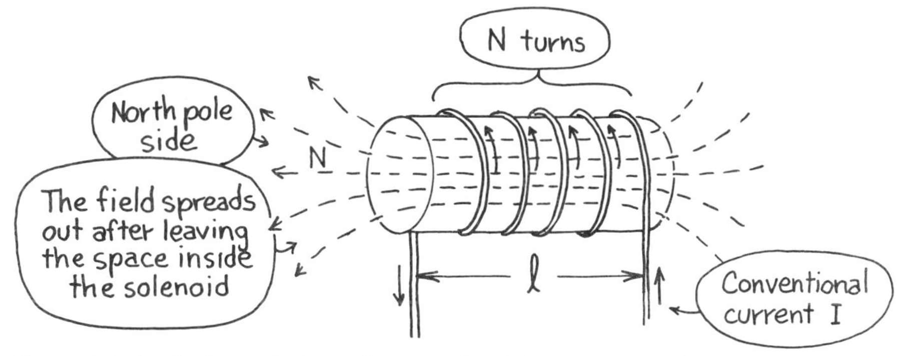

Figure 8-5. A winding called a coil or a solenoid concentrates the magnetic field. The strength of this field depends on the current, the number of turns, and the length of the space in which the field is concentrated.

The strength of this field depends upon the current and the number of turns. It also depends on the length ℓ of the space where the field is concentrated and on the material upon which the coil is wound. If the material is nonmagnetic, the equation for the magnetic field is

$$B = \mu_0 \frac{NI}{\ell} \tag{8-3}$$

B Magnetic Field Strength	N Number of Turns	I Current	ℓ Length
tesla (T)	—	ampere (A)	metre (m)

The constant μ_0 is called the *permeability* of a vacuum. Its value in SI units is

$$\mu_0 = 1.257 \times 10^{-6}\ \text{T} \cdot \text{m/A}$$

The coil length and the length of the space in which the field is concentrated are the same in this case. The number of turns N is a dimensionless quantity. Examples of nonmagnetic materials upon which the coil may be wound are plastics, wood, and brass.

The direction of the magnetic field can be found by grasping the coil in your right hand with your fingers coiled in the direction of the conventional current. Your thumb then points in the direction of the field. The arrows (flux lines) of the field by definition leave the north pole and enter the south pole. A current-carrying coil is called an *electromagnet.*

Example 8-2 A 100-turn coil is wound uniformly from one end to the other on a bakelite (plastic) bobbin. The bobbin is 0.10 m long and has a cross-sectional area of $1.0 \times 10^{-4}\ \text{m}^2$. The current is 2.0 A.
(a) What is the magnetic field strength within the bakelite bobbin?
(b) What is the value of the magnetic flux within the bobbin?

Solution to (a)

$$\left\{\begin{aligned} N &= 100 \\ \ell &= 0.10\ \text{m} \\ I &= 2.0\ \text{A} \end{aligned}\right\}$$

$$B = \mu_0 \frac{NI}{\ell} = \frac{(1.257 \times 10^{-6}\text{T} \cdot \text{m/A})(100)(2.0\ \text{A})}{0.10\ \text{m}} = 2.514 \times 10^{-3}$$

$$\rightarrow \boxed{2.5 \times 10^{-3}\ \text{T}}$$

Solution to (b)

$$\{A = 1.0 \times 10^{-4}\ \text{m}^2\}$$

$$B = \frac{\phi}{A} \Rightarrow \phi = BA = (2.5 \times 10^{-3}\text{T})(1.0 \times 10^{-4}\ \text{m}^2) = \boxed{2.5 \times 10^{-7}\ \text{Wb}}$$

Magnetic Fields in Ferromagnetic Materials

Magnetic field strength increases dramatically if the coil is wound on a *ferromagnetic* material like iron. Equation (8-3) still applies, but we have to be careful about the permeability μ and the length ℓ.

$$B = \mu \frac{NI}{\ell} = \mu H \qquad (8\text{-}4)$$

Magnetic intensity

where

$$H = \frac{NI}{\ell} \qquad (8\text{-}5)$$

The quantity H defined by Eq. (8-5) is called *magnetic intensity*. The units of H are amperes/metre (A/m).

Let's look at the length ℓ in Eq. (8-4) by referring to Fig. 8-6. In each case the magnetic lines of flux concentrate inside the ferromagnetic material. This length where the magnetic lines are concentrated is the length that should be used in Eq. (8-4). As you can see, the length is different for each situation.

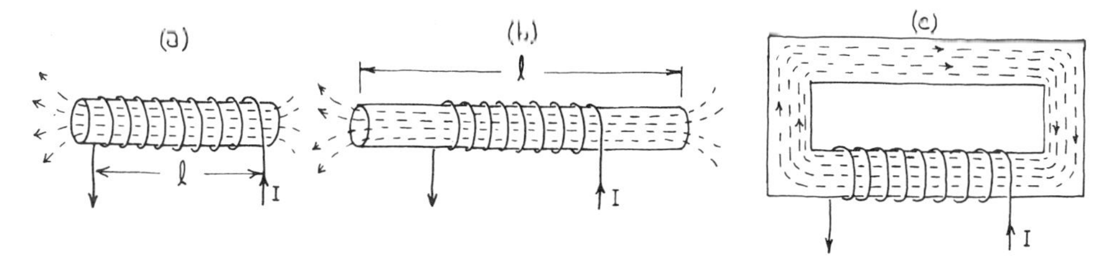

Figure 8-6. The length ℓ in each ferromagnetic core is different even though the coil length is the same. In (c) the length is the average perimeter of the ferromagnetic rectangle.

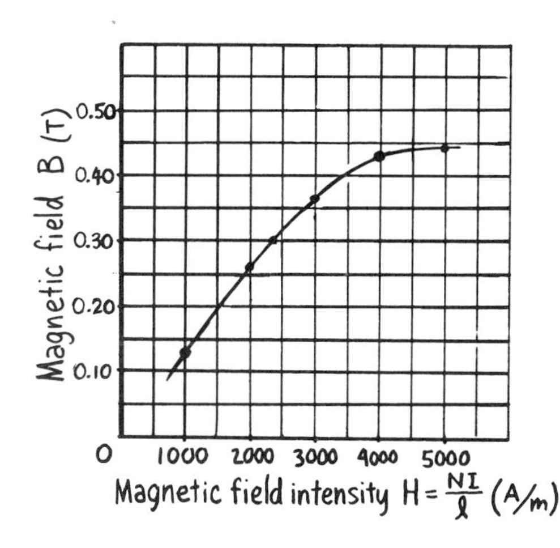

Figure 8-7. *B* versus *H* curve for soft iron.

The permeability is not constant for a ferromagnetic material. It changes when the coil current changes. Experiments are performed on ferromagnetic materials to determine the values of μ for different currents. The results are published in the form of B versus H curves. A typical B versus H curve is shown in Fig. 8-7 for soft iron.

Since $\mu = B/H$ from Eq. (8-4), we can calculate μ for a particular value of H using the B versus H curve. The value of μ depends upon the point's location along the curve, which in turn depends upon the current (Fig. 8-8).

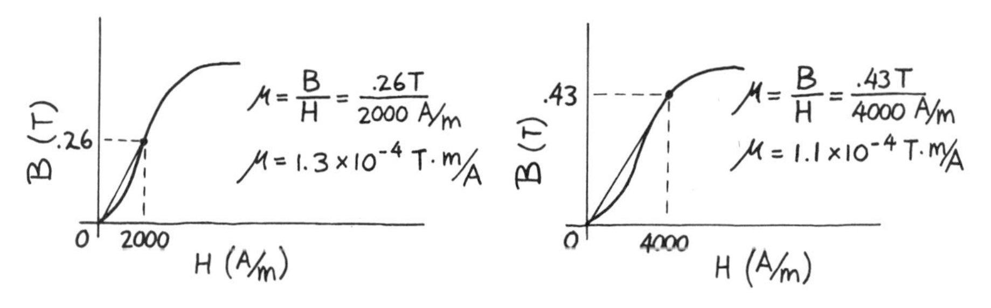

Figure 8-8. The value of $\mu = B/H$ depends on the value of H and is the slope of a line drawn from the origin to the point specified by H. In (a), $H = 2000$ A/m; in (b), $H = 4000$ A/m. The calculated values of μ, 1.3×10^{-4} and 1.1×10^{-4} T · m/A, are different. The material is soft iron.

Example 8-3 Suppose that the bobbin in Example 8-2 is made of soft iron.
(a) What is the magnetic field strength inside the iron?
(b) What is the permeability of the iron under these conditions?
(c) How does the iron's permeability compare with that of the bakelite?

Solution to (a)

$$\left\{\begin{array}{l} N = 100 \\ \ell = 0.10\text{ m} \\ I = 2.0\text{ A} \end{array}\right\}$$

B = 0.26T

(Fig. 8-7)

H = 2000 A/m

$$H = \frac{NI}{\ell} = \frac{100(2.0\text{ A})}{0.10\text{ m}} = 2000\text{ A/m}$$

From Fig. 8-7, $\boxed{B = 0.26\text{ T}}$

Solution to (b)

$$\mu = \frac{B}{H} = \frac{0.26\text{ T}}{2000\text{ A/m}} = \boxed{1.3 \times 10^{-4}\,\frac{\text{T} \cdot \text{m}}{\text{A}}}$$

Solution to (c)

The permeability of bakelite is very close to μ_0:

$$\frac{\mu}{\mu_0} = \frac{1.3 \times 10^{-4}}{1.257 \times 10^{-6}} = 103 \rightarrow \boxed{100}$$

The permeability of the iron is 100 times greater than the permeability of bakelite.

We must go back to the atomic nature of matter (Sec. 3-6) to explain the ferromagnetic behavior of materials like iron and steel. An electron of an atom rotates about the nucleus and spins about its own axis like a top. Both motions contribute to the creation of magnetic fields, because charge is in motion. Usually, the fields of paired electrons in an atom tend to cancel, just as the external fields of two bar magnets tend to cancel when placed next to each other with their north poles at opposite ends.

In ferromagnetic materials, however, certain electrons in each atom are unpaired, and they interact with neighboring atoms to produce a localized magnetic field in part of the material. The regions where this happens are called *domains*. They are quite large by atomic standards, large enough to be seen under a microscope. When unmagnetized, the domains are arranged in a random manner, and there is no net magnetic field (Fig. 8-9). Under the urging of an external field, such as from a current-carrying solenoid,

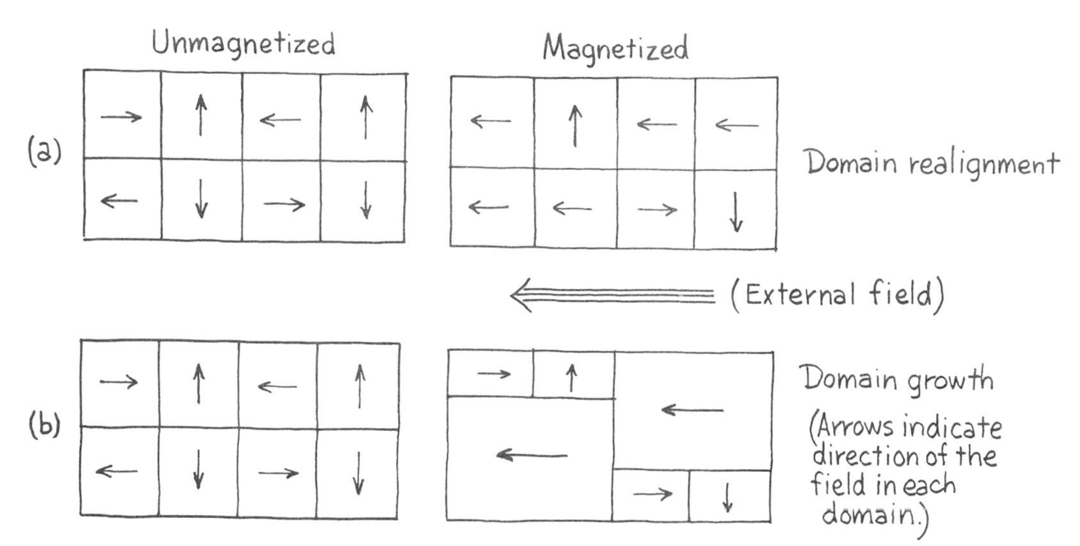

Figure 8-9. There are two ways in which a ferromagnetic material becomes magnetized when placed inside a current-carrying solenoid. In (a) the domains (represented by little boxes) realign themselves with the external field. In (b) the originally aligned domains grow in size at the expense of others.

1. The domains align themselves with the external field;

2. Those domains originally aligned with the external field grow in size at the expense of the others.

Once the domains have realigned and grown, no further increase in the field is possible and the material is said to be *saturated.*

8-3 COMPARISON OF MAGNETIC AND ELECTRIC CIRCUITS

Magnetic circuits bear some resemblance to electric circuits. The resemblance can be enhanced if we pick certain magnetic quantities to represent the unified-concept notions of force and parameter rate. The force-like quantity becomes the product of turns times current. Such a product is actually called the *magnetomotive force* F_m, commonly abbreviated mmf.

$$F_m = NI \tag{8-6}$$

We will choose magnetic flux ϕ as parameter rate for magnetic circuits.

The ratio of magnetomotive force to flux is called the *reluctance* R_m of the magnetic circuit.

$$R_m = \frac{F_m}{\phi} = \frac{NI}{\phi} \tag{8-7}$$

R_m *Reluctance*	F_m *Magnetomotive Force*	ϕ *Magnetic Flux*	N *Coil Turns*	I *Coil Current*
ampere/weber (A/Wb)	ampere (A)	weber (Wb)	—	ampere (A)

Working with Eqs. (8-1) and (8-3),

$$B = \underbrace{\frac{\phi}{A}}_{\text{Eq. (8-1)}} = \mu \underbrace{\frac{NI}{\ell}}_{\text{Eq. (8-3)}} \Rightarrow \frac{NI}{\phi} = \frac{1}{\mu}\frac{\ell}{A}$$

But

$$\frac{NI}{\phi} = R_m$$

Therefore,

$$R_m = \frac{1}{\mu}\frac{\ell}{A} \tag{8-8}$$

R_m *Reluctance*	μ *Permeability*	ℓ *Length*	A *Area*
ampere/weber (A/Wb)	$\dfrac{\text{tesla} \cdot \text{metre}}{\text{ampere}}$ (T · m/A)	metre (m)	square metre (m^2)

Notice that the reluctance equation $R_m = (1/\mu)(\ell/A)$ has the same form as the electrical resistance equation $R = \rho\ell/A$ in Chapter 7.

Table 8-1 shows these relations between electrical and magnetic circuits using the unified concepts of force, parameter rate, and resistance.

TABLE 8-1 UNIFIED CONCEPT RELATIONS BETWEEN ELECTRICAL AND MAGNETIC CIRCUITS

	Electric Circuit	Magnetic Circuit
Force symbol and units	Electromotive force (emf) E volt (V)	Magnetomotive force (mmf) $F_m = NI$ ampere (A)
Parameter rate symbol and units	Electric current I ampere (A)	Magnetic flux ϕ weber (Wb)
Resistance symbol, equations and units	Electrical resistance $R = E/I = \rho\ell/A$ ohm (Ω)	Reluctance $R_m = \dfrac{F_m}{\phi} = \dfrac{1}{\mu}\dfrac{\ell}{A}$ ampere/weber (A/Wb)

We can imagine that the magnetomotive force $F_m = NI$ drives magnetic flux ϕ around a magnetic circuit whose length is ℓ, area is A, and reluctance is R_m. There are some glaring inconsistencies in relating the flux and the reluctance to unified concepts:

1. The magnetic flux is not a real parameter rate. It does not have the units of a parameter divided by time.

2. The reluctance is not a resistance in the sense that resistors remove energy from the system. No energy is being lost in the magnetic circuit. In fact, the magnetic circuit acts as a storage device for the energy of the magnetic field.

Nevertheless, unified concepts can serve as a mechanism to aid in the transfer of knowledge from systems we already know something about to the magnetic circuit. As an example, can you identify in Fig. 8-10 which magnetic

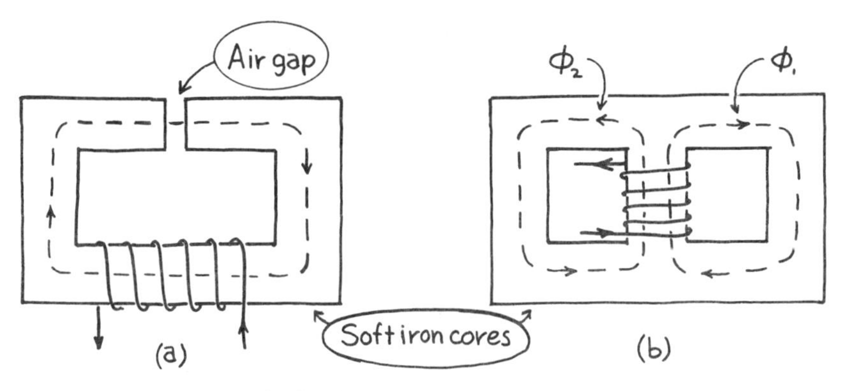

Figure 8-10. Two magnetic circuits.

circuit is a series circuit and which is parallel? Circuit (a) is a series circuit. The same flux flows through two materials with different reluctances, the iron and the air. The total reluctance is the sum of the individual reluctances. Circuit (b) in Fig. 8-10 is a symmetrical parallel circuit. The total flux flowing through the center splits up. Half of the flux goes to the path on the right and the rest to the path on the left. An equivalent electrical circuit is a battery connected to two equal resistors in parallel. The same techniques for finding battery voltage, currents, and resistances in electrical circuits can be used to find their unified-concept counterparts in the magnetic circuits.

Example 8-4 Suppose that the length of the soft iron path in the magnetic circuit of Fig. 8-10a is 0.10 m and the length of the air gap is 0.0010 m. The cross-sectional area is $1.0 \times 10^{-4}\ \text{m}^2$. A current of 3.0 A flows in the 100-turn coil.

(a) What is the reluctance of the magnetic circuit?

(b) What magnetic flux exists in the air gap?

Solution to (a)

$$\left\{ \begin{aligned} \mu_1 &= \mu_o = 1.257 \times 10^{-6}\ \text{T}\cdot\text{m/A} \\ \ell_1 &= 1.0 \times 10^{-3}\ \text{m} \\ A &= 1.0 \times 10^{-4}\ \text{m}^2 \end{aligned} \right\}$$

If resistors add in an electrical series circuit, then reluctances add in a series magnetic circuit. The reluctance R_{m1} of the air gap is

$$R_{m1} = \frac{1}{\mu_1}\frac{\ell_1}{A} = \left(\frac{1}{1.257 \times 10^{-6}\ \text{T}\cdot\text{m/A}}\right)\left(\frac{1.0 \times 10^{-3}\ \text{m}}{1.0 \times 10^{-4}\ \text{m}^2}\right)$$

$$= 7.96 \times 10^{6}\ \text{A/Wb}$$

The reluctance R_{m2} of the soft iron is

$$\left\{\begin{array}{l} \ell_2 = 1.0 \times 10^{-1}\ \text{m} \\ A = 1.0 \times 10^{-4}\ \text{m}^2 \\ N = 100 \\ I = 3.0\ \text{A} \end{array}\right\}$$

(Fig. 8-7)

$$H = \frac{NI}{\ell_2} = \frac{100(3.0\ \text{A})}{1.0 \times 10^{-1}\ \text{m}} = 3000\ \text{A/m} \Rightarrow B = 0.37\ \text{T}$$

$$\mu_2 = \frac{B}{H} = \frac{0.37\ \text{T}}{3000\ \text{A/m}} = 1.23 \times 10^{-4}\ \frac{\text{T}\cdot\text{m}}{\text{A}}$$

$$R_{m2} = \frac{1}{\mu_2}\frac{\ell_2}{A} = \left(\frac{1}{1.23 \times 10^{-4}\ \text{T}\cdot\text{m/A}}\right)\left(\frac{1.0 \times 10^{-1}\ \text{m}}{1.0 \times 10^{-4}\ \text{m}^2}\right)$$

$$= 8.13 \times 10^{6}\ \text{A/Wb}$$

The total reluctance R_m is the sum of R_{m1} and R_{m2}:

$$R_m = R_{m1} + R_{m2} = 7.96 \times 10^6\ \text{A/Wb} + 8.13 \times 10^6\ \text{A/Wb} = 1.61 \times 10^7$$

$$\rightarrow \boxed{1.6 \times 10^7\ \text{A/Wb}}$$

Solution to (b)

$$\phi = \frac{NI}{R_m} = \frac{100(3.0\ \text{A})}{1.61 \times 10^7\ \text{A/Wb}} = \boxed{1.9 \times 10^{-5}\ \text{Wb}}$$

8-4 INDUCED ELECTROMOTIVE FORCE

So far we have learned a couple of interesting things about magnetic fields:

1. A conductor placed within and perpendicular to a magnetic field experiences a force when a current passes through the conductor (Sec. 8-1).

2. A magnetic field is generated in the vicinity of a current-carrying conductor (Sec. 8-2).

A third interesting thing about magnetic fields is that they can generate or *induce* electromotive forces in conductors. This is true provided that (1) there is some motion of the conductor perpendicular to the field, or (2) the field near the conductor is not constant but changes with time. Let's discuss these two possibilities.

Induced Electromotive Force in a Moving Conductor

If a wire is forced to move across magnetic flux lines, a potential difference appears across the wire. Figure 8-11 shows the relation between the motion of the wire, expressed by its constant velocity v, the direction of the magnetic field B, and the direction of the induced emf E. All three quantities are mutually perpendicular to each other.

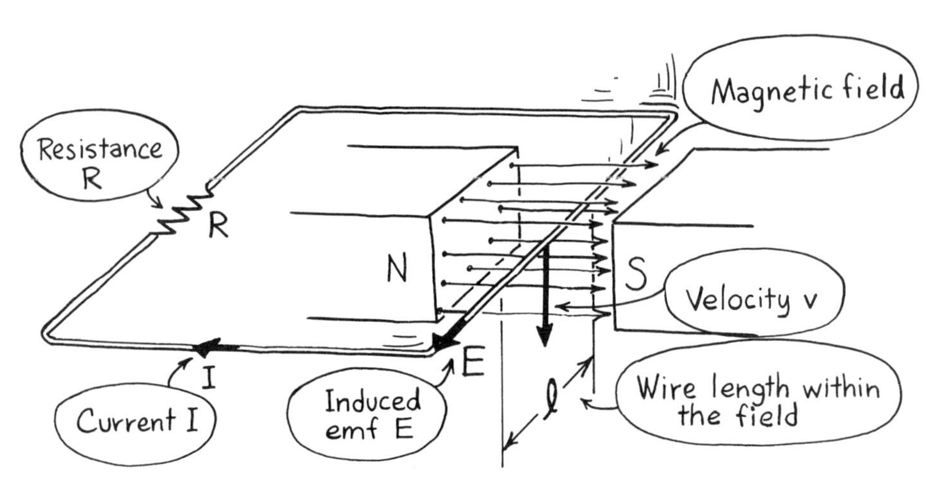

Figure 8-11. The length of wire ℓ within the magnetic field B experiences an induced emf E when forced downward at a constant velocity v. The current I depends on the induced emf and the resistance R.

The emf E is the product of the field B, the velocity, and the length of wire within the field.

$$E = B\ell v \tag{8-9}$$

E *Induced emf*	B *Magnetic Field Strength*	ℓ *Length*	v *Velocity*
volt (V)	tesla (T)	metre (m)	metre/second (m/s)

Conventional current I flows as shown in Fig. 8-11 if the ends of the wire are connected to a resistor. The current's magnitude depends on the induced emf E and the resistance R. It can be calculated using Ohm's law, $I = V_R/R = E/R$.

The wire's motion may not be perpendicular to the field. When this happens, only the perpendicular component of the velocity is used in Eq. (8-9). No emf is induced if the wire moves parallel to the field (Fig. 8-12). Wires moving in magnetic fields are used to generate sinusoidal voltages and to detect mechanical motion.

Example 8-5 The length of wire within the field in Fig. 8-11 is 0.050 m, the strength of the magnetic field is 0.20 T, and the resistance is 60 Ω. The wire moves downward with a velocity of 30 m/s.

(a) What is the induced emf?
(b) What is the current in the resistor?

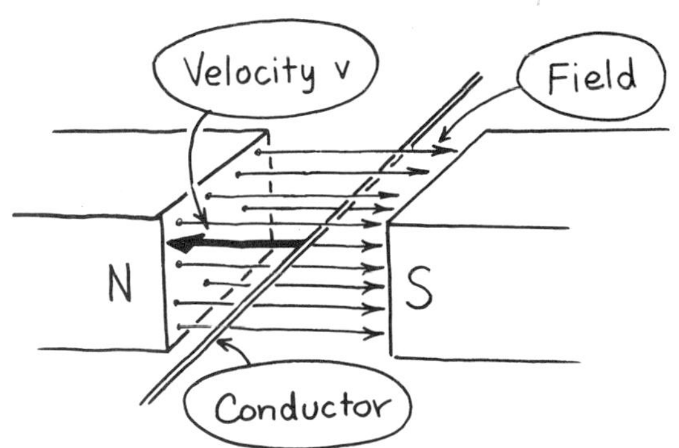

Figure 8-12. There is no induced voltage if the velocity is parallel to the direction of the magnetic field.

Solution to (a)

$\left\{\begin{array}{l} B = 0.20\text{ T} \\ \ell = 0.050\text{ m} \\ v = 30\text{ m/s} \end{array}\right\}$

$$E = B\ell v = 0.20\text{ T}(0.050\text{ m})(30\text{ m/s}) = \boxed{0.30\text{ V}}$$

Solution to (b)

$\{R = 60\ \Omega\}$

$$I = \frac{E}{R} = \frac{0.30\text{ V}}{60\ \Omega} = \boxed{5.0 \times 10^{-3}\text{ A}}$$

Induced Electromotive Force from a Changing Magnetic Field

We have just seen that voltage is induced when a conductor moves across or cuts magnetic flux lines. Voltage can also be induced with no apparent physical motion at all. One illustration of this is in Fig. 8-13.

In Fig. 8-13a the switch is open. No current flows in the first coil, no flux lines are present in the iron core, and the voltmeter connected across the second coil of N turns reads zero. The voltmeter deflects momentarily when the switch is closed (Fig. 8-13b), but returns to zero when the current in the first coil becomes constant at its maximum value (Fig. 8-13c). The voltmeter deflection measures an induced emf E. The induced emf comes from a *changing* magnetic flux through each of the coil's N turns.

Experiments show that the induced emf E is

$$E = N\frac{\Delta\phi}{\Delta t} \qquad (8\text{-}10)$$

E *Induced Voltage*	N *Number of Turns*	$\Delta\phi$ *Change in Magnetic Flux*	Δt *Change in Time*
volt (V)	—	weber (Wb)	second (s)

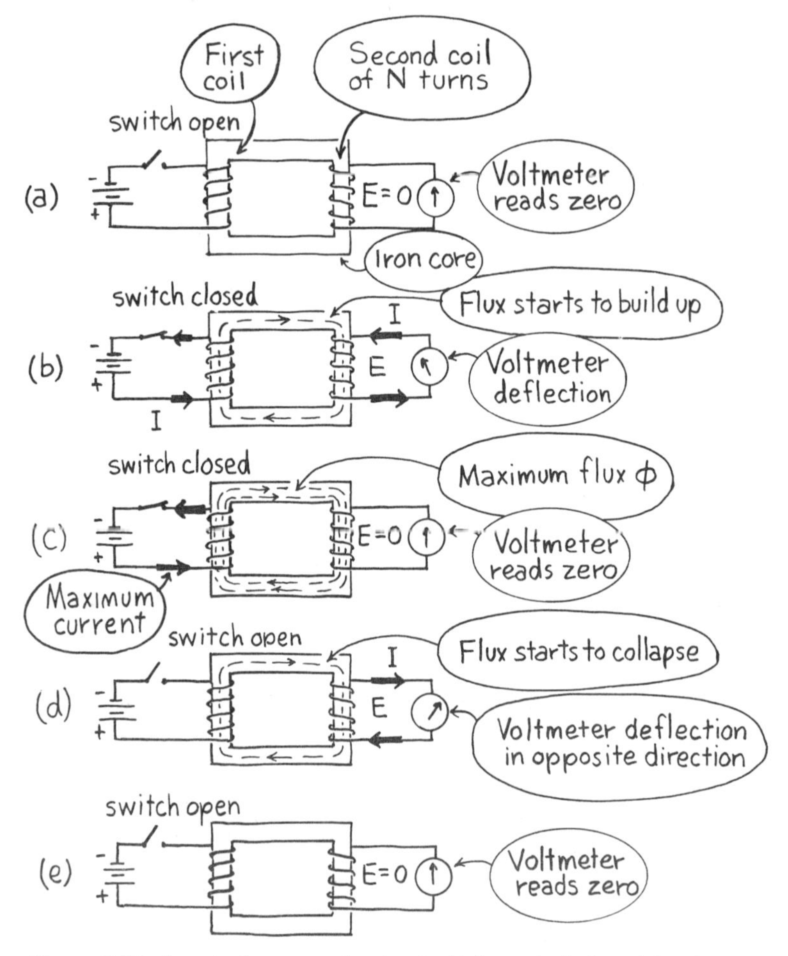

Figure 8-13. An electromagnetic circuit. Voltage is induced in the second coil whenever the flux through the turns of this coil changes.

This important relation is called *Faraday's law.* Induced emf is due to a *changing* magnetic field. The magnetic flux through a coil must change with time in order for there to be an induced voltage. $\Delta\phi$ represents the change in flux through each turn during the time interval Δt. In Fig. 8-13a and 8-13c, the flux is constant at either zero or its maximum value and there is no induced voltage. Only when the field changes do we get an induced emf.

Another interesting thing about this experiment is the direction of the induced emf. The induced current I in the second coil sets up a magnetomotive force (NI) that tries to oppose the buildup of magnetic flux from the first coil. Use the right-hand rule discussed in Sec. 8-2 to verify this. This is true of all induced voltages and is a consequence of the law of conservation of energy.

Logically, if the induced emf drives a current that aids the original flux buildup, the field increases indefinitely, which is not possible. Magnetic fields represent energy in one of its many forms.

Induced voltages always oppose whatever causes them.

This statement is referred to as *Lenz's law*. In our example, the cause of the induced emf is an increasing magnetic field. The induced emf drives a current. The resulting mmf of this current opposes the flux buildup.

In Fig. 8-13d the switch is opened, the flux in the iron core collapses toward zero, and the changing flux through the turns of the second coil induces a voltage as measured by the voltmeter deflection in the opposite direction. Can you explain why the voltmeter deflects the other way?

Example 8-6 The magnetic flux through each of the 200 turns of a coil changes steadily from 0 to 0.016 Wb in 40 ms. What voltage is induced in the coil?

Solution

$$\left\{\begin{aligned} N &= 200 \\ \Delta\phi &= 0.016 - 0 \\ &= 0.016 \text{ Wb} \\ \Delta t &= 0.040 - 0 \\ &= 0.040 \text{ s} \end{aligned}\right\}$$

$$E = N\frac{\Delta\phi}{\Delta t} = \frac{200(0.016 \text{ Wb})}{0.040 \text{ s}} = \boxed{80 \text{ V}}$$

8-5 SELF-INDUCTION

The discussion in the previous section centers around what happens to the second coil in Fig. 8-13. When the switch closes, magnetic flux builds up with time through the turns of the second coil. A changing flux induces a voltage that opposes the flux buildup.

But aren't the same kinds of things happening to the first coil, too? Isn't magnetic flux changing with time through its turns? Isn't there an induced voltage in the first coil, and doesn't this induced voltage oppose the cause (current from battery) producing it? The answer to these questions is—yes.

Figure 8-14a shows a coil of negligible resistance connected through a switch to a battery. Your first thought might be that the current surges quickly to infinity when the switch closes. This does not happen. Experiments show that the current increases uniformly with time (Fig. 8-14b).

Remember that magnetic flux (caused by the magnetomotive force of the coil current) builds up as the coil current increases. The changing flux induces a voltage to oppose the very current responsible for it. This self-induced voltage is called the *back emf*. It is oriented such that the coil looks to the battery like another battery in the circuit trying to force charge in the opposite direction. If it bothers you that there should be any change of current at all with two equal

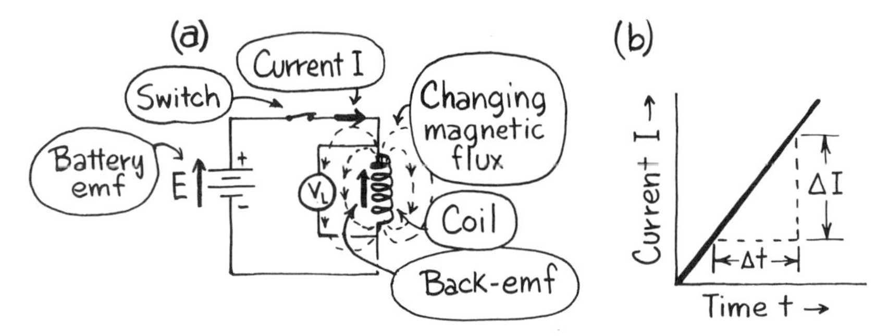

Figure 8-14. When the switch is closed, the current builds up at a constant rate instead of surging quickly to infinity. The changing magnetic flux due to the coil current generates a back emf that limits the rate of current buildup.

and opposing sources of emf in the circuit, look ahead to Sec. 10-5. The process is called *self induction.*

Experiments show that the uniform rate of change of current and the coil voltage V_L are related. Equation (8-11) shows this relation:

$$V_L = L\frac{\Delta I}{\Delta t} \tag{8-11}$$

V_L Coil Voltage	L Inductance	ΔI Change of Current	Δt Change of Time
volt (V)	henry (H)	ampere (A)	second (s)

Inductance L as defined by Eq. (8-11) is a measure of a coil's opposition to change. A coil acts like a mass in a mechanical system. An object's mass opposes change of velocity (Sec. 3-2), just as a coil's inductance opposes change of current. This unified-concept idea is discussed further in Sec. 10-2. The unit of inductance, the *henry* (H), is a volt/(ampere/second).

A coil's inductance L depends on the number of turns N, the length ℓ of the confined magnetic flux path, the cross-sectional area A and the permeability of the core μ.

$$L = \frac{\mu N^2 A}{\ell} \tag{8-12}$$

L *Inductance*	μ *Permeability*	N *Number of Turns*	ℓ *Length*	A *Area*
henry (H)	$\frac{\text{tesla} \cdot \text{metre}}{\text{ampere}}$ (T · m/A)	—	metre (m)	square metre (m^2)

Example 8-7 may help in getting you to understand more about this fairly difficult concept called self-induction.

Example 8-7 The current in a coil of negligible resistance changes from 0 to 0.10 A in 10 ms when connected through a switch to the terminals of a 6.0-V battery.

(a) What is the coil voltage?
(b) What is the rate of change of current?
(c) What is the coil's inductance?
(d) What is the current 50 ms after the switch is thrown?
(e) What is the rate of change of current if the same coil is connected to a 12-V battery in place of the 6.0-V battery?

Solution to (a)

$\{E = 6.0 \text{ V}\}$

$$\text{Coil voltage } V_L = E = \boxed{6.0 \text{ V}}$$

Solution to (b)

$\left\{\begin{array}{l}\Delta I = 0.10 - 0 = 0.10 \text{ A} \\ \Delta t = 10 \text{ ms} = 10 \times 10^{-3} \text{ s}\end{array}\right\}$

$$\frac{\Delta I}{\Delta t} = \frac{0.10 \text{ A}}{10 \times 10^{-3} \text{ s}} = \boxed{10 \text{ A/s}}$$

Solution to (c)

$$V_L = L\frac{\Delta I}{\Delta t} \Rightarrow L = \frac{V_L}{\Delta I/\Delta t} = \frac{6.0 \text{ V}}{10 \text{ A/s}} = \boxed{0.60 \text{ H}}$$

Solution to (d)

$\{t = 50 \text{ ms} = 50 \times 10^{-3} \text{ s}\}$

$$\frac{\Delta I}{\Delta t} = \frac{I - 0}{50 \times 10^{-3} \text{ s}} = 10 \text{ A/s}$$

Solution to (b)

$$I = 10 \text{ A/s}(50 \times 10^{-3} \text{ s}) = \boxed{0.50 \text{ A}}$$

Solution to (e)

$\{E = 12 \text{ V}\}$

$$V_L = E = 12 \text{ V}$$

$$= L\frac{\Delta I}{\Delta t} \Rightarrow \frac{\Delta I}{\Delta t} = \frac{V_L}{L} = \frac{12 \text{ V}}{0.60 \text{ H}} = \boxed{20 \text{ A/s}}$$

Solution to (c)

8-6 SUMMARY

Here is a summary of the important ideas in this chapter.

1. Magnetic forces are exerted on *moving* charged particles in a magnetic field. Charges moving in a conductor constitute an electric current I, and the force F on the length ℓ of the current-carrying conductor within the field is a measure of the magnetic field strength B. The magnetic field can be thought of as lines of magnetic flux ϕ coming from the north pole side of a magnet with cross-sectional area A (Fig. 8-15).

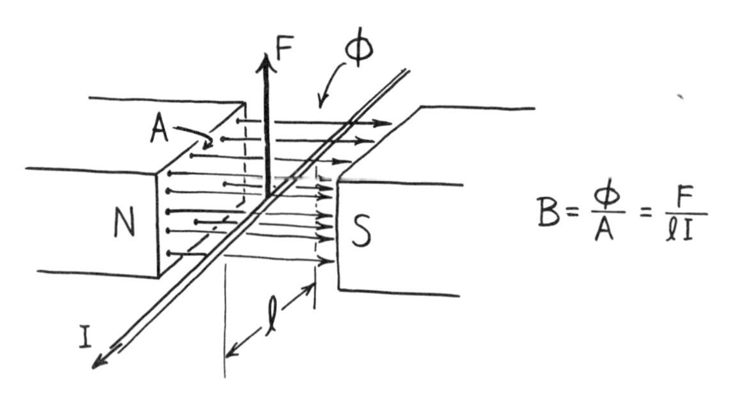

Figure 8-15. Magnetic forces are exerted on moving charges in a magnetic field.

2. Moving charges generate their own magnetic field. If moving charge in the form of an electric current I passes through the N turns of a solenoid, the magnetic field strength B confined within the length ℓ can be calculated. The permeability μ is essentially constant for most nonmagnetic materials. A B versus H curve is needed to find the value of μ for ferromagnetic materials (Fig. 8-16).

3. Simple magnetic and electric circuits can be compared using unified concepts. Table 8-1 shows the relationships.

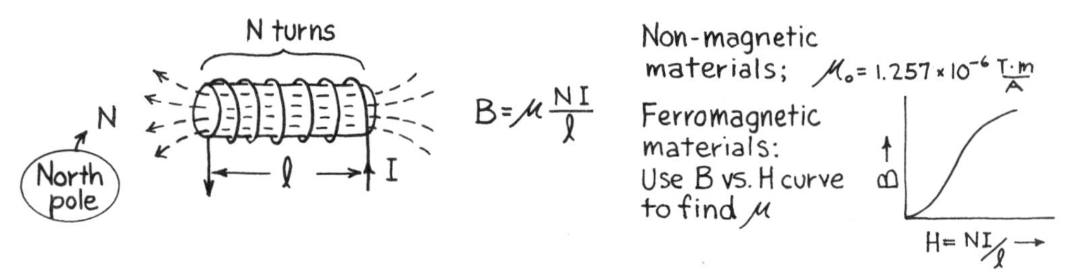

Figure 8-16. Moving charges generate their own magnetic field.

4. Magnetic fields induce electromotive forces in conductors. Two situations are discussed. In the first, a wire moves downward with a velocity v perpendicular to a magnetic field of strength B. The wire length ℓ within the field experiences an induced emf E. In the other situation a second coil experiences an induced emf E when the magnetic flux through the coil's N turns *changes* with time $(\Delta\phi/\Delta t)$. The changing magnetic field is caused by a changing current in the first coil when the switch is closed (Fig. 8-17).

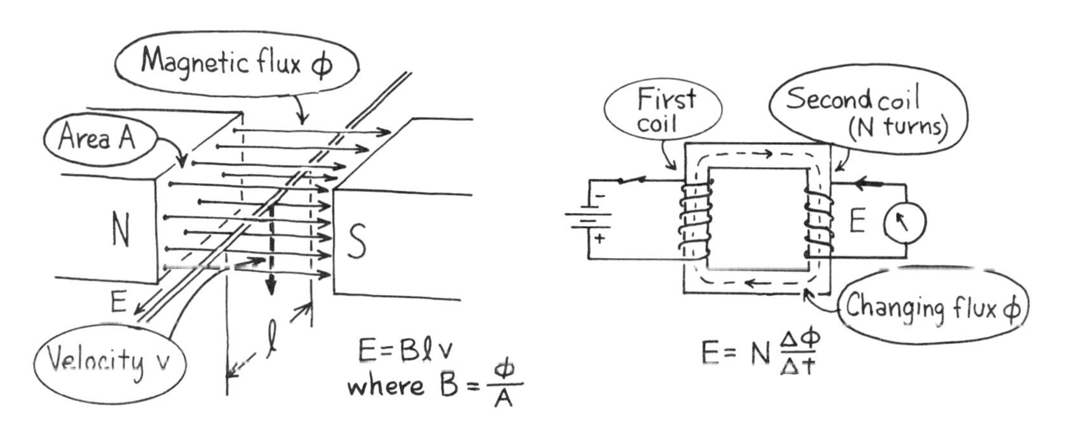

Figure 8-17. Magnetic fields induce electromagnetic forces in conductors.

5. If a constant voltage V_L is applied to a coil, the current through the coil changes at a constant rate $\Delta I/\Delta t$. The proportionality constant between V_L and $\Delta I/\Delta t$ is called the inductance L of the coil. The coil's inductance depends on the number of turns N, the length ℓ of the confined flux path, the cross-sectional area A, and the permeability μ of the core (Fig. 8-18).

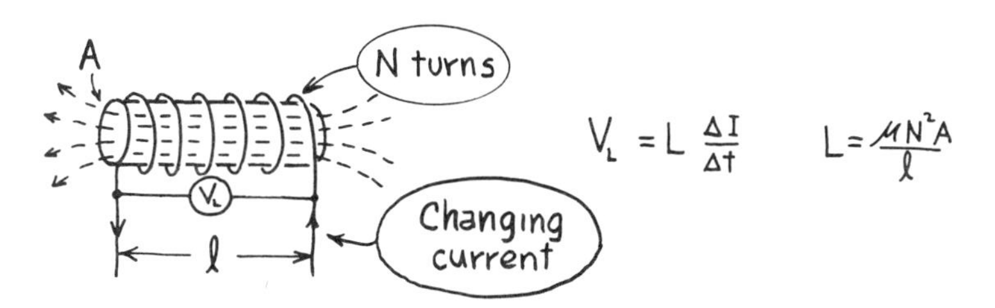

Figure 8-18. A coil opposes any change of current through it.

The inductance of a coil is like the mass of an object in a mechanical system. This is another unified-concept idea that is discussed further in Sec. 10-2.

PROBLEMS

8-1. The magnetic field between the poles of a magnet is 0.15 T. The magnet's face area is 2.0×10^{-4} m^2. What is the total magnetic flux in the space between the poles?

8-2. The magnetic flux in the space between the poles of a magnet is 1.8×10^{-5} Wb. The cross-sectional area of the magnet's face is in the form of a circle with a radius of 0.60 cm. What is the magnetic field strength in tesla?

8-3. The magnetic field strength B in the space between the poles of the magnet is 0.50×10^{-3} T.

(a) What is the total magnetic flux in the space between the magnet's poles?

(b) Draw arrows representing the direction of the flux.

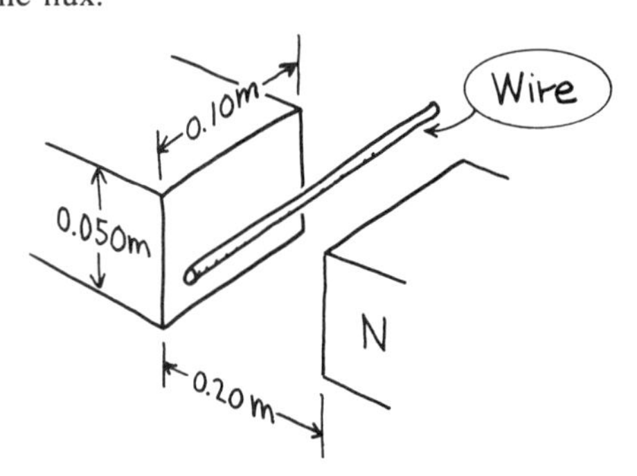

Problem 8-3

8-4. Show the direction of the force exerted on the current-carrying conductor placed as shown between the poles of a permanent magnet.

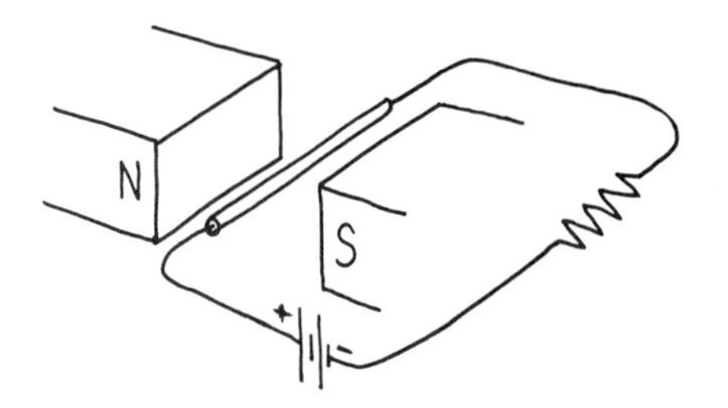

Problem 8-4

8-5. Which end of the electromagnet becomes the north pole when the switch is closed?

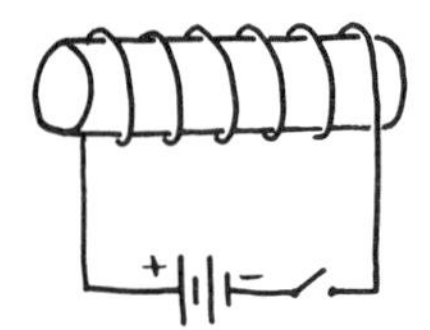

Problem 8-5

8-6. The wire between the poles of the magnet in Problem 8-3 carries a conventional current of 40 A. The current is in a direction which is out of the paper.

(a) What is the magnitude of the force exerted on the wire by the magnetic field?

(b) In what direction does the force point?

8-7. A current of 15 A in a conductor perpendicular to the poles of a magnet causes a magnetic force of 6.0×10^{-2} N to act on the conductor. The length of conductor within the field is 5.0×10^{-3} m. What is the strength B of the magnetic field causing the force?

8-8. A solenoid of 200 turns carries a current of 2.5 A. The magnetic flux within the solenoid is 7.5×10^{-4} Wb. What is the solenoid's reluctance?

8-9. A coil is wound on a plastic nonmagnetic bobbin of cross-sectional area 4.0×10^{-4} m^2. The coil has 150 turns and a length of 0.20 m. A current of 1.2 A flows in the coil.

(a) What is the reluctance of the coil?

(b) What is the value of the magnetic flux inside the bobbin?

8-10. A coil of 80 turns and a resistance of 6.0 Ω is connected across the terminals of a 1.5-V battery. The coil is wound around a soft iron bar whose length is 2.0×10^{-2} m. What is the permeability of the bar under these conditions?

8-11. The plastic bobbin in Problem 8-9 is replaced by a soft iron core.

(a) What is the reluctance of the coil?

(b) What is the value of the magnetic flux inside the core?

8-12. What is the direction of the induced conventional current in the resistor (up or down) if

(a) The bar magnet moves toward the coil?

(b) The bar magnet moves away from the coil?
(c) What causes the induced currents in both cases?

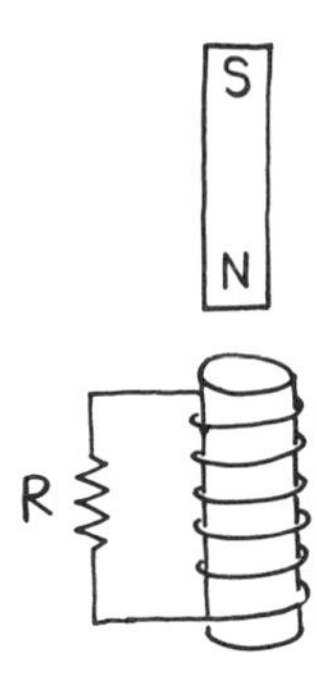

Problem 8-12

8-13. An airplane with a 30-m wingspread flies from east to west at a speed of 120 m/s. What induced emf appears in the wings if the vertical component of the earth's magnetic field is 3.0×10^{-5} T?
8-14. A 10.0-cm wire moves perpendicular to a 0.20-T magnetic field. What velocity must the wire have in order to have an emf of 0.50 V induced across its ends?
8-15. A magnetic flux changes uniformly from 0.020 to 0.060 Wb through 400 turns of a solenoid in 0.20 s. What emf is induced in the solenoid?
8-16. The units for induced emf E in the equation $E = N(\Delta\phi/\Delta t)$ are webers/second. Prove that the units for the induced emf E in the equation $E = B\ell v$ are also webers/second.
8-17. The magnetic flux through the turns of a solenoid changes uniformly at the rate of 0.060 Wb/s. How many turns must the solenoid have so that the induced emf is 18 V?
8-18. What is the direction of the induced conventional current in R_2 (up or down) if
(a) Resistance R_1 steadily decreases?
(b) Resistance R_1 steadily increases?
(c) Resistance R_1 remains constant?

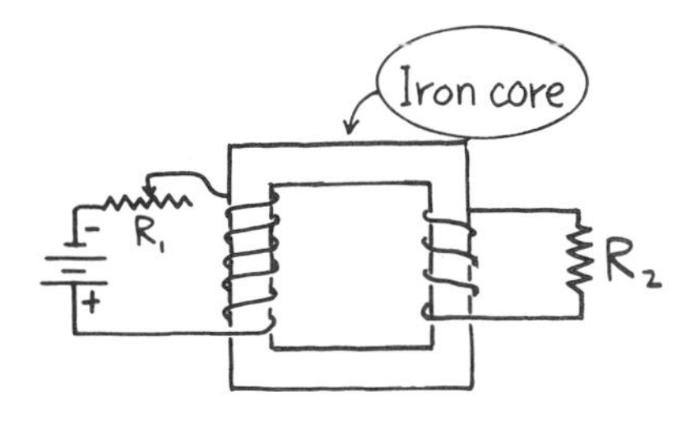

Problem 8-18

8-19. Current in a 0.50-H coil changes at a steady rate of 40 A/s. What back emf is induced in the coil?
8-20. A 12-V battery is connected by a switch to a 1.2-H coil of negligible resistance. What is the current in the circuit 50 ms after the switch is closed?
8-21. What is the inductance of an air-core, 200-turn coil? Its cross-sectional area is 1.0×10^{-4} m^2 and its length is 0.10 m.
8-22. What is the inductance of the coil in Problem 8-21 if the core is made of soft iron and the current is 2.0 A?
8-23. The length of the soft iron core is 0.20 m and the cross-sectional area is 2.0×10^{-4} m^2. The 200-turn coil carries 2.4 A.
(a) What is the reluctance of the iron–air path?
(b) What is the value of the magnetic flux in the air gap?

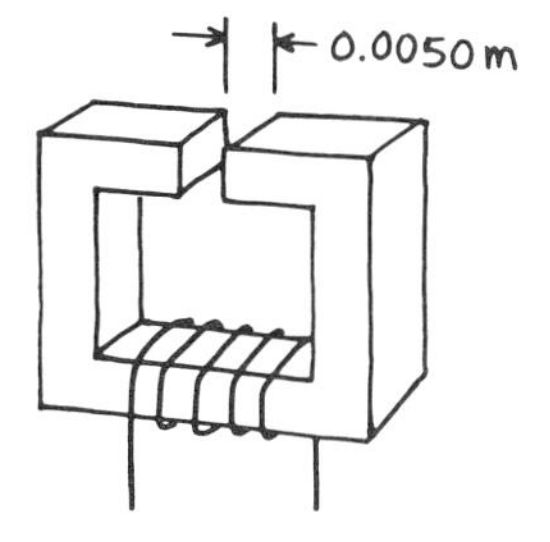

Problem 8-23

8-24. Two soft iron cores are joined together. The cores have the same geometry (area and perimeter). The same magnetic flux is needed in both the left and right sections of the core. This flux value should be 1.6×10^{-4} Wb. What current would you recommend in the 100-turn coil?

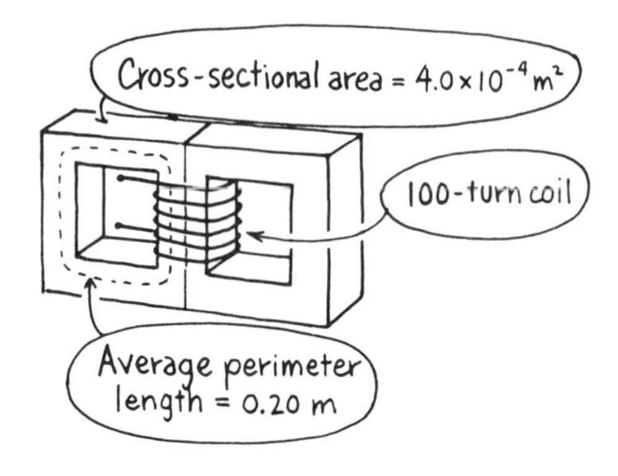

Problem 8-24

ANSWERS TO ODD-NUMBERED PROBLEMS

8-1. 3.0×10^{-5} Wb
8-3. 2.5×10^{-6} Wb
8-5. North pole is at the right end.
8-7. 0.80 T
8-9. (a) 4.0×10^{8} A/Wb
(b) 4.5×10^{-7} Wb
8-11. (a) 4.1×10^{6} A/Wb
(b) 4.4×10^{-5} Wb
8-13. 0.11 V
8-15. 80 V
8-17. 300
8-19. 20 V
8-21. 5.0×10^{-5} H
8-23. (a) 2.8×10^{7} A/Wb
(b) 1.7×10^{-5} Wb

9 Capacitance and Potential Energy Storage

The ideas of work and energy have been dealt with most recently in Section 7-4. There we saw that work done by the resistive force F_R was transferred outside the system and effectively lost. Work done by other forces may not get away. If a force puts energy into a system to stay, at least temporarily, we say that the energy is ***stored*** *in the system. Either way, "doing work" involves transfer of mechanical energy. Thus work and energy are the same kind of thing and have the same units. It is simply two ways of discussing the same topic.*

Doing work (transferring energy) is like depositing money in a checking account. The bank agrees to take care of the money, but inevitably the banker takes a cut for his services. That money is lost so far as the depositor is concerned. The rest of the money is stored by being locked up in a vault or by circulating it in some safe enterprise. Either way the stored money is available for withdrawal and use by the depositor.

When the auditors come to check the books, all cash involved must be accounted for. The net input must equal the amount stored plus the amount "lost" to the banker's fee.

The same thing applies to energy. Energy is not something that just goes away without leaving a trace. Whatever work is done on a physical system must be accounted for. Some of the work can be lost through resistors and some can be stored in one way or another, but the work done (energy put in) must equal the energy lost plus energy stored. If the books don't balance, we know something is wrong. This idea is called the principle or law of ***conservation of energy****. The principle is useful because it is an equation, and thus it allows one to solve for some unknown value.*

$$\text{work done} = \text{energy lost} + \text{energy stored} \quad \text{(conservation of energy)}$$

Chapter 7 was devoted to resistive forces through which energy is lost. The rest of this chapter (and the next, too) will deal with energy that is stored within systems and which, therefore, is available for withdrawal.

This chapter discusses ***potential energy****, the energy stored by virtue of position. Devices such as springs, reservoirs, and heated objects store potential energy. These devices are called* ***capacitors****. Capacitive force does the work that is transferred and stored as potential energy in capacitors. A property called* ***capacitance*** *describes how the capacitor reacts to the capacitive force. The purpose of this chapter is to study capacitance and potential energy storage in capacitors of the various physical systems.*

9-1 TWO WAYS TO STORE ENERGY

Suppose that a car is sitting motionless beside a roadway. We will say that by definition the car has no energy as it sits there, and we will be concerned with the car's energy *relative* to that condition.

When the car is put into gear, an excitation force F_E acts on it. As the car moves a distance s, an amount of work $W = F_E s$ is done on the car (Fig. 9-1).

Figure 9-1. The excitation force F_E puts energy into the system. Friction force f is responsible for energy lost from the system. The car accelerates if F_E is greater than f.

While this is happening, the friction force f shows up. Part of the excitation force must be used to overcome friction. The work done by that part—the resistive force F_R—is the energy E_R lost from the system. That energy is now largely in the form of slightly warmer air, tires, and roadway.

The rest of the energy transferred to the car must have been stored in the system somehow, but can this be proved?

Suppose that we quickly rig up a block and tackle (Fig. 9-2) and throw a hook over the car's rear bumper as it goes by. If the rope doesn't break, the car will raise the load attached to the other end of the rope. In other words, the car is able to do some work.

Figure 9-2. The car is able to lift the load (do work) because it is moving. The ability to do work is a result of the car's motion. The car had kinetic energy, which was transferred to it by the excess excitation force.

If we try the same thing with a car that is not moving, nothing happens. A stationary car is not able to do the work. Thus the moving car contained energy (was able to do work) purely because it was moving. In mechanical systems this is called *kinetic energy*—energy of motion.

Now let us look a bit farther along the road. It seems that there is a hill up which the car moves until it is well above its original elevation. The car, having just made it up the hill (Fig. 9-3), teeters on the edge for a moment and then drops off into space. Luckily, we are standing by with our hook rigged as shown. We snag the car by its front bumper and lower it gently to its original level by means of a counterweight on the other end of the rope. As the car slowly falls, work is done to lift the counterweight. The car could not have

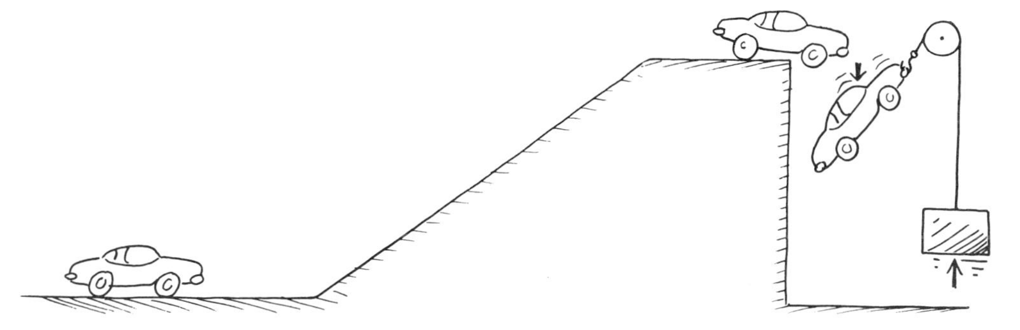

Figure 9-3. The car is able to lift the load (do work) because of its position. The car had potential energy.

done this from the bottom of the hill. Its ability to do the work was strictly because of its position. This sort of energy is called *potential energy*—energy of position or configuration. Both kinetic and potential energy have their equivalents in other systems.

9-2 CAPACITIVE FORCE, CAPACITORS, AND CAPACITANCE

Section 4-6 introduced capacitive force through which energy is stored in a capacitor. The energy of interest here is potential energy, so let's look further into these ideas.

Mechanical Translational System

It takes work (force × extension) to stretch an elastic object such as the spiral spring of Fig. 9-4. Thus the spring is a *capacitor*, a potential energy

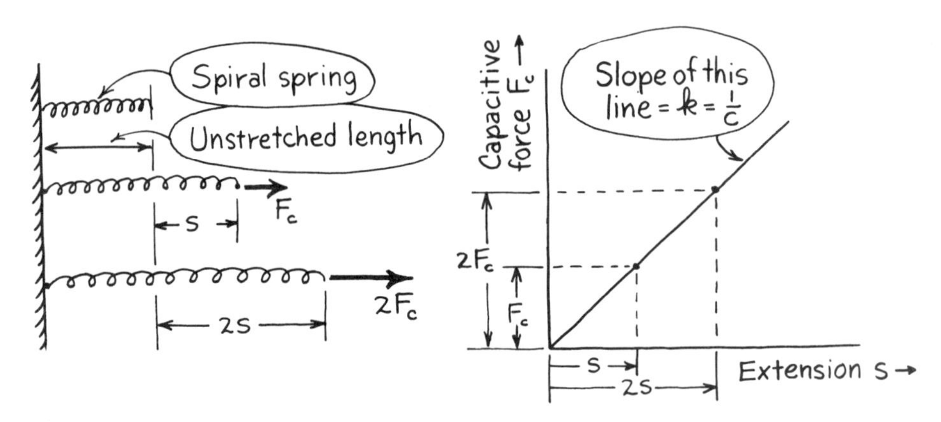

Figure 9-4. A spiral spring is a mechanical capacitor. Unless it has been stretched too far, force is proportional to extension. The constant ratio of extension to force is called capacitance.

storage device. Until it gets stretched too much, the spring's extension is directly proportional to the force applied. Pull twice as hard and you get twice as much extension. The spring is called a *linear* capacitor, because the graph of force versus extension is a straight line passing through the origin. The graph's slope is the constant ratio of capacitive force F_C to extension s. Its value describes the "stiffness" of the spring and is called the *spring constant* k. The ratio of extension to force, the reciprocal of spring constant, is called *capacitance* C.

$$C = \frac{s}{F_C} \tag{9-1}$$

where

$$k = \frac{1}{C} \tag{9-2}$$

C *Capacitance*	F_C *Capacitive Force*	s *Extension*	k *Spring Constant*
metre/newton (m/N)	newton (N)	metre (m)	newton/metre (N/m)
foot/pound (ft/lb)	pound (lb)	foot (ft)	pound/foot (lb/ft)

Capacitance is the inverse of stiffness. It can be thought of as a measure of how easily a spring can be stretched, or in general how easily parameter can be changed in an energy-storage device.

If the value of k or C is constant, we say that a capacitor is *linear*. Not all capacitors are linear, but enough of them are so that the concept is important. It is essential that you understand this sort of device and be able to recognize it in the various systems.

Example 9-1 It takes a force of 26 N to stretch the spring a distance of 0.37 m.

(a) What is the spring constant k?

(b) What is the capacitance C?

(c) How much force will be needed to stretch the spring a total of 0.50 m?

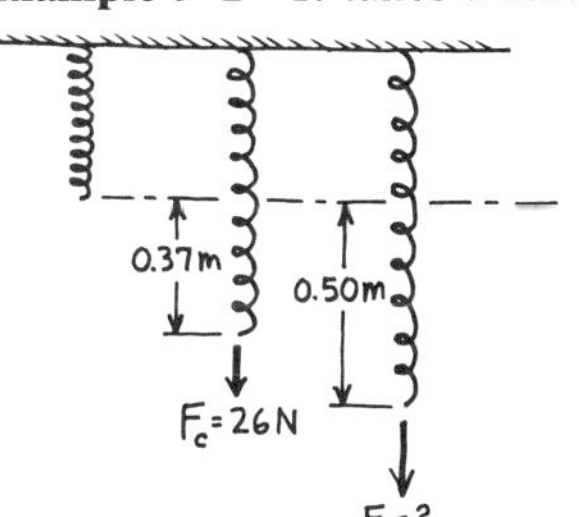

Solution to (a)

$\left\{ \begin{array}{l} s = 0.37 \text{ m} \\ F_C = 26 \text{ N} \end{array} \right\}$

$$k = \frac{F_C}{s} = \frac{26 \text{ N}}{0.37 \text{ m}} = \boxed{70 \frac{\text{N}}{\text{m}}}$$

Solution to (b)

$$C = \frac{1}{k} = \frac{s}{F_C} = \frac{0.37 \text{ m}}{26 \text{ N}} = \boxed{1.4 \times 10^{-2} \frac{\text{m}}{\text{N}}}$$

Solution to (c)

$\{s = 0.50 \text{ m}\}$

$$F_C = \frac{s}{C} = \frac{0.50 \text{ m}}{1.4 \times 10^{-2} \text{ m/N}} = \boxed{35 \text{ N}}$$

Capacitors and capacitance exist in the other systems, too. We will formally define capacitance for all systems as the ratio of parameter Q to capacitive force F_C.

$$\underset{\text{Capacitance}}{C} = \frac{\overset{\text{Parameter}}{Q}}{\underset{\text{Capacitive force}}{F_C}} \tag{9-3}$$

Linear capacitors in these systems have a force versus parameter curve that is a straight line passing through the origin, as in Fig. 9-4. Before moving on to the other systems, we want to examine an important exception in the mechanical translational system.

The Gravitational "Spring"—a Special Case

A very familiar capacitive system is illustrated in Fig. 9-5. Work is done and potential energy is stored whenever an object is lifted from one elevation to another. In order to get the block to move upward, we need an excitation force F_E at least as great as the weight of the block. F_E may need to be larger to provide a resistive force F_R to overcome friction. If F_E is large

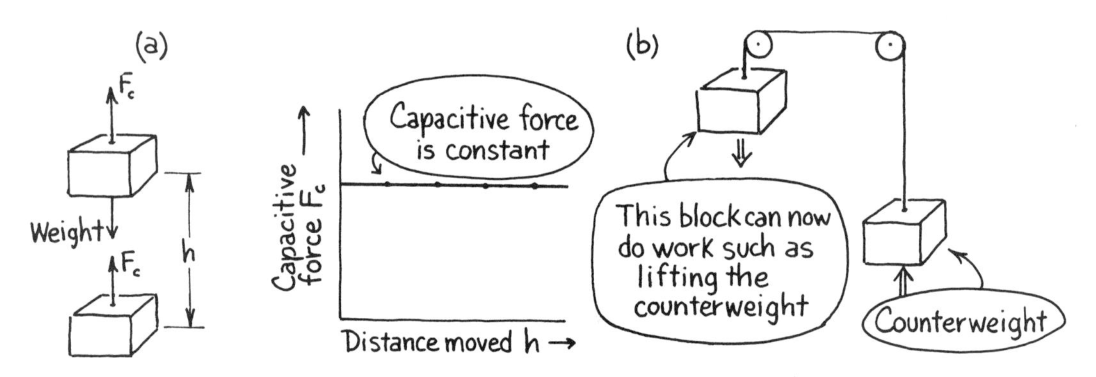

Figure 9-5. (a) Lifting or capacitive force F_c has to be at least as large as the block's weight in order to raise it off the ground. (b) Once up to height h, the block is able to do work. Gravitational potential energy is stored in it.

enough, there may be some force left over to accelerate the block. Whatever the size of F_E, a part of it, the capacitive force F_C, is needed to overcome the earth's downward pull. It is as though a gravitational "spring" were being stretched as the block rises. It is not the ordinary kind of spring, for in most springs capacitive force increases with extension (distance moved). Gravitational force, on the other hand, actually becomes less as the distance increases. It is inversely proportional to the square of the distance from the center of the earth. The earth's radius is about 4000 miles (6400 km), and a change of elevation of a few metres or even kilometres will not make much difference in the gravitational force (weight) of an object. It is a reasonable approximation to say that an object's weight is constant near the earth's surface.

Thus the capacitive force (force needed to overcome weight) is constant in this system. Although the graph of Fig. 9-5 is a straight line, it does not pass through the origin, and gravitation is not a linear capacitor. The ratio of force to distance moved (spring constant) or the reciprocal (capacitance) is not constant. The situation is analogous to the special case of dry friction discussed in Sec. 7-2.

Mechanical Rotational System

Where space or other physical requirements dictate, elastic potential energy storage devices are used that wind up instead of stretching. Examples include the torsion fiber employed in sensitive metering devices such as galvanometers and the torsion bars used instead of leaf or coil springs on some cars. All the generalizations about stretching of springs apply to twisting of elastic objects. In the rotational case, angle is parameter and capacitive force is torque. Figure 9-6 is similar to Fig. 9-4, and the *rotational spring constant* k_r is defined in the same way. Again, capacitance is the reciprocal of spring

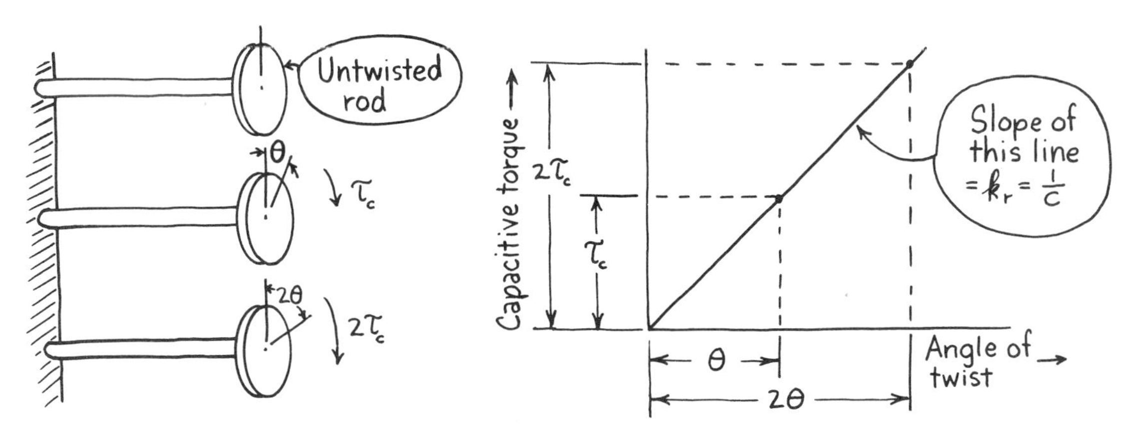

Figure 9-6. A twisted rod acts as a mechanical capacitor. Capacitive torque is proportional to angular deflection.

constant. The spring's characteristics may be expressed either way, although you are more likely to hear about k or k_r (stiffness) than about C in these mechanical systems.

$$C = \frac{\theta}{\tau_C} \qquad (9\text{-}4) \qquad \text{where} \qquad k_r = \frac{1}{C}$$

C *Capacitance*	τ_C *Capacitive Torque*	θ *Angle*	k_r *Rotational Spring Constant*
$\frac{\text{radian}}{\text{newton} \cdot \text{metre}}$ $\left(\frac{\text{rad}}{\text{N} \cdot \text{m}}\right)$	newton · metre (N · m)	radian (rad)	$\frac{\text{newton} \cdot \text{metre}}{\text{radian}}$ $\left(\frac{\text{N} \cdot \text{m}}{\text{rad}}\right)$
$\frac{\text{radian}}{\text{pound} \cdot \text{foot}}$ $\left(\frac{\text{rad}}{\text{lb} \cdot \text{ft}}\right)$	pound · foot (lb · ft)		$\frac{\text{pound} \cdot \text{foot}}{\text{radian}}$ $\left(\frac{\text{lb} \cdot \text{ft}}{\text{rad}}\right)$

Example 9-2 A torque of 5.0 lb · ft twists a metal rod through an angle of 60°.
(a) What is the rotational spring constant?
(b) What is the capacitance of this rotational system?
(c) Through what angle will a torque of 12.5 lb · ft twist the rod?

Solution to (a)

$$\left\{\begin{aligned} \tau_C &= 5.0 \text{ lb} \cdot \text{ft} \\ \theta &= 60° \times \left(\frac{1 \text{ rad}}{57.3°}\right) \\ &= 1.05 \text{ rad} \end{aligned}\right\}$$

$$k_r = \frac{1}{C} = \frac{\tau_C}{\theta} = \frac{5.0 \text{ lb} \cdot \text{ft}}{1.05 \text{ rad}} = \boxed{4.8 \frac{\text{lb} \cdot \text{ft}}{\text{rad}}}$$

Solution to (b)

$$C = \frac{1}{k_r} = \frac{\theta}{\tau_C} = \frac{1.05 \text{ rad}}{5.0 \text{ lb} \cdot \text{ft}} = \boxed{0.21 \frac{\text{rad}}{\text{lb} \cdot \text{ft}}}$$

Solution to (c)

$$\{\tau_C = 12.5 \text{ lb} \cdot \text{ft}\}$$

$$\theta = \tau_C C = 12.5 \text{ lb} \cdot \text{ft} \times 0.21 \frac{\text{rad}}{\text{lb} \cdot \text{ft}} = \boxed{2.6 \text{ rad}}$$

Fluid System (Liquid)

A typical city water supply system includes a number of storage tanks in high places around town. Water is pumped into the tanks and later flows to consumers in the area. That way, relatively constant hydraulic head (pressure) is maintained. Also, the work needed to store the water can be spread out over times of high and low demand. Energy put into the system late at night will be available the next morning. The tanks are potential energy storage devices; they are capacitors.

If a tank is to be filled through an inlet at the bottom (Fig. 9-7), the first bit of liquid to enter the tank encounters no opposition, for the depth h is zero. There is no hydraulic head to work against. As the depth increases, however, it becomes increasingly harder to get liquid into the tank. When the depth is h, the pump must exert a capacitive pressure p_C equal to $h\rho g$ to force liquid into the tank. Also, at that point the volume of liquid stored is $V = hA$, where A is

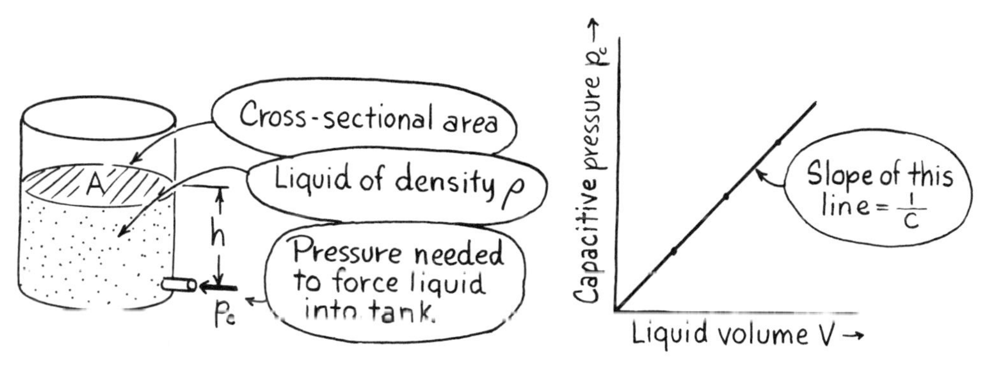

Figure 9-7. A straight-sided tank is a linear fluid capacitor when filled from the bottom. Capacitive gauge pressure is proportional to volume. Note that, if the sides of the tank are not parallel, the graph would not be straight and C would not be constant.

the cross-sectional area of the tank (assumed to be straight-sided). The graph of force (pressure) versus parameter (volume) is once again a straight line. Capacitive pressure is really a gauge pressure, which is zero when the tank contains no liquid (Sec. 3-3).

$$C = \frac{V}{p_C} \tag{9-5}$$

C Liquid Capacitance	V Liquid Volume	p_C Capacitive Pressure
cubic metre / pascal (m^3/Pa)	cubic metre (m^3)	pascal (Pa) ← Note that a pascal is a newton/square metre
cubic foot / pound/square foot [$ft^3/(lb/ft^2)$]	cubic foot (ft^3)	pound / square foot (lb/ft^2)

Example 9-3 illustrates the idea of fluid capacitance.

Example 9-3 A gauge pressure of 1.0×10^4 Pa at the bottom of a straight-sided tank is able to force $0.63\ m^3$ of water into a tank.

(a) What is the tank's capacitance?
(b) What is the water depth?
(c) What is the tank's cross-sectional area?
(d) What gauge pressure is required to force $2.0\ m^3$ of water into the tank?

Solution to (a)

$\left\{ \begin{matrix} p_C = 1.0 \times 10^4\ \text{Pa} \\ V = 0.63\ m^3 \end{matrix} \right\}$

$$C = \frac{V}{p_C} = \frac{0.63\ m^3}{1.0 \times 10^4\ \text{Pa}} = \boxed{0.63 \times 10^{-4} \frac{m^3}{\text{Pa}}}$$

Solution to (b)

$\{\rho = 1.0 \times 10^3\ kg/m^3\}$
App. B, Table 1

$$p_C = \rho g h \Rightarrow h = p_C/\rho g$$

$$h = \frac{1.0 \times 10^4\ \text{Pa}}{(1.0 \times 10^3\ kg/m^3)(9.81\ m/s^2)} = \boxed{1.0\ m}$$

Solution to (c)

$$A = \frac{V}{h} = \frac{0.63\ m^3}{1.02\ m} = \boxed{0.62\ m^2}$$

Solution to (d)

$\{V = 2.0\ m^3\}$

$$p_C = \frac{V}{C} = \frac{2.0\ m^3}{0.63 \times 10^{-4}\ m^3/\text{Pa}} = \boxed{3.2 \times 10^4\ \text{Pa}}$$

Capacitive pressure may be expressed as the height of a column of liquid rather than a force per unit area. In these situations capacitance is just the cross-sectional area of the reservoir, since

$$C = \frac{V}{p_C} = \frac{Ah}{h} = A$$

A large tank such as a lake behind a dam has more capacitance, stores more potential energy, and is a large capacitor compared to a small tank, such as a glass of water.

Fluid System (Gas)

Usually, gases are not dense enough for the pressure $h\rho g$ to be significant relative to the pressure involved in compressing the gas in a tank. Unless the container is extremely tall (like the earth's atmosphere), the pressures at the bottom and at the top are very nearly the same. Thus it matters little where the inlet is located or what the tank's shape is when the fluid is a gas.

Gases are by definition compressible and thereby are not the same as liquids. When a gas compresses or expands, its temperature must change as detailed in Sec. 3-3. For slow processes, when the gas temperature is allowed to continuously adjust to the surroundings, gaseous (pneumatic) systems behave like liquid systems in some respects.

When a compressor forces gas into a pressure tank, the change in absolute pressure is proportional to the amount of gas added. "Amount" of gas in this case must represent the number of molecules of gas in the tank and not the tank's volume, which is constant. The amount might be expressed as the *mass* of gas or as the *volume* of gas measured at some standard temperature and pressure. For example, it is common practice for natural gas to be measured and sold in terms of cubic feet.

If "amount" of gas is expressed as a volume, a graph of tank pressure versus volume (Fig. 9-8) looks just like that for a liquid system, provided changes are made slowly so that the temperature remains constant. Capacitance C for a pneumatic system is defined in the same way as for a liquid system. Equation (9-5) applies to gases as well as to liquids.

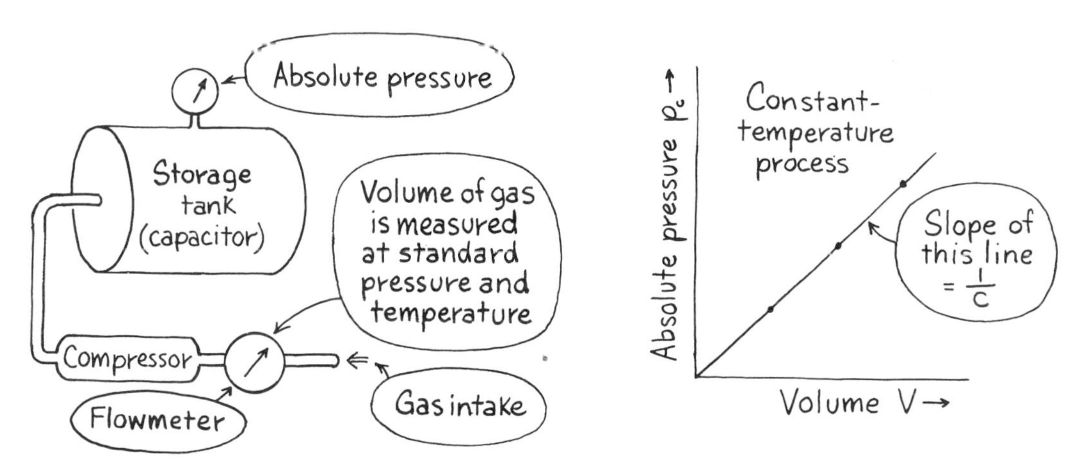

Figure 9-8. A pneumatic system acts just like a liquid storage system if changes are made slowly, and volume is measured under standard conditions of temperature and pressure.

$$C = \frac{V}{p_C} \qquad (9\text{-}5)$$

where V = Volume, p_C = Absolute pressure

A subtle difference is that the tank pressure is absolute rather than gauge. Absolute pressure is used for gases because at atmospheric pressure (zero gauge pressure) the tank still contains some gas. Only at zero absolute pressure is the tank really "empty."

In many cases, the mass of gas remains constant, and the volume is changed by means of a piston that either compresses the gas or allows it to expand, depending upon the pressure exerted by the piston. A graph of absolute pressure versus volume in such a situation is not linear. A typical compression curve is shown in Fig. 9-9. The piston starts in position 1 and

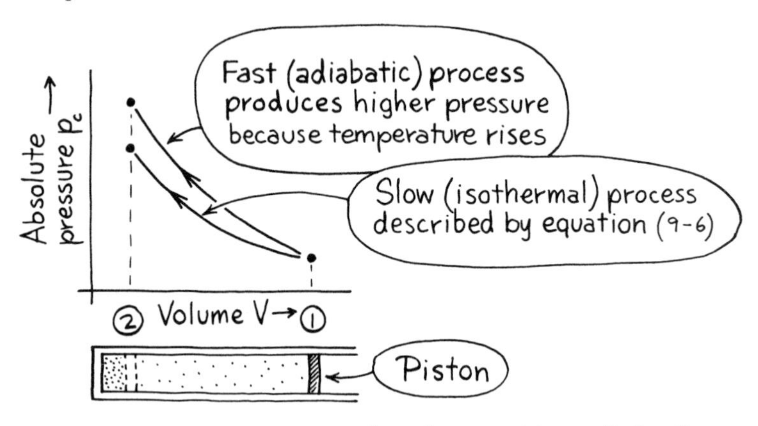

Figure 9-9. If a constant amount of gas is trapped in a cylinder, the pressure versus volume graph is *not* a straight line and it certainly does not look like the graph of Fig. 9-8.

compresses the gas to position 2. The pressure rise is nonlinear. The shape of the curve is also affected by how fast the compression occurs. If the process is very slow, the gas temperature does not change enough to matter. This is called an *isothermal* (constant-temperature) process. If the change of volume is very fast, the heat does not have time to escape and the gas warms up, causing higher pressure. This called an *adiabatic* (no heat added or removed) process.

There are special equations for these processes. The easiest is for the slow isothermal process in which the product of pressure and volume remains constant.

$$p_1 V_1 = p_2 V_2 \quad \text{(isothermal process)} \tag{9-6}$$

p *Absolute Pressure*	V *Volume*	*Note*: Subscripts 1 and 2 refer to values at different piston positions
pascal (Pa)	cubic metre (m^3)	
pound/square foot (lb/ft^2)	cubic foot (ft^3)	

Equation (9-6) is called *Boyle's law*.

Example 9-4 Two hundred cubic feet of air at atmospheric pressure are forced isothermally into an evacuated 16-ft^3 tank.

(a) What will the absolute pressure be?
(b) What is the tank's capacitance?
(c) What pressure is needed to force an additional 100 ft^3 of air into the tank?

Solution to (a)

Conversion factor in Appendix A

$$\left\{\begin{aligned} p_1 &= 1.0\text{ atm} \checkmark \\ &= 2.12 \times 10^3\text{ lb/ft}^2 \\ V_1 &= 200\text{ ft}^3 \\ V_2 &= 16\text{ ft}^3 \end{aligned}\right\}$$

$$p_1 V_1 = p_2 V_2 \Rightarrow p_2 = p_1 \frac{V_1}{V_2}$$

$$p_2 = \left(2.12 \times 10^3 \frac{\text{lb}}{\text{ft}^2}\right)\left(\frac{200\text{ ft}^3}{16\text{ ft}^3}\right) = \boxed{2.7 \times 10^4 \frac{\text{lb}}{\text{ft}^2}}$$

Solution to (b)

$$C = \frac{V}{p_C}$$

$$= \frac{200\text{ ft}^3}{2.7 \times 10^4\text{ lb/ft}^2} = \boxed{7.5 \times 10^{-3} \frac{\text{ft}^3}{\text{lb/ft}^2}}$$

Solution to (c)

$\{V_1 = 200 + 100 = 300\text{ ft}^3\}$

Using capacitance,

$$C = \frac{V_1}{p_C} \Rightarrow p_C = \frac{V_1}{C} = \frac{300\text{ ft}^3}{7.5 \times 10^{-3}\text{ ft}^3/(\text{lb/ft}^2)} = \boxed{4.0 \times 10^4\text{ lb/ft}^2}$$

or using Boyle's law

$$p_2 = p_1 \frac{V_1}{V_2} = \left(2.12 \times 10^3 \frac{\text{lb}}{\text{ft}^2}\right)\left(\frac{300\text{ ft}^3}{16\text{ ft}^3}\right) = 4.0 \times 10^4\text{ lb/ft}^2$$

Electrical System

An electrical capacitor normally consists of two large plates of conducting material. The plates are close to each other but separated by a layer of insulating material. In one kind of capacitor the plates are made of metal foil with a layer of paper or plastic in between. To get a large plate area into a small package, the whole thing is rolled up and sealed (Fig. 9-10).

If a capacitor is connected to a source of emf (Fig. 9-11), charge will not flow continuously because the circuit is not complete. When the switch is closed, however, there is a crowding of charge onto the plates until the potential difference V_C (capacitor voltage) becomes equal to the excitation force, which is the battery's emf E. The work done by the battery to produce the imbalance of charge remains in the capacitor as it does in a stretched spring or a filled tank. The stored potential energy can be reclaimed in one form or another when the switch is thrown to position 2.

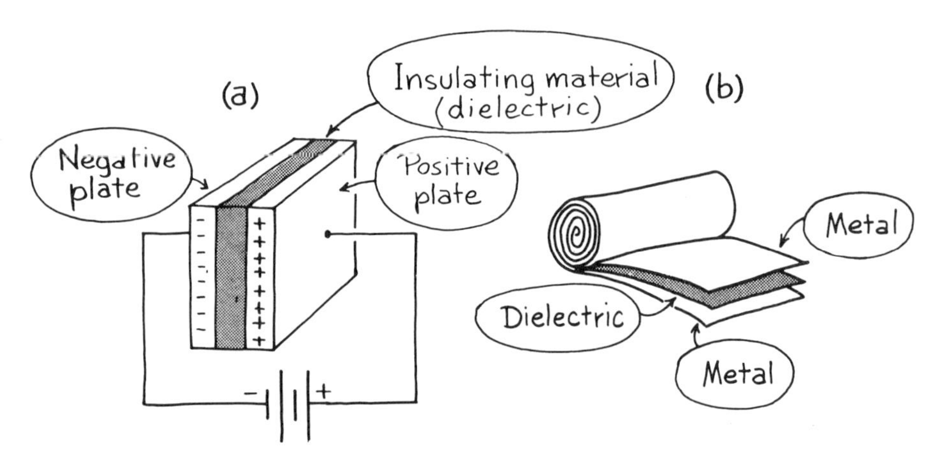

Figure 9-10. (a) A simple capacitor consists of two conducting plates with a layer of insulating material between. (b) Capacitors may be rolled or otherwise folded to save space.

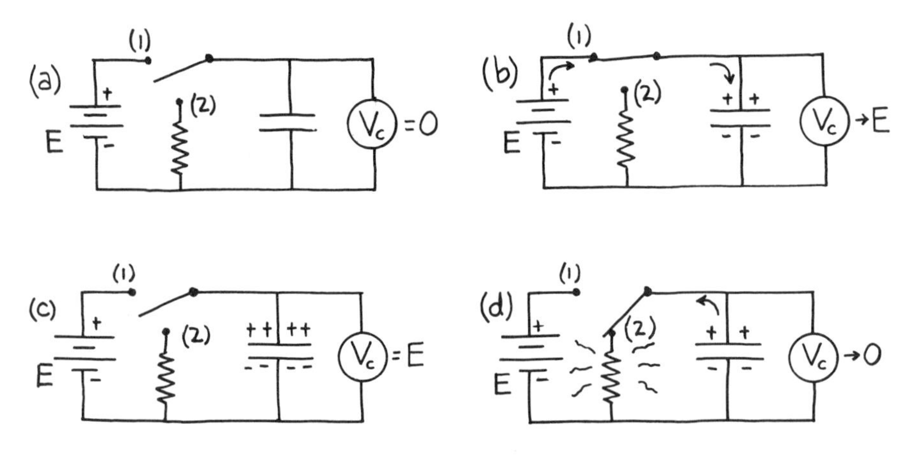

Figure 9-11. (a) Capacitor, battery, and voltmeter to measure V_C. Switch in neutral position. (b) Switch thrown to position 1. Battery does work by forcing excess charge onto capacitor plates. Charge movement stops when $V_C = E$. (c) Switch in neutral position. Charge remains on plates. Energy is stored in the capacitor. (d) Switch thrown to position 2. Capacitor becomes driving force like a wound up spring. Stored energy is reclaimed as heat or stored in another form.

If the amount of charge (parameter) stored is directly proportional to the capacitor voltage (capacitive force), the capacitor is linear in the same sense that springs or straight-sided tank capacitors are linear (Fig. 9-12). If a spring is stretched too hard, it may pull out of shape or even break. If too much air is forced into a pressure tank, it may explode. If too much potential difference is

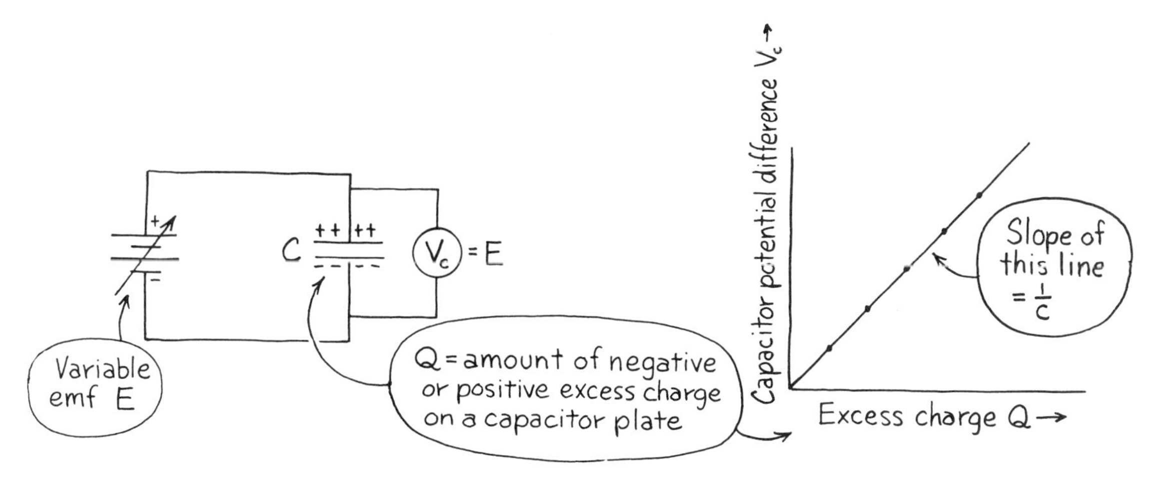

Figure 9-12. Many electrical capacitors act like straight-sided tanks. Charge Q is proportional to potential difference V_C. Such a capacitor is said to be linear.

imposed across an electrical capacitor, the insulation between plates may not be enough to prevent arcing and resultant breakdown of the capacitor.

Electrical capacitance is defined in exactly the same way as for the other systems. In fact, electrical capacitance is the original model selected for the unified concept of capacitance.

$$C = \frac{Q}{V_C} \tag{9-7}$$

C *Electrical Capacitance*	Q *Excess Charge*	V_C *Capacitor Voltage*
farad (F)	coulomb (C)	volt (V)

Also note the same symbol V is used for voltage (potential difference) and the unit volt.

Note: The same symbol C is used for capacitance and coulomb. Don't mix them up.

The unit of electrical capacitance is the *farad* (F). By definition, one farad = one coulomb/volt. A farad is a very large unit, and real capacitors are never close to being a full farad. Commonly used capacitors may be rated in terms of microfarads ($1\ \mu F = 10^{-6}$ F) or picofarads ($1\ pF = 10^{-12}$ F).

Example 9-5 A 24-V battery is connected to a 0.10-μF capacitor.

(a) How much charge flows onto a capacitor plate?
(b) How many electrons does this represent?
(c) What voltage is needed to force 1.0×10^{-5} C onto a capacitor plate?

Solution to (a)

$$\begin{Bmatrix} V_C = E = 24\text{ V} \\ C = 0.10\ \mu F \\ = 0.10 \times 10^{-6}\text{ F} \end{Bmatrix}$$

$$Q = CV_C = (0.10 \times 10^{-6}\text{ F})(24\text{ V}) = \boxed{2.4 \times 10^{-6}\text{ C}}$$

Solution to (b)

Appendix A

$$2.4 \times 10^{-6}\text{ C}\left(\frac{6.25 \times 10^{18}\text{ electrons}}{1\text{ C}}\right) = \boxed{1.5 \times 10^{13}\text{ electrons}}$$

Solution to (c)

$$\{Q = 1.0 \times 10^{-5}\ C\}$$

$$V_C = \frac{Q}{C} = \frac{1.0 \times 10^{-5}\text{ C}}{0.10 \times 10^{-6}\text{ F}} = \boxed{100\text{ V}}$$

Larger plates will provide more room for charge to accumulate in a capacitor, just as large cross-sectional area permits more liquid to be stored in a tank. The amount of charge that accumulates for a given voltage depends also on the distance between plates and on the insulating properties of the material between the plates. Such a material is called a *dielectric*, and its presence can change the capacitance. Experiments show that material, area, and distance are related according to the equation

$$C = \varepsilon_0 K \frac{A}{s} \tag{9-8}$$

where

$$\varepsilon_0 = 8.85 \times 10^{-12} \frac{\text{C} \cdot \text{m}}{\text{V} \cdot \text{m}^2}$$

C *Capacitance*	K *Dielectric Constant*	A *Area*	s *Separation Distance*
farad (F)	(dimensionless ratio)	square metre (m^2)	metre (m)

The constant ε_0 (epsilon) is called the *permittivity* of a vacuum. The *dielectric constant* K shows how different dielectric materials increase the capacitance if the material replaces a vacuum (or air) between the plates of a capacitor. For example, the dieletric constant of paraffin is about 2.2. A capacitor made with paraffin between the plates has 2.2 times the capacitance of an identical capacitor with only air between its plates or with a vacuum. Appendix B, Table 10, lists the dielectric constants of various materials used in capacitors.

Example 9-6 What is the capacitance of a capacitor made of two metallic plates separated by a 1.0-mm sheet of paraffin? The plate area is $1.0 \times 10^{-2}\ \text{m}^2$.

Solution

$$\left\{\begin{aligned} s &= 1.0\ \text{mm} \\ &= 1.0 \times 10^{-3}\ \text{m} \\ A &= 1.0 \times 10^{-2}\ \text{m}^2 \end{aligned}\right\}$$

$$C = \varepsilon_0 K \frac{A}{s}$$

Appendix B, Table 10, for paraffin (points to 2.2)

$$= \left(8.85 \times 10^{-12}\ \frac{\text{C} \cdot \text{m}}{\text{V} \cdot \text{m}^2}\right) \frac{(2.2)(1.0 \times 10^{-2}\ \text{m}^2)}{1.0 \times 10^{-3}\ \text{m}} = \boxed{1.9 \times 10^{-10}\ \text{F}}$$

Note: A farad (F) is a coulomb (C) per volt (V).

Heat System

Equation (3-14) incorporates the definition of a quantity called specific heat:

$$Q = mc(T_f - T_o) \tag{3-14}$$

Amount of heat added or removed (Q); Specific heat (c); Mass of object (m); Change of temperature ($T_f - T_o$)

The product mc (mass × specific heat) is the ratio of heat to temperature difference: $mc = Q/(T_f - T_o)$. But this ratio of heat (parameter) to temperature (force) is, by definition, capacitance, or thermal capacitance as it is called in heat systems.

$$C = \frac{Q}{T_f - T_o} \tag{9-9}$$

where

$$C = mc \tag{9-10}$$

C *Thermal Capacitance (Heat Capacity)*	m *Mass (or Weight)*	c *Specific Heat*	Q *Heat Energy Added or Removed*	T_f *Final Temperature*	T_o *Starting Temperature*
$\frac{\text{calorie}}{\text{Celsius degree}}$ (cal/°C)	gram (g)	$\frac{\text{calorie}}{\text{gram} \cdot \text{degree Celsius}}$ $\left(\frac{\text{cal}}{\text{g} \cdot °\text{C}}\right)$	calorie (cal)	Celsius degree (°C)	
$\frac{\text{British thermal unit}}{\text{Fahrenheit degree}}$ (Btu/°F)	pound (lb)	$\frac{\text{British thermal unit}}{\text{pound} \cdot \text{degree Fahrenheit}}$ $\left(\frac{\text{Btu}}{\text{lb} \cdot °\text{F}}\right)$	British thermal unit (Btu)	Fahrenheit degree (°F)	

As in the other systems, a graph of force (temperature change) versus parameter (heat added) is very nearly a straight line for solid materials (Fig. 9-13). For liquids and gases, however, the line is not nearly as straight.

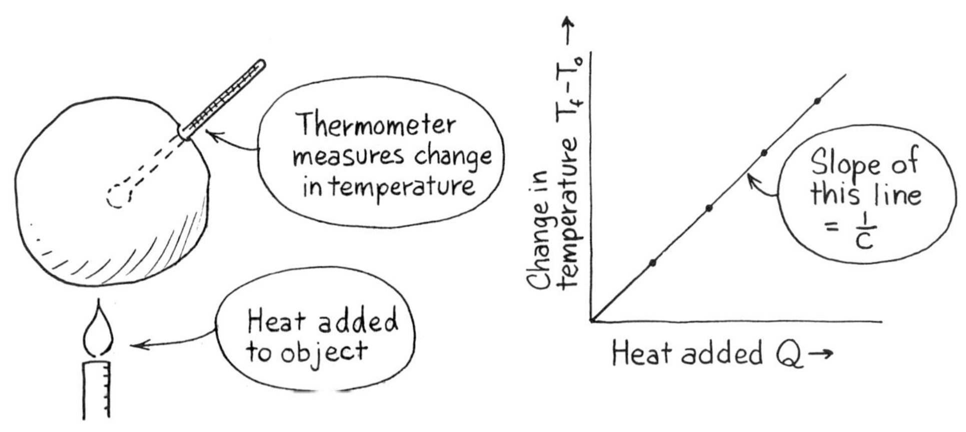

Figure 9-13. When heat is added to an object, the temperature rises correspondingly. Thermal capacitance (heat capacity) C is the ratio of parameter to force as in all the other systems.

Thermal capacitance or *heat capacity C* of a particular object describes how much heat is involved in changing the temperature of that object. *Specific heat c* applies to a particular kind of substance. It tells how much heat is involved in changing the temperature of a given amount of the substance. The heat capacity of a 20-kg brass monkey is twice as much as that of a 10-kg brass monkey. But according to Table 3, Appendix B, it requires 0.094 cal to raise the temperature of 1 g of brass from either monkey by 1°C.

Example 9-7 A 330-lb cast-iron automobile engine operates at a temperature of 180°F.

(a) What is the thermal capacitance of the engine?

(b) If the engine starts at 60°F, how much energy must be added to bring it up to operating temperature?

Solution to (a)

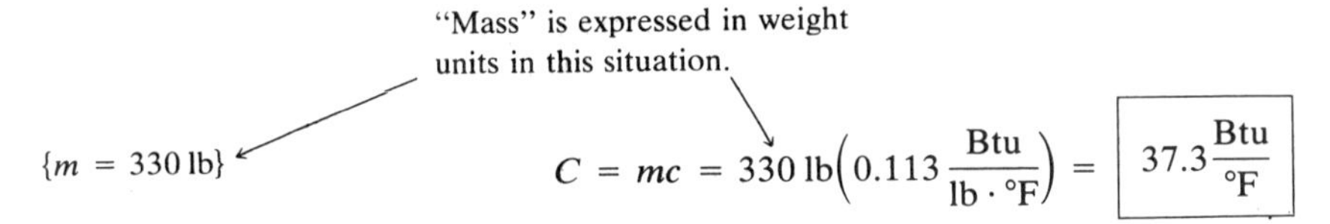

Solution to (b)

$$\left\{\begin{matrix} T_f = 180^\circ\text{F} \\ T_o = 60^\circ\text{F} \end{matrix}\right\} \qquad Q = C(T_f - T_o) = \left(37.3\frac{\text{Btu}}{{}^\circ\text{F}}\right)[(180 - 60)^\circ\text{F}] = \boxed{4.47 \times 10^3\text{ Btu}}$$

9-3 POTENTIAL ENERGY STORAGE IN CAPACITORS

The capacitors that we have looked at so far are devices in which a change of parameter depends upon what we have called capacitive force. When the capacitive force F_C does work on the system, energy is transferred to the capacitor and that energy is stored in the capacitor. The amount of energy stored depends upon the amount of parameter and on the characteristics of the capacitor as described by its capacitance C.

Energy storage in a capacitor does not depend on motion or "flow" of any sort. A spring-loaded mouse trap can be cocked and left motionless until the unfortunate mouse comes along to release the energy. Work done to pump water into a storage tank remains there until a valve is opened below to drain the water and the energy from the tank. Thus we are not dealing with kinetic energy E_K but potential energy E_P. Except for heat systems in which parameter is energy, we say by definition that

work done by capacitive force = potential energy (E_P) stored in the capacitor

Section 7-4 showed that energy can be represented as the area under a graph of force versus parameter. This applies just as well to potential energy, as shown in Fig. 9-14. Figure 9-14a illustrates the rather special case in which F_C is constant. The most familiar example is of the nonlinear gravitational system (Fig. 9-5). The constant capacitive force F_C needed to lift an object is equal to the object's weight W, and the distance moved is the vertical height h.

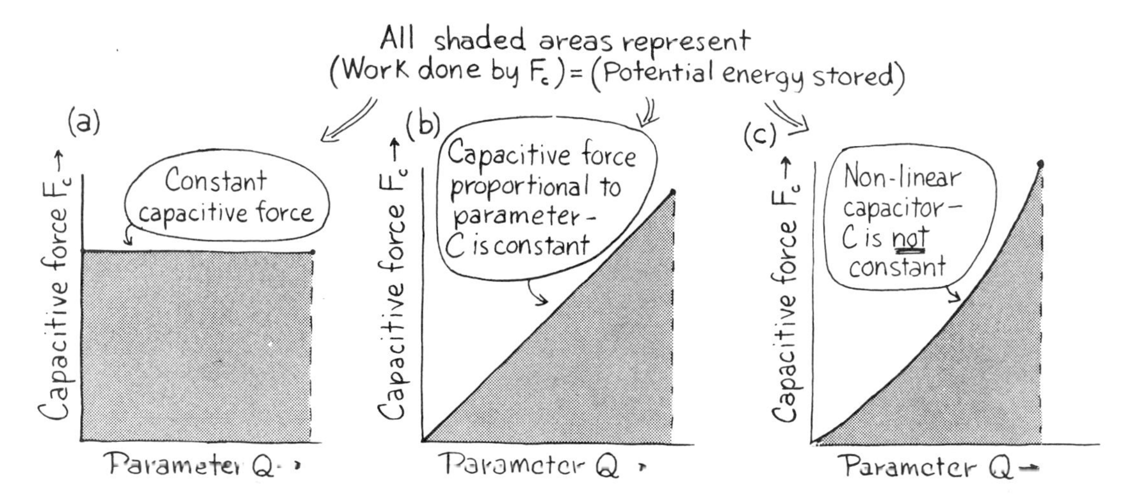

Figure 9-14. Regardless of the shape of the graph, work done by capacitive force is equal to potential energy stored in the capacitor. The amount of energy stored is represented by the area under the graph.

The work done is force times distance, which is transferred as stored potential energy:

$$\text{Gravitational potential energy} \rightarrow E_P = \overset{\text{Weight}}{W} \underset{\text{Change of elevation}}{h} \tag{3-11}$$

The product Wh is the rectangular area of Fig. 9-14a.

Example 9-8 Last summer I went to the mountains. I started at sea level and camped at 8000-ft elevation in my 4000-lb camper.

(a) How much work did my engine have to do just to make the elevation change?

(b) While there I collected 15 kg of rocks and brought them home. How much potential energy did the rocks lose?

Solution to (a)

$$\left\{\begin{array}{l} h = 8000\ \text{ft} \\ W = 4000\ \text{lb} \end{array}\right\}$$

$$E_P = Wh = 4000\ \text{lb}(8000\ \text{ft}) = \boxed{3.2 \times 10^7\ \text{ft} \cdot \text{lb}}$$

This is about the amount of energy from one third of a gallon of gasoline

Solution to (b)

$$\left\{\begin{array}{l} h = 8000\ \text{ft}\left(\dfrac{1\ \text{m}}{3.28\ \text{ft}}\right) \\ \quad = 2.44 \times 10^3\ \text{m} \\ m = 15\ \text{kg} \end{array}\right\}$$

$$W = mg$$

Change mass (kg) to weight (N)

$$E_P = Wh = mgh$$

$$= 15\ \text{kg}\left(9.81\ \frac{\text{m}}{\text{s}^2}\right)(2.44 \times 10^3\ \text{m}) = \boxed{3.6 \times 10^5\ \text{J}}$$

Figure 9-14b represents the behavior of a linear capacitor—springs, straight-sided tanks, heated objects, or electrical capacitors. Stored potential energy is represented by the area of the triangle with altitude F_C and base Q. The area and the stored potential energy is therefore

$$E_P = \tfrac{1}{2}F_C Q \tag{9-11}$$

But capacitance C is defined as the ratio of parameter to capacitive force:

$$C = \frac{Q}{F_C} \tag{9-3}$$

Substituting Eq. (9-3) into the energy equation (9-11) results in several equations for stored potential energy E_P as represented by the triangular area in Fig. 9-14b. These are

$$E_P = \tfrac{1}{2}CF_C^2 \tag{9-12}$$

$$= \frac{1}{2}\frac{Q^2}{C} \tag{9–13}$$

Also, since spring constant $k = 1/C$ [Eq. (9-2)],

$$E_P = \tfrac{1}{2}kQ^2 \tag{9-14}$$

These equations are summarized for the various systems in Table 9-1.

TABLE 9-1 POTENTIAL ENERGY STORAGE EQUATIONS

Systems	*Energy Equations*	E_P *Potential Energy*	F_C *Capacitive Force*	Q *Parameter*	C *Capacitance*	k *Spring Constant*
Mechanical translational	$\frac{1}{2}F_Cs$ $\frac{1}{2}CF_C^2$ $\frac{1}{2}s^2/C$ $\frac{1}{2}ks^2$	(J) (ft · lb)	F_C Capacitive force (N) (lb)	s Distance (m) (ft)	(m/N) (ft/lb)	(N/m) (lb/ft)
Mechanical rotational	$\frac{1}{2}\tau_C\theta$ $\frac{1}{2}C\tau_C^2$ $\frac{1}{2}\theta^2/C$ $\frac{1}{2}k_r\theta'$	(J) (ft · lb)	τ_C Capacitive torque (N · m) (lb · ft)	θ Angle (rad)	(rad/N · m) (rad/lb · ft)	(N · m/rad) (lb · ft/rad)
Fluid (for *liquids*; *not* for gases)	$\frac{1}{2}p_CV$ $\frac{1}{2}Cp_C^2$ $\frac{1}{2}V^2/C$	(J) (ft · lb)	p_C *Pressure* (Pa) (lb/ft²)	V *Volume* (m³) (ft³)	$\left(\frac{m^3}{Pa}\right)$ $\left(\frac{ft^3}{lb/ft^2}\right)$	
Electrical	$\frac{1}{2}V_CQ$ $\frac{1}{2}CV_C^2$ $\frac{1}{2}Q^2/C$	(J)	V_C Capacitive voltage (V)	Q Charge (C)	(F)	
Heat	Equations do not apply					

The examples that follow illustrate applications of these equations in solving potential energy storage problems.

Example 9-9 The spring of Example 9-1 is stretched 0.37 m with a force of 26 N. Its spring constant was found to be 70 N/m. How much work was done to stretch it?

Solution

$$\left\{\begin{aligned} s &= 0.37\text{ m} \\ F_C &= 26\text{ N} \\ k &= 70\text{ N/m} \end{aligned}\right\}$$

$$E_P = \tfrac{1}{2}F_Cs = \tfrac{1}{2}(26\text{ N})(0.37\text{ m}) = \boxed{4.8\text{ J}}$$

or

$$E_P = \tfrac{1}{2}ks^2 = \tfrac{1}{2}(70\text{ N/m})(0.37\text{ m})^2 = 4.8\text{ J}$$

Example 9-10 The metal rod of Example 9-2 twisted 60° under a torque of 5.0 lb · ft. Its rotational spring constant k_r was found to be 4.8 lb · ft/rad. How much energy is stored in the rod with the 60° twist?

Solution

$$\left\{\begin{aligned} \theta &= 60°\left(\frac{1\text{ rad}}{57.3°}\right) \\ &= 1.05\text{ rad} \\ \tau_C &= 5.0\text{ lb}\cdot\text{ft} \\ k_r &= 4.8\text{ lb}\cdot\text{ft/rad}\end{aligned}\right\}$$

$$E_P = \tfrac{1}{2}\tau_C\theta = \tfrac{1}{2}(5.0\text{ lb}\cdot\text{ft})(1.05\text{ rad}) = \boxed{2.6\text{ ft}\cdot\text{lb}}$$

or

$$E_p = \tfrac{1}{2}k_r\theta^2 = \tfrac{1}{2}(4.8\text{ lb}\cdot\text{ft/rad})(1.05\text{ rad})^2 = 2.6\text{ ft}\cdot\text{lb}$$

Example 9-11 The tank in Example 9-3 contained 0.63 m^3 of water when a gauge pressure of 1.0×10^4 Pa was applied at the bottom. Its capacitance was found to be 0.63×10^{-4} m^3/Pa. How much potential energy was stored in the tank?

Solution

$$\left\{\begin{aligned} V &= 0.63\text{ m}^3 \\ p_C &= 1.0\times 10^4\text{ Pa} \\ C &= 0.63\times 10^{-4}\text{ m}^3/\text{Pa}\end{aligned}\right\}$$

$$E_P = \tfrac{1}{2}p_C V = \tfrac{1}{2}(1.0 \times 10^4\text{ Pa})(0.63\text{ m}^3) = \boxed{3.2 \times 10^3\text{ J}}$$

or

$$E_P = \tfrac{1}{2}Cp_C^2 = \tfrac{1}{2}(0.63 \times 10^{-4}\text{ m}^3/\text{Pa})(1.0 \times 10^4\text{ Pa})^2 = 3.2 \times 10^3\text{ J}$$

or

$$E_p = \frac{1}{2}\frac{V^2}{C} = \frac{\tfrac{1}{2}(0.63\text{ m}^3)^2}{0.63 \times 10^{-4}\text{ m}^3/\text{Pa}} = 3.2 \times 10^3\text{ J}$$

The equations in Table 9-1 do not apply to gas systems. Energy stored in a gas due to work done by mechanical forces is complicated mainly because a gas is compressible. The correct equations are developed in a branch of the physical sciences called thermodynamics, but this is beyond the scope of our text. Mechanical work done on a gas is still, however, the area under a pressure versus volume graph.

Example 9-12 The 0.10-μF electrical capacitor of Example 9-5 held 2.4×10^{-6} C of charge when the potential difference was 24 V.

(a) How much potential energy was stored in the capacitor?
(b) How much additional energy is added when the capacitor is charged to 100 V?

Solution to (a)

$$\left\{\begin{aligned} C &= 0.10\ \mu\text{F} \\ &= 0.10\times 10^{-6}\text{ F} \\ V_C &= 24\text{ V}\end{aligned}\right\}$$

$$E_P = \tfrac{1}{2}CV_C^2 = \tfrac{1}{2}(0.10 \times 10^{-6}\text{ F})(24\text{ V})^2 = \boxed{2.9 \times 10^{-5}\text{ J}}$$

Solution to (b)

$$\{V_C = 100\text{ V}\}$$

$$E_P = \tfrac{1}{2}CV_C^2 = \tfrac{1}{2}(0.10 \times 10^{-6}\text{ F})(100\text{ V})^2 = 50 \times 10^{-5}\text{ J}$$

$$\text{Additional energy} = (50 - 2.9)10^{-5} = \boxed{47 \times 10^{-5}\text{ J}}$$

The equations in Table 9-1 do not apply to heat systems. It is possible, however, to calculate heat added or removed using $Q = mc(T_f - T_o)$ [Eq.

(3-14)]. This was illustrated in Example 9-7 where a temperature change of 180°F − 60°F = 120°F to a 330-lb iron engine required the addition of

$$Q = mc(T_f - T_o) = 330\text{ lb}\left(\underset{\text{Specific heat of iron}}{0.113}\,\frac{\text{Btu}}{\text{lb}\cdot{}^\circ\text{F}}\right)(120^\circ\text{F}) = 4.47 \times 10^3\text{ Btu}$$

*9-4 PROPERTIES OF ELASTIC MATERIALS

An object's dimensions will change if it is stretched (tension), squeezed (compression), or twisted (shear). Potential energy is stored in the deformed object in these situations. If the force is not too large, the object returns to its original condition when the force is removed. Let's look at tension, compression, and shear separately.

Tension Causes Change of Length

Figure 9-15 shows a solid object being stretched by a tensional force F. The change in length $\Delta\ell$ due to the force depends upon the original length

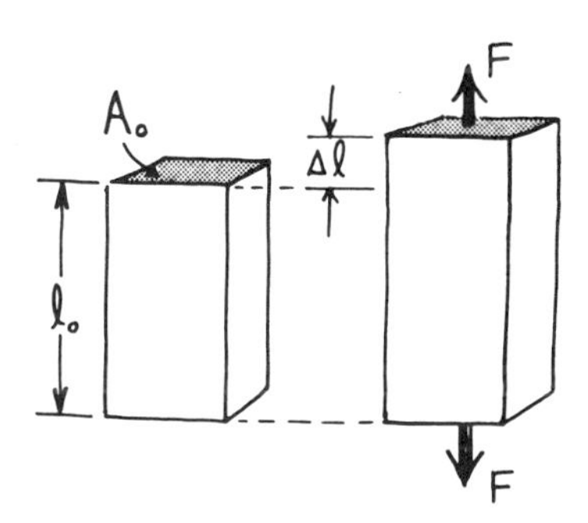

Figure 9-15. A tensional force F stretches the solid. The change in length is $\Delta\mathscr{L}$, the original length is ℓ_o, and cross-sectional area is A_o.

ℓ_o, the original cross-sectional area A_o, and the kind of material the solid is made of. Experiments show that if the force is not too large all these factors are related by Eq. (9-15):

$$Y = \frac{\sigma}{\varepsilon} \tag{9-15}$$

where the ratio of force to original cross-sectional area is called engineering *stress* σ, and the ratio of change in length $\Delta\ell$ to original length ℓ_o is called engineering *strain* ε:

$$\sigma = \frac{F}{A_o} \tag{9-16}$$

$$\varepsilon = \frac{\Delta\ell}{\ell_o} \tag{9-17}$$

Y *Young's Modulus*	σ *Stress*	ε *Strain*	F *Force*	A_o *Original Area*	ℓ_o *Original Length*	$\Delta\ell$ *Change of Length*
newton / square metre (N/m^2)	newton / square metre (N/m^2)	metre / metre (m/m)	newton (N)	square metre (m^2)	metre (m)	metre (m)
pound / square inch ($lb/in.^2$)	pound / square inch ($lb/in.^2$)	inch / inch (in./in.)	pound (lb)	square inch ($in.^2$)	inch (in.)	inch (in.)

Appendix B, Table 7, lists values of elastic moduli for different materials. The modulus for tensile stress defined in Eq. (9-15) is commonly called *Young's modulus.* Like other elastic moduli, it is a ratio of stress to strain. It is characteristic of a material and is constant for moderate forces that do not permanently deform an object.

Curves of force versus change in length are shown in Fig. 9-16 for two different materials. The original cross-sectional area A_o is 1 in.^{2} and the

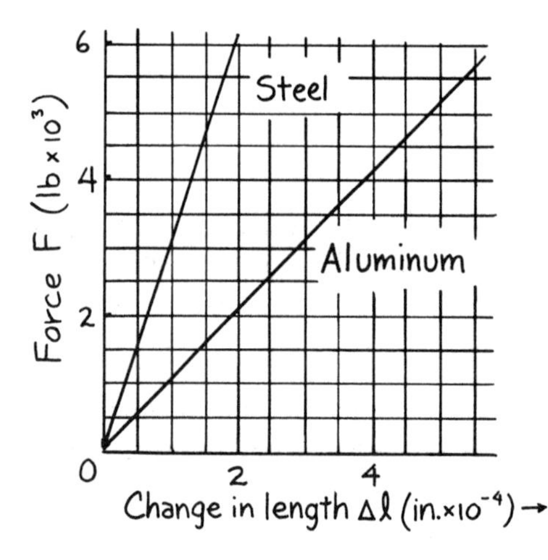

Figure 9-16. Graphs of force versus change in length for 1-in. cubes of steel and aluminum. Steel is a stiffer material and its graph is steeper. More force is needed to stretch the steel than the aluminum for a given deflection. The deflections are extremely small.

original length ℓ_o is 1 in. for both the steel and aluminum blocks. Note that the changes in length are small. Solids may be shaped in the form of a spring, bellows, or diaphragm to allow for a greater change in length.

Compression Causes Change of Volume

When a material is subjected to an increase in pressure, it is squeezed down so that it occupies less volume (Fig. 9-17). The decrease in

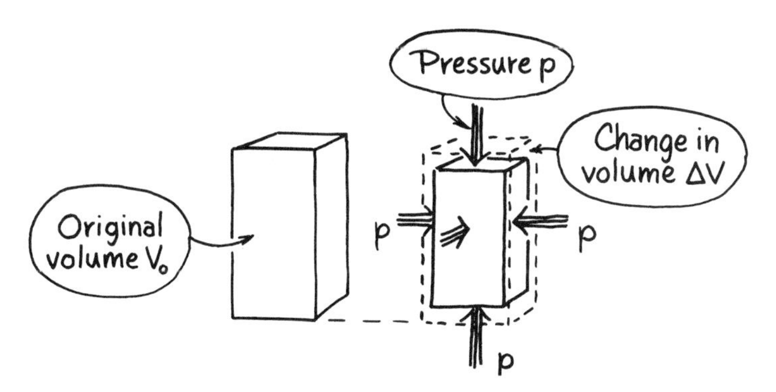

Figure 9-17. A pressure increase p reduces the volume by an amount ΔV. Solids and liquids are difficult to compress. Gases compress with relative ease.

volume of liquids and solids is small relative to that of gases. The change, however, is real. For example, rocks deep in the earth are noticeably denser than at the surface, and deep ocean water is measurably denser than surface water.

The relation between pressure p and change in volume ΔV looks just like Eq. (9-15). The pressure p is the volume "stress," and $\Delta V/V_o$ is the volume strain. Again, stress is proportional to strain; that is, their ratio is a constant characteristic of the material. Equation (9-18) defines that constant called the *bulk modulus*.

$$B = \frac{p}{\Delta V/V_o} \tag{9-18}$$

B Bulk Modulus	p Pressure	ΔV Change in Volume	V_o Original Volume
pascal (Pa)		cubic metre (m^3)	
pound/square inch ($lb/in.^2$)		cubic inch ($in.^3$)	

The bulk modulus B compares pressure to volume strain for a given substance. Values of bulk modulus appear in Appendix B, Table 7.

Gases compress more easily than liquids or solids. Volume changes are complicated by possible changes in temperature, which can occur when the gas is expanded or compressed. Gases cool if expanded and become warmer if compressed.

Shear Causes Change of Shape

Figure 9-18 shows a solid block acted on by forces that are parallel but not on the same line. The block tends to be smeared out like a deck of cards. This sort of deformation is called *shear.* Shear stress is defined as F/A_o, and shear strain is the tangent of the deflection angle θ. Experiments show that shear stress is proportional to shear strain. The *shear modulus G* compares shear stress to shear strain for a given material.

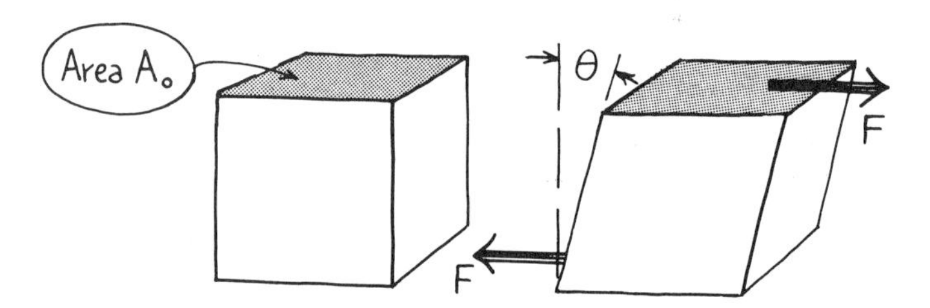

Figure 9-18. The block is twisted through a small angle θ by the parallel forces. The shear stress is F/A_o and shear strain is the tangent of θ.

$$G = \frac{F/A_o}{\tan\theta} \tag{9-19}$$

G *Shear Modulus*	*F* *Shear Force*	A_o *Area*	θ *Shear Angle*
$\frac{\text{newton}}{\text{square metre}}$ (N/m^2)	newton (N)	square metre (m^2)	degree (°)
$\frac{\text{pound}}{\text{square inch}}$ ($lb/in.^2$)	pound (lb)	square inch ($in.^2$)	

Like Young's modulus and the bulk modulus, it is constant for moderate forces. Values of shear modulus appear in Appendix B, Table 7, for various solid materials. There is no such thing as a shear modulus for a fluid. Fluids offer virtually no opposition to shearing forces.

9-5 SUMMARY

1. Energy has to be accounted for. If work is done on a system, it must show up as energy lost E_R or as energy stored within the system. This is a

statement of the principle of *conservation of energy*. The energy that is stored (not lost) may be in the form of *potential energy* E_P or *kinetic energy* E_K, or some of each.

2. *Capacitive force*, the second of our three basic forces, does work on *capacitors* to store potential energy in them. The characteristics of a capacitor are described by the quantity *capacitance*, the ratio of parameter to capacitive force:

$$C = \frac{Q}{F_C} \tag{9-3}$$

One exception to this is the gravitational "spring" discussed in Sec. 9-2. Table 9-2 summarizes the equations for capacitance in the various systems.

TABLE 9-2 CAPACITORS AND CAPACITANCE IN THE VARIOUS SYSTEMS

System	*Capacitor*	*Parameter*	*Capacitive Force-like Quantity*	*Capacitance*
Mechanical translational	Linear spring	Extension s	Capacitive force F_C	$C = \frac{s}{F_C}$
Mechanical rotational	Rotational spring	Angle θ	Capacitive torque	$C = \frac{\theta}{\tau_C}$
Fluid (liquid)	Reservoir	Volume V	Capacitive (gauge) pressure p_C	$C = \frac{V}{p_C}$
Fluid (gas)	Tank	Volume added V	Capacitive (absolute) pressure p_C	$C = \frac{V}{p_C}$
Electrical	Capacitor	Charge Q	Capacitive voltage V_C	$C = \frac{Q}{V_C}$
Heat	Heated object	Heat energy Q	Temperature change $T_f - T_o$	$C = \frac{Q}{T_f - T_o}$

3. A capacitor stores potential energy. In all but heat systems the amount of potential energy stored in a capacitor is represented by the area under a graph of capacitive force versus parameter. Various equations used to calculate potential energy storage are located in Table 9-1.

Chapter 10 investigates inertive force and systems that store kinetic energy. A quantity called inertance will describe this kinetic energy storage capability. We will look at the relations between inertance, inertive force, and motion or flow in systems containing inertance.

PROBLEMS

9-1. A 1700-lb elevator is lifted a vertical distance of 64 ft by an electric motor. Sketch a graph of capacitive force versus height and calculate the potential energy added to and stored in the elevator.

9-2. The horse lifts the 100-lb bale of hay 20 ft by applying a 110-lb force using the setup illustrated. Assume that there is no energy loss due to friction.

(a) How much work is done by the horse?

(b) What is the change in potential energy of the hay?

(c) What became of the energy that was put into the system but was not stored as potential energy?

Problem 9.2

9-3. A spring stretches 0.15 m when a load of 1.0 kg is hung vertically from one end.

(a) What is the spring constant in newtons/metre?

(b) What is the total extension when the spring supports a load of 4.0 kg?

(c) How much potential energy is stored in the spring when supporting the 4.0-kg load?

9-4. A linear spring requires a force of 30.0 oz to stretch it 0.250 in.

(a) What is the spring constant in units of pounds/inch?

(b) What is the spring's capacitance?

(c) What force is needed to stretch the spring to a total deflection of 0.125 in.?

9-5. A steel drive shaft twists 8.6° (one end relative to the other) when 786 lb · ft of torque is applied.

(a) What is the shaft's capacitance?

(b) How much torque is needed to twist the shaft 25.3°? Assume linear behavior.

9-6. A pointer in a simple electrical meter winds up a small spring when it rotates. Suppose the pointer is 5.0 cm long and a 0.75-N force, acting at a right angle to the pointer at its end, rotates the spring through an angle of 30°.

(a) What torque is exerted on the coil spring?

(b) What is the spring's capacitance?

(c) How much energy is stored in the spring?

9-7. A pump gauge pressure of 3.00×10^4 Pa will force 10.0 m^3 of water into a tank.

(a) What is the tank's capacitance in SI units?

(b) How much work does the pump have to do in order to force 50.0 m^3 of water into the tank?

9-8. A water tank has a cross-sectional area of 25 ft^2 and a height of 10 ft. The tank is filled through a pipe at its base.

(a) Sketch a graph of pressure p at the bottom of the tank versus volume V of water in the tank as it is filled.

(b) How much potential energy is stored in the tank when it is full?

9-9. (a) What volume of air measured at atmospheric pressure will a 9.0 ft^3 compressor tank hold if the air is pumped slowly into the tank until the tank's absolute pressure is 100 psi?

(b) What is the tank's capacitance in units of $ft^3/(lb/ft^2)$?

9-10. According to my calculations, a tire on my car holds 1.7 ft^3 of compressed air. I run them at 30-psi *gauge* pressure.

(a) What volume of air, measured at atmospheric pressure, is contained in the inflated tire? Assume that there is no change in temperature of the air during the compression.

(b) What is the capacitance of the tire in units of $ft^3/(lb/ft^2)$?

9-11. An electrical capacitor of 1.5 μF is connected to a 6.0-V battery.

(a) How much negative charge is stored on one of the capacitor plates?

(b) How much potential energy is stored in the capacitor?

9-12. The charge on a capacitor is 3.6×10^{-4} C when the potential difference across its plates is 80 V.

(a) How much energy is stored?

(b) What is the capacitance?

(c) How much energy is stored if the potential difference is raised to 120 V?

9-13. A block of aluminum has a mass of 1.0 kg.

(a) What is the thermal capacitance of the block?

(b) How much heat energy must be added to the block to change its temperature from 10 to 75°C?

9-14. (a) What is the thermal capacitance of 25 lb of water? Give your answer in units of Btu/°F.

(b) How much heat must be added to (stored in) the water to change its temperature from 32 to 75°F?

9-15. How much *more* energy is needed to stretch the spring in Problem 9-3 from 0.30 to 0.60 m?

9-16. Energy is stored in a spring as it is stretched. Values of force and deflection are as follows:

Force F (N)	0	2.5	5.0	7.5	10.0
Deflection s (m)	0	0.20	0.40	0.60	0.80

(a) Plot a graph of force versus deflection with the force on the vertical axis.

(b) Calculate the spring constant and the capacitance.

(c) How much energy is stored in the spring when the deflection is 0.64 m?

9-17. The illustrated torsion bar is at a wheel that supports one quarter of the total weight of a 3200-lb car. The car's suspension system allows the car to settle 3.0 in. due to the weight of the car alone. The length of the support arm is 1.0 ft. What is the rotational spring constant of the torsion bar?

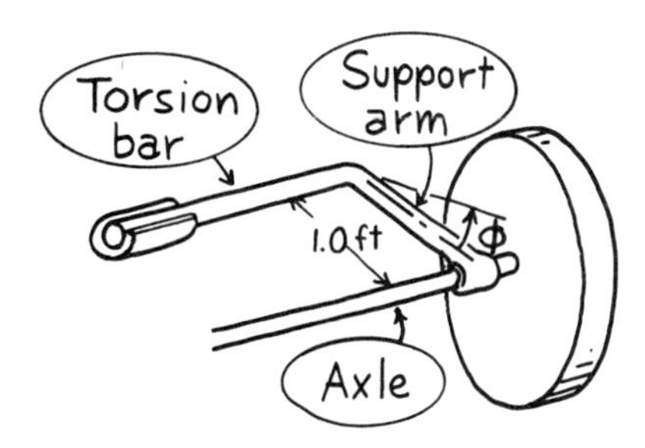

Problem 9.17

9-18. How many degrees must the shaft in Problem 9-5 be twisted in order to store 800 ft · lb of energy?

9-19. In some locations it is necessary to put city water supply tanks on towers to maintain hydraulic head. One such tank has its base situated so that it is 40 ft above the ground. The tank itself is 20 ft in diameter and 15 ft tall. Water is pumped from ground level through a pipe into the base of the tank. Plot an accurate graph of pump pressure versus amount of water in the tank—enough to fill the tank. Indicate areas on the graph that represent energy needed to raise water to the base of the tank and the additional energy to fill the tank once the water is 40 ft off the ground. Calculate the total stored energy (relative to ground level) when the tank is full of water.

9-20. A water storage tank is 26 ft in diameter and 20 ft deep.

(a) Calculate the tank's capacitance.

(b) How much potential energy is stored when the tank is full?

9-21. A straight-sided tank has a cross-sectional area of 12.0 ft^2. Initially, it contains water to a depth of 15.0 ft.

(a) How much water is needed to change the level from 15.0 to 18.0 ft?

(b) How much additional energy is stored in the water added?

9-22. A photographic flash unit consists of a capacitor that is slowly charged by a 600-V battery and then discharged quickly through a xenon glow tube. The tube requires 10 J of energy for a succesful discharge.
(a) What must be the capacitance of the capacitor?
(b) How much charge is stored when the capacitor is fully charged?
9-23. A capacitor holds 2.0×10^{-3}C of charge when a potential difference of 240 V is applied.
(a) Draw a graph of force versus parameter for this linear capacitor and find its capacitance.
(b) How much additional energy is needed to charge the capacitor from 240 to 300 V? Show this energy on the force versus parameter graph.
9-24. Lake Tahoe (California and Nevada) has a surface area of 495 km^2 (nearly 200 mi^2). The upper 10 m of water in the lake is warmed by the sun from about 5°C in the winter to perhaps 18°C by late summer.
(a) What is the thermal capacitance of the 10-m surface layer of water?
(b) How much heat (in calories) is added to the water to produce the temperature change?
(c) One gallon of gasoline is equivalent to 3.2×10^7 cal of heat energy when burned. How many gallons of gasoline would have to be burned to produce the energy calculated in part (b)?
9-25. A heat sink for a transistor is a copper plate with a thickness of 0.25 cm and area of 9.0 cm^2.
(a) What is the thermal capacitance of the heat sink in cal/°C?
(b) How much energy in joules is required to change the temperature of the copper from 20 to 100°C?
9-26. A force of 100 N pushes a 140-N block 5.0 m along a plane that is inclined 30° to the horizontal. Assume that there is no friction between block and plane.
(a) How much work is done by the force?
(b) How much of that work is stored as potential energy?
(c) How much of the work is transferred as kinetic energy?
9-27. A steel bar with cross-sectional area of 0.500 in.^{2} is placed in tension, and a foil strain gauge measures a mechanical strain of 50×10^{-6} in./in. when this happens.
(a) What is the stress in the steel bar?
(b) What force is needed to obtain this stress?
9-28. What pressure in pounds/square inch is needed to change the volume of lubricating oil from 150.00 to 149.88 in.^{3}?
9-29. A shear stress of 25×10^6 N/m^2 acts on a cubic block of material 0.100 m on a side. The upper surface is deflected 0.61×10^{-4} m with respect to the bottom surface. What is one possible material that would exhibit this behavior?
9-30. A variable capacitor detects the position of a neoprene rubber probe. The capacitance is 2.0×10^{-12} F with the rubber fully withdrawn as in (A). What does the capacitance become if the probe is fully inserted as in (B)?

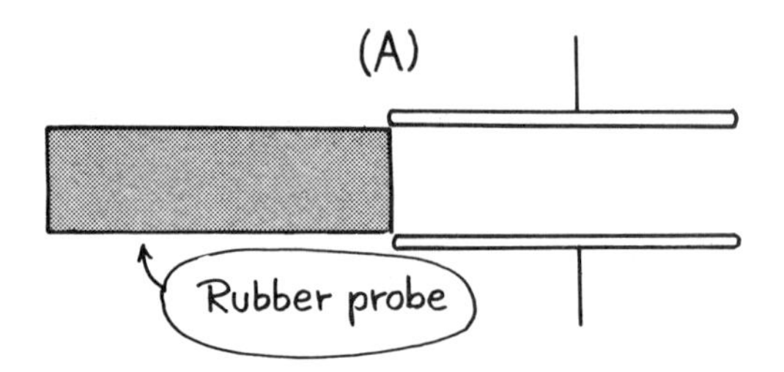

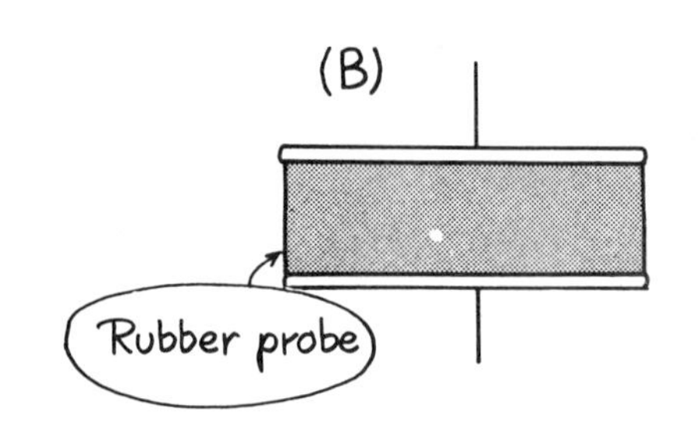

Problem 9.30

ANSWERS TO ODD-NUMBERED PROBLEMS

9-1. 1.1×10^5 ft · lb
9-3. (a) 65 N/m
(b) 0.60 m
(c) 12 J
9-5. (a) 1.91×10^{-4} rad/(lb · ft)
(b) 2.31×10^3 lb · ft
9-7. (a) 3.33×10^{-4} m^3/Pa
(b) 3.75×10^6 J
9-9. (a) 61 ft^3
(b) 4.2×10^{-3} $ft^3/(lb/ft^2)$
9-11. (a) 9.0×10^{-6} C
(b) 2.7×10^{-5} J
9-13. (a) 2.2×10^2 cal/°C
(b) 1.4×10^4 cal
9-15. 8.8 J
9-17. 3.3×10^3 lb · ft/rad
9-19. 1.4×10^7 ft · lb
9-21. (a) 36 ft^3
(b) 3.7×10^4 ft · lb
9-23. (a) 8.3×10^{-6} F
(b) 0.13 J
9-25. (a) 1.9 cal/°C
(b) 6.2×10^2 J
9-27. (a) 1.5×10^3 lb/in.^{2}
(b) 7.5×10^2 lb
9-29. $G = 4.1 \times 10^{10}$ N/m^2 is the value for copper

10

Inertance and Kinetic Energy Storage

We come now to the last of the three styles of energy—kinetic energy. A moving car in Figure 9-2 was able to perform useful work by lifting a counterweight. The moving car possessed kinetic energy, the energy of motion.

This chapter studies kinetic energy by first examining the net force responsible for accelerating objects in mechanical systems. Objects have inertia or a property that we will call ***inertance****, which describes how the net force is able to accelerate the object. Examples of inertance are mass and moment of inertia. The net or* ***inertive*** *force does work that is stored as kinetic energy in moving objects.*

Net force and inertance in mechanical systems have analogues in electrical and fluid systems where these quantities are identified using unified-concept ideas. Heat systems, however, do not exhibit inertive properties.

10-1 NET FORCE, MASS AND NEWTON'S SECOND LAW

In the late 1600's, an Englishman, Isaac Newton, worked out a series of statements and equations that described (not explained!) what it takes to make things move. Some of what he had to say can be summarized in what are commonly called *Newton's laws of motion.* The first of these goes like this: "An object remains at rest, or if in motion it remains in motion at constant speed in a straight line unless it is acted upon by an unbalanced external force." More simply, *Newton's first law* says that the motion of an object (speed or direction) will not change unless there is something to make it change. That sounds reasonable, doesn't it?

Newton's second law takes the first law and puts it in terms of mathematics. Newton reasoned that there is a definite relation between the amount of force acting on an object and the rate at which its velocity changes (the acceleration). It turns out that the greater the force, the more the acceleration. An object's acceleration is directly proportional to the net force acting on it. The resultant or net force is actually the vector sum of all forces acting on an object, and the resulting acceleration is in the same direction as the resultant force.

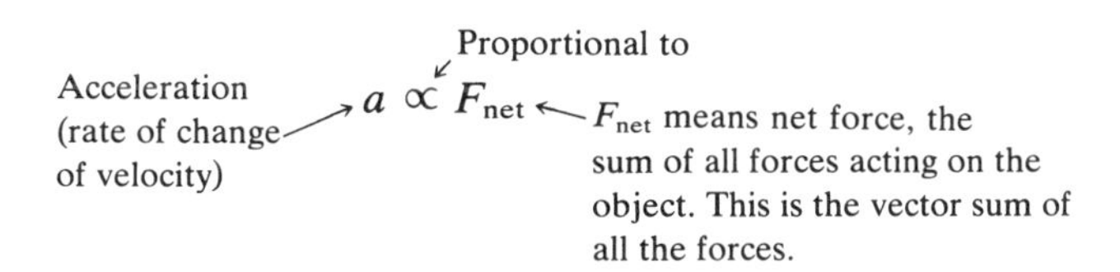

It is said that you can put any engine into any car—if you have the money. People interested in quick starts install large engines in cars and, sure enough, the greater force produces larger acceleration. Another way of maximizing acceleration is to reduce the amount of material that is to be accelerated. Drag strip cars for example consist of very powerful engines in only the most rudimentary framework of a car. Careful experiment shows that acceleration increases as the amount of substance decreases. Acceleration is inversely proportional to the amount of material in an object as represented by its *mass.*

$$a \propto \frac{1}{m}$$

Acceleration → a; m ← Mass of object = amount of material

These two relations can be put together to produce the mathematical statement of Newton's second law:

$$F_{net} = ma \quad \text{(Newton's second law, for mechanical translational system)} \qquad (3\text{-}3)$$

F_{net} *Resultant (net) Force*	m *Mass of Object*	a *Acceleration*
newton (N)	kilogram (kg)	$\frac{\text{metre/second}}{\text{second}}$ (m/s^2)
pound (lb)	slug (slug)	$\frac{\text{foot/second}}{\text{second}}$ (ft/s^2)

The equation $F_{net} = ma$ says that acceleration is proportional to net force ($a \propto F_{net}$) and, if rearranged to $a = F_{net}/m$, acceleration is inversely proportional to mass ($a \propto 1/m$). The equation $F_{net} = ma$ can also be solved for m:

$$m = \frac{F_{net}}{a}$$

which serves as a definition for mass. It is useful to think of mass as the ratio of net force to acceleration. Mass is a measure of how much force it takes to produce a certain amount of acceleration for a given object. A graph of net force versus acceleration for an object is a straight line passing through the origin (Fig. 10-1). The slope of this line is F_{net}/a, the object's mass.

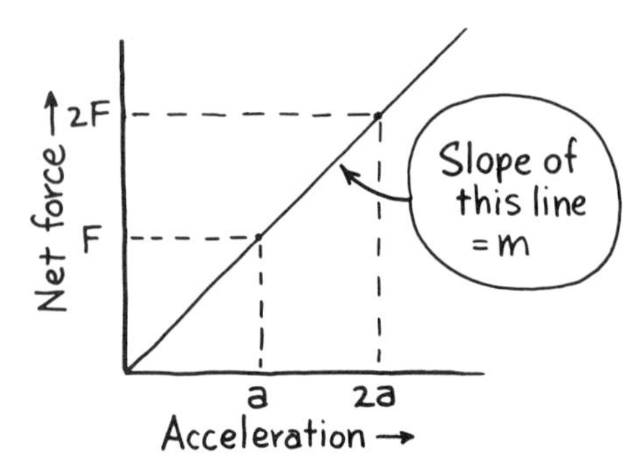

Figure 10-1. A net force acting on an object causes it to accelerate. Doubling the net force doubles the acceleration. Force is proportional to acceleration. The slope of the graph is the object's mass *m*.

For most practical situations the mass of an object is constant. However, the mass of an object increases as its velocity approaches the speed of light.

An alternative and important way to look at mass is as a measure of an object's ability to store kinetic energy. An obect with a large mass needs to have more work done on it to attain a certain velocity. Once this velocity is attained, more energy must be lost before motion ceases. Thus objects with greater mass are harder to start and stop.

Chapter 6 looked at the topic of kinematics, the study of motion. Newton's second law is the main equation of *dynamics*, the study of the relation between force and motion. As with any equation, $F_{net} = ma$ is useful because it allows you to solve for an unknown value. Here are a couple of examples that illustrate its use.

Example 10-1 A 1000-kg car is lifted by a crane that exerts an upward force of 1.50×10^4 N. What is the car's acceleration?

Solution

Whenever such a problem is encountered, a *free-body* diagram should be drawn. The free-body diagram isolates the object in space. All forces acting *on* the object, represented by vector arrows, are drawn on the diagram. The net or resultant force can then be calculated.

The crane pulls upward with a force $P = 1.50 \times 10^4$ N, and the earth pulls downward with a force equal to the car's weight W:

P

$$W = mg = 1000 \text{ kg}(9.81 \text{ m/s}^2) = 0.981 \times 10^4 \text{ N}$$

The net force F_{net} is the vector sum of P and W. Since both forces are vertical, the sum is a simple subtraction:

W

Free-body diagram

W is negative (points downward) — Net force is positive (upward)

$$F_{net} = \vec{P} + \vec{W} = 1.50 \times 10^4 \text{ N} - 0.981 \times 10^4 \text{ N} = 0.52 \times 10^4 \text{ N}$$

Consider P to be positive because it points upward — The acceleration is upward in the same direction as the net force

$$a = F_{net}/m = \frac{0.52 \times 10^4 \text{ N}}{1000 \text{ kg}} = \boxed{5.2 \text{ m/s}^2}$$

Example 10-2 In Example 4-1, an object weighing 4.0 N falls to the ground on a windy day. The horizontal wind force is 3.0 N to the right. What is the object's acceleration when first released?

Solution

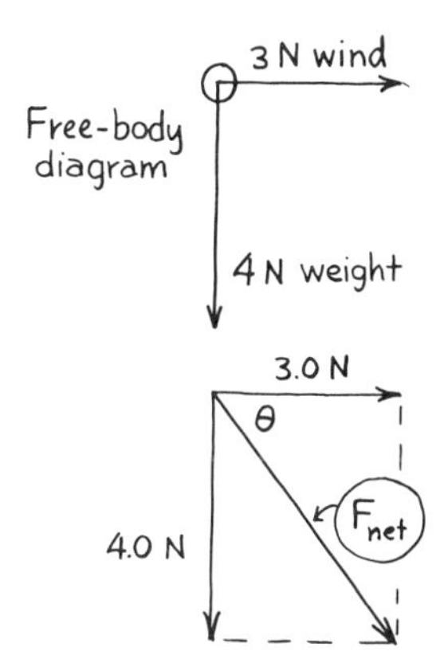

Draw a free-body diagram of the object showing the two forces. Add these forces by constructing a parallelogram and finding the diagonal. The diagonal is the vector sum or resultant force F_{net} (Sec. 2-5).

$$\tan\theta = \frac{4.0}{3.0} = 1.33 \Rightarrow \theta = 53°$$

$$F_{net} = \sqrt{(3.0 \text{ N})^2 + (4.0 \text{ N})^2} = 5.0 \text{ N}$$

The mass

$$m = \frac{W}{g} = \frac{4.0 \text{ N}}{9.81 \text{ m/s}^2} = 0.41 \text{ kg}$$

and the acceleration

$$a = \frac{F_{net}}{m} = \frac{5.0 \text{ N}}{0.41 \text{ kg}} = \boxed{12 \text{ m/s}^2 \ \searrow 53°}$$

Most of our work in finding net force will involve simple vector additions of the type illustrated in the first of these two examples.

10-2 INERTANCE

In unified-concepts terms, mass is the ratio of a force-like quantity to rate of change of parameter rate. We will soon see that quantities analogous to mass can be defined and used in all but heat systems. Since mass is a measure of the inertia of an object, its opposition to acceleration, we will use the term *inertance M* and *inertive force F_M* for net force when writing Newton's second law in its general form.

$$F_M = M\ddot{Q} \qquad \begin{pmatrix}\text{Newton's second law,}\\ \text{unified-concepts form;}\\ \text{not for heat systems}\end{pmatrix} \qquad (10\text{-}1)$$

(Labels: F_M — Inertive force; M — Inertance; $\ddot{Q}$ — Rate of change of parameter rate)

Inertance M is a measure of the tendency of a system to oppose change of parameter rate $\dot{Q}$. Said in another way, inertance is a measure of the force needed to accelerate the system at a given rate. Figure 10-2 shows that it can be represented as the slope of a graph of inertive force F_M versus rate of change of parameter rate $\ddot{Q}$.

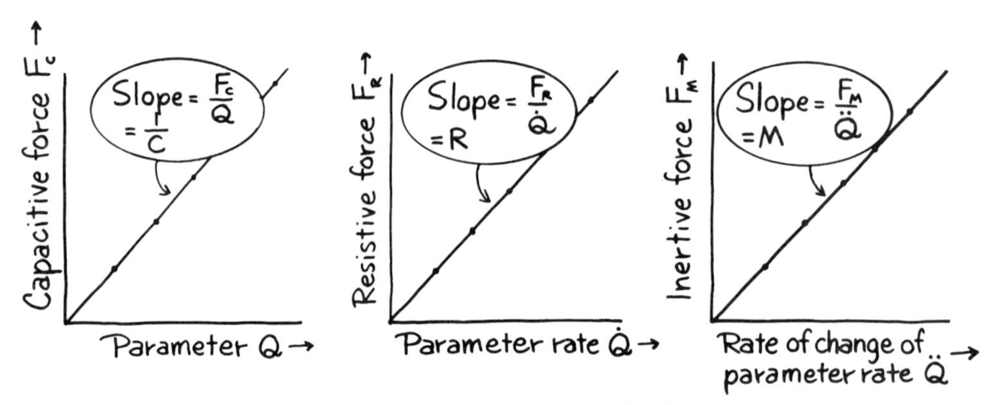

Figure 10-2. Capacitance, resistance, and inertance all describe how much force is needed to do certain things to a system. Together these three values say a lot about how a system behaves.

Figure 10-2 also shows slope relations for capacitance and resistance. Capacitance is the reciprocal of the slope of a graph of capacitive force versus parameter. It is a measure of how much a capacitor's parameter will change in response to a given change of capacitive force. Resistance is the slope of a resistive force versus parameter rate curve. It is a measure of how much force is needed to maintain a given parameter rate (velocity, volume flow rate, current, etc.) in a resistor.

We will find that the three quantities, capacitance, resistance, and inertance, taken together describe a system remarkably well and allow us to predict what is going to happen when a system is disturbed in various ways.

Mechanical Rotational System

Newton's second law adapts easily to rotational systems. Rotational force is torque τ, and acceleration gives way to angular acceleration α. *Moment of inertia I* is the common name for rotational inertance. It is defined by writing Newton's second law in rotational terms:

$$\tau_{\text{net}} = I\alpha \quad \left(\begin{array}{l}\text{Newton's second law,}\\ \text{for mechanical}\\ \text{rotational system}\end{array}\right) \tag{10-2}$$

τ_{net} *Net Torque*	I *Moment of Inertia*	α *Angular Acceleration*
newton · metre (N · m)	kilogram · metre^2 (kg · m^2)	$\dfrac{\text{radian/second}}{\text{second}}$ (rad/s^2)
pound · foot (lb · ft)	slug · foot^2 (slug · ft^2)	

Equation (10-2) can be derived for a simple rotational system where a small concentrated object of mass m accelerates tangentially at a distance r about an axis of rotation. The accelerating force F is

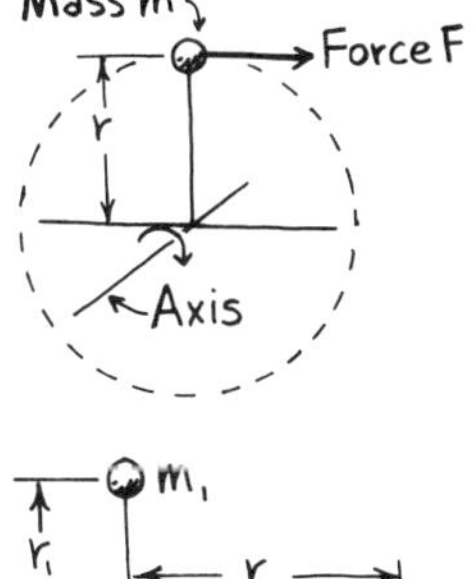

$$F = ma$$

Multiply both sides by the radius r:

$$Fr = mra$$

But acceleration $a = r\alpha$ [Eq. (6-14)] and torque $\tau = Fr$ [Eq. (4-1)]. Substituting,

$$\tau = (mr^2)\alpha$$

The term mr^2 must be the moment of inertia I.

If more than one mass is involved, the moment of inertia becomes the sum of all the mr^2 terms:

$$I = m_1r_1^2 + m_2r_2^2 + \cdots \tag{10-3}$$

The units of moment of inertia are (mass) × (distance)^2. See if you can show that these units are consistent with the units of I as defined in Eq. (10-2), where $I = \tau_{net}/\alpha$. Rotational inertance depends not only on the mass of a rotating object but also on the distribution of the mass relative to the axis of rotation. A large-diameter wheel has more rotational inertance than a small-diameter wheel of the same mass. Figuring out moment of inertia is usually a job involving calculus. Formulas for a number of common shapes are given in Appendix B, Table 6.

Example 10-3 A "kinetic energy wheel" propulsion system is considered for use in trolley buses. The system is based upon a specially designed super flywheel that is accelerated to angular speeds of as much as 20,000 rpm by an electric motor when the trolley is connected to its overhead wire. The energy in the wheel is then available to drive the bus, allowing it to be free of overhead wires much of the time. The wheel is equivalent to a 220-kg disc (solid cyclinder) of 1.0-m diameter.

(a) What is the wheel's moment of inertia?

(b) What constant net torque is needed to bring the wheel up to 20,000 rpm in 25 min?

Solution to (a)

$$\left\{ \begin{array}{l} m = 220 \text{ kg} \\ r = 0.50 \text{ m} \end{array} \right\}$$

Appendix B, Table 6

$$I = \tfrac{1}{2}mr^2 = \tfrac{1}{2}(220 \text{ kg})(0.50 \text{ m})^2 = \boxed{28 \text{ kg} \cdot \text{m}^2}$$

Solution to (b)

$$\left\{ \begin{array}{l} t = 25 \text{ min} \left(\dfrac{60 \text{ s}}{1 \text{ min}} \right) \\ \quad = 1.5 \times 10^3 \text{ s} \\ \omega_f = 20{,}000 \text{ rpm} \left(\dfrac{0.1047 \text{ rad/s}}{1 \text{ rpm}} \right) \\ \quad = 2.09 \times 10^3 \text{ rad/s} \end{array} \right\}$$

First calculate the angular acceleration:

$$\omega_f = \omega_o + \alpha t, \quad \text{[Eq. (6-1)]}$$

$$\alpha = \frac{\omega_f - \omega_o}{t} = \frac{(2.09 \times 10^3 - 0) \text{ rad/s}}{1.5 \times 10^3 \text{ s}} = 1.4 \frac{\text{rad}}{\text{s}^2}$$

Then calculate the net torque:

$$\tau_{net} = I\alpha = (28 \text{ kg} \cdot \text{m}^2)(1.4 \text{ rad/s}^2) = \boxed{39 \text{ N} \cdot \text{m}}$$

Moment of inertia, defined by Eq. (10-2) as the ratio of torque to angular acceleration, $I = \tau_{net}/\alpha$, becomes the slope of a net torque versus angular acceleration curve (Fig. 10-3). Experimentally, an unknown I can be found by measuring the torque needed to accelerate an object about its rotational axis.

Fluid System

Fluid flowing in a pipe is matter in motion and opposes any attempt to change its flow rate. *Fluid inertance M* is best defined by writing Newton's

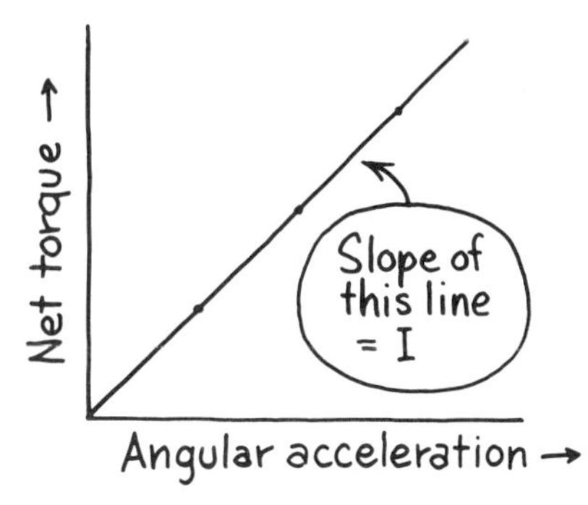

Figure 10-3. Moment of inertia I is the slope of a net torque versus angular acceleration curve.

second law in terms of fluid things:

$$p_{\text{net}} = M\ddot{V} \quad \text{(Newton's second law, for fluid system)} \tag{10-4}$$

p_{net} *Net Pressure*	M *Fluid Inertance*	$\ddot{V}$ *Rate of Change of Volume Flow Rate*
pascal (Pa)	$\frac{\text{pascal}}{\text{cubic metre/second}^2}$ $\left(\frac{\text{Pa}}{\text{m}^3/\text{s}^2}\right)$	cubic metre/second^2 (m^3/s^2)
pound/square foot (lb/ft^2)	$\frac{\text{pound/square foot}}{\text{cubic foot/second}^2}$ $\left(\frac{\text{lb/ft}^2}{\text{ft}^3/\text{s}^2}\right)$	cubic foot/second^2 (ft^3/s^2)

Note: A pascal is a newton/square metre.

Equation (10-4) can be derived for a liquid of density ρ accelerated in a pipe of length ℓ and cross-sectional area A (Fig 10-4). Look at the liquid

Figure 10-4. Liquid in a pipe can be analyzed like a mechanical translational system. The net pressure responsible for accelerating the liquid is $p_{\text{net}} = p_1 - p_2$.

contained in the pipe as though it were an object in a mechanical translational system. The mass m of this liquid is $m = \rho V$, and its volume V is $V = A\ell$. Hence

$$m = \rho A \ell$$

If friction forces are negligible, the net force

$$F_{net} = F_1 - F_2 = p_1 A - p_2 A = (p_1 - p_2)A = p_{net} A$$

Hence

$$F_{net} = p_{net} A$$

The liquid acceleration $a = \Delta v/\Delta t$, where v is the liquid velocity. But since volume flow rate $\dot{V} = Av$ [Eq. (6-17)],

$$a = \frac{\Delta(\dot{V}/A)}{\Delta t} = \frac{1}{A}\frac{\Delta \dot{V}}{\Delta t} = \frac{1}{A}\ddot{V}$$

Combining all this into $F_{net} = ma$,

$$p_{net} A = \rho A \ell \left(\frac{\ddot{V}}{A}\right)$$

$$p_{net} = \left(\frac{\rho \ell}{A}\right) \ddot{V}$$

Since $p_{net} = M\ddot{V}$ [Eq. (10-4)], the quantity $\rho\ell/A$ must be liquid inertance:

$$M = \rho \frac{\ell}{A} \tag{10-5}$$

M *Liquid Inertance*	ρ *Liquid Density*	ℓ *Pipe Length*	A *Cross-sectional Area*
$\frac{\text{kilogram}}{\text{metre}^4}$ (kg/m^4)	$\frac{\text{kilogram}}{\text{cubic metre}}$ (kg/m^3)	metre (m)	square metre (m^2)
slug/foot^4 (slug/ft^4)	slug/cubic foot (slug/ft^3)	foot (ft)	square foot (ft^2)

Notice that units of M given here are quite different from those for M in Eq. (10-4). You might try as a quick exercise to prove to yourself that the units are equivalent.

Liquid inertance can be defined as the slope of a graph showing net pressure versus rate of change of volume flow rate (Fig. 10-5). Under normal operating conditions an applied pressure difference is balanced by friction in

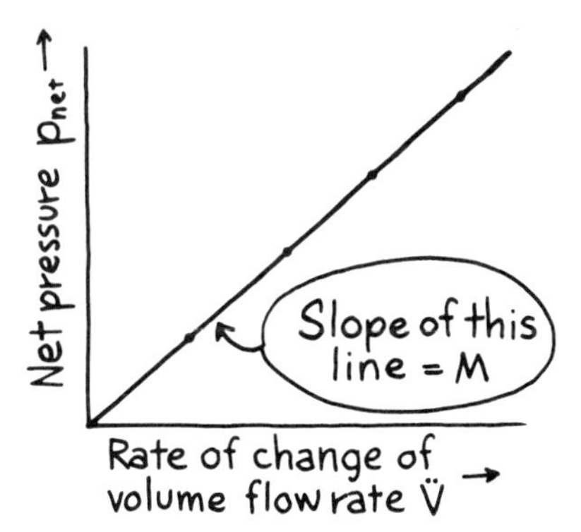

Figure 10-5. In keeping with analogies to the other systems, liquid inertance M is the slope of a p_{net} versus $\ddot{V}$ curve.

the system. In that case the net pressure is zero and the system is in the steady state. Liquid inertance (as with any inertance) comes into play only when some change occurs, such as opening or closing a valve.

Calculation of fluid inertance for gases is more complicated because a gas can be compressed. Equation (10-5) applies to gases if the net pressure responsible for accelerating the gas is small.

Example 10-4 Water from a kitchen sink faucet has to flow through 50 ft of pipe from the main valve. The pipe is $3.0 \times 10^{-3}\ \text{ft}^2$ in cross-sectional area, and the gauge pressure at the main valve is 75 psi. The pipe is horizontal.

(a) Calculate the inertance of the water in that 50-ft length of pipe.

(b) What is the initial value of $\ddot{V}$ when the faucet is suddenly opened?

Solution to (a)

$$\left\{\begin{aligned} \ell &= 50\ \text{ft} \\ A &= 3.0 \times 10^{-3}\ \text{ft}^2 \\ \rho &= \frac{\rho g}{g} = \frac{62.4}{32.2} \\ &= 1.94\ \text{slug/ft}^3 \end{aligned}\right\}$$

$$M = \rho\frac{\ell}{A} = 1.94\frac{\text{slug}}{\text{ft}^3}\left(\frac{50\ \text{ft}}{3.0 \times 10^{-3}\ \text{ft}^2}\right) = \boxed{3.2 \times 10^4\ \text{slug/ft}^4}$$

Solution to (b)

When the faucet is first opened, there is no flow ($\dot{V} = 0$) and so there is no resistive force. The net pressure for accelerating the water is the pressure at the main valve minus the atmospheric pressure at the faucet. This pressure difference is simply the main valve gauge pressure.

$$p_{net} = 75\ \text{psi}\left(\frac{2.12 \times 10^3\ \text{lb/ft}^2}{14.7\ \text{psi}}\right) = 10.8 \times 10^3\ \text{lb/ft}^2$$

Appendix A (arrow to 14.7 psi)

$$p_{net} = M\ddot{V} \Rightarrow \ddot{V} = \frac{p_{net}}{M} = \frac{10.8 \times 10^3\ \text{lb/ft}^2}{3.2 \times 10^4\ \text{slug/ft}^4} = \boxed{0.33\ \frac{\text{ft}^3/\text{s}}{\text{s}}}$$

Electrical System

Whenever an electrical excitation force causes charge to flow in a conductor, the energy supplied to the circuit meets one or more of three fates:

1. It can be lost as heat in a resistor.
2. It can be stored as potential energy in a capacitor.
3. It can be stored as the energy of a magnetic field that appears when charge is in motion (Sec. 8-2).

The above paragraph is a statement of the principle of conservation of energy in electrical terms (see Sec. 9-1). Item 1 is energy lost, and item 2 is stored potential energy; but what is item 3? This energy is not "kinetic" in the sense that it is not contained in moving particles, for the mechanical energy of moving electrons is negligible compared to the energy of the magnetic field. However, the magnetic field exists only as long as charge is in motion. In addition, as we will see (Sec. 10-3), the formula for magnetic field energy is remarkably like that for kinetic energy in the other systems considered in this chapter. Thus, for comparison with other systems, we will designate the energy of a magnetic field about a coil (inductor) as stored kinetic energy.

When the switch is closed in the circuit of Fig. 10-6, the excitation force E makes charge flow through the circuit. There is a tendency inherent in the coil to oppose the buildup of current. This is analogous to the effect of the mass of

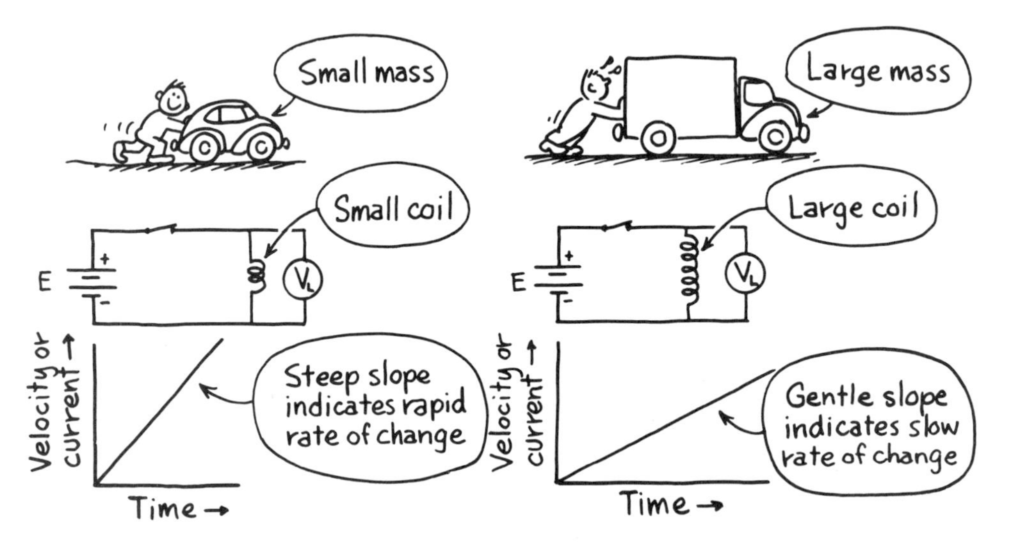

Figure 10-6. Electrical inductance has the effect of slowing down changes in an electrical system, just as mass does for a mechanical system. The greater the inertance, the more the systems oppose change.

the accelerating car, which opposes the increase in velocity. A coil acts like a mass. Mass m opposes change in velocity just like a coil's inductance L opposes change in current (Sec. 8-5). The ratio of net force to velocity change (acceleration) is mass in the mechanical system. The ratio of coil voltage V_L to rate of current change $\dot{I}$ is *inductance* L in the analogous electrical system.

$$V_L = L\dot{I} \quad \text{(Newton's second law, for electrical system)} \tag{10-6}$$

V_L *Coil Voltage*	L *Inductance*	$\dot{I}$ *Rate of Change of Current*
volt (V)	henry (H)	ampere/second (A/s)

Take some time to go back and review Sec. 8-5, which discusses inductance. Equation (10-6), $V_L = L\dot{I}$, is the same as Eq. (8-11), $V_L = L\,\Delta I/\Delta t$, since $\dot{I} = \Delta I/\Delta t$. The unit of inductance, the henry (H), is a volt/(ampere/second). Equation (8-12), $L = \mu N^2 A/\ell$, describes what affects a coil's inductance.

A graph of coil voltage V_L versus rate of current change $\dot{I}$ has the same form as a graph of force versus rate of velocity change (acceleration). Figure 10-7 shows this. The slope of the graphs is mass in the mechanical system and inductance L in the electrical system.

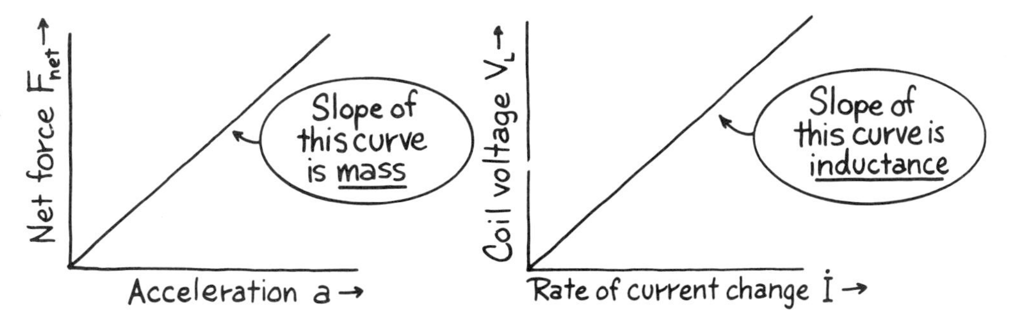

Figure 10-7. Coil inductance is analogous to mass. Both are slopes of force versus rate of change of parameter rate.

The analogy between inductance and mass is remarkably consistent and applies not only when the velocities and currents are increasing, but also when they are decreasing. Such systems inherently oppose change either way.

Example 10-5 A coil of wire that acts as an inductor is connected through a switch to a 24-V battery.

(a) What is the coil's inductance if the rate of change of current $\dot{I}$ is 4.0×10^3 A/s just after the switch is closed?

(b) What is the current 5.0 μs after the switch is closed?

Solution to (a)

$$\left\{\begin{array}{l} V_L = E = 24 \text{ V} \\ \dot{I} = 4.0 \times 10^3 \text{ A/s} \end{array}\right\}$$

$$V_L = L\dot{I} \rightarrow L = \frac{V_L}{\dot{I}} = \frac{24 \text{ V}}{4.0 \times 10^3 \text{ A/s}} = \boxed{6.0 \times 10^{-3} \text{ H}}$$

Solution to (b)

$$\left\{\begin{array}{l} \Delta t = 5.0 \ \mu\text{s} \\ \quad = 5.0 \times 10^{-6} \text{ s} \end{array}\right\}$$

$$\dot{I} = \frac{\Delta I}{\Delta t} \rightarrow \Delta I = \dot{I}\,\Delta t = \left(4.0 \times 10^3 \frac{\text{A}}{\text{s}}\right)(5.0 \times 10^{-6} \text{ s})$$

$$= \boxed{2.0 \times 10^{-2} \text{ A}}$$

Heat System

Thermal inertance apparently does not exist, so heat systems do not fit into this scheme. Do not despair, for thermal systems will make a comeback in Chapter 11. Table 10-2 at the end of the chapter summarizes the important facts about inertance discussed so far in the various systems.

10-3 KINETIC ENERGY STORAGE

In each of the cases we have look at so far, net force (inertive force) did work on the system to change the rate at which something moves or flows. The work done is stored in the system as kinetic energy—the energy of motion. According to our scheme, there are three kinds of forces and three kinds of energy. Resistive force F_R does work to produce energy E_R that is lost through a resistor of some sort. Capacitive force F_C does work that is stored as potential energy E_P. The inertive or net force F_M is responsible for the storage of kinetic energy E_K:

Work done by net (inertive) force = kinetic energy stored in system

Consider the simple system of Fig. 10-8. The net force responsible for accelerating the object, moves a distance s and does work $W = F_{\text{net}} \cdot s$. But according to Newton's second law, $F_{\text{net}} = ma$. Substituting,

$$\text{Work} \nearrow W = F_{\text{net}} \cdot s = ma \cdot s$$

Use Eq. (6-3) for constant acceleration

$$v_f^2 = v_o^2 + 2as \tag{6-3}$$

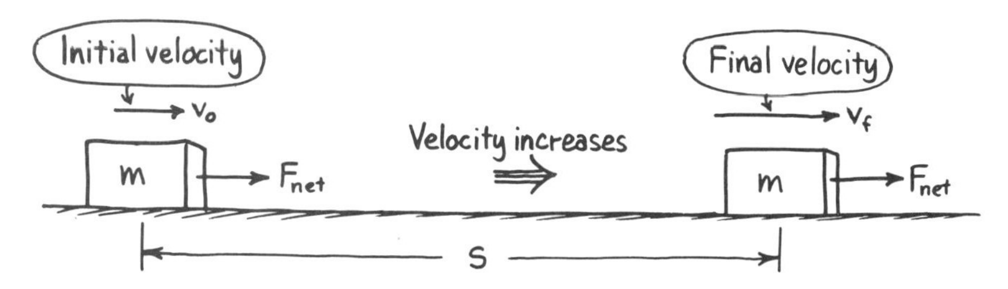

Figure 10-8. The object of mass m is accelerated (velocity increases from v_o to v_f) by net force F_{net} while moving a distance s. F_{net} does work $F_{net} \cdot s$ that is stored as kinetic energy.

and solve for distance s:

$$s = \frac{v_f^2 - v_o^2}{2a}$$

When this is substituted into the work equation, we get

$$W = mas = ma\left(\frac{v_f^2 - v_o^2}{2a}\right) = \frac{1}{2}mv_f^2 - \frac{1}{2}mv_o^2.$$

But the work done equals the change in kinetic energy ΔE_K:

$$\Delta E_K = \tfrac{1}{2}mv_f^2 - \tfrac{1}{2}mv_o^2 \quad \text{(mechanical translational system)} \qquad (10\text{-}7)$$

ΔE_K *Change in Kinetic Energy*	m *Mass*	v_f *Final Velocity*	v_o *Initial Velocity*
joule (J)	kilogram (kg)	metre/second (m/s)	
foot · pound (ft · lb)	slug (slug)	foot/second (ft/s)	

The quantities $\tfrac{1}{2}mv_o^2$ and $\tfrac{1}{2}mv_f^2$ represent the original and final values of kinetic energy. This equation can be stated in words: *The work done by the net force on a body is equal to the change of kinetic energy of the body.* The statement is called the *work–energy theorem* and is a useful thing to know. In many situations the initial velocity v_o is zero, and so the initial value of kinetic energy is zero. In that case the work done by the net force is just $\tfrac{1}{2}mv^2$:

$$E_K = \tfrac{1}{2}mv^2 \qquad (3\text{-}12)$$

Example 10-6 In Example 10-1, a 1000-kg car, starting from rest, was lifted by a 1.5×10^4 N force. Acceleration was calculated to be 5.2 m/s². What is the car's kinetic energy after it has been lifted a distance of 8.0 m?

Solution

$\left\{ a = 5.2\ \text{m/s}^2 \text{ (calculated in Example 10-1)}, \; m = 1000\ \text{kg} \right\}$

This problem can be solved in a couple of ways. The first method uses the car's acceleration to find the velocity and its resulting kinetic energy.

$$v_f^2 = v_o^2 + 2\ as = 0 + 2(5.2\ \text{m/s}^2)(8.0\ \text{m}) = 83.2\ \text{m}^2/\text{s}^2$$

$$E_K = \tfrac{1}{2}mv_f^2 - \tfrac{1}{2}mv_o^2 = \tfrac{1}{2}(1000\ \text{kg})(83.2\ \text{m}^2/\text{s}^2) = \boxed{4.2 \times 10^4\ \text{J}}$$

$\left\{ F_{net} = P - W = 0.52 \times 10^4\ \text{N} \text{ (calculated in Example 10-1)} \right\}$

The second method calculates the work done by the net force, which, according to the work–energy theorem, equals the kinetic energy.

$$E_K = F_{net} \cdot s = (0.52 \times 10^4\ \text{N})8.0\ \text{m} = \boxed{4.2 \times 10^4\ \text{J}}$$

By the magic of unified concepts, you now know about kinetic energy in rotational and fluid systems. The electrical equivalent is the energy stored in the magnetic field about a coil. In general terms, we can say that

$$E_K = \tfrac{1}{2}M\dot{Q}^2 \tag{10-8}$$

where E_K is the kinetic energy of a system with inertance M and parameter rate $\dot{Q}$. The same approach is used in deriving kinetic energy equations for the other systems as was just used in deriving the equation for the mechanical translational system. Results of such derivations, in agreement with unified-concepts Eq. (10-8), are presented in Table 10-1.

The following examples illustrate how these equations can be used to solve kinetic-energy problems in the various systems.

Example 10-7 Example 10-3 involves a "kinetic energy wheel" that spins at 20,000 rpm. Its moment of inertia was found to be 28 kg · m². How much energy is stored in the wheel?

Solution

$\left\{ \omega = 2.09 \times 10^3\ \text{rad/s}, \; I = 28\ \text{kg} \cdot \text{m}^2 \text{ From Example 10-3} \right\}$

$$E_K = \tfrac{1}{2}I\omega^2 = \tfrac{1}{2}(28\ \text{kg} \cdot \text{m}^2)(2.09 \times 10^3\ \text{rad/s})^2$$

$$= \boxed{6.1 \times 10^7\ \text{J}}$$

$\left\{ t = 1.5 \times 10^3\ \text{s}, \; \alpha = 1.4\ \text{rad/s}^2, \; \omega_f = 2.09 \times 10^3\ \text{rad/s}, \; \omega_o = 0 \text{ From Example 10-3} \right\}$

This problem can also be solved using the work–energy theorem. The constant torque that accelerates the wheel is 39 N · m, as calculated in Example 10-3, part (b). The angle through which the wheel rotates in 25 min is

$$\theta = \frac{\omega_f^2 - \omega_o^2}{2\alpha} = \frac{(2.09 \times 10^3\ \text{rad/s})^2 - 0}{2(1.4\ \text{rad/s}^2)}$$

$$= 1.56 \times 10^6\ \text{rad}$$

TABLE 10-1 KINETIC ENERGY EQUATIONS IN THE VARIOUS SYSTEMS

$$E_K = \tfrac{1}{2}M\dot{Q}^2$$

System	E_K Kinetic Energy	M Inertance	$\dot{Q}$ Parameter Rate	Equation
Mechanical translational	(J) (ft · lb)	m Mass (kg) (slug)	v Velocity (m/s) (ft/s)	$E_K = \frac{1}{2}mv^2$
Mechanical rotational	(J) (ft · lb)	I Moment of inertia (kg · m²) (slug · ft²)	ω Angular velocity (rad/s)	$E_K = \frac{1}{2}I\omega^2$
Fluid	(J) (ft · lb)	M Fluid inertance (kg/m⁴) (slug/ft⁴)	$\dot{V}$ Volume flow rate (m³/s) (ft³/s)	$E_K = \frac{1}{2}M\dot{V}^2$
Electrical	(J)	L Inductance (H)	I Electric current (A)	$E_K = \frac{1}{2}LI^2$
Heat	No inertance and no kinetic energy in heat systems			

and the work done is the product of net torque and angle.

$$E_K = \tau_{\text{net}} \cdot \theta = (39\ \text{N} \cdot \text{m})(1.56 \times 10^6\ \text{rad}) = \boxed{6.1 \times 10^7\ \text{J}}$$

Example 10-8 The hydraulic system of Example 10-4 was found to have inertance M of $3.2 \times 10^4\ \text{slug/ft}^4$. How much kinetic energy is in the water when the faucet is on and delivering $0.50\ \text{ft}^3$ of water in 10 s?

Solution

$$\left\{\begin{aligned} M &= 3.2 \times 10^4\ \text{slug/ft}^4 \\ \Delta V &= 0.50\ \text{ft}^3 \\ \Delta t &= 10\ \text{s} \end{aligned}\right\}$$

$$\dot{V} = \frac{\Delta V}{\Delta t} = 0.50\ \text{ft}^3/10\ \text{s} = 5.0 \times 10^{-2}\ \text{ft}^3/\text{s}$$

$$E_K = \tfrac{1}{2}M\dot{V}^2 = \tfrac{1}{2}(3.2 \times 10^4\ \text{slug/ft}^4)(5.0 \times 10^{-2}\ \text{ft}^3/\text{s})^2$$

$$\boxed{= 40\ \text{ft} \cdot \text{lb}}$$

Example 10-9 The coil of Example 10-5 has inductance L of 6.0 mH. How much energy is stored in the coil's magnetic field when the current is 0.020 A?

Solution

$$\begin{Bmatrix} L = 6.0\,\text{mH} = 6.0 \times 10^{-3}\,\text{H} \\ I = 0.020\,\text{A} \end{Bmatrix}$$

$$E_K = \tfrac{1}{2}LI^2 = \tfrac{1}{2}(6.0 \times 10^{-3}\ \text{H})(2.0 \times 10^{-2}\ \text{A})^2$$

$$\boxed{= 1.2 \times 10^{-6}\ \text{J}}$$

10-4 CONSERVATION OF ENERGY

We are now in a position to work some more with the conservation of energy. The law states that whatever work is done on a system must be accounted for. Energy put into a system can be lost in a resistor, stored as potential energy in a capacitor, or as kinetic energy in an inertive element.

$$\begin{pmatrix}\text{work done by}\\ \text{excitation}\\ \text{force } F_E\end{pmatrix} = \begin{pmatrix}\text{work done by}\\ \text{resistive}\\ \text{force } F_R\end{pmatrix} + \begin{pmatrix}\text{work done by}\\ \text{capacitive}\\ \text{force } F_C\end{pmatrix} + \begin{pmatrix}\text{work done by}\\ \text{inertive}\\ \text{force } F_M\end{pmatrix}$$

$$\begin{pmatrix}\text{energy put into}\\ \text{a system}\end{pmatrix} = \begin{pmatrix}\text{energy lost}\\ E_R\end{pmatrix} + \begin{pmatrix}\text{potential energy}\\ \text{stored } E_P\end{pmatrix} + \begin{pmatrix}\text{kinetic energy}\\ \text{stored } E_K\end{pmatrix} \qquad (10\text{-}9)$$

The law, as stated by the two word equations, gives you a way to order your thoughts on the subject. As you know, equations are your friends. They let you solve for something you don't already know and that is quite often worthwhile.

We have been making use of the law in this chapter, perhaps without realizing it. The solution to Example 10-6, for instance, could have been laid out like this:

Energy put into system by crane's force:

$$P \cdot s = (1.5 \times 10^4\ \text{N})(8.0\ \text{m}) = 12 \times 10^4\ \text{J}$$

Energy lost to friction:

$$E_R = 0\ \text{J}$$

Stored gravitational potential energy:

$$E_P = mgh = 1000\ \text{kg}(9.81\ \text{m/s}^2)(8.0\ \text{m})$$

$$= 7.8 \times 10^4\ \text{J}$$

Energy put into system $= E_R + E_P + E_K$:

$$12 \times 10^4\ \text{J} = 0 + 7.8 \times 10^4\ \text{J} + E_K$$

$$E_K = 4.2 \times 10^4\ \text{J}$$

Try your hand at the next example.

Example 10-10 A 50.0-N object is pulled 12.0 m up a 20.0° incline by a constant 25.0-N force parallel to the incline. A constant 3.00-N friction force opposes the motion. Identify the forces F_E, F_R, F_C, and F_M, and calculate the work done by each. Is the conservation of energy law satisfied?

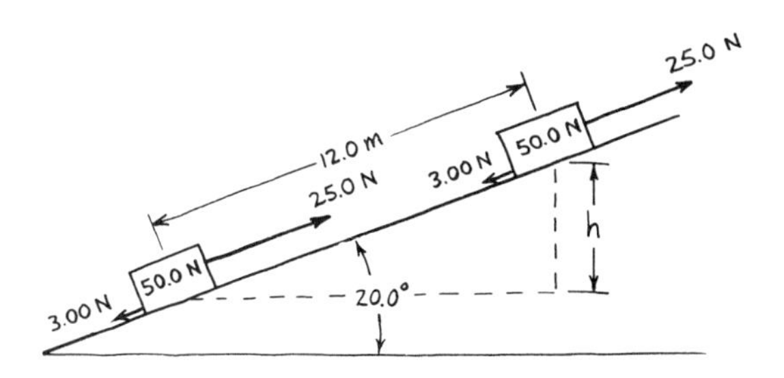

Solution

F_E is the 25.0 N force that pulls the block up the incline. The work done by F_E is

$$W = F_E \cdot s = 25.0\text{ N}(12.0\text{ m}) = \boxed{300\text{ J}}$$

F_R overcomes the friction force of 3.00 N. The work done by F_R, which is lost as heat, is

$$E_R = F_R \cdot s = 3.00\text{ N}(12.0\text{ m}) = \boxed{36.0\text{ J}}$$

F_C lifts the object to a height h:

$$h = 12.0(\sin 20.0°) = 12.0(0.342) = 4.10\text{ m}$$

Its value equals the weight of 50 N. The increase in gravitational potential energy is

$$E_P = Wh = 50.0\text{ N}(4.10\text{ m}) = \boxed{205\text{ J}}$$

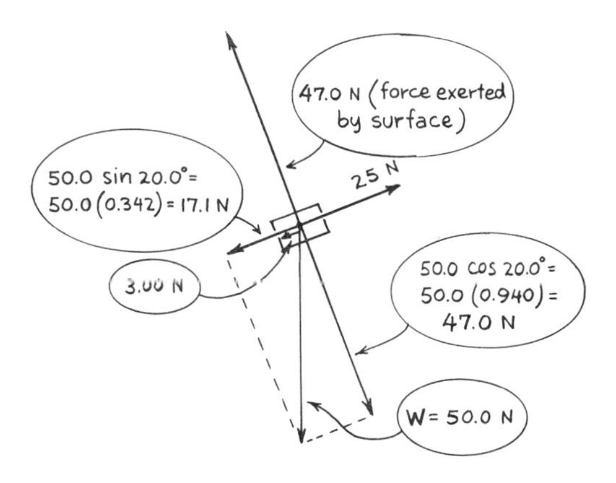

F_M or F_{net} is the resultant of all forces acting on the object and points in the direction of movement. A free-body diagram shows that

$$F_M = 25.0\text{ N} - 3.0\text{ N} - 17.1\text{ N} = 4.9\text{ N}$$

The work done by F_M is kinetic energy

$$E_K = F_M \cdot s = 4.9\text{ J}(12.0\text{ m}) = \boxed{59\text{ J}}$$

The conservation of energy law is satisfied since

$$\begin{pmatrix}\text{work done}\\ \text{by } F_E\end{pmatrix} = \begin{pmatrix}\text{work done}\\ \text{by } F_R\end{pmatrix} + \begin{pmatrix}\text{work done}\\ \text{by } F_C\end{pmatrix} + \begin{pmatrix}\text{work done}\\ \text{by } F_M\end{pmatrix}$$

$$300\text{ J} = 36\text{ J} + 205\text{ J} + 59\text{ J} = 300\text{ J}$$

10-5 NEWTON'S THIRD LAW

So far in this chapter we have applied Newton's first and second laws to various systems. Newton is generally credited with three laws of motion. The third one sounds like it is the easiest to understand, but it can be misleading. The law can be stated as

> For every action there is an equal but opposite reaction.

The hard part of this law is realizing where forces are coming from and where they are going (Fig. 10-9). When object A exerts a force (action force) on

Figure 10-9. A seeming paradox. If there is a reaction force that appears equal and opposite to every action, how can there ever be a net force to cause acceleration?

object B, object B exerts an equal (magnitude) but opposite (direction) reaction force on object A.

Many people have asked the question, "How can you get a net accelerating force if for every action force there is an equal but opposite reaction force?" The car in Fig. 10-9 apparently cannot accelerate because the forces seem to balance! The figure is misleading and Fig. 10-10 does a better job. The action force is exerted *by* the person *on* the car. The reaction force is exerted *by* the car *on* our friend. The reaction force does indeed affect the way he will move, but it doesn't have a thing to do with the car's motion. Acceleration results if the action force is greater than all the other forces (wind, friction, etc.) acting on the car.

The same seeming paradox shows up in other systems and for the same reason. In various places in this book terms like "back pressure" and "back

Figure 10-10. This emphasizes what third law really says. The reaction force does not act on *B* and in fact has no effect on what *B* is going to do.

emf" appear without identification. The term "back pressure" refers to forces exerted *by* a moving fluid *on* the pump that pushes it through the pipe. This is a reaction pressure because it is exerted *by* the fluid and not *on* the fluid.

A "back emf" appears across a coil when the current changes. The back emf is a reaction *by* the coil to the excitation emf of the battery. Battery emf and back emf are equal (but opposite in direction) if the coil is the only element in the circuit. Taking this view of things, only the battery emf affects matters in the coil. The battery emf is the action force and, like the force on the car of Fig. 10-10, it is devoted solely to "accelerating" the system or, in this case, to changing the current.

10-6 SUMMARY

So ends our direct examination of what happens to energy that is put into a system. The important points in this chapter are summarized below.

1. *Newton's second law* of motion defines mass as the ratio of *net force* to acceleration. The ratio is constant for a given object. Mass can be thought of as a measure of an object's *inertia* or of the amount of matter in an object.

2. Newton's second law can be expanded to cover mechanical rotational, fluid, and electrical systems. In each of these the *inertive force* F_M is proportional to acceleration $\ddot{Q}$. $F_M = M\ddot{Q}$, where M is the system *inertance*, analogous to mass. Table 10-2 summarizes inertance for the various systems. Heat systems do not have inertance.

3. The work done by the inertive force equals the change of *kinetic energy* stored in a system. Table 10-1 summarizes the kinetic energy storage equations for the various systems.

4. Any work done on a system by an excitation force must be accounted for as the sum of the energy lost through a resistor, potential energy stored in a capacitor, and kinetic energy stored by an inertive device.

Starting with Chapter 11, we put resistance, capacitance, and inertance together in various combinations to see how they behave.

The table of unified concepts in Appendix A summarizes many of the definitions, formulas, and units associated with R, M, and C systems.

TABLE 10-2 INERTANCE IN THE VARIOUS SYSTEMS

System	*Inertive Force* F_M	*Rate of Change of Parameter Rate* $\ddot{Q}$	*Inertance* M	*Equations* $F_M = M\ddot{Q}$
Mechanical translational	Net force F_{net}	Acceleration $a = \dfrac{\Delta v}{\Delta t}$	Mass m	$F_{net} = ma$
Mechanical rotational	Net torque τ_{net}	Angular acceleration $\alpha = \dfrac{\Delta\omega}{\Delta t}$	Moment of inertia I	$\tau_{net} = I\alpha$ $I = m_1r_1^2 + m_2r_2^2 + \cdots$
Fluid	Net pressure difference p_{net}	Rate of change of volume flow rate $\ddot{V} = \dfrac{\Delta\dot{V}}{\Delta t}$	Fluid inertance M	$p_{net} = M\ddot{V}$ $M = \rho\ell/A$ ← liquids
Electrical	Coil voltage V_L	Rate of change of current $\dot{I} = \dfrac{\Delta I}{\Delta t}$	Inductance L	$V_L = L\dot{I}$ $L = \dfrac{\mu N^2 A}{\ell}$
Heat	No inertance in heat systems			

PROBLEMS

10-1. A 1000-kg car accelerates along a horizontal roadway from 0 to 30 m/s in 10 s.
(a) What net force was acting on the car?
(b) How much kinetic energy does the car have when its velocity is 30 m/s?

10-2. A 10-kg box is lifted vertically to a height of 6.5 m by a constant upward 200-N force. The box initially is at rest.
(a) What is the acceleration of the box as it rises?
(b) What is the velocity of the box when it reaches the 6.5-m height?

10-3. In Problem 10-2, how much of the work done by the 200-N upward force is stored as potential energy and as kinetic energy?

10-4. A 180-lb mouse sits on a scale in an elevator.
(a) What is the scale reading if the elevator moves upward at a constant speed of 15 ft/s?
(b) What is the scale reading if the elevator accelerates upward at a constant 4.5 ft/s^2?

10-5. How much work does it take to get a frictionless flywheel, starting from rest, to rotate at 650 rpm? The wheel is a solid disc of 1.00-m diameter and 80.0-kg mass. (*Hint:* Refer to Appendix B, Table 6, for moment of inertia equations.)

10-6. An airplane propeller is approximated as a uniform rod 8.0 ft long, pivoted at its center. If it weighs 35 lb, how much kinetic energy is stored when it spins at 4000 rpm? (*Hint:* Refer to Appendix B, Table 6, for moment of inertia equations.)

10-7. What is the moment of inertia of the system of masses about the axis of rotation? Neglect any effects of the thin arms and hub.

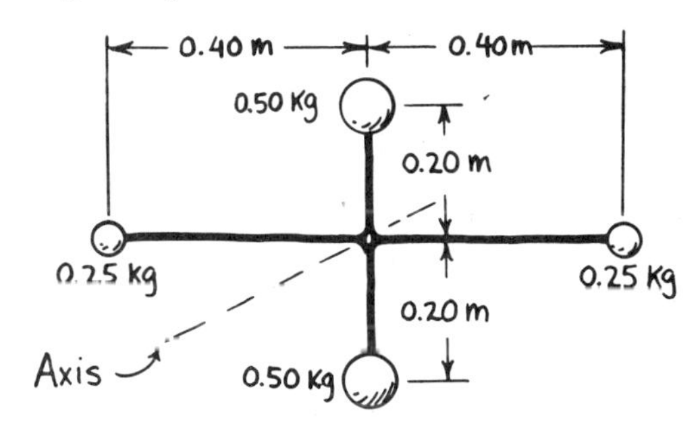

Problem 10-7

10-8. A constant torque of 500 lb · ft is applied to a solid 400-lb flywheel that rotates on frictionless bearings. The wheel's radius is 2.0 ft.

(a) What is the moment of inertia of the flywheel?

(b) What is its angular acceleration?

10-9. Water flows in a fire hose. The cross-sectional area of the hose is $0.050\ \text{ft}^2$ and it is 200 ft long.

(a) What is the mass of water in the hose?

(b) What is the inertance of the water in the hose?

10-10. The water in the fire hose of Problem 10-9 flows at the rate of $720\ \text{ft}^3/\text{min}$.

(a) Calculate the water's kinetic energy using $E_K = \frac{1}{2}mv^2$, where m and v are the mass and velocity of the water.

(b) Calculate the water's kinetic energy using $E_K = \frac{1}{2}M\dot{V}^2$, where M and $\dot{V}$ are the inertance and volume flow rate of the water.

10-11. A 20-m length of garden hose with a cross-sectional area of $3.0 \times 10^{-4}\ \text{m}^2$ delivers water at the rate of $45 \times 10^{-4}\ \text{m}^3/\text{s}$.

(a) What is the mass of water inside the hose?

(b) What is the inertance of the water inside the hose?

10-12. How much kinetic energy is stored in the water of Problem 10-11?

10-13. A 10-V battery is connected through a switch to a 0.020-H coil. What is the initial rate of change of current when the switch is first closed?

10-14. A coil connected to a battery produces a magnetic field with 1.0×10^{-3} J of stored energy when the current is 200 mA. What is the inductance of the coil?

10-15. A 0.25-H coil is connected to a 12-V battery through a switch. Assuming that there is no resistance in the circuit, what is the current 2.0 ms after the switch is closed?

10-16. A 200-Ω resistor and a 0.50-H coil are connected in series to a 10-V battery. The resistor limits the final *steady* current to a value determined by Ohm's law. How much energy is stored in the magnetic field around the coil when the current reaches its final steady value?

10-17. The values given are for distance and time for a 2.0-kg mass accelerated along a frictionless horizontal surface by a constant force.

Distance (m)	0	1.0	4.0	9.0
Time (s)	0	1.0	2.0	3.0

(a) What is the object's acceleration? Is it constant?

(b) What is the constant force?

10-18. A spring gun shoots a 10-g (0.010-kg) projectile at a velocity of 50 m/s. To cock the gun, the spring is compressed 0.20 m. What is the spring constant? (*Hint:* The projectile's kinetic energy must originally be stored as potential energy in the spring.)

10-19. A bicycle wheel and tire, essentially a hollow cylinder, have a diameter of 0.72 m, a mass of 2.0 kg, and spin at 200 rpm.

(a) How much kinetic energy is stored in the wheel?

(b) How much frictional torque will it take to stop the wheel in 10 revolutions?

10-20. A solid flywheel rotates at a constant angular speed of 3600 rpm. Its moment of

inertia is 4.0 slug · ft^2 about its rotational axis.
(a) How much kinetic energy is stored in the flywheel?
(b) How much torque in pound · feet is required to uniformly decelerate the flywheel so that it comes to rest in 30 s?

10-21. When a liquid passes from a larger tube to a smaller one, its velocity must increase if the tubes are to remain full. Now take a look at the energy of the liquid in those tubes. Consider a capillary tube of 1.0-mm inside diameter being fed by (in series with) rubber tubing of 5.0-mm inside diameter. Each tube is 15 cm long. The volume flow rate of water for both tubes is 1.5 cm^3/s.
(a) Calculate the kinetic energy of the water flowing in the capillary tube and the kinetic energy of the water in the rubber tube.
(b) Why is the energy larger in the capillary tube?

10-22. An air duct is 28 m long and has a rectangular cross section of 1.0 m by 0.50 m. A fan moves air at a rate of 6.0 m^3/s. How much kinetic energy is stored in the moving air inside the duct? Assume standard conditions of temperature and pressure. Assume also that the air is only slightly compressed so that the equation for liquid inertance is applicable.

10-23. A 60-V battery is connected through a switch to a 1000-Ω resistor and a 50-mH coil. When the switch is closed, current builds up from zero to its final steady value as determined by the resistance of *R* and Ohm's law.
(a) What is the initial rate at which current changes with time?
(b) What is the final steady current?
(c) When is the coil's back emf greatest?

10-24. An electron gun in a TV tube increases the speed and energy of electrons before they hit the screen. Each electron passes through a 3000-V potential difference.
(a) How much kinetic energy (in joules) does each electron gain?
(b) The mass of an electron is 9.1×10^{-31} kg. Find the speed of an electron after passing through the above potential difference.

10-25. A 0.50-kg block hung vertically from a spring causes the spring to stretch 0.20 m. The block is pulled downward an additional 0.10 m, held momentarily, and released.
(a) What is the net force acting on the block just after release?
(b) What is the block's acceleration just after release? Give both magnitude and direction of the acceleration.

10-26. A spring lies parallel to a frictionless horizontal surface. One end of the spring is fixed and a 0.50-kg block is attached to the other end. The spring stretches 0.10 m when a horizontal force of 15 N is applied.
(a) How much potential energy is stored in the spring in the stretched position?
(b) After release, what is the block's velocity as it passes through the position where the spring is neither stretched nor compressed? (*Hint:* The stored potential energy of the spring is transferred as kinetic energy to the mass.)

10-27. A 2.0-kg object is pulled 3.0 m along a plane inclined at 30°. The excitation force responsible for this is a constant 15 N. There is a constant 2.0-N friction force opposing the motion.
(a) How much work is done by the excitation force?
(b) How much energy is lost to friction?
(c) How much potential energy is stored by raising the block?
(d) What is the block's velocity after passing the 3.0-m mark? (*Hint:* Use the conservation of energy to simplify this calculation.)

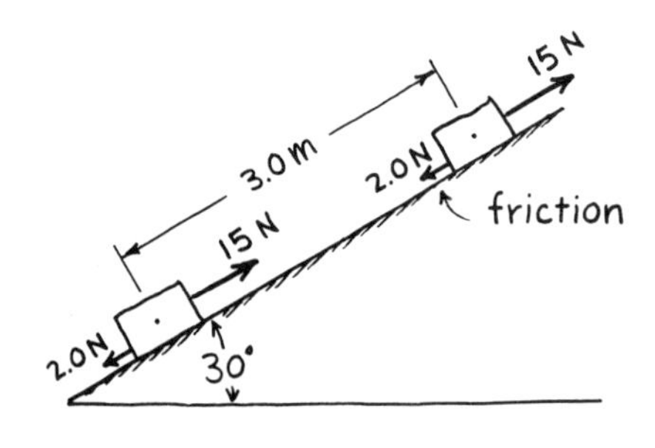

Problem 10-27

ANSWERS TO ODD-NUMBERED PROBLEMS

10-1. (a) 3.0×10^3 N
(b) 4.5×10^5 J

10-3. $E_P = 6.4 \times 10^2$ J
$E_K = 6.6 \times 10^2$ J

10-5. 2.32×10^4 J

10-7. 0.12 kg · m^2

10-9. (a) 19 slug
(b) 7.8×10^3 slug/ft^4

10-11. (a) 6.0 kg
(b) 6.7×10^7 kg/m^4

10-13. 5.0×10^2 A/s

10-15. 96 mA

10-17. (a) 2.0 m/s^2; yes
(b) 4.0 N

10-19. (a) 57 J
(b) 0.90 N · m

10-21. (a) 2.1×10^{-4} J
(b) 8.6×10^{-6} J

10-23. (a) 1.2×10^3 A/s
(b) 0.060 A
(c) When switch is first closed

10-25. (a) 2.4 N
(b) 4.9 m/s^2 upward

10-27. (a) 45 J
(b) 6.0 J
(c) 29 J
(d) 3.1 m/s

11

Energy Transfer and Storage—*RC* Systems

In many practical situations it is useful to know how fast an energy-storage system will give up or accept whatever it is storing. For example, how long does it take to drain the water from a tank? Does a certain thermocouple react to temperature changes quickly enough to keep up with the changing temperature? How fast does the suspension of an automobile return the car to equilibrium after going over a bump?

The answers to such questions are a little more complicated than they may at first seem. Neither parameter rate (velocity, current, etc.) nor rate of change of parameter rate (acceleration, current rate, etc.) is constant. Contrast this with the kinematics of Chapter 6, in which at least acceleration remained constant. But not all is lost, especially if the systems contain linear storage elements and linear resistors. Fairly simple mathematical equations describe changes such as charging a capacitor, stretching a spring, or filling a reservoir through a resistor. The equations apply to all our systems and represent a unifying basis for their study.

Chapter 7 dealt with systems in which energy lost depends on resistance R. Chapters 9 and 10 considered two types of energy-storage elements described by capacitance C and inertance M. Beginning in this chapter and continuing through Chapter 16, we will look at systems that both lose and store energy. First we will study RC systems containing resistance and potential energy storage elements. Chapter 12 studies RM systems, those containing resistance and kinetic energy storage. In Chapter 13, all three elements, resistance, capacitance, and inertance, are put together into a single system with interesting results.

11-1 SYSTEMS WITH CAPACITANCE ONLY

A system with capacitance but no resistance or inertance reacts quickly to a sudden application of force. Any sudden change (*step change*) of force produces instantaneous results (Fig. 11-1) because there is nothing to slow it down.

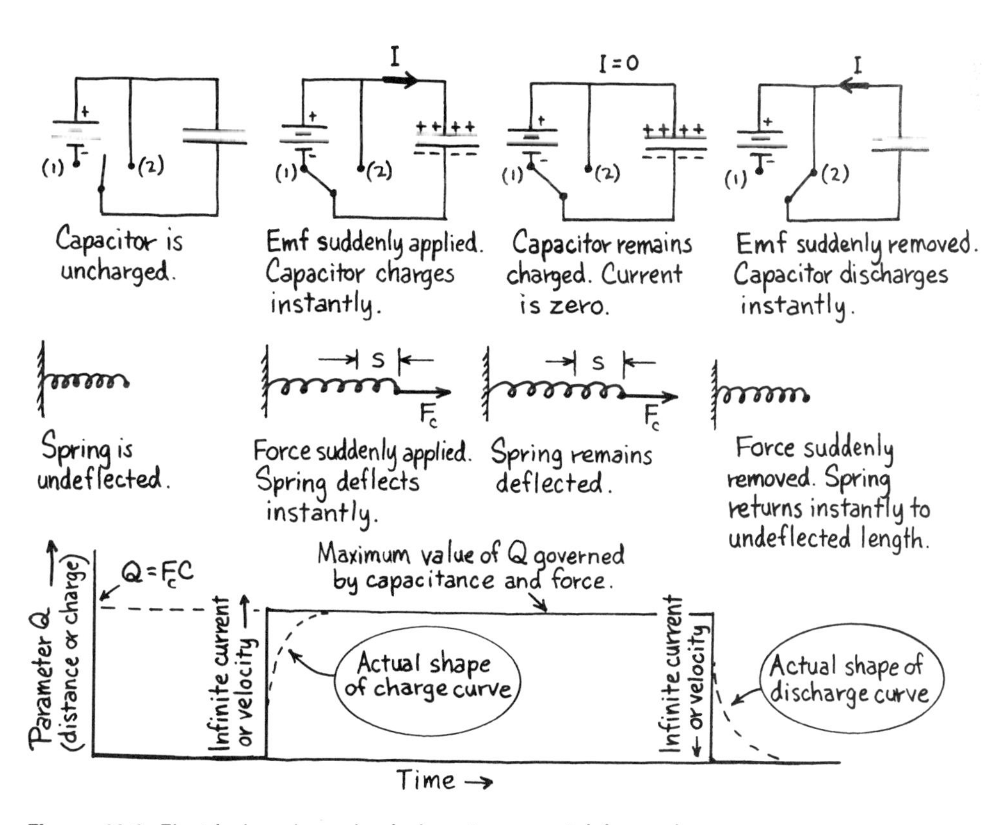

Figure 11-1. Electrical and mechanical systems containing only a capacitor react instantly to step changes of force. The parameter's maximum value (charge, deflection) is determined by the capacitance and the size of the force.

When the switch in the electrical circuit is thrown to position 1, the capacitor charges instantly to its maximum value Q governed by the emf E and the capacitance C ($Q = EC$). The curve goes straight up. The current, represented by the slope of the charge versus time curve, is infinite for a very brief time. Charge remains on the capacitor plates until the switch is thrown to position 2. The capacitor then discharges instantly and the current is again infinite, but in the opposite direction for the smallest imaginable time.

The same thing happens to the mechanical spring. A sudden applied force causes the massless, frictionless spring to deflect instantly a distance s governed by the force F_C and the spring's capacitance C ($s = F_C C$). The deflection remains as long as the force remains, but changes instantly to zero upon removal of the force.

In real systems, some time is needed to charge the capacitor or stretch the spring. Real systems react according to the dashed line in Fig. 11-1. The shape of the dashed line is what this chapter is all about. The shape depends upon system capacitance and resistance.

11-2 RESISTANCE AND CAPACITANCE TOGETHER

In most practical situations, resistance cannot be ignored. Neither can inertance be eliminated from real systems, but in many cases inertance can be small enough to ignore. It is such systems, containing only R and C for all practical purposes, that will occupy our attention for the rest of this chapter.

A resistor connected in series with a capacitor (Fig. 11-2) acts to limit the parameter rate. Current or velocity can only get up to a value given by Ohm's law:

Current or velocity → $\dot{Q}$; Resistive Force → F_R; Resistance → R

$$\dot{Q} = \frac{F_R}{R} \qquad (7\text{-}2)$$

Thus the capacitor's charge or the spring's extension changes at a rate governed by R. At the start when the capacitor has no charge, the entire potential drop is across the resistor ($V_R = E$), and the initial current is $I_o = E/R$. When the spring's extension is still zero, none of the applied excitation force goes toward stretching the spring. All the force is used up in the dashpot resistor ($F_R = F_E$). The initial velocity is $v_o = F_E/R$.

But what then? Just as soon as some charge has accumulated on the capacitor plates or as the spring stretches a bit, the capacitor or spring starts "pushing back." Some of the excitation force now has to be used to keep that charge in the capacitor or to keep the spring stretched. Less of the excitation force is available to maintain flow through the resistor. The result must be a decrease of parameter rate (current or velocity) as represented by the decreasing slope of the graph in Fig. 11-2. The graph must level off to zero slope at a

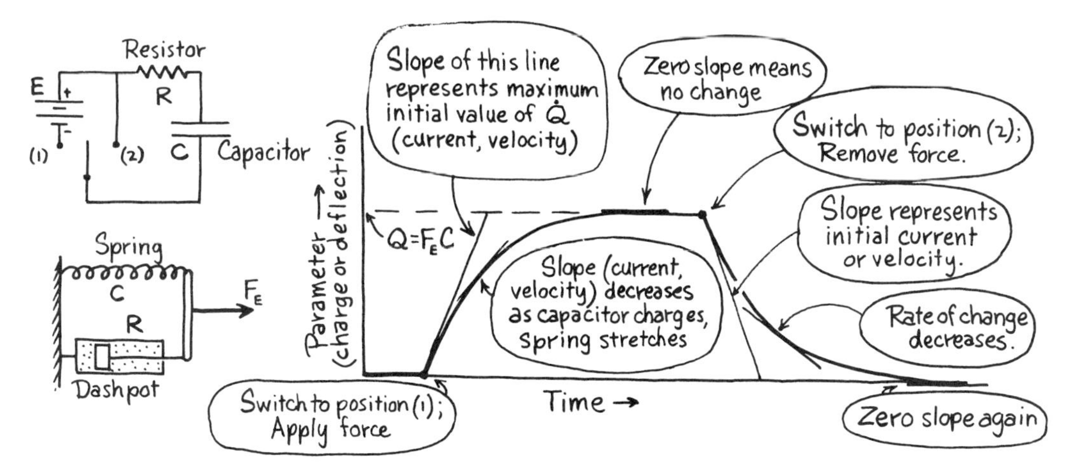

Figure 11-2. In series-connected systems the capacitor and spring increase their opposition to change as parameter (charge or extension) increases. Rate of change decreases as parameter increases. Finally the capacitor's opposition completely balances the applied force and change stops.

steady value of parameter determined by capacitance and size of the applied force, $Q = F_E C$. Zero slope means no more change. The capacitor is fully charged and the spring is fully extended.

Compare the graphs of Figs. 11-1 and 11-2. The addition of resistance to a system has slowed down the system's response to change. By sketching a couple of other graphs (Figs. 11-3 and 11-4), we can make some guesses about what affects the reaction time of an RC system. Since parameter rate is limited

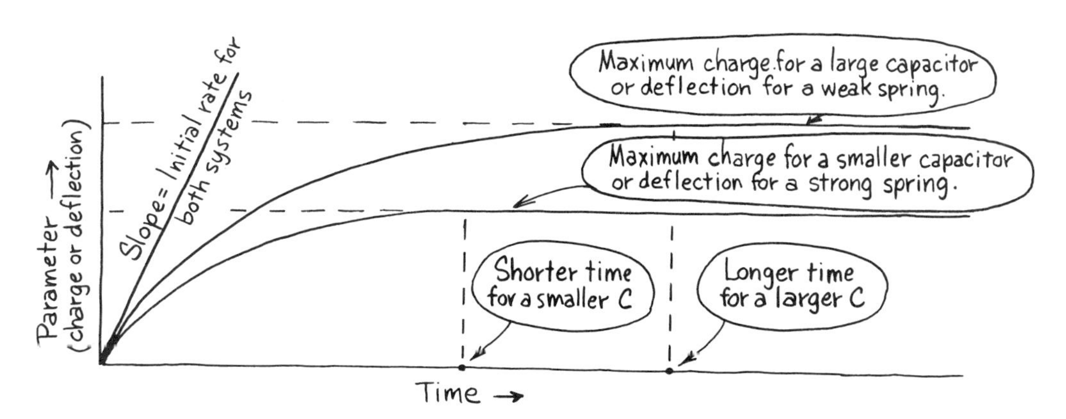

Figure 11-3. Two *RC* systems with the same *R* and different values of *C*. Initial values of parameter rate are the same so it must take longer to charge up a larger capacitor, all else being equal.

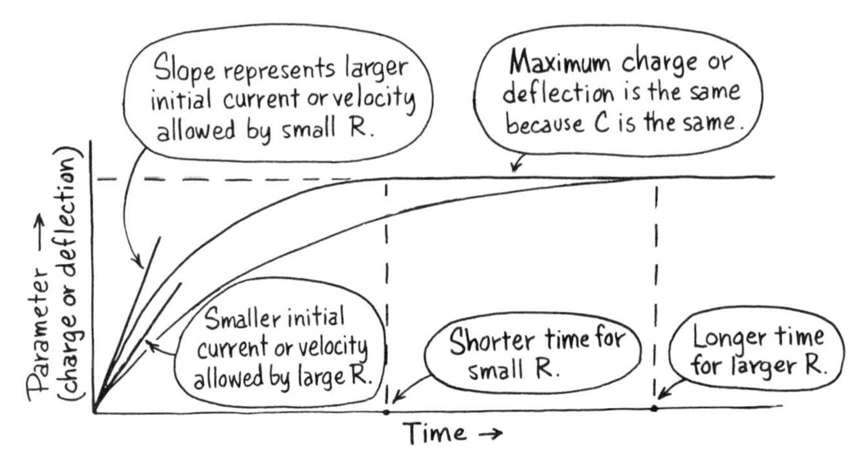

Figure 11-4. Two *RC* systems with the same *C* and different values of *R*. When *R* is large, the same amount of parameter is delivered more slowly. The system with greater *R* takes longer to charge or to stretch, all else being equal.

by the value of *R*, charge cannot initially be delivered any faster to a large capacitor than to a small one (Fig. 11-3). The initial rate at which a stiff spring stretches (small *C*) will be no different than for a weak spring (large *C*) if resistance and applied force are all the same. Figure 11-3 shows that a system with a larger electrical capacitor or with a weaker spring takes longer to come to equilibrium. Reaction time depends on the value of *C*. Think of it in terms of fluids or heat systems. It takes longer to fill a bucket than a coffee cup at the same rate because the bucket holds more water. It takes longer to heat up a large object than a smaller one because the large object holds more heat.

Figure 11-4 compares two *RC* systems with the same *C* but different *R*. A given capacitor will fill up with charge sooner or the spring will stretch more quickly to a given length if lower resistance allows larger current or greater velocity. Again, in terms of fluids, a bucket can be filled faster with a fire hose than with a garden hose. Or, in terms of heat, it is easier to warm up something with a small thermal resistance than something that is well insulated.

It appears then that the response time of an *RC* system is related to both resistance and capacitance. It must be a direct relationship. A longer time is needed to come to equilibrium for greater values of *R* or *C*. Now we will take a closer, more mathematical look at a system to see if our reasoning is correct.

11-3 EXPONENTIAL FUNCTIONS

The following derivations of Eqs. (11-1) and (11-3) are something that you can skim over or swallow whole if it suits you. They are included here to show where the exponential charge and discharge equations, (11-2) and (11-4), come from.

***Derivation of the Exponential Functions**

These derivations apply to any series RC system in which inertance and inertive force are small enough to be neglected. We use a mechanical system because it may be a bit easier to visualize.

When a constant, steady excitation force F_E is applied (Fig. 11-5a), its

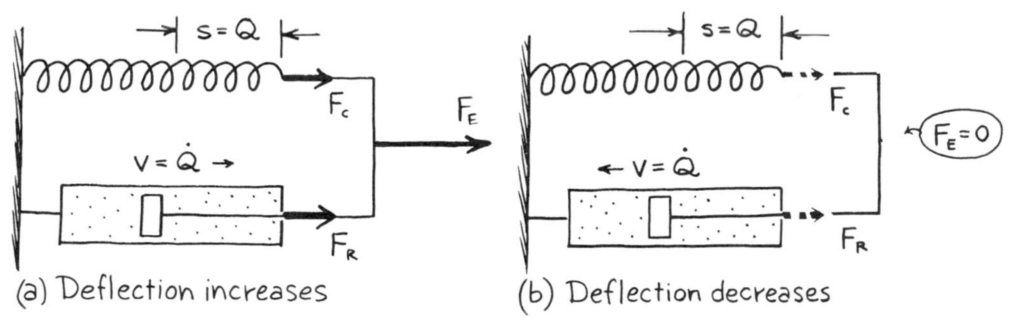

Figure 11-5. (a) A sudden excitation force F_E extends the spring and the dashpot. The final deflection depends upon F_E and the spring capacitance C. (b) The excitation force is suddenly removed ($F_E = 0$). The spring and dashpot contract, coming to rest at the spring's equilibrium position (no deflection).

value at any instant must equal the sum of resistive force F_R on the dashpot and capacitive force F_C on the spring:

$$F_R + F_C = F_E$$

Resistive and capacitive forces can be written in terms of the definitions of R and C:

$$F_R = \dot{Q}R \tag{7-2}$$

$$F_C = \frac{Q}{C} \tag{9-3}$$

Q represents deflection s, and $\dot{Q}$ velocity v. Substituting,

$$\dot{Q}R + \frac{Q}{C} = F_E \tag{11-1}$$

Equation (11-1) is a thing called a *differential equation*, which can be solved using the methods of calculus. Its solution depends upon the final maximum steady spring deflection $s_f = Q_f$.

$$Q = Q_f(1 - e^{-t/RC}) \tag{11-2}$$

Q is the parameter (deflection) at any time t. The letter e represents a number (2.71828 . . .) that keeps cropping up in situations like this.

If the excitation force F_E is suddenly removed, the extended spring force, equal to F_C, causes the system to contract (Fig. 11-5b). Mathematically,

$$F_R + F_C = 0$$

$$\dot{Q}R + \frac{Q}{C} = 0 \qquad (11\text{-}3)$$

Here is another slightly different differential equation, which can be solved using calculus. Its solution depends upon the initial steady deflection $s_o = Q_o$.

$$Q = Q_o e^{-t/RC} \qquad (11\text{-}4)$$

Equations for Charging and Discharging Systems

Equations (11-2) and (11-4) allow us to calculate values of Q at various times if we know R and C and the original or final value of Q. The equations are typical *exponential functions*, and their graphs are exponential curves (Fig. 11-6).

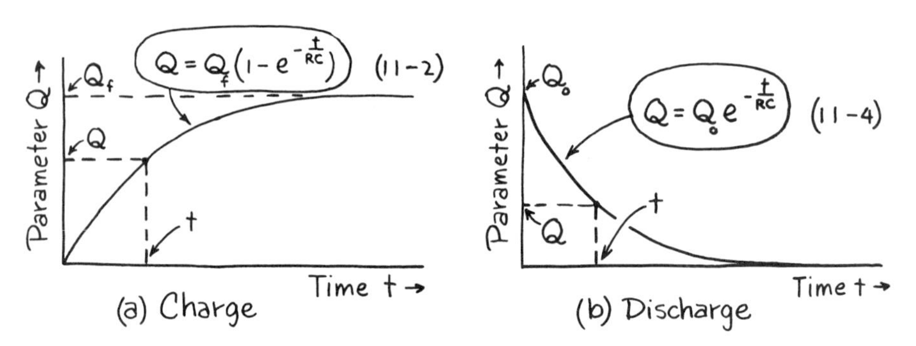

Figure 11-6. Charging and discharging curves for *RC* series systems.

Figure 11-6a describes a charging situation characterized by increasing parameter as the curve flattens out to its steady value of Q_f. A spring stretches, a reservoir fills up, an object gets hotter, or a capacitor charges.

Figure 11-6b describes a discharging situation characterized by decreasing parameter as the curve flattens out to its final steady value of zero. The spring contracts, the reservoir empties, the object cools, or the capacitor discharges.

Table 11-1 presents the charge and discharge exponential equations for the various systems. All *RC* series systems exhibit the same behavior as described by Eqs. (11-2) and (11-4).

TABLE 11-1 CHARGE AND DISCHARGE EXPONENTIAL EQUATIONS FOR *RC* SERIES SYSTEMS

Charging System	*Discharging System*	
$Q = Q_f(1 - e^{-t/RC})$	$Q = Q_o e^{-t/RC}$	Q_o = *original value* Q_f = *final value* $e = 2.71828\ldots$

System	*Equations*	Q *Parameter*	t *Time*	R *Resistance*	C *Capacitance*
Mechanical translational	$s = s_f(1 - e^{-t/RC})$ $s = s_o e^{-t/RC}$	s Deflection (m) (ft)	↕	$R = F_R/v$ $\left(\frac{\text{N}}{\text{m/s}}\right)$ $\left(\frac{\text{lb}}{\text{ft/s}}\right)$	$C = s/F_C$ (m/N) (ft/lb)
Mechanical rotational	$\theta = \theta_f(1 - e^{-t/RC})$ $\theta = \theta_o e^{-t/RC}$	θ Angle (rad)		$R = p_R/\omega$ $\left(\frac{\text{N}\cdot\text{m}}{\text{rad/s}}\right)$ $\left(\frac{\text{lb}\cdot\text{ft}}{\text{rad/s}}\right)$	$C = \theta/p_C$ $\left(\frac{\text{rad}}{\text{N}\cdot\text{m}}\right)$ $\left(\frac{\text{rad}}{\text{lb}\cdot\text{ft}}\right)$
Fluid	$V = V_f(1 - e^{-t/RC})$ $V = V_o e^{-t/RC}$	V Volume (m^3) (ft^3)	(s)	$R = p_R/\dot{V}$ $\left(\frac{\text{Pa}}{\text{m}^3/\text{s}}\right)$ $\left(\frac{\text{lb/ft}^2}{\text{ft}^3/\text{s}}\right)$	$C = V/p_C$ $\left(\frac{\text{m}^3}{\text{Pa}}\right)$ $\left(\frac{\text{ft}^3}{\text{lb/ft}^2}\right)$
Electrical	$Q = Q_f(1 - e^{-t/RC})$ $Q = Q_o e^{-t/RC}$	Q Charge (C)		$R = V_R/I$ (Ω)	$C = Q/V_C$ (F)
Heat	$Q = Q_f(1 - e^{-t/RC})$ $Q = Q_o e^{-t/RC}$	Q Heat energy (cal) (Btu)		$R = \frac{T_2 - T_1}{\dot{Q}}$ $\left(\frac{°\text{C}}{\text{cal/s}}\right)$ $\left(\frac{°\text{F}}{\text{Btu/s}}\right)$	$C = \frac{Q}{T_2 - T_1}$ $\left(\frac{\text{cal}}{°\text{C}}\right)$ $\left(\frac{\text{Btu}}{°\text{F}}\right)$

11-4 *RC* TIME CONSTANT

The product *RC* appears in the charge and discharge equations (11-2) and (11-4). It represents a very important *time* used to express how fast a system reacts to a step change. That amount of time is called the time constant τ

(Greek letter tau). Since we are dealing here with *RC* systems, we will call it the *RC time constant* τ_{RC}.

$$\tau_{RC} = RC \tag{11-5}$$

The unit of τ_{RC} is the *second* if you use the units for R and C in Table 11-1. For example, the units of RC in the mechanical translational system are $[N/(m/s)](m/N) = s$. The units of the product RC in the heat system are $[°C/(cal/s)](cal/°C) = s$.

Let us now do a bit of arithmetic to see how τ_{RC} helps to describe the charge and discharge curves of Fig. 11-6. Take the discharge curve first.

$$Q = Q_o e^{-t/RC} \tag{11-4}$$

When $t = RC = \tau_{RC}$ (i.e., after one time constant),

$$Q = Q_o e^{-RC/RC} = Q_o e^{-1} = Q_o \frac{1}{2.71828} = 0.37Q_o$$

The value of Q is 37 percent of its original value. Or, saying it another way, τ_{RC} is the time it takes for the system to lose 63 percent of its parameter. It doesn't matter what value of parameter you start with. Whatever the value of Q at a given time, it will be only 37 percent of that value (will have lost 63 percent) τ_{RC} seconds later. If you begin with 100 units of parameter (Fig. 11-7), after τ_{RC} seconds, 63 will have gone and 37 will remain. After $2\tau_{RC}$ seconds, 63 percent of those 37 will have gone, and only 37 percent of 37 or about 13.5 units are left. After three time constants, 5 percent is left. After

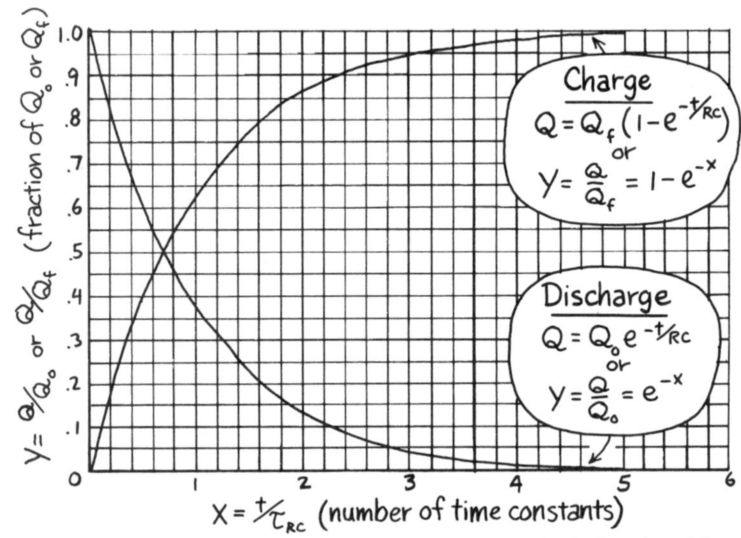

Figure 11-7. Accurately plotted graphs of charging and discharging *RC* series systems. Values of parameter *Q* or time can be calculated using these graphs.

four time constants, 2 percent remains, and after $5\tau_{RC}$ seconds, less than 1 percent of the original amount of parameter remains in the system. Mathematically, it looks as though the parameter will never completely disappear. Practically, however, it is assumed that a system is drained after five time constants.

The same sort of thing happens using the charging equation.

$$Q = Q_f(1 - e^{-t/RC}) \tag{11-2}$$

When $t = RC = \tau_{RC}$ (i.e., after one time constant),

$$Q = Q_f(1 - e^{-RC/RC}) = Q_f(1 - e^{-1}) = 0.63Q_f$$

The value of Q is 63 percent of its final value. After five time constants, Q almost equals the final value Q_f.

It is interesting to note that the definition of time constant is not sacred. It would be alright to define τ as the time needed to decay or rise to some other percentage of the original or final value. Such a time constant is in common use. The time constant for decay of radioactive materials is defined on a basis of 50 percent remaining. This time constant is called the *half-life* of the material and it works perfectly well.

In an *RC* series system, it is more convenient to use the 37 percent mark so that $\tau_{RC} = RC$. With this definition, τ_{RC} can be expressed in another very useful form. Take a look at Fig. 11-8. The *RC* time constant can be written in terms of $\dot{Q}$ and Q, where $\dot{Q}$ is the slope of the curve.

$$\textit{Charge}: \quad \dot{Q} = \frac{Q_f - Q}{\tau_{RC}} \quad \text{or} \quad \tau_{RC} = \frac{Q_f - Q}{\dot{Q}} \tag{11-6}$$

$$\textit{Discharge}: \quad \dot{Q} = \frac{-Q}{\tau_{RC}} \quad \text{or} \quad \tau_{RC} = \frac{-Q}{\dot{Q}} \tag{11-7}$$

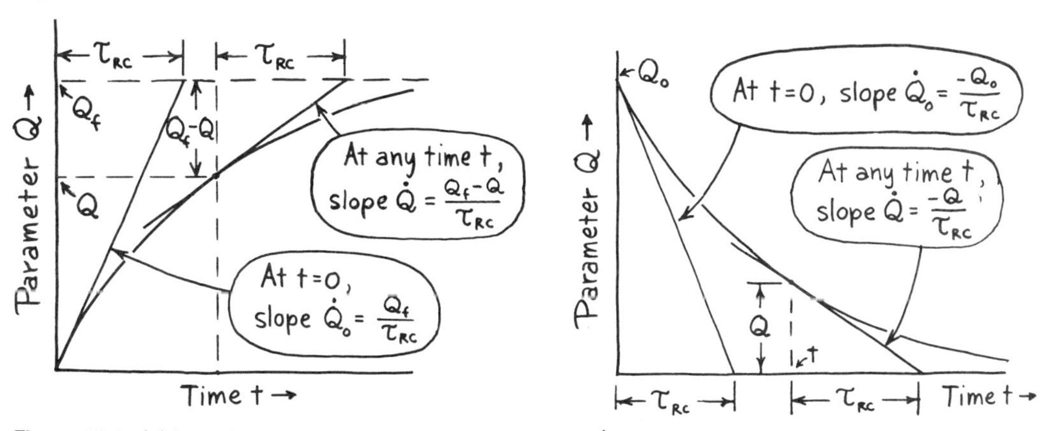

Figure 11-8. (a) In a charging *RC* series system the slope $\dot{Q}$ is the ratio $(Q_f - Q)/\tau_{RC}$. (b) In a discharging system the slope $\dot{Q}$ is the ratio of Q to τ_{RC} and it is negative.

If the slope of the charge curve (Fig. 11-8a) were to suddenly stop changing, it would take the same time τ_{RC} for the parameter to reach the final steady value Q_f, no matter where you are on the curve. If the slope of the discharge graph (Fig. 11-8b) were to stop changing, it would take the same time τ_{RC} for the parameter to go to zero at that rate from any point on the curve. The minus sign in Eq. (11-7) means that Q gets smaller as time goes on.

We can summarize the important facts about time constant and its relation to the charge and discharge curves:

1. Time constant τ_{RC} is the product of R and C:

$$\tau_{RC} = RC \tag{11-5}$$

2. In one time constant of time, a system charges to 63 percent of its final value or discharges to 37 percent of its original value.

3. Practically speaking, systems are completely charged or discharged after five time constants.

4. The parameter rate depends upon the time constant and the amount of parameter remaining in the system.

Let's apply our knowledge about *RC* systems to solving some specific examples.

Electrical System

Series *RC* systems were introduced at the beginning of the chapter using an electrical system. Here is a real application.

Example 11-1 A photographic flash unit is fired by the sudden discharge of a capacitor. The 10-μF capacitor is slowly charged by a 450-V battery through a 300-kΩ resistor.

(a) Calculate the time constant for the charging system.
(b) How long will it take the capacitor to charge completely?
(c) How long does it take to accumulate 50 percent of the maximum charge?

Solution to (a)

$$\left\{ \begin{matrix} C = 10\ \mu\text{F} = 10 \times 10^{-6}\ \text{F} \\ R = 300\ \text{k}\Omega = 300 \times 10^{3}\ \Omega \end{matrix} \right\}$$

$$\tau_{RC} = RC = (3.0 \times 10^5\ \Omega)(10 \times 10^{-6}\ \text{F}) = \boxed{3.0\ \text{s}}$$

Solution to (b)

The charging will be complete for all practical purposes after five time constants.

$$5\tau_{RC} = 5(3.0\ \text{s}) = \boxed{15\ \text{s}}$$

Solution to (c)

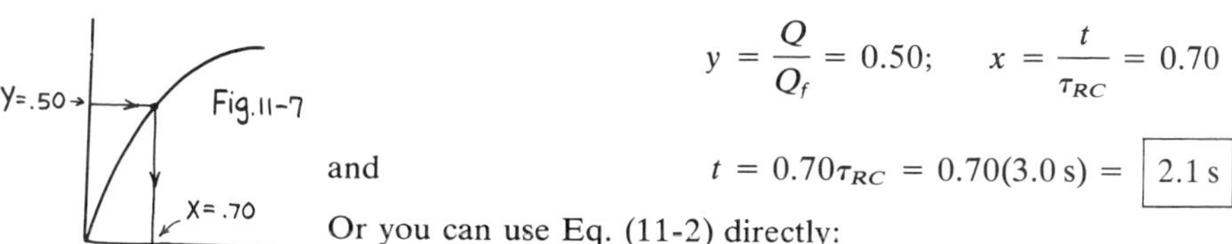

One way to do this is to consult Fig. 11-7. From the graph

$$y = \frac{Q}{Q_f} = 0.50; \qquad x = \frac{t}{\tau_{RC}} = 0.70$$

and

$$t = 0.70\tau_{RC} = 0.70(3.0\text{ s}) = \boxed{2.1\text{ s}}$$

Or you can use Eq. (11-2) directly:

$$\frac{Q}{Q_f} = 1 - e^{-t/RC} \quad \text{or} \quad 0.50 = 1 - e^{-t/3.0}$$

and solve for t.

Mechanical Translational System

The suspension system of a car includes resistors and capacitors in the form of shock absorbers and springs, respectively (see Fig. 11-9 on p. 238). If the wheel goes quickly over a bump, the wheel and axle move while the bulk of the car attached to the frame remains nearly stationary. The mass of the wheel and axle can be ignored to reduce the complexity of the system. A heavy-duty suspension system is one with stiff springs (small C) and stiff shock absorbers (large R). The ride can be made softer by increasing C and decreasing R together to maintain the same response time.

Although certain aspects of automobile suspension can be understood in these terms, the topic is far from simple. Example 11-2 is of a similar but less complicated system.

Example 11-2 A lightweight platform is designed to isolate the vibrations of machinery placed upon it. The mountings include springs and shock absorbers arranged in series as illustrated. A 20-kg piece of machinery is suddenly lifted from the platform. The platform rises 5.0 cm in 1.25 s before effectively coming to rest.

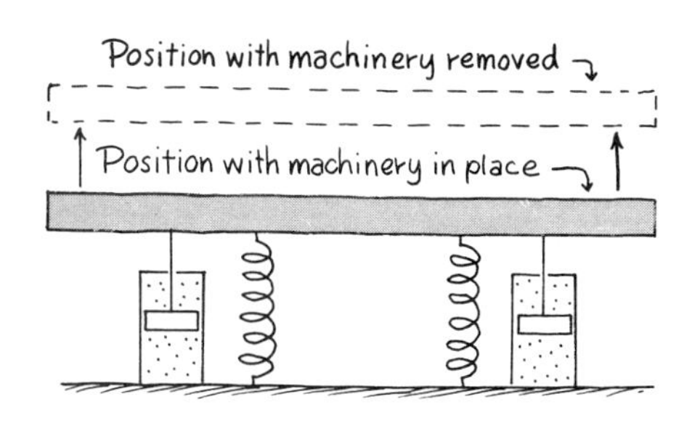

(a) Calculate the time constant of the system.

(b) What is the initial velocity of the platform just as it starts to rise?

(c) How far does the platform move in 0.50 s?

(d) What is the effective resistance of the shock absorbers?

Solution to (a)

The time constant is one fifth of the total rise time.

$$5\tau_{RC} = 1.25\text{ s} \Rightarrow \tau_{RC} = 1.25\text{ s}/5 = \boxed{0.25\text{ s}}$$

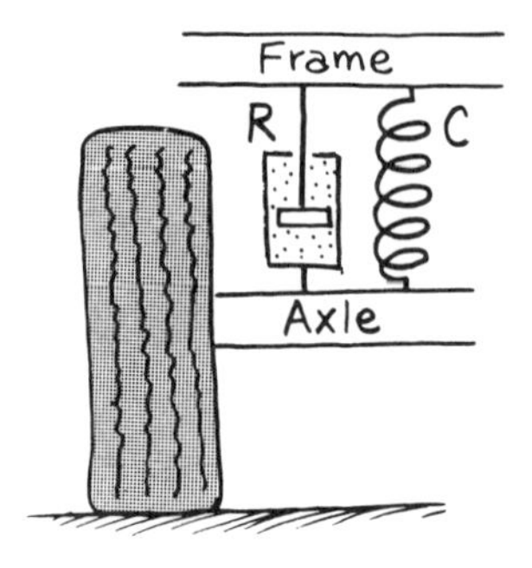

Figure 11-9. An automobile suspension system can be considered an *RC* system. The spring is the capacitor and the shock absorber (dashpot) is the resistor.

Solution to (b)

{$s_o = 5.0\text{ cm} = 5.0 \times 10^{-2}\text{ m}$} This is a discharging system.

$$v_o = \frac{s_o}{\tau_{RC}} = \frac{5.0 \times 10^{-2}\text{ m}}{0.25\text{ s}} = \boxed{0.20\text{ m/s}}$$

Solution to (c)

{$t = 0.50\text{ s}$} Use Fig. 11-7. From the graph,

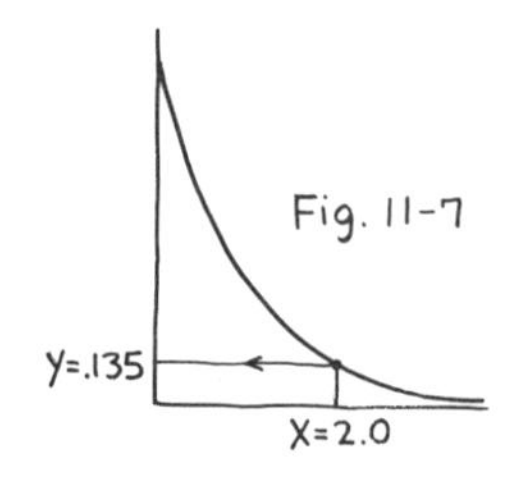

$$x = \frac{t}{\tau_{RC}} = \frac{0.50}{0.25} = 2.0; \qquad y = \frac{s}{s_o} = 0.135$$

$$s = 0.135(5.0\text{ cm}) = \boxed{0.67\text{ cm}}$$

Or, use Eq. (11-4) directly.

$$s = s_o e^{-t/RC} = 5.0e^{-0.50/0.25}$$

$$= 5.0e^{-2} = 5.0(0.135) = 0.67\text{ cm}$$

Solution to (d)

{$m = 2000\text{ kg}$} First calculate C for the springs and then get R from $\tau_{RC} = RC$.

$$F_C = mg = 20\text{ kg}(9.81\text{ m/s}^2) = 1.96 \times 10^2\text{ N}$$

$$C = \frac{s_o}{F_C} = \frac{5.0 \times 10^{-2}\text{ m}}{1.96 \times 10^2\text{ N}} = 2.55 \times 10^{-4}\text{ m/N}$$

$$\tau_{RC} = RC$$

$$R = \frac{\tau_{RC}}{C} = \frac{0.25\text{ s}}{2.55 \times 10^{-4}\text{ m/N}} = \boxed{9.8 \times 10^2 \frac{\text{N}}{\text{m/s}}}$$

Mechanical Rotational System

A mechanical door closer may have all the elements of an *RC* system. A coiled spring stores energy as the door is opened and returns the energy when the door is closed. A viscous damping device controls the rate at which the energy is released. If R and C are constant, we are dealing with a system that we know something about (Fig. 11-10). You may wonder how, if

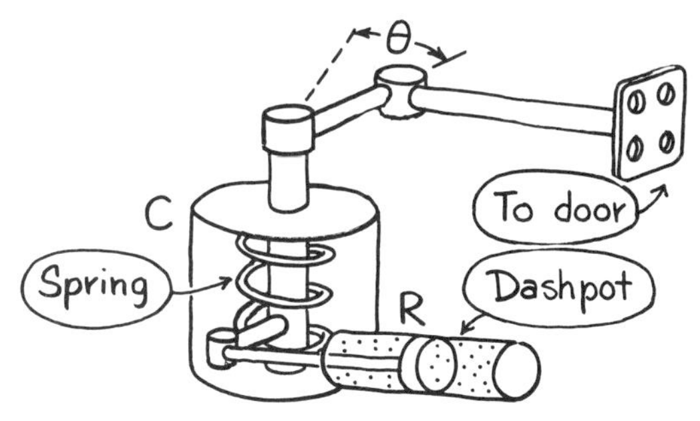

Figure 11-10. A door closer is a mechanical rotational RC system.

torque is proportional to angle θ, the door ever gets closed. Good question! It would theoretically not close completely if the spring were completely unwound when the door is shut. Some energy must be stored in the spring (i.e., it is partially wound even when the door is closed). The rotational capacitor is never allowed to "discharge" completely.

Example 11-3 Suppose for the sake of simplicity that the spring of a door closer is completely unwound when the door is closed. The door opens 90° when a 70-N · m torque is applied. When the door is released, it closes completely in 3.0 s.

(a) Calculate the time constant, the capacitance, and the resistance of the system.
(b) What will the door's angular velocity be just after it is released?
(c) At what angle is the door open 1.0 s after it is released?

Solution to (a)

Notice that τ_{RC} is *time* and τ_C is capacitive *torque*.

$$5\tau_{RC} = 3.0\text{ s}$$
$$\tau_C = 70\text{ N}\cdot\text{m}$$
$$\theta_o = 90^\circ\left(\frac{1\text{ rad}}{57.3^\circ}\right) = 1.57\text{ rad}$$

$$5\tau_{RC} = 3.0\text{ s} \Rightarrow \tau_{RC} = \frac{3.0\text{ s}}{5} = \boxed{0.60\text{ s}}$$

$$C = \frac{\theta_o}{\tau_C} = \frac{1.57\text{ rad}}{70\text{ N}\cdot\text{m}}$$

$$= 2.24\times10^{-2} \rightarrow \boxed{2.2\times10^{-2}\,\frac{\text{rad}}{\text{N}\cdot\text{m}}}$$

$$\tau_{RC} = RC$$

$$R = \frac{\tau_{RC}}{C} = \frac{0.60\text{ s}}{2.24\times10^{-2}\text{ rad/N}\cdot\text{m}}$$

$$= 26.7 \rightarrow \boxed{27\,\frac{\text{N}\cdot\text{m}}{\text{rad/s}}}$$

Solution to (b)

$$\tau_{RC} = \frac{Q}{\dot{Q}} = \frac{\theta}{\omega} \Rightarrow \omega = \frac{\theta}{\tau_{RC}}$$

$$= \frac{1.57 \text{ rad}}{0.60 \text{ s}} = 2.62 \rightarrow \boxed{2.6 \text{ rad/s}}$$

Solution to (c)

$$\theta = \theta_o e^{-t/\tau_{RC}} = (90°)e^{-1.0/0.60} = 17.00 \rightarrow \boxed{17° \quad \text{or} \quad 0.30 \text{ rad}}$$

Fluid System

The simplest potential energy storage device in a hydraulic system is a straight-sided tank. When the tank is filled from the bottom, it is like charging a capacitor. If the hose through which it is filled has constant resistance, the system will fill and drain according to the exponential time constant curves.

Example 11-4 A tank of 5.0-m depth and 24-m^2 cross-sectional area is drained through a hose at the bottom. The volume flow rate is measured as 0.60 litre/s (1 litre = 1×10^{-3} m^3) when the water depth is down to 1.0 m. Assume that the hose resistance remains constant.

(a) Find the system *RC* time constant.
(b) How long does it take to drain the tank?
(c) What is the hose resistance?

Solution to (a)

$$\dot{V} = 0.60 \frac{\text{litre}}{\text{s}}\left(\frac{1 \times 10^{-3} \text{ m}^3}{1 \text{ litre}}\right)$$

$$V = Ah = 24 \text{ m}^2(1.0 \text{ m}) = 24 \text{ m}^3$$

This again is a discharging system. Use $\dot{Q} = Q/\tau_{RC}$ [Eq. (11-7)] to calculate the time constant.

$$\tau_{RC} = \frac{Q}{\dot{Q}} = \frac{24 \text{ m}^3}{6.0 \times 10^{-4} \text{ m}^3/\text{s}} = \boxed{4.0 \times 10^4 \text{ s}}$$

Solution to (b)

After five time constants, the tank has less than 1 percent of its original contents.

$$5\tau_{RC} = 5(4.0 \times 10^4 \text{ s}) = \boxed{20 \times 10^4 \text{ s}}$$

About 55 hr

Solution to (c)

$\left\{\begin{array}{l} A = 24\text{ m}^2 \\ h = 5.0\text{ m} \end{array}\right\}$

First calculate the tank's capacitance, and then use $\tau_{RC} = RC$ to calculate the hose resistance.

$$C = \frac{V}{p_C} = \frac{Ah}{h\rho g} = \frac{A}{\rho g} = \frac{24\text{ m}^2}{(1000\text{ kg/m}^3)(9.81\text{ m/s}^2)} = 2.4 \times 10^{-3}\,\frac{\text{m}^3}{\text{Pa}}$$

Appendix B, Table 1

$$R = \frac{\tau_{RC}}{C} = \frac{4.0 \times 10^4\text{ s}}{2.4 \times 10^{-3}\text{ m}^3/\text{Pa}} = \boxed{1.6 \times 10^7\,\frac{\text{Pa}}{\text{m}^3/\text{s}}}$$

Heat System

Response times for temperature-sensing devices such as thermocouples are commonly given in terms of an *RC* time constant. Other thermal systems, including heating and cooling of buildings, can be described in this way, too. Calculations of *R* and *C* become very complex, and theoretical estimates of time constants often don't agree well with experimental values. Thermal time constants are more likely to be determined experimentally than computed from theory.

Example 11-5 An ice chest contains 10 kg of melted ice (water) at 0°C. It is placed in a room where the temperature is 20°C. The chest has a thermal resistance of 4.2°C/(cal/s).

(a) Calculate the thermal time constant of the system.

(b) How long does it take for the water inside to come to the outside room temperature?

(c) How much heat in calories enters the chest in 15 hr?

Solution to (a)

$\left\{\begin{array}{l} R = 4.2°\text{C}/(\text{cal/s}) \\ m = 10\text{ kg}\left(\frac{10^3\text{ g}}{1\text{ kg}}\right) \\ \quad = 10 \times 10^3\text{ g} \end{array}\right\}$

First calculate the thermal capacitance of the water and then use the time-constant equation.

App. B, Table 3

$$C = mc = (10 \times 10^3\text{ g})\left(1.0\,\frac{\text{cal}}{\text{g}\cdot°\text{C}}\right) = 1.0 \times 10^4\,\frac{\text{cal}}{°\text{C}}$$

$$\tau_{RC} = RC = \left(4.2\,\frac{°\text{C}}{\text{cal/s}}\right)\left(1.0 \times 10^4\,\frac{\text{cal}}{°\text{C}}\right) = \boxed{4.2 \times 10^4\text{ s}}$$

Solution to (b)

It takes about five time constants for the water to warm up to room temperature.

$$5\tau_{RC} = 5(4.2 \times 10^4\text{ s}) = \boxed{2.1 \times 10^5\text{ s}}$$

← About 58 hr

Solution to (c)

$$\left\{\begin{aligned} T_f &= 20°\text{C} \\ T_o &= 0°\text{C} \\ t &= 15\text{ hr}\left(\frac{3600\text{ s}}{1\text{ hr}}\right) \\ &= 5.4 \times 10^4\text{ s} \end{aligned}\right\}$$

This is a charging system. Heat is being added to the chest. First calculate Q_f, the total heat needed to raise the water temperature to 20°C. Then use Fig. 11-7 to get the heat Q added after 15 hr.

$$Q_f = mc(T_f - T_o) = (10 \times 10^3\text{ g})\left(1.0\frac{\text{cal}}{\text{g}\cdot°\text{C}}\right)(20 - 0)°\text{C} = 2.0 \times 10^5\text{ cal}$$

$$x = \frac{t}{\tau_{RC}} = \frac{5.4 \times 10^4\text{ s}}{4.2 \times 10^4\text{ s}} = 1.29$$

$$y = 0.73 = \frac{Q}{Q_f}$$

$$Q = 0.73Q_f = 0.73(2.0 \times 10^5\text{ cal})$$

$$Q_f = \boxed{1.5 \times 10^5\text{ cal}}$$

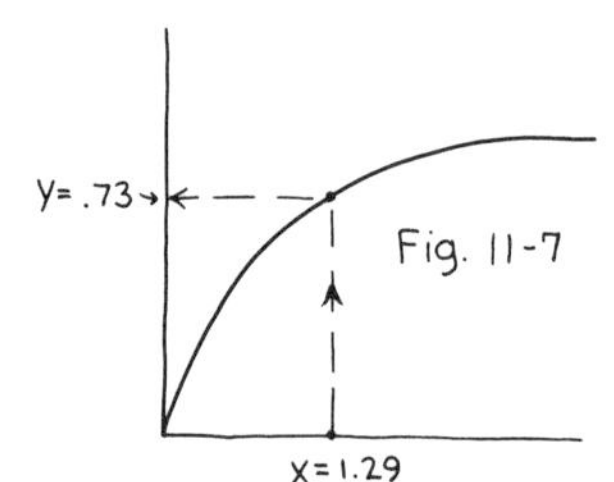

Or use Eq. (11-2):

$$Q = Q_f(1 - e^{-t/RC}) = (2.0 \times 10^5\text{ cal})(1 - e^{-1.29}) = (2.0 \times 10^5)(1 - 0.275)$$
$$= 1.4 \times 10^5\text{ cal}$$

11-5 SUMMARY

This chapter has given you ammunition to attack a whole new class of problems that are common in engineering practice. If a system contains constant resistance and capacitance but negligible inertance, its response to sudden changes is now quite predictable. Some of the more important points made in this chapter are outlined below.

1. If a system has only capacitance (negligible resistance and inertance), it will change instantaneously to a final steady value in response to a suddenly applied or suddenly removed force.

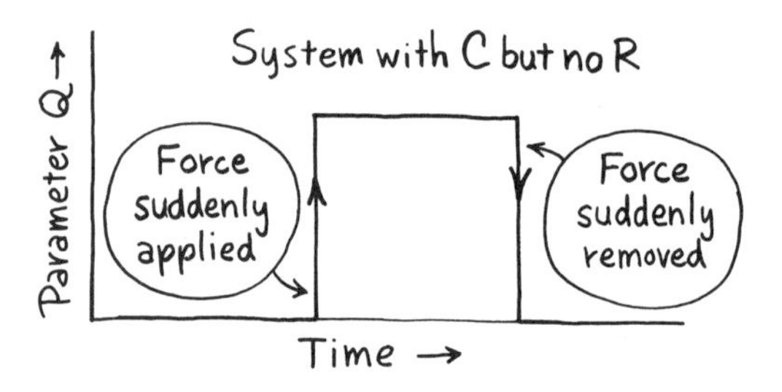

2. If resistance is added, it acts to limit the system's rate of change. The response time for a step application of force increases as both R and C increase.

3. Equations describing the change of parameter with time in an RC series system are exponential functions. The rate of change of Q depends upon the value of Q at any time. As time goes on, the rate of change $\dot{Q}$ approaches zero, and Q theoretically never quite reaches its ultimate goal.

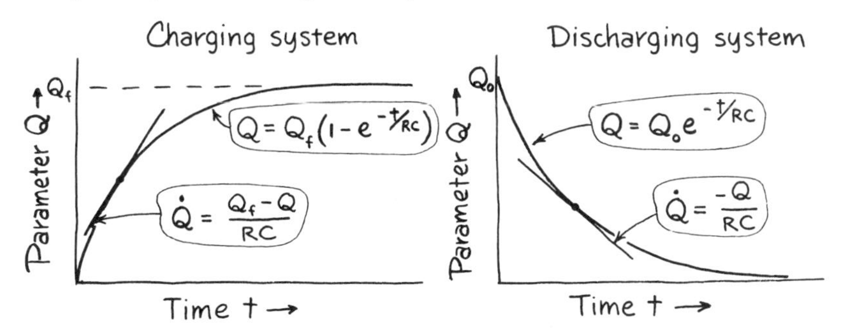

4. If Q of such a system were to suddenly stop changing, the final value of Q would be reached in a time called the system's time constant. In an RC series system the *RC time constant* τ_{RC} is the product of R and C:

$$\tau_{RC} = RC$$

After one time constant, Q will have lost or gained 63 percent of its original or final value. After five time constants, the system is essentially fully charged or completely discharged.

5. Table 11-1 on page 233 shows how the exponential equations apply to the various systems.

If you have a grasp of the material presented in this chapter, Chapter 12 should be easy. We will be doing the same sort of thing, but on systems that contain resistance and inertance—*RM* series systems.

PROBLEMS

11-1. What is the time constant of the mechanical system illustrated?

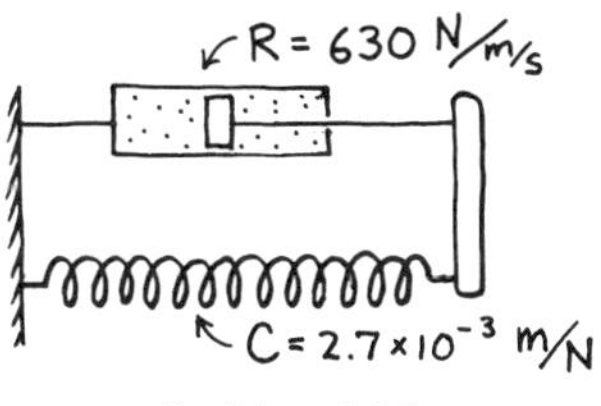

Problem 11-1

11-2. The suddenly applied force P produces an initial velocity of 2.3 m/s and ultimately stretches the spring 0.30 m from its equilibrium position. How long does it take the spring to fully stretch?

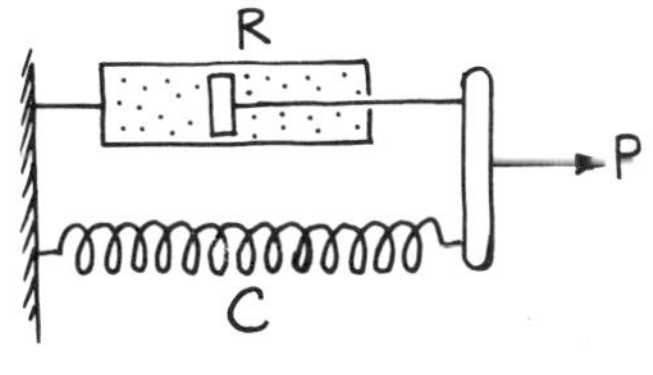

Problem 11-2

11-3. A torque of 12 lb · ft winds a door closer. The door closer is similar to the one described in

Sec. 11-4 (Fig. 11-10). When released, the system unwinds completely in 15 s. What is the constant resistance of the rotary viscous damper if the door rotates through a total angle of 90°? Ignore the door's moment of inertia.

11-4. What is the initial angular velocity of the system in Problem 11-3?

11-5. Water drains through a pipe at the base of a straight-sided tank of cross-sectional area $2.0\ m^2$. What is the time constant of the water tank if the volume flow rate is $1.0 \times 10^{-4}\ m^3/s$ when the depth equals 0.54m?

11-6. A water tank has a cross-sectional area of $1.0\ ft^2$. Initially, the depth is 2.0 ft and the volume flow rate is $2.5 \times 10^{-2}\ ft^3/s$. Estimate how long it takes for the tank to drain completely. Assume that R and C are constant.

11-7. A 30×10^3-Ω resistor and a 2.0 μF capacitor are connected in series through a switch to a 12.0-V battery. About how long does it take for the capacitor to fully charge once the switch is closed?

11-8. What is the time constant of the system as it discharges after the switch is thrown to position 2?

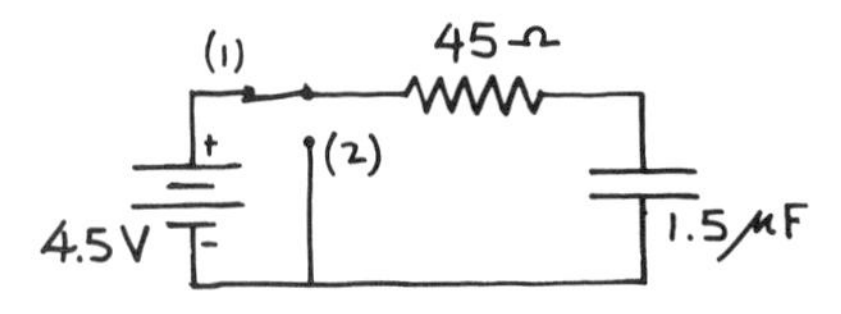

Problem 11-8

11-9. An ice box with thermal resistance of 3.5°C/(cal/s) contains 5.0 kg of water. What is the time constant of this *RC* system?

11-10. The time constant of a very small temperature-sensing device as specified by the manufacturer is 5.0 s. It is used to measure the temperature of a liquid at 80°C. How long does it take before a steady temperature reading can be expected if the initial temperature of the device is 20°C prior to immersion in the liquid?

11-11. A system like the spring–dashpot system of Problem 11-1 has a time constant of 0.45 s. If it is stretched to 0.15 m and released, how far will it be from its final equilibrium position after 0.25 s?

11-12. A mechanical system consists of a linear shock absorber and a spring. A 40-N force applied to the shock absorber by itself causes it to expand at a constant velocity of 0.20 m/s. The same force applied to the spring by itself causes the spring to extend 0.15 m. The shock absorber and spring are then connected in series.

(a) How long does it take for the spring to stop stretching for all practical purposes if a 100-N force is suddenly applied to the system?

(b) After a long period of time, the 100-N force is suddenly removed. How far from its final no-load equilibrium position will the system be 1.0 s after release?

11-13. Suppose that the resistance of the hose in Problem 11-6 is changed so that the system *RC* time constant is 55 s. How much water remains in the tank 110 s after the tank starts to discharge water?

11-14. A water tank, draining from its base, has an *RC* time constant of 105 s. How long does it take to drain the tank if the original depth is 3.0 ft? How long does it take to drain the tank if the original depth is 2.0 ft? Having answered these questions, what does "empty" mean? Will there be the same amount of water in the tank when it is "empty" in each case?

11-15. A 2000-Ω resistor and a capacitor are connected in series through a switch to a 24-V battery. The capacitor charges fully in 50 ms after the switch is closed. What is the value of capacitance C?

11-16. What is the initial charging rate of the capacitor in Problem 11-15? (A charging rate is an electric current.)

11-17. A mass of lava hardens into volcanic rock and cools to the temperature of its surroundings in 5.0 days. At one point it was losing heat at the rate of 400 cal/s. How much heat in calories (relative to the surroundings) did the rock contain at that time? Assume constant R and constant C for this heat system.

11-18. Refer to the temperature-sensing device of Problem 11-10. What is the temperature of the device after it is immersed in the 80°C liquid for 8.0 s?

11-19. A spring with constant $k = 50$ lb/in. is

series connected with a linear shock absorber. The system comes to rest about 2.0 s after the sudden application of a constant force. What is the resistance of the shock absorber? Give your answer in units of lb/(in./s).

11-20. The illustrations show a spring in combination with a shock absorber. They are arranged in two different ways.

(a) Draw electrical circuits equivalent to the two mechanical systems.

(b) Describe what happens to the systems if a sudden constant force is applied. Use the electrical circuits to help reason out your answers.

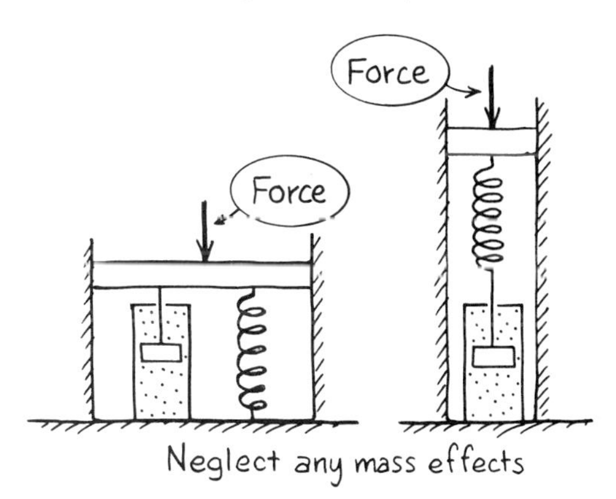

Problem 11-20

11-21. The data show height versus time for the discharge of a burette through a horizontal capillary tube.

Height (cm)	40.0	21.4	11.5	6.1	3.3	1.8	0.9	0.5	0.3
Time (s)	0	5.0	10.0	15.0	20.0	25.0	30.0	35.0	40.0

(a) How long does it take for the volume of water to reach 37 percent of its starting volume? (*Hint*: Plot the data.)

(b) Does the resistance of the capillary tube remain constant?

11-22. The data show gauge pressure versus time for the discharge of a compressor tank through a long thin hose.

Pressure (psi)	100	43	15	4	0
Time (s)	0	10	20	30	40

(a) How long does it take for the gauge pressure to reach 37 percent of its starting pressure? (*Hint*: From a practical viewpoint, the tank is "empty" when the gauge pressure is zero. Plot gauge pressure versus time.)

(b) Is this a linear RC system?

11-23. An electrical circuit has a resistor and capacitor connected in series through a switch to a 12-V battery. The RC time constant of the combination is 1.0 ms. Calculate the capacitor voltage 0.50 ms after the switch is closed.

11-24. An electrical RC series circuit has a time constant of 60 μs. How long after the switch is moved to position 2 does it take for the capacitor to lose half of its original charge?

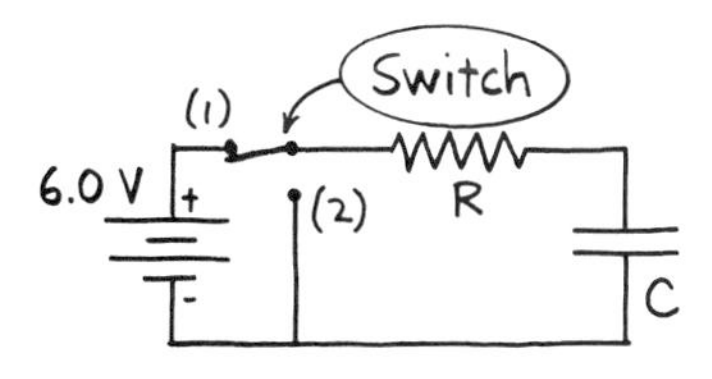

Problem 11-24

11-25. The heating coil of a hair dryer is suddenly turned off but the fan is kept on. A thermocouple measures the temperature of the exit air.

Time (s)	0	5	10	15	20	25
Temperature (°F)	121	116	107	100	95	91
Time (s)	30	40	50	60	70	80
Temperature (°F)	87	83	80	78	77	76

Room temperature is 75°F.

(a) How much time is needed for the temperature to fall 63 percent of the total temperature drop possible? (*Hint*: Plot the data using room temperature as reference.)

(b) Do the data plotted in part (a) form a true exponential curve?

11-26. A 150-lb aluminum engine block cools from 150 to 100°F in 15 min. Assume that the temperature drops exponentially with time. Surrounding air is 70°F.

(a) What is the time constant of the block?

(b) What is its temperature 20 min after cooling starts?

11-27. An object cools from 100 to 70°C in 20 s. Room temperature is 20°C. How long will it take for the object to cool from 70 to 40°C if temperature decreases exponentially with time?

11-28. A small thermocouple cools from 100°C to room temperature in about 35 s. If the thermocouple is heated in a flame to 500°C, what would its temperature be 14 s after the flame is removed? Room temperature is 20°C.

ANSWERS TO ODD-NUMBERED PROBLEMS

11-1. 1.7 s
11-3. 23 lb · ft/(rad/s)
11-5. 1.1×10^4 s
11-7. 30 ms
11-9. 1.8×10^4 s
11-11. 8.6×10^{-2} m
11-13. 0.27 ft^3
11-15. 5.0×10^{-6} F
11-17. 3.5×10^7 cal
11-19. 20 lb/(in./s)
11-21. (a) 8.0 s
(b) Yes
11-23. 4.7 V
11-25. (a) 24 s
(b) Yes
11-27. 39 s

12

Energy Transfer and Storage—*RM* Systems

When a force acts on a real object, the object's mass prevents the velocity from changing instantaneously to a huge value. Even if there is no friction, mass opposes any attempt to change the object's velocity. But there is acceleration (i.e., change in velocity). The acceleration depends upon the net force and the mass (Newton's second law). The methods of Chapter 6 allow us to find the position or velocity of the object at any time when the acceleration is constant.

But what if acceleration is not constant? If the system contains a resistor that provides a changing frictional force, the net force and thus acceleration will also change. If the resistor is linear, we are in luck. It is fairly easy to deal with the kinematics of such a system and its analogs in other systems. That is the subject of this chapter—the behavior of RM series systems.

The format for the study of RM systems will be similar to that used for RC systems in the last chapter. First we look at how parameter rate (velocity, current, etc.) changes with time in a system containing inertance only, when a constant force is suddenly applied or removed. Then resistance is added to form an RM system. Resistance puts a limit on the maximum parameter rate possible in a system. The equations relating parameter rate and time are the same as those developed in Chapter 11 for RC systems except that the time constant expression is different.

With this as background we can study how RM systems behave when a constant force is suddenly applied or removed. The one exception is the heat system. It doesn't follow the RM exponential equations because it has no inertance.

12-1 SYSTEMS WITH INERTANCE ONLY

A system with inertance but no resistance or capacitance reacts in a predictable manner to the sudden application of a constant force. Newton's second law [Eq. (3-3)] says that the parameter rate (velocity, current, etc.) changes at a rate that depends upon the net force F_{net} and the inertance M. $F_{net} = M\ddot{Q} = M(\Delta\dot{Q}/\Delta t)$, where $\dot{Q}$ is the parameter rate and $\ddot{Q}$ is the rate of change of parameter rate. A graph of $\dot{Q}$ versus time (Fig. 12-1) is a straight line passing through the origin. The slope of the graph is $\ddot{Q}$.

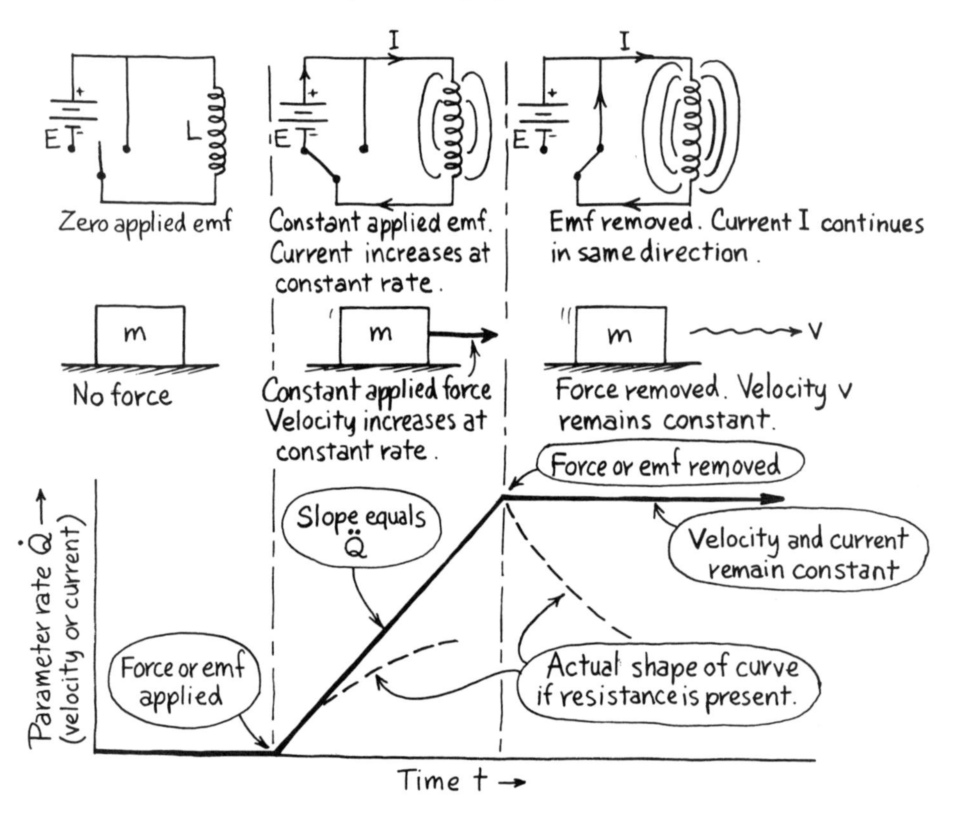

Figure 12-1. An applied force produces a constant rate of change of $\dot{Q}$ in a system with inertance only. $\dot{Q}$ remains constant if the force is removed.

A constant force changes the velocity of the object at a constant rate. Velocity increases uniformly with time. A constant emf changes the current in the coil at a constant rate. The current builds up uniformly with time. Current does not jump instantaneously to infinity, because the changing magnetic field surrounding the turns of the coil opposes change in current. Likewise, velocity does not change quickly to infinity because of the object's inertia. Both velocity and current increase indefinitely until force or emf is removed. Velocity or current must then remain constant. The presence of resistance, however, changes system behavior as indicated by the dashed lines in Fig. 12-1.

12-2 RESISTANCE AND INERTANCE TOGETHER

Now let us introduce some resistance into the systems. Figure 12-2 shows a mass–dashpot mechanical system and a coil–resistor electrical system. Both are series-connected systems.

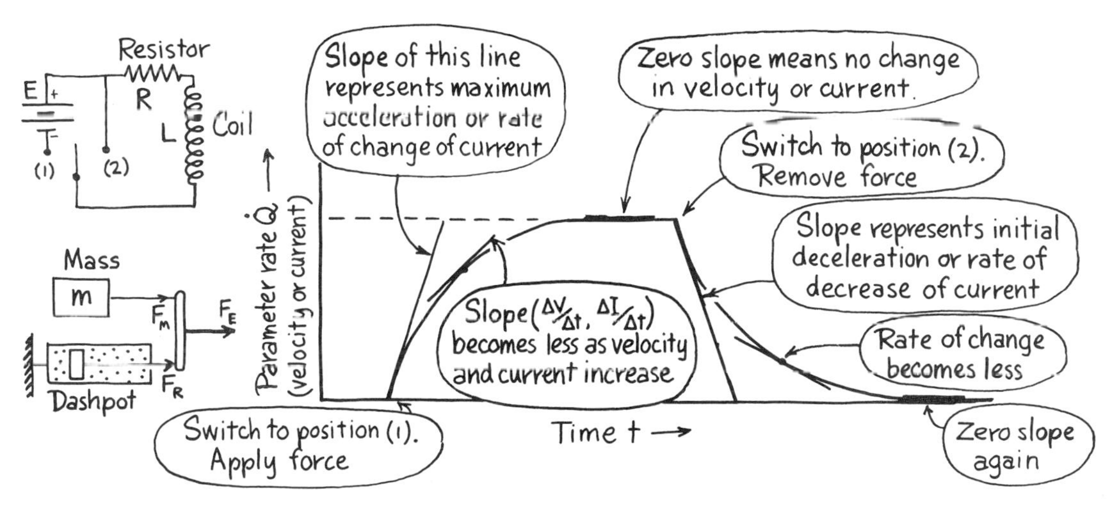

Figure 12-2. In series connected *RM* systems the resistor's opposition increases as the parameter rate (velocity, current) increases. Finally, the resistor's opposition completely balances the applied force and change stops. Velocity or current stays at its final steady value. When the force is removed, resistors take energy away from the mass or the coil's magnetic field. Velocity or current decreases and eventually becomes zero.

When a constant excitation force F_E is first applied to the mechanical system, it goes entirely into accelerating the mass. There is no resistive force F_R because the velocity to start with is zero ($F_R = Rv$). There is no capacitive force because the system does not include a capacitor. The acceleration (rate of change of velocity $\Delta v/\Delta t$), as determined by Newton's second law, is $a = \Delta v/\Delta t = F_E/m$. This acceleration is the initial slope of the velocity versus time curve.

When a constant emf E is first applied to the electrical system, the emf goes entirely into changing the current and building up the magnetic field around the coil. The current to start with is zero, and the potential difference across the resistor is zero, too ($V_R = IR$). The coil voltage V_L equals E at this instant, and the current rate of change $\dot{I}$ is

$$\dot{I} = \underbrace{\frac{\Delta I}{\Delta t} = \frac{V_L}{L}}_{\text{Eq. (8-11)}} = \frac{E}{L}$$

This rate at which current changes is the initial slope of the current versus time curve.

As soon as there is movement, a resistive force shows up in the mechanical system and less of the excitation force is available to accelerate the mass. The system accelerates at a lower rate. Eventually, all the excitation force is used up in the dashpot and the acceleration is zero. Both dashpot piston and mass move at a constant *terminal* velocity $v_f = F_E/R$. When the excitation force is suddenly removed, the inertia of the mass tries to maintain the terminal velocity. But the dashpot resistor drains kinetic energy from the mass, and the system eventually comes to rest.

By analogy, the same kinds of things happen to the electrical system. As soon as there is current, a potential difference appears across the resistor. Less battery emf is available to change current through the coil. Rate of change of current decreases until eventually all the battery emf is across the resistor. The current stops changing, but maintains a final steady value. When the battery emf is suddenly removed (switch in position 2), the coil tries to maintain the current. But the resistor drains away energy stored in the magnetic field. The field collapses and the current is reduced eventually to zero. The graphs of Fig. 12-2 look like the *RC* exponential curves of Chapter 11.

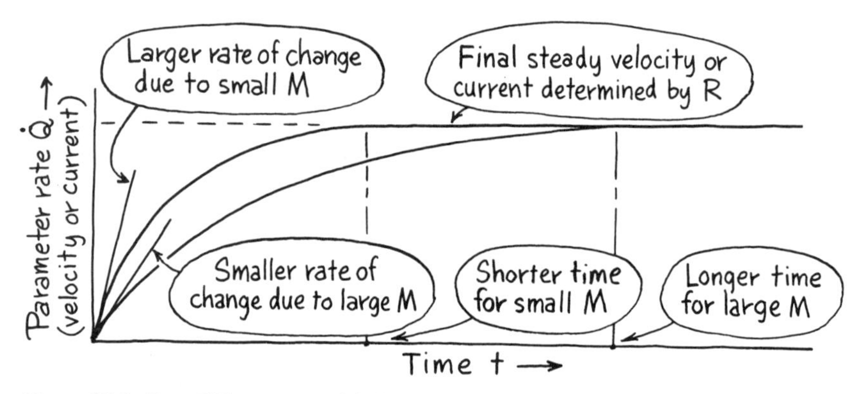

Figure 12-3. Two *RM* systems with the same *R* and different values of *M*. Systems with large *M* change at slower rates and take longer times to reach equilibrium.

We can make some guesses about what effect inertance (mass, inductance) and resistance have on the reaction time of an *RM* system. Inertance determines the initial rate of change of parameter rate. The larger the mass or inductance, the slower is the change of velocity or the change of current. It takes a longer time to get large things moving. Large inertance means slow reaction, all else being equal (Fig. 12-3).

Resistance determines the final steady parameter rate (velocity or current). Large resistance means small parameter rate and less time to reach the smaller final parameter rate, all else being equal (Fig. 12-4).

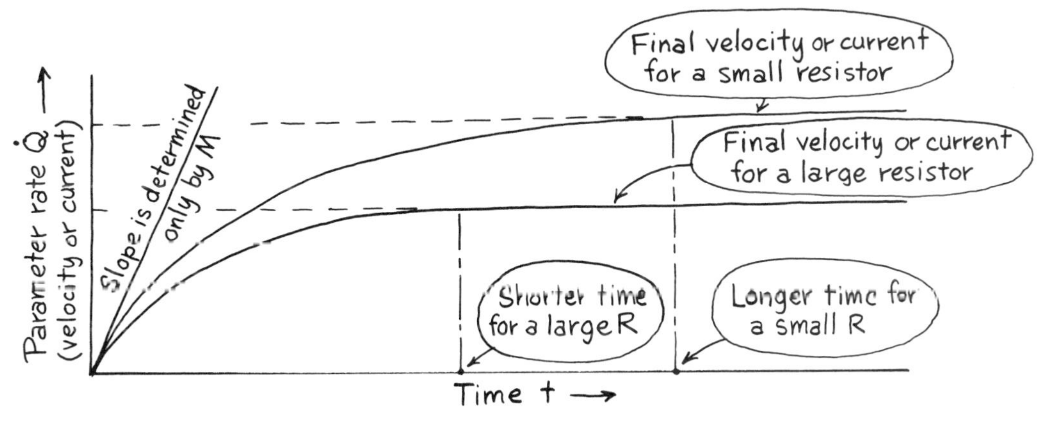

Figure 12-4. Two *RM* systems with the same *M* but different values of *R*. Systems with large *R* attain smaller velocities and currents. Less time is needed to reach these lower values.

It appears then that the response time of an *RM* system is related to both *R* and *M*. Systems take a longer time to come to equilibrium the greater the value of *M* and the smaller the value of *R*. Again, as in Chapter 11, we take a closer more mathematical look to see if our guesses are correct.

12-3 A QUANTITATIVE LOOK AT *RM* SYSTEMS

The following derivations may be skimmed over lightly if you wish. They are put here to point out the similarity between *RM* systems and the *RC* systems of Chapter 11.

*Derivation of the Exponential Functions

We use an electrical system in the derivations that follow. The results, however, apply to any *RM* system. Consider the electric circuit of Fig. 12-5 with a resistor *R* and an inductor *L* in series. Throw the switch to position 1 and allow the system to come to steady state. As the current builds, "kinetic"

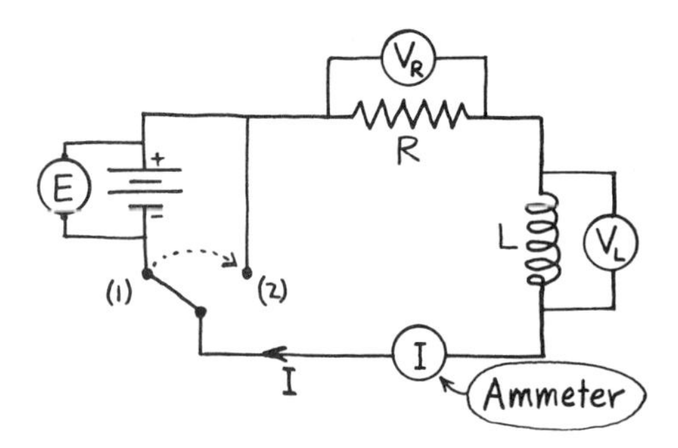

Figure 12-5. Current builds up when the switch is thrown to position 1 and collapses when the switch is thrown to position 2.

energy is stored in the magnetic field about the inductor. At any instant the sum of voltages across R and L must equal the emf E:

$$V_L + V_R = E$$

But V_R can be written as IR using Ohm's law, and V_L can be written using the definition of inductance (Newton's second law for electrical systems).

$$V_R = IR \qquad (7\text{-}2)$$

$$V_L = L\dot{I} \qquad (8\text{-}11)$$

Making these substitutions,

$$\dot{I}L + IR = E \qquad (12\text{-}1)$$

Equation (12-1) is a differential equation very similar to Eq. (11-1). Its solution is

$$I = I_f[1 - e^{-t/(L/R)}]$$

or in unified-concept terms, with $I = \dot{Q}$ and $L = M$,

$$\dot{Q} = \dot{Q}_f[1 - e^{-t/(M/R)}] \qquad (12\text{-}2)$$

$\dot{Q}$ is the parameter rate (velocity, current, etc.) at any time t. $\dot{Q}_f$ is the final steady value of the parameter rate.

Now move the switch to position 2 in Fig. 12-5 to remove the excitation emf. Current as observed in the ammeter does not come to a sudden stop. Mathematically,

$$V_L + V_R = 0$$

$$\dot{I}L + IR = 0 \qquad (12\text{-}3)$$

Equation (12-3) is another differential equation very similar to Eq. (11-3). Its solution is

$$I = I_o e^{-t/(L/R)}$$

In unified-concept terms,

$$\dot{Q} = \dot{Q}_o e^{-t/(M/R)} \tag{12-4}$$

where $\dot{Q}_o$ is the initial parameter rate.

Equations for Increasing and Decreasing Parameter Rates

Equations (12-2) and (12-4) allow us to calculate values of $\dot{Q}$ (velocity, current, etc.) at various times if we know R, M, and the initial or final value of $\dot{Q}$. Graphs of these equations (Fig. 12-6) are identical to those of the charging and discharging equations for RC systems, except that $\dot{Q}$ replaces Q and M/R replaces RC.

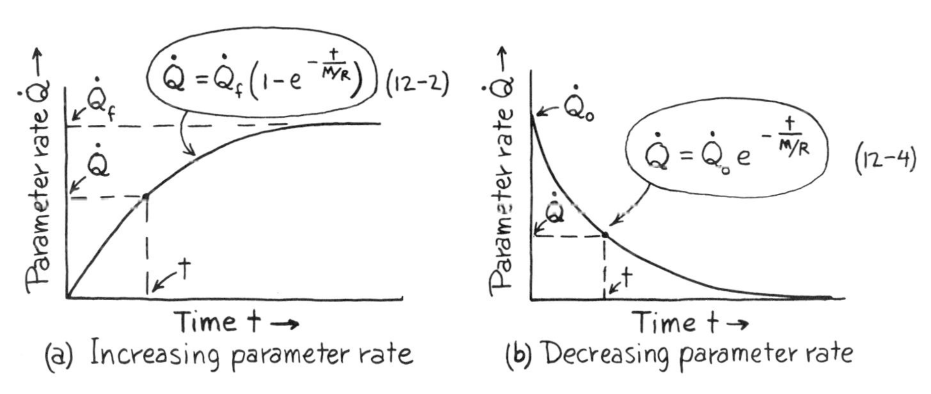

Figure 12-6. Curves of *RM* systems with increasing and decreasing parameter rates.

Figure 12-6a describes a parameter rate increase in an *RM* system. An object's velocity increases, a flywheel speeds up, volume flow rate increases, or a coil current increases. As time goes on, these quantities approach their final steady value $\dot{Q}_f$. Motion still continues in the steady state, as contrasted with a charging *RC* system in which the final parameter rate is zero.

Figure 12-6b describes a parameter rate decrease in an *RM* system. The velocity of an object decreases, a flywheel slows up, volume flow rate decreases, or coil current decreases. As time goes on, these quantities approach their final value of zero.

Another interesting thing about these curves is that the area underneath them represents parameter (Sec. 6-2)—distance traveled, angle turned, volume transferred, or charge moved.

Table 12-1 presents the exponential equations for increasing and decreasing parameter rate in the various systems. All *RM* systems exhibit the same behavior as described by Eqs. (12-2) and (12-4). Heat systems are an exception because they lack inertance. They cannot be *RM* systems.

TABLE 12-1 EXPONENTIAL EQUATIONS FOR *RM* SYSTEMS

Increasing Parameter Rate	*Decreasing Parameter Rate*	
$\dot{Q} = \dot{Q}_f[1 - e^{-t/(M/R)}]$	$\dot{Q} = \dot{Q}_o e^{-t/(M/R)}$	$\dot{Q}_o$ = *original value* $\dot{Q}_f$ = *final value* $e = 2.71828\cdots$

System	*Equations*	$\dot{Q}$ *Parameter Rate*	t *Time*	R *Resistance*	M *Inertance*
Mechanical translational	$v = v_f[1 - e^{-t/(m/R)}]$ $v = v_o e^{-t/(m/R)}$	v Velocity (m/s) (ft/s)	↑	$R = F_R/v$ $\left(\frac{\text{N}}{\text{m/s}}\right)$ $\left(\frac{\text{lb}}{\text{ft/s}}\right)$	$m = F_M/a$ (kg) (slug)
Mechanical rotational	$\omega = \omega_f[1 - e^{-t/(I/R)}]$ $\omega = \omega_o e^{-t/(I/R)}$	ω Angular velocity (rad/s)	(s)	$R = \tau_R/\omega$ $\left(\frac{\text{N}\cdot\text{m}}{\text{rad/s}}\right)$ $\left(\frac{\text{lb}\cdot\text{ft}}{\text{rad/s}}\right)$	$I = \tau_M/\alpha$ $(\text{kg}\cdot\text{m}^2)$ $(\text{slug}\cdot\text{ft}^2)$
Fluid	$\dot{V} = \dot{V}_f[1 - e^{-t/(M/R)}]$ $\dot{V} = \dot{V}_o e^{-t/(M/R)}$	$\dot{V}$ Volume flow rate (m^3/s) (ft^3/s)		$R = p_R/\dot{V}$ $\left(\frac{\text{Pa}}{\text{m}^3/\text{s}}\right)$ $\left(\frac{\text{lb/ft}^2}{\text{ft}^3/\text{s}}\right)$	$M = p_M/\ddot{V}$ (kg/m^4) (slug/ft^4)
Electrical	$I = I_f[1 - e^{-t/(L/R)}]$ $I = I_o e^{-t/(L/R)}$	I Current (A)	↓	$R = V_R/I$ (Ω)	$L = V_L/\dot{I}$ (H)

12-4 *RM* TIME CONSTANT

Again we can define a time constant. This one is the *RM time constant* τ_{RM}, and it plays the same role as the *RC* time constant τ_{RC} in Chapter 11.

$$\tau_{RM} = \frac{M}{R} \tag{12-5}$$

The unit of τ_{RM} is the *second* if the units for resistance R and inertance M are as in Table 12-1. For example, the units of M/R in the mechanical translational system are

$$\frac{\text{kg}}{\text{N}/(\text{m/s})} = \frac{\text{kg}\cdot\text{m}}{\text{N}\cdot\text{s}}$$

But from $F = ma$, a newton (N) has the units kg · m/s². Substituting,

$$\frac{\text{kg} \cdot \text{m}}{(\text{kg} \cdot \text{m/s}^2) \cdot \text{s}} = \text{s}$$

$\tau_{RM} = M/R$ occupies the same place in *RM* systems as $\tau_{RC} = RC$ does in *RC* systems, and the exponential functions that describe such systems are very similar. Thus we can summarize what was learned in Chapter 11 about *RC* systems and apply it to *RM* systems.

1. In one time constant of time ($\tau_{RM} = M/R$), a system accelerates to 63 percent of its final parameter rate (velocity, current, etc.) or decelerates to 37 percent of its original parameter rate.

2. Practically speaking, steady-state conditions are reached after five time constants. Parameter rate is steady either at its final value $\dot{Q}_f$ or at zero.

3. The rate of change of parameter rate $\ddot{Q}$ (acceleration, rate of change of current, etc.) at any particular instant depends on the time constant and the parameter rate at that time (Fig. 12-7):

$$\left.\begin{aligned} \textit{Increasing parameter rate:} \quad \ddot{Q} &= \frac{\dot{Q}_f - \dot{Q}}{\tau_{RM}} \\ \tau_{RM} &= \frac{\dot{Q}_f - \dot{Q}}{\ddot{Q}} \end{aligned}\right\} \tag{12-6}$$

$$\left.\begin{aligned} \textit{Decreasing parameter rate:} \quad \ddot{Q} &= \frac{-\dot{Q}}{\tau_{RM}} \\ \tau_{RM} &= \frac{-\dot{Q}}{\ddot{Q}} \end{aligned}\right\} \tag{12-7}$$

If the slope of the increasing parameter rate curve (Fig. 12-7a) were to stop changing, it would take the same time τ_{RM} for the parameter rate $\dot{Q}$ to reach its final steady value of $\dot{Q}_f$ from any point on the curve. If the slope of the decreasing parameter rate curve (Fig. 12-7b) were to stop changing, it would take the same time τ_{RM} for the parameter rate to go to zero from any point on the curve. The minus sign in Eq. (12-7) means that $\dot{Q}$ gets smaller as time goes on.

Figure 12-8 shows accurately plotted graphs of Eqs. (12-2) and (12-4). The graphs are equivalent to Fig. 11-7 for *RC* systems and can be used to solve *RM* problems directly.

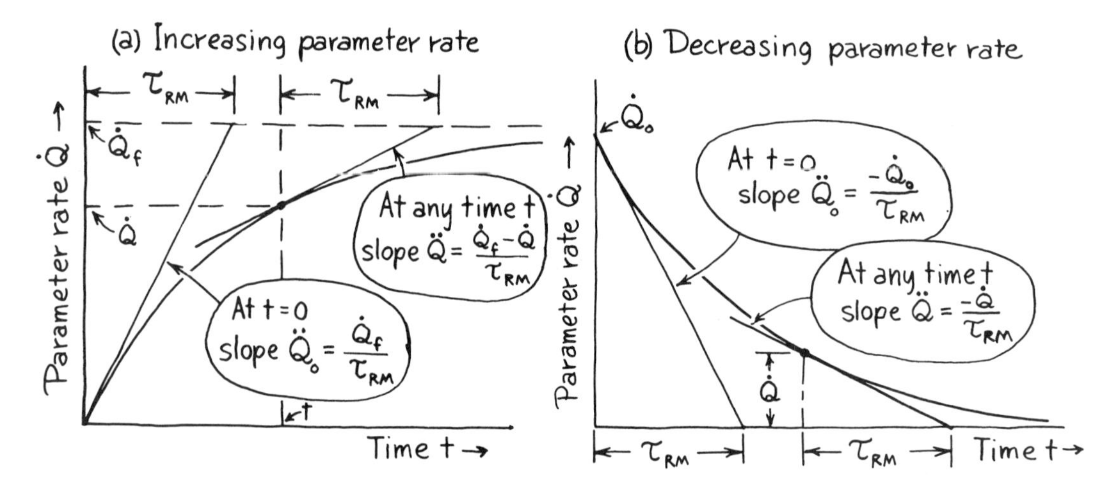

Figure 12-7. Exponential curves for *RM* series systems have all the same properties as those for *RC* series systems. All you have to do is substitute $\dot{Q}$ for Q, $\ddot{Q}$ for $\dot{Q}$, and M/R for RC.

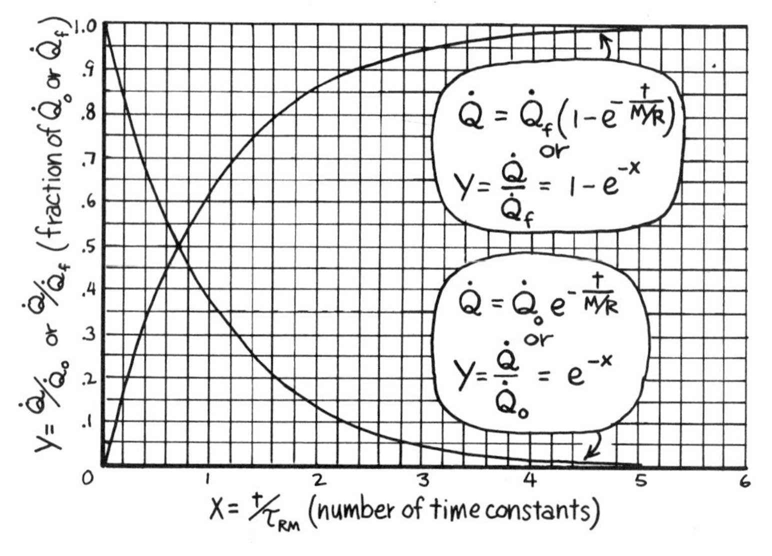

Figure 12-8. Accurately plotted graphs for *RM* systems. Values of parameter rate $\dot{Q}$ and time t can be calculated using these graphs.

Mechanical Translational System

In an ideal system involving only dry friction, any force large enough to overcome static friction on an object would continue to accelerate the object, however gradually, until some other physical effect took over to change things. The object would have to melt from the heat generated or smash into a wall or any number of other possible disasters before acceleration stops.

Systems in which the resistive force increases along with velocity are different in that they have a built in governor that limits the velocity to some maximum value proportional to excitation force. Free fall of an object in a viscous fluid is a practical illustration of such an *RM* system. Take the case of Crazy Harry who bails out of a hovering helicopter with a parachute that will remain closed for the moment. Figure 12-9 illustrates the situation.

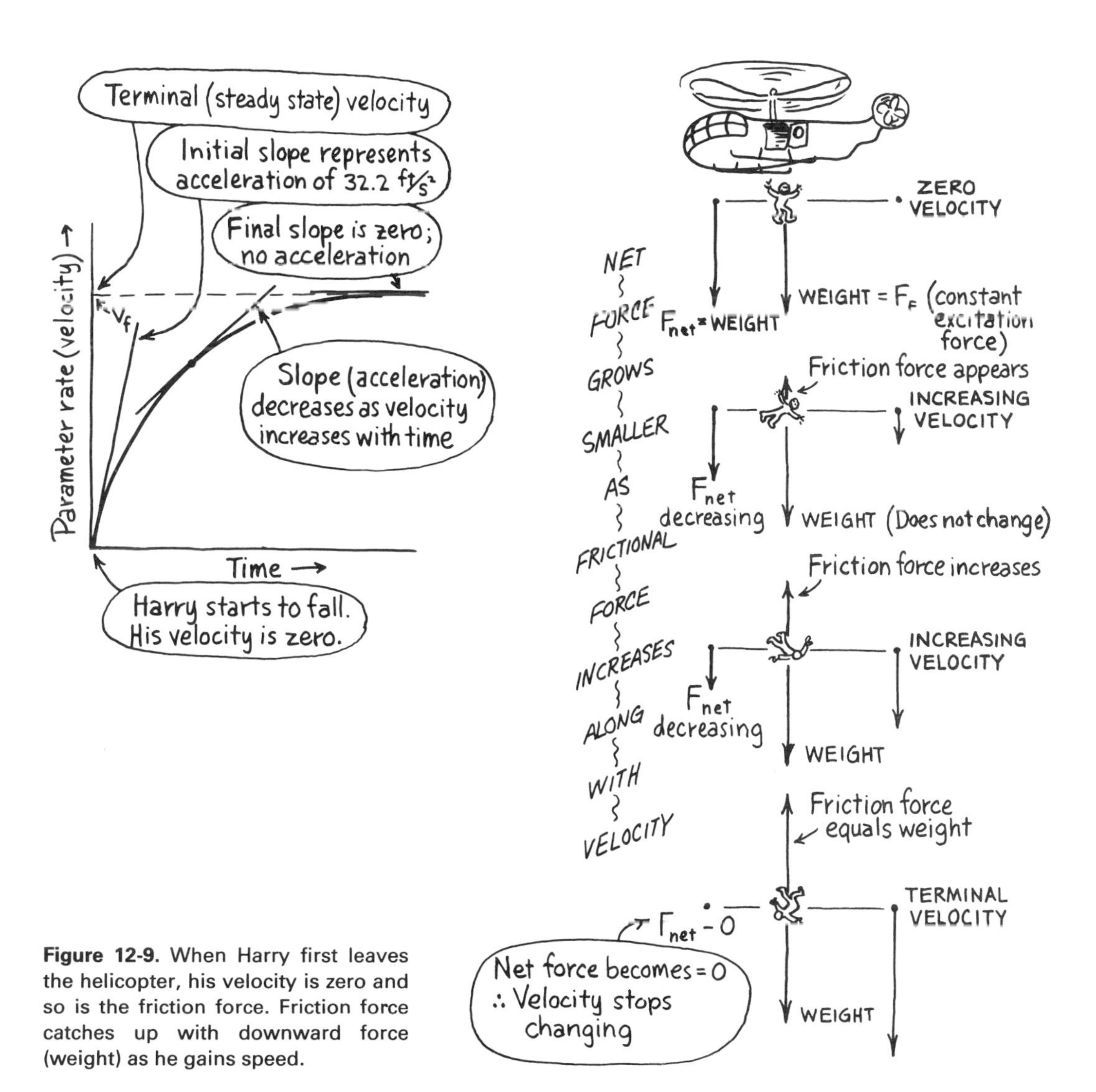

Figure 12-9. When Harry first leaves the helicopter, his velocity is zero and so is the friction force. Friction force catches up with downward force (weight) as he gains speed.

Just as he steps out the door, his vertical velocity is zero. Since frictional force depends upon speed, there is no friction at this point. The only force acting on him is his weight (the earth is pulling him downward), which is the excitation force F_E. He begins to accelerate downward, initially at 32.2 ft/s^2. As our hero picks up speed, an upward friction forcc appears to oppose motion. The net force on him is reduced correspondingly, and so is his acceleration. Eventually (assuming that the helicopter was high enough to begin with), the friction force becomes equal to his weight. The excitation force is now used up completely to overcome friction, the net force is zero, and no further acceleration takes place. The system reaches steady state. The speed at which this happens is called the *terminal velocity*—aptly named if the parachute doesn't open.

Note that terminal velocity depends upon the value of the excitation force, in this case the weight. If Crazy Harry had just eaten a heavy meal, his weight would have been larger, and he would need to move faster in order to produce correspondingly larger friction forces to balance his weight. If (a large if) friction force is directly proportional to speed (i.e., if R is constant), terminal velocity must be directly proportional to weight, all else being equal. Let's take a quantitative look at Crazy Harry as he hurtles through the atmosphere.

Example 12-1 Harry's weight, including parachute, is 205 lb. His terminal velocity (with parachute closed) is 176 ft/s (120 mi/hr).

(a) What is the resistance (assumed constant) of Harry in the air?
(b) What is Harry's inertance?
(c) What is the *RM* time constant?
(d) What was Harry's acceleration just as he began to fall?

Solution to (a)

$\left\{\begin{array}{l} W = 205 \text{ lb} \\ v_f = 176 \text{ ft/s} \end{array}\right\}$ When $\quad v = v_f, \quad F_R = W,$

$$R = \frac{F_R}{v} = \frac{W}{v_f} = \frac{205 \text{ lb}}{176 \text{ ft/s}} = \boxed{1.16 \text{ lb/(ft/s)}}$$

Solution to (b)

Inertance in this system is mass.

$$m = \frac{W}{g} = \frac{205 \text{ lb}}{32.2 \text{ ft/s}^2} = \boxed{6.37 \text{ slug}}$$

Solution to (c)

$$\tau_{RM} = \frac{M}{R} = \frac{m}{R} = \frac{6.37 \text{ slug}}{1.16 \text{ lb/(ft/s)}} = \boxed{5.47 \text{ s}}$$

Solution to (d)

We know that the initial acceleration is 32.2 ft/s^2, but let's make a calculation using the slope equation (12-6). $\dot{Q}$ is increasing.

$$\ddot{Q} = \frac{\dot{Q}_f - \dot{Q}}{\tau_{RM}} \quad \text{or} \quad a = \frac{v_f - 0}{\tau_{RM}} = \frac{176\text{ ft/s}}{5.47\text{ s}} = \boxed{32.2\text{ ft/s}^2}$$

Example 12-2 Continuing with the flight of Crazy Harry:

(a) What was his velocity when he was accelerating at 16.1 ft/s^2 (half the acceleration of gravity)?

(b) Harry pulled his rip cord 15.0 s after he jumped. How fast was he going at the time?

Solution to (a)

$$\left\{ \begin{aligned} \ddot{Q} &= a = 16.1\text{ ft/s}^2 \\ v_f &= 176\text{ ft/s} \\ \tau_{RM} &= 5.47\text{ s} \end{aligned} \right\}$$

$$a = \frac{v_f - v}{\tau_{RM}}$$

$$v = v_f - a\tau_{RM} = 176\text{ ft/s} - (16.1\text{ ft/s}^2)(5.47\text{ s})$$

$$= \boxed{87.9\text{ ft/s}}$$

Solution to (b)

$$\left\{ \begin{aligned} t &= 15.0\text{ s} \\ \tau_{RM} &= 5.47\text{ s} \\ v_f &= 176\text{ ft/s} \end{aligned} \right\}$$

Using Fig. 12-8,

Y = 0.94

Fig. 12-8

X = 2.74

$$x = \frac{t}{\tau_{RM}} = \frac{15.0\text{ s}}{5.47\text{ s}} = 2.74; \; y = 0.94 = \frac{v}{v_f}$$

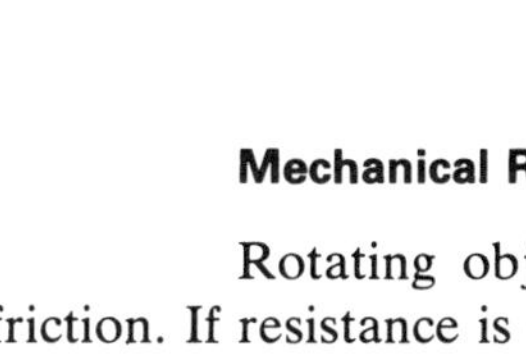

$$v = 0.94v_f = 0.94(176\text{ ft/s}) = \boxed{165\text{ ft/s}}$$

or using Eq. (12-2),

$$v = v_f[1 - e^{-t/(m/R)}] = (176\text{ ft/s})(1 - 2.718^{-2.74})$$

$$= (176\text{ ft/s})(0.936) = 165\text{ ft/s}$$

↑ This can best be done with a calculator

Mechanical Rotational System

Rotating objects may speed up or slow down in the presence of friction. If resistance is constant, estimates can be made of how long it takes to reach either a terminal angular velocity or to come to rest. Let us look at a *viscosimeter* used to find the viscosity (frictional properties) of liquids.

A viscosimeter consists of one metal cylinder rotating inside another (Fig. 12-10). The space between cylinders is filled with the liquid, and the bearings are such that almost no friction is exerted except by the liquid. The system acts as an almost perfectly linear rotational resistor. *R* really is constant in this device.

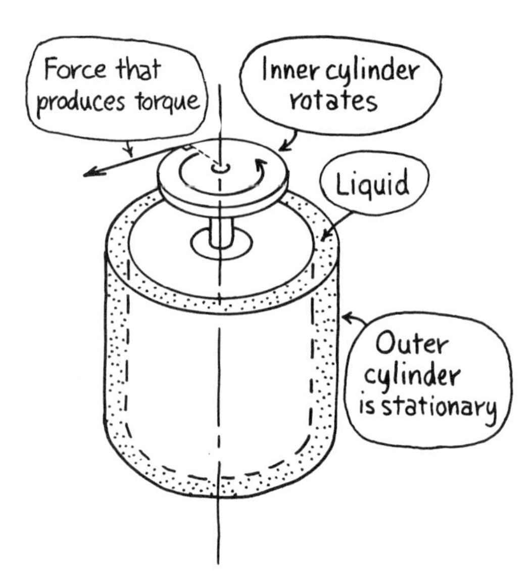

Figure 12-10. A viscosimeter consists of one metal cylinder rotating inside another. Space between cylinders is filled with the liquid whose frictional properties are to be determined.

Example 12-3 A constant excitation torque of 7.8×10^{-4} N · m is applied to the inner cylinder of a viscosimeter whose moment of inertia is 4.3×10^{-5} kg · m^2. The system's resistance for a mixture of glycerin and water is 3.4×10^{-5} N · m/(rad/s).

(a) Calculate the system's terminal angular velocity.

(b) How long, starting from rest, does it take to reach terminal velocity?

Solution to (a)

$\left\{ \begin{array}{l} \tau_E = 7.8 \times 10^{-4}\ \text{N} \cdot \text{m} \\ R = 3.4 \times 10^{-5}\ \text{N} \cdot \text{m/(rad/s)} \end{array} \right\}$

When $\omega = \omega_f$, $\tau_E = \tau_R$; (τ means "torque" here)

$$R = \frac{\tau_R}{\omega} = \frac{\tau_E}{\omega_f} \Rightarrow \omega_f = \frac{\tau_E}{R} = \frac{7.8 \times 10^{-4}\ \text{N} \cdot \text{m}}{3.4 \times 10^{-5}\ \text{N} \cdot \text{m/(rad/s)}} = \boxed{23\ \text{rad/s}}$$

Solution to (b)

$\{I = 4.3 \times 10^{-5}\ \text{kg} \cdot \text{m}^2\}$

τ means "time constant" here

$$\tau_{RM} = \frac{M}{R} = \frac{I}{R} = \frac{4.3 \times 10^{-5}\ \text{kg} \cdot \text{m}^2}{3.4 \times 10^{-5}\ \text{N} \cdot \text{m/(rad/s)}} = 1.26\ \text{s}$$

$$5\tau_{RM} = 5(1.26\ \text{s}) = \boxed{6.3\ \text{s}}$$

Reminder: The symbol τ (tau) stands for both torque and time constant.

Example 12-4 Continuing with the viscosimeter problem of Example 12-3:

(a) What is the angular velocity 2.0 s after the excitation torque is applied? What is the angular acceleration at this time?

(b) How long does it take to reach an angular velocity of 7.0 rad/s?

Solution to (a)

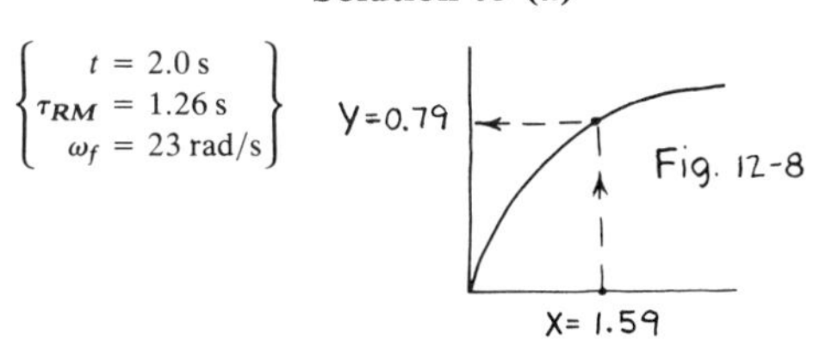

Warning: Do *not* use the equations of Chapter 6 for a problem like this. Acceleration is not constant.

Use Fig. 12-8:

$$x = \frac{t}{\tau_{RM}} = \frac{2.0\text{ s}}{1.26\text{ s}} = 1.59;\ y = 0.79 = \frac{\omega}{\omega_f}$$

$$\omega = 0.79\omega_f = 0.79(23\text{ rad/s}) = \boxed{18\text{ rad/s}}$$

Or use Eq. (12-2):

$$\omega = \omega_f(1 - e^{-t/\tau_{RM}}) = (23\text{ rad/s})(1 - 2.718^{-1.59}) = (23\text{ rad/s})(0.795) = 18\text{ rad/s}$$

Use Eq. (12-6) to calculate angular acceleration:

$$\ddot{Q} = \frac{\dot{Q}_f - \dot{Q}}{\tau_{RM}} \Rightarrow \alpha = \frac{\omega_f - \omega}{\tau_{RM}} = \frac{(23 - 18)\text{ rad/s}}{1.26\text{ s}} = \boxed{4.0\text{ rad/s}^2}$$

Solution to (b)

$\omega = 7.0$ rad/s
$\omega_f = 23$ rad/s

Use Fig. 12-8:

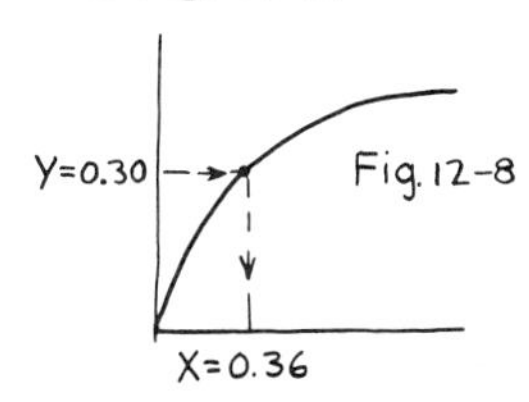

$$y = \frac{\omega}{\omega_f} = \frac{7.0\text{ rad/s}}{23\text{ rad/s}} = 0.30;\ x = 0.36 = \frac{t}{\tau_{RM}}$$

$$t = 0.36\tau_{RM} = 0.36(1.26\text{ s}) = \boxed{0.45\text{ s}}$$

Or use Eq. (12-2):

$$\omega = \omega_f(1 - e^{-t/\tau_{RM}}) \Rightarrow e^{-t/\tau_{RM}} = 1 - \frac{\omega}{\omega_f}$$

$$e^{-t/1.26} = 1 - \frac{7.0\text{ rad/s}}{23\text{ rad/s}} = 0.696$$

← Find the natural log (ln) of both sides of this equation. Use a calculator.

$$\frac{t}{1.26} = 0.363 \quad \text{and} \quad t = 1.26(0.363) = 0.46\text{ s}$$

Fluid System

Studies of reaction time in fluid systems are more likely to be concerned with liquids than with gases because liquids are denser. More time is needed to get them in motion. Liquid resistance is often assumed to be constant, so that estimates can be made of how much time is needed to either accelerate a liquid to its final steady-state volume flow rate or to stop the flow of a moving liquid.

Example 12-5 A quick-acting pump suddenly applies a pressure of 1.3×10^4 Pa to water at rest in a hose 2.0 m long and $3.0 \times 10^{-6}\ \text{m}^2$ in cross-sectional area. The viscosity of the water is $1.0 \times 10^{-3}\ \text{kg/(m} \cdot \text{s)}$. Assume constant resistance, laminar flow conditions.

(a) How long does it take for the water to reach its final steady-state flow rate?
(b) What is the final volume flow rate?

Solution to (a)

$$\left\{\begin{aligned} \eta &= 1.0 \times 10^{-3}\ \text{kg/(m} \cdot \text{s)} \\ \ell &= 2.0\ \text{m} \\ A &= 3.0 \times 10^{-6}\ \text{m}^2 \end{aligned}\right\}$$

First calculate the resistance R using Eq. (7-13) and the inertance M using Eq. (10-5). Then calculate the time constant τ_{RM} and multiply it by 5.

Eq. (7-13)

$$R = 8\pi\eta\frac{\ell}{A^2} = 8\pi\left(1.0 \times 10^{-3}\frac{\text{kg}}{\text{m} \cdot \text{s}}\right)\frac{2.0\ \text{m}}{(3.0 \times 10^{-6}\ \text{m}^2)^2}$$

$$= 5.58 \times 10^9 \rightarrow 5.6 \times 10^9 \frac{\text{Pa}}{\text{m}^3/\text{s}}$$

Eq. (10-5)

$$M = \frac{\rho\ell}{A} = \left(1000\frac{\text{kg}}{\text{m}^3}\right)\frac{2.0\ \text{m}}{(3.0 \times 10^{-6}\ \text{m}^2)} = 6.67 \times 10^8\ \frac{\text{kg}}{\text{m}^4}$$

App. B, Table 1

$$\tau_{RM} = \frac{M}{R} = \frac{6.67 \times 10^8\ \text{kg/m}^4}{5.58 \times 10^9\ \text{Pa/(m}^3\text{/s)}} = 0.119\,\text{s}$$

$$5\tau_{RM} = 5(0.119\ \text{s}) = \boxed{0.60\ \text{s}}$$

Solution to (b)

$\{p_E = 1.3 \times 10^4\ \text{Pa}\}$

When $\dot{V} = \dot{V}_f$, $p_R = p_E$,

$$R = \frac{p_R}{\dot{V}} = \frac{p_E}{\dot{V}_f}$$

$$\dot{V}_f = \frac{p_E}{R} = \frac{1.3 \times 10^4\ \text{Pa}}{5.58 \times 10^9\ \text{Pa/(m}^3\text{/s)}} = \boxed{2.3 \times 10^{-6}\ \text{m}^3/\text{s}}$$

Electrical System

An electrical system, used at the start of the chapter to develop the *RM* exponential equations, behaves just like the mechanical system we've looked at. A changing current produces a changing magnetic field around a coil, and the field opposes further change by generating a back emf. The coil's inductance is the analog of mass or moment of inertia. Let's look at an example in which parameter rate decreases.

Example 12-6 The circuit illustrated includes a 3.0-kΩ resistor and a 5.0-mH coil. The switch is left in position 1 for at least five time constants and then it is thrown to position 2.

(a) What is the initial current with the switch in position 2?

(b) How long does it take for the current to stop?

(c) What is the current after 3.0 μs? What is the rate of change of current at this time?

Solution to (a)

$\left\{ \begin{array}{l} E = 64\text{ V} \\ R = 3.0 \times 10^3\ \Omega \end{array} \right\}$ When $t = 0$ s, $V_R = E$,

$$I_o = \frac{V_R}{R} = \frac{E}{R} = \frac{64\text{ V}}{3.0 \times 10^3\ \Omega} = \boxed{21 \times 10^{-3}\text{ A}}$$

Solution to (b)

$\{L = 5.0 \times 10^{-3}\text{ H}\}$

$$\tau_{RM} = \frac{L}{R} = \frac{5.0 \times 10^{-3}\text{ H}}{3.0 \times 10^{+3}\ \Omega} = 1.67 \times 10^{-6}\text{ s}$$

$$5\tau_{RM} = 5(1.67 \times 10^{-6}) = \boxed{8.3 \times 10^{-6}\text{ s}}$$

Solution to (c)

$\left\{ \begin{array}{l} t = 3.0 \times 10^{-6}\text{ s} \\ \tau_{RM} = 1.67 \times 10^{-6}\text{ s} \end{array} \right\}$

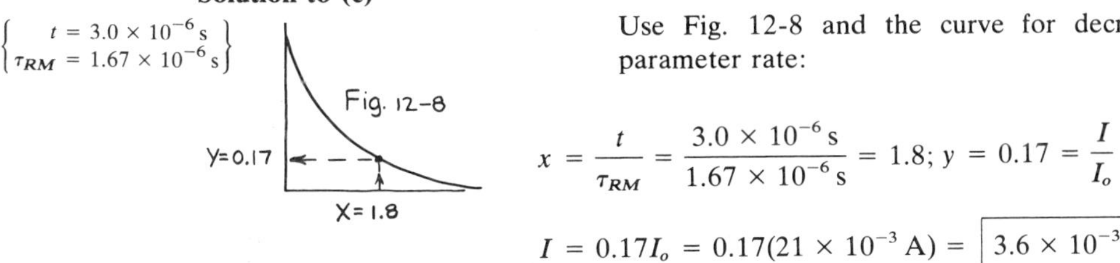

Use Fig. 12-8 and the curve for decreasing parameter rate:

$$x = \frac{t}{\tau_{RM}} = \frac{3.0 \times 10^{-6}\text{ s}}{1.67 \times 10^{-6}\text{ s}} = 1.8;\ y = 0.17 = \frac{I}{I_o}$$

$$I = 0.17 I_o = 0.17(21 \times 10^{-3}\text{ A}) = \boxed{3.6 \times 10^{-3}\text{ A}}$$

Or use Eq. (12-4):

$$\dot{Q} = \dot{Q}_o e^{-t/(M/R)} \Rightarrow I = I_o e^{-t/(L/R)} = (21 \times 10^{-3}\text{ A})(e^{-1.8})$$
$$= (21 \times 10^{-3}\text{ A})(0.166) = 3.5 \times 10^{-3}\text{ A}$$

↑ Use a calculator

Use Eq. (12-7) to find the rate of current change:

$$\ddot{Q} = \frac{\dot{Q}}{\tau_{RM}} \Rightarrow \dot{I} = \frac{-I}{\tau_{RM}} = \frac{-3.6 \times 10^{-3}\text{ A}}{1.67 \times 10^{-6}\text{ s}} = \boxed{-2.1 \times 10^3\text{ A/s}}$$

The negative sign means that I is decreasing with time

12-5 SUMMARY

This chapter has to a great extent been a replay of Chapter 11. We have seen that the exponential equations of *RC* systems have very close analogues in the *RM* series systems of this chapter. The difference is that we talk about $\dot{Q}$ instead of Q, $\ddot{Q}$ instead of $\dot{Q}$, and M/R instead of RC. The major points are summarized below.

1. A system containing only inertance will accelerate gradually and uniformly if a constant excitation force is suddenly applied. If the force acts opposite to the motion, the object decelerates gradually and uniformly. Resistance, if added to the system, gradually stops the acceleration or deceleration. System parameter rates assume steady values as time goes on.

2. Equations describing changes in *RM* systems are exponential functions. The rate of change of parameter rate $\ddot{Q}$ depends upon the value of $\dot{Q}$ in the system. As time goes on, $\ddot{Q}$ (the slope of the exponential curve) approaches zero, and $\dot{Q}$ theoretically never reaches its ultimate goal.

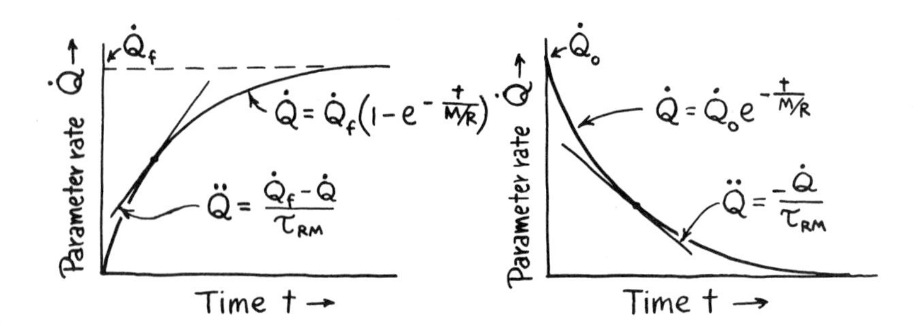

3. If the acceleration $\ddot{Q}$ were to suddenly stop changing, the final value of $\dot{Q}$ would be reached in time τ_{RM}, the system's *RM* time constant.

$$\tau_{RM} = \frac{M}{R}$$

After one time constant, $\dot{Q}$ will have gained or lost 63 percent of its original or final value. After five time constants, the system is essentially in the steady state at some final value of $\dot{Q}_f$ or at zero.

4. Table 12-1 on page 254 shows how the exponential equations apply to the various systems.

The study of *RC* and *RM* systems is important if we are interested in what happens just after the system experiences a step change of excitation force. System response at the start is governed by values of resistance, capacitance, and inertance. This usually small time period is referred to as the *transient* time period. Time delays due to combinations of *RM* and *RC* have an important bearing on how energy is controlled in systems. After five time constants, these systems are considered to be in the steady state. The *RC* system is either fully charged or fully discharged. The *RM* system maintains a constant parameter rate, either $\dot{Q}_f$ or zero.

There is one more way to put *R*, *M*, and *C* together in pairs. Chapter 13 will investigate systems that combine *M* and *C*. We will find that *MC* systems act differently. With no resistor to dissipate potential or kinetic energy, the *MC* system repeatedly overshoots its equilibrium position. For the first time we will be dealing with vibrating or *oscillating* systems. Resistance, if added to the system, damps out the oscillations.

PROBLEMS

12-1. A force of 6.0 N moves the 2.0-kg block at a maximum speed of 0.30 m/s. What is the system's *RM* time constant? The block and damper are in series.

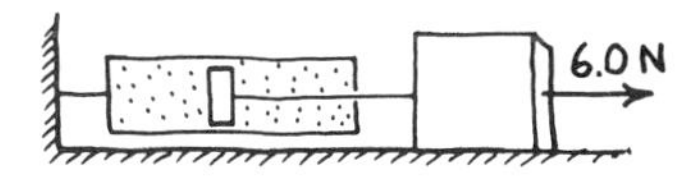

Problem 12-1

12-2. The block and damper are series connected. What is the *RM* time constant?

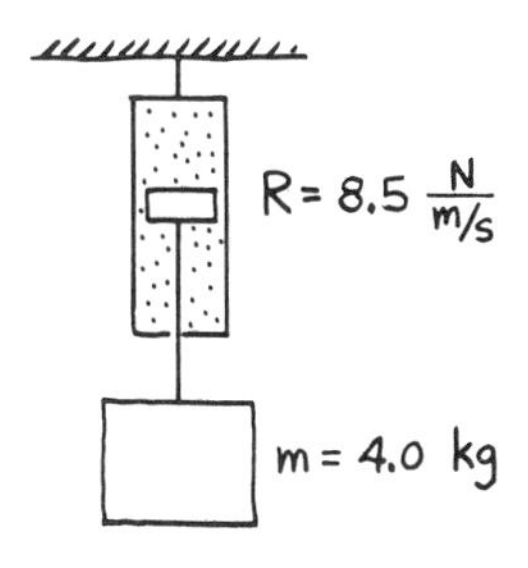

Problem 12-2

12-3. A 1000-kg car accelerates from rest under a constant 900-N force. Motion is opposed by frictional force *f* that is proportional to velocity. The car reaches a velocity of 30 m/s before it stops accelerating. How long does it take to reach top speed?

12-4. A 100-g steel sphere falls through a tube of water with a terminal velocity of 10 cm/s (0.10 m/s). Assuming that the water provides a constant resistance, how long after release does it take the ball to attain 99+ percent of its terminal velocity?

12-5. Refer to the viscosimeter of Example 12-3. If motor oil replaces the glycerin–water mixture, the system's resistance increases to 6.1×10^{-5} N · m/(rad/s). How long does it take to get the inner cylinder to maximum speed?

12-6. An airplane propeller with a moment of inertia of 5.8 slug · ft^2 attains its maximum angular velocity 3.5 s after it begins to spin. Assume that the plane's engine exerts constant torque and that the system's resistance is constant. What is the system's resistance?

12-7. A system of water pipes has an inertance of 8.3×10^7 kg/m^4 and resistance of 1.8×10^8 Pa/(m^3/s). How long does it take to reach maximum volume flow rate after a step change of pressure is applied?

12-8. What is the maximum flow rate in Problem 12-7 if the step change of pressure applied is 2.0×10^4 Pa?

12-9. A 40-Ω resistor and a 0.35-mH coil are connected in series through a switch to a 12-V battery. How long will it take for the current to stop changing after the switch is closed?

12-10. The resistor voltage reaches its maximum value 2.0 μs after the switch is closed. What is the value of the coil's inductance *L*?

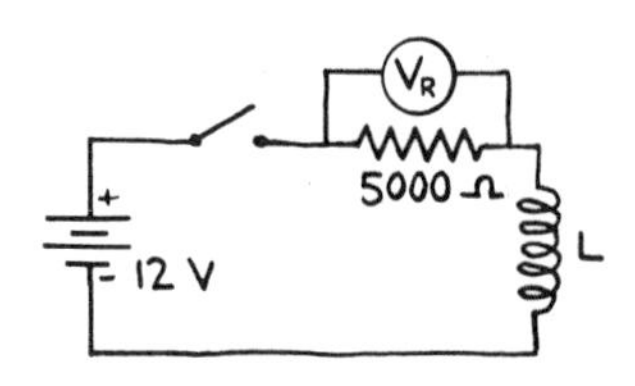

Problem 12-10

12-11. A potential difference of 100 V is suddenly impressed across a 0.50-H coil that has a resistance of 20 Ω. How long does it take for the current to reach steady state?

12-12. The switch is placed in position 1 for a while and then suddenly thrown to position 2. How long does it take for current to stop flowing?

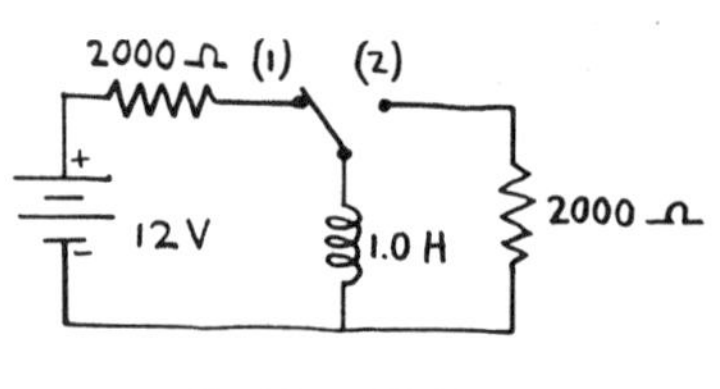

Problem 12-12

12-13. The frictional drag on the hull of a boat is very nearly proportional to the speed of the boat for low speeds. Suppose that the engine stops when the boat is traveling 10 ft/s, and the initial deceleration (negative acceleration) is 2.0 ft/s^2. Estimate how long it takes for the boat to stop.

12-14. If the boat in Problem 12-13 were moving at 5.0 ft/s when the engine quit, approximately how long would it take before the boat "stopped"? Would it be the same time as if it started decelerating while moving 10 ft/s? What does "stopped" mean in this situation?

12-15. An airplane propeller accelerates exponentially from zero to a steady value of 3000 rpm in 3.5 s. What is the angular acceleration as the angular velocity passes 1000 rpm?

12-16. A flywheel has an angular acceleration of 3.0 rad/s^2 when its angular velocity is 12 rad/s. How long does it take for the flywheel, starting from rest, to reach its maximum angular velocity of 25 rad/s? Assume that angular velocity changes exponentially with time.

12-17. A 50-ft length of garden hose with inside diameter 0.50 in. delivers water at a maximum rate of 1.5×10^{-2} ft^3/s when a gauge pressure equivalent to 25 ft of water is applied across its ends. Assume that the hose resistance remains constant. How long does it take to reach the maximum flow rate after the pressure is suddenly applied?

12-18. In Example 12-5, is it a good assumption that the resistance of the hose remains constant? (*Hint*: Check the value of the Reynolds number.)

12-19. An electrical circuit has a time constant of 2.5 μs and carries a steady-state current of 22 mA with the switch in position 1. What is the initial rate of change of current when the switch is thrown to position 2?

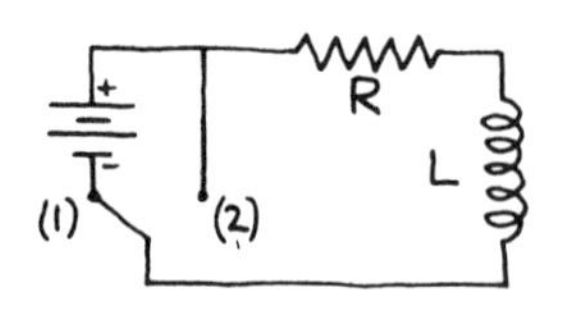

Problem 12-19

12-20. An electrical circuit similar to that of Problem 12-19 has a time constant of 15 μs. It carries a current of 75 mA when the switch is in position 1. What is the current 7.0 μs after the switch is thrown to position 2?

12-21. A car such as the one in Problem 12-3 has a top speed of 30 m/s and a time constant of 20 s. What is its speed 30 s after accelerating from rest?

12-22. How long would it take for the car in Problem 12-21 to attain a speed of 12 m/s?

12-23. A flywheel in the form of a solid cylinder 0.50 m in diameter has a mass of 160 kg. A constant torque keeps the flywheel rotating at 200 rad/s. The flywheel decelerates and comes to rest 180 s after the torque is removed. What is the constant resistance of this rotational system?

12-24. Refer to Problem 12-23.

(a) What constant excitation torque kept the flywheel rotating at 200 rad/s?

(b) What is the angular velocity 50 s after the excitation torque is removed?

12-25. A sphere falls through a liquid with a time constant of 0.35 s. Its final velocity is 2.7 m/s. What is its velocity 0.50 s after it starts to fall? Assume that resistance remains constant.

12-26. A boat weighs 12,000 lb. Energy supplied at the rate of 1.0 hp causes the boat to move at a constant speed of 3.0 ft/s. Assume that the resistance is constant for small speeds. The engine suddenly quits.

(a) What is the initial deceleration?

(b) What is the boat's speed 10 s after the engine quits?

12-27. Experiments using small oil drops to measure the charge of an electron were first done by American physicist Robert A. Millikan. The oil drops contained excess electrons and could be stopped in space by establishing an electric field in the space where the drops were falling. Terminal velocity measurements were needed to calculate the radius of the drop and its weight.

One of these drops has a radius of 1.0×10^{-5} m and a density of 800 kg/m^3. It is dropped, starting from rest, into air, which is a viscous fluid. The oil drop quickly attains a steady terminal velocity of 0.010 m/s. Neglect any effects of the buoyancy of air in the following calculations.

(a) What is the value of the excitation force acting on the falling drop?

(b) What is the value of the resistance?

Problem 12-27

12-28. Refer to Problem 12-27.

(a) What is the drop's initial velocity and initial acceleration?

(b) How long does it take for the drop to reach terminal velocity?

ANSWERS TO ODD-NUMBERED PROBLEMS

12-1. 0.10 s
12-3. 1.70×10^2 s
12-5. 3.5 s
12-7. 2.3 s
12-9. 4.4×10^{-5} s
12-11. 0.13 s
12-13. 25 s
12-15. 3.0×10^2 rad/s^2
12-17. 3.4 s
12-19. -8.8×10^3 A/s
12-21. 23 m/s
12-23. 0.14 N · m/(rad/s)
12-25. 2.1 m/s
12-27. (a) 3.3×10^{-11} N
(b) 3.3×10^{-9} N/(m/s)

13

Energy Transfer and Storage—*MC* and *RMC* Systems

So far we have been looking at the transient behavior of systems with resistance and either potential energy storage (Chapter 11) or kinetic energy storage (Chapter 12). A time constant $\tau_{RC} = RC$ or $\tau_{RM} = M/R$ expresses how quickly such systems react when an excitation force is suddenly applied or removed. The final effect has been a steady-state condition in which the parameter or the parameter rate remains at a constant value.

This chapter will first look at a system that contains both capacitance and inertance but no resistance. Such systems oscillate when an excitation force is suddenly applied or removed. The parameter describing the system does not remain constant but surges back and forth, first in one direction and then the other, producing an oscillation that theoretically never stops. If resistance is added, the oscillations eventually die out.

Our strategy in developing the material in this chapter is to first examine a mechanical MC system consisting of a mass attached to a spring. The oscillatory movement of the system is called simple harmonic motion. The motion is examined and the effects of mass and the spring's capacitance on the motion are studied. Sinusoidal equations describing system movement are developed. Ideas developed in the mechanical system are applied by analogy to fluid and electrical systems. Finally, resistance is added to a system already containing capacitance and inertance. Resistance draws energy from the system to damp out the oscillations. If resistance is large enough, it may completely prevent oscillations. Such systems are said to be critically damped or overdamped.

13-1 SYSTEMS WITHOUT RESISTANCE (*MC* SYSTEMS)

Consider the mechanical system illustrated in Fig. 13-1. It includes a spring of capacitance C and a block with inertance M in the form of mass m. The block slides back and forth on a frictionless surface. A black dot on top of the block serves as a reference for making distance measurements. The spring is unstretched when the block is in position 1. The spring stretches a distance S_m when the block is moved to position 2. At position 2 the system's energy is stored in the spring as potential energy $E_p = \frac{1}{2}(1/C)S_m^2$ [Eq. (9-12)].

The block is released and moves to the left, gaining velocity v. The spring returns to its natural length losing potential energy in the process. At position 1 the system's energy is stored in the block as kinetic energy $E_K = \frac{1}{2}mv^2$ [Eq. (3-12)]. There has been no loss of energy during all this because the surface is frictionless. The potential energy of the spring at position 2 is the same as the kinetic energy of the block at position 1. What happens after this? The block does not crash into the wall. It starts to slow up because the spring compresses. The block comes to rest at position 3, exactly a distance S_m from position 1. Potential energy in the compressed spring at position 3 is the same as potential energy stored in the stretched spring at position 2. There is no kinetic energy because the block is momentarily at rest, and there has been no energy loss due to friction.

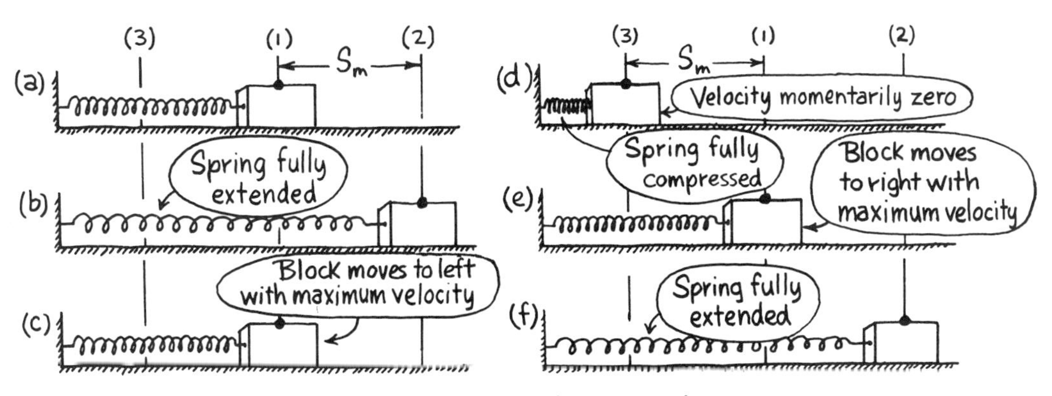

Figure 13-1. In (a) the block and the unstretched spring rest on the frictionless surface. In (b) the block is moved to the right and held. The spring deflects a distance S_m. When the block is released, it moves back to its original position, overshoots and retraces its path (c), (d), (e), and (f).

Now the block starts moving to the right, picking up speed, passing position 1, slowing up and momentarily coming to rest at position 2 to complete one *cycle* of motion (i.e., one round trip). The block's velocity is greatest at position 1. It is at this point that all the spring's potential energy has been changed to kinetic energy stored in the block.

This kind of motion repeats itself. The block oscillates back and forth between positions 2 and 3. It is momentarily at rest at the end points and is moving fastest at the midpoint. Such oscillatory motion is very common in nature.

Simple Harmonic Motion

Our next task is to study the motion of the block in more detail. We will do this by placing a camera far above the frictionless surface. A time exposure of the motion shows a black strip made by the dot on the block as it moves back and forth (Fig. 13-2a). The length of the strip is $2S_m$.

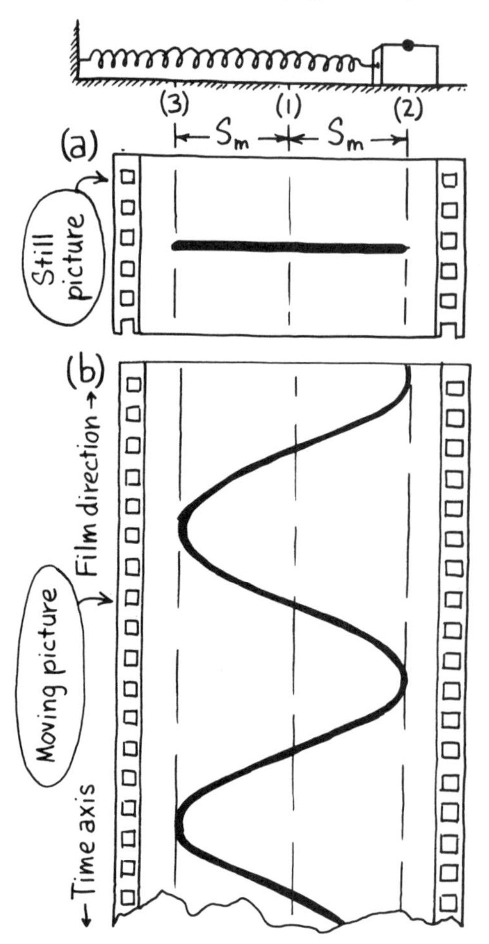

Figure 13-2. A camera mounted far above the oscillating block records the position of the black dot. In (a) a still picture shows a straight line whose length is $2S_m$. In (b) the film has been set in motion at a constant speed. The black dot follows a sinusoidal curve.

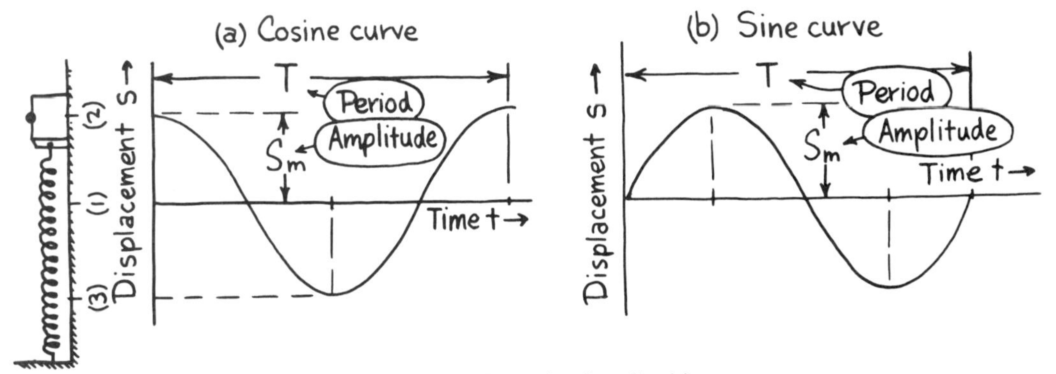

Figure 13-3. The motion of the oscillating block can be described by either a cosine curve (a) or a sine curve (b). The amplitude S_m is the maximum displacement. It is positive if s is to the right of position 1. Period T is the time needed for one complete cycle.

If we set the camera's film in motion at a constant speed, successive pictures of the dot show it meandering back and forth across the strip of film. The familiar picture made by the black dot on the block (Fig. 13-2b) is called a *cosine curve* or a *sine curve*, depending on your point of reference in the time sequence.

We have to introduce some new terms to talk about this sort of curve (see Fig. 13-3). The *amplitude* S_m of the motion is the largest displacement of the block from its equilibrium position. When the subscript $_m$ is added to a symbol, it will mean the *maximum* value of that changing quantity (i.e., its amplitude). If we are talking about no specific value of a changing quantity, the lowercase letter will be used. For example, the changing displacement is referred to in equations by the letter s. The maximum value (amplitude) is S_m. The uppercase without subscript (S) has yet another meaning, which we will get to later.

The *period* T is the time to complete one cycle of motion. The *frequency* f of the motion is the number of cycles of motion made in a 1-s time interval.

Frequency is just the reciprocal of period. Its unit is cycles/second, which is called hertz (Hz).

$$f = \frac{1}{T} \tag{13-1}$$

f *Frequency*	T *Period*
hertz (Hz)	second (s)

A system that oscillates sinusoidally with a constant amplitude and frequency is said to exhibit *simple harmonic motion*. Our mass–spring system exhibits simple harmonic motion in the absence of friction. Equations (13-2) and (13-3) are for finding the block's displacement s from its reference position 1 at any time t.

$$s = S_m \cos(2\pi ft) \tag{13-2}$$

or

$$s = S_m \cos(360ft) \tag{13-3}$$

s *Displacement*	S_m *Amplitude*	f *Frequency*	t *Time*
Any unit of length		hertz (Hz)	second (s)

The equations differ only slightly. The product $2\pi ft$ is an angle measured in radians. This is the form the equation takes when it is derived using calculus. The product $360ft$ is an angle measured in degrees. This form is easier to use in calculations. The displacement s is positive if the block is to the right of position 1 and negative if the block is to the left of position 1. Both equations describe the cosine curve of Fig. 13-3a.

Example 13-1 The block in a system like that of Fig. 13-1 is displaced 0.10 m to the right of its equilibrium position and released. The block first returns to its original position 2.0 s later. Assume that no friction is present.

(a) Find the amplitude, period, and frequency of the resulting simple harmonic motion.

(b) What is the block's displacement 1.2 s after it is released?

Solution to (a)

$$S_m = \text{amplitude} = \text{maximum displacement} = \boxed{0.10\ \text{m}}$$

$$T = \text{period} = \text{time for one complete cycle of motion} = \boxed{2.0\ \text{s}}$$

$$f = \text{frequency} = \frac{1}{T} = \frac{1}{2.0\ \text{s}} = \boxed{0.50\ \text{Hz}}$$

Solution to (b)

Use Eq. (13-2):

$$s = S_m \cos(360ft) = (0.10\ \text{m}) \cos[360(0.50\ \text{Hz})(1.2\ \text{s})]$$

$$= (0.10\ \text{m}) \cos 216° = (0.10\ \text{m})(-0.809)$$

$$= \boxed{-8.1 \times 10^{-2}\ \text{m}}$$

The minus sign means that the mass is to the left of the equilibrium position.

13-2 A QUANTITATIVE LOOK AT *MC* SYSTEMS

The movement of our mechanical mass–spring system is a periodic, cyclic kind of motion described by a sinusoidal equation containing amplitude, frequency, and time. Both mass and the spring's capacitance affect the system's motion. The frequency should be greater if the mass is smaller, because the motion of a smaller mass can be more easily changed. A stiffer spring provides more accelerating force for a given displacement than does a weaker one. Stiffer springs with smaller capacitance should increase the frequency of oscillation, all else being equal. The amplitude, on the other hand, should not depend upon either mass or spring capacitance, but only upon the initial spring displacement.

The next section lays the groundwork for deriving Eq. (13-2). You can skip over this lightly if you wish.

*Derivation of the Sinusoidal Function

Figure 13-4a shows our mass–spring system in its equilibrium position with the spring unstretched. An excitation force F_E stretches the spring a distance S_m (Fig. 13-4b). We observe the system t seconds after this force is suddenly removed (Fig. 13-4c). The spring displacement s creates a

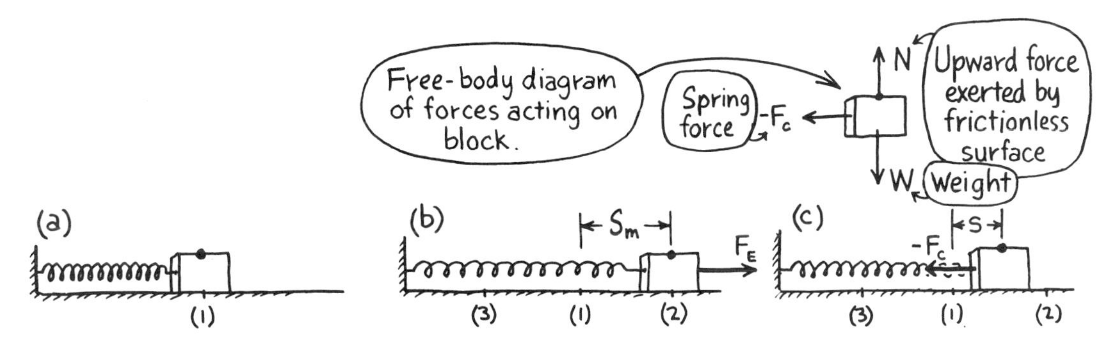

Figure 13-4. A mass–spring system in its equilibrium position 1 is displaced a maximum distance S_m to the right by excitation force F_E. When F_E is removed, the spring force $-F_C$ acting on the block accelerates it to the left.

spring force $-F_C$ that acts on the block and accelerates it to the left. The spring force is the net force F_{net} acting on the mass. Using Newton's second law, $F_{net} = ma = -F_C$. The minus sign means that F_C is to the left. Rearranging terms, $ma + F_C = 0$. But $F_C = s/C$ [Eq. (9-1)], and $a = \ddot{s}$, using the dot notation of Sec. 6-4. Substituting,

$$m\ddot{s} + \left(\frac{1}{C}\right)s = 0 \qquad (13\text{-}4)$$

The solution to this differential equation is either a sine curve or a cosine curve, depending on how the motion originally begins. In our case the motion starts when displacement is at a maximum S_m. This gives us the cosine form of the solution:

$$s = S_m \cos\left(\frac{1}{\sqrt{mC}} \cdot t\right) \tag{13-5}$$

Comparing Eq. (13-5) to Eq. (13-2),

$$s = S_m \cos(2\pi f t) \tag{13-2}$$

we conclude that

$$2\pi f = \frac{1}{\sqrt{mC}}$$

or

$$f = \frac{1}{2\pi\sqrt{mC}} \tag{13-6}$$

The frequency f is indeed larger if either the mass m or the spring capacitance C is made smaller.

System Equations

The frequency equation (13-6) can be generalized and used for other oscillating MC systems. Just substitute inertance M for mass m:

$$f = \frac{1}{2\pi\sqrt{MC}} \tag{13-7}$$

The displacement equation (13-3) can also be generalized for the oscillating systems. Since s represents the parameter q,

$$q = Q_m \cos(360 f t) \tag{13-8}$$

Table 13-1 summarizes the form of the frequency and parameter equations for the various MC systems.

The unit of frequency is the hertz (Hz) if you use the units for capacitance and inertance given in Table 13-1. Since 1 Hz is 1 cycle/second, the units are 1/s (a cycle is dimensionless) and $\sqrt{MC}$ should have the units of seconds. Let's check this out for the mechanical system. The units of $\sqrt{MC}$ are $\sqrt{\text{kg} \cdot (\text{m/N})}$. But using $F = ma$, the unit newton (N) is equivalent to $\text{kg} \cdot \text{m/s}^2$. Substituting,

$$\sqrt{\text{kg} \cdot (\text{m/N})} = \sqrt{\text{kg} \cdot \frac{\text{m}}{\text{kg} \cdot \text{m/s}^2}} = \sqrt{\text{s}^2} = \text{s}$$

TABLE 13-1 SUMMARY OF FREQUENCY AND PARAMETER EQUATIONS FOR THE VARIOUS OSCILLATING *MC* SYSTEMS

$$q = Q_m \cos(360ft) \qquad f = \frac{1}{2\pi\sqrt{MC}}$$

System	*Equations*	f *Frequency*	t *Time*	q *Parameter*	C *Capacitance*	M *Inertance*	R* *Resistance*
Mechanical translational	$s = S_m \cos(360ft)$ $f = \frac{1}{2\pi\sqrt{mC}}$	Hertz (Hz)	Second (s)	s Distance (m) (ft)	C Spring capacitance (m/N) (ft/lb)	m Mass (kg) (slug)	R $\left(\frac{N}{m/s}\right)$ $\left(\frac{lb}{ft/s}\right)$
Mechanical rotational	$\theta = \theta_m \cos(360ft)$ $f = \frac{1}{2\pi\sqrt{IC}}$	Hertz (Hz)	Second (s)	θ Angle (rad)	C Spring capacitance $\left(\frac{rad}{N \cdot m}\right)$ $\left(\frac{rad}{lb \cdot ft}\right)$	I Moment of inertia $(kg \cdot m^2)$ $(slug \cdot ft^2)$	R $\left(\frac{N \cdot m}{rad/s}\right)$ $\left(\frac{lb \cdot ft}{rad/s}\right)$
Fluid	$V = V_m \cos(360ft)$ $f = \frac{1}{2\pi\sqrt{MC}}$	Hertz (Hz)	Second (s)	V Volume (m^3) (ft^3)	C Reservoir capacitance $\left(\frac{m^3}{Pa}\right)$ $\left(\frac{ft^3}{lb/ft^2}\right)$	M Fluid inertance (kg/m^4) $(slug/ft^4)$	R $\left(\frac{Pa}{m^3/s}\right)$ $\left(\frac{lb/ft^2}{ft^3/s}\right)$
Electrical	$q = Q_m \cos(360ft)$ $f = \frac{1}{2\pi\sqrt{LC}}$	Hertz (Hz)	Second (s)	q Charge (C)	C Electrical capacitance (F)	L Coil inductance (H)	R (Ω)
Heat	Equations do not apply because there is no thermal inertance						

* *Note*: Resistance is included here to cover the elements included in Eq. (13-12).

13-3 OSCILLATING SYSTEMS

Now for a look at some other oscillating *MC* systems by using the knowledge we have just acquired.

Mechanical Translational System

An oscillating mechanical translational system consists of a mass connected to a spring. This system was used as an example in deriving the equations for frequency and parameter in *MC* systems. A common illustration of such a system is a mass suspended by a vertical spring. Figure 13-5a shows the spring by itself. A mass placed on the spring displaces it downward a distance x until the upward spring force equals the weight of the object (Fig. 13-5b). If the mass is pulled down farther and released (Fig. 13-5c), it oscillates about its equilibrium position with simple harmonic motion.

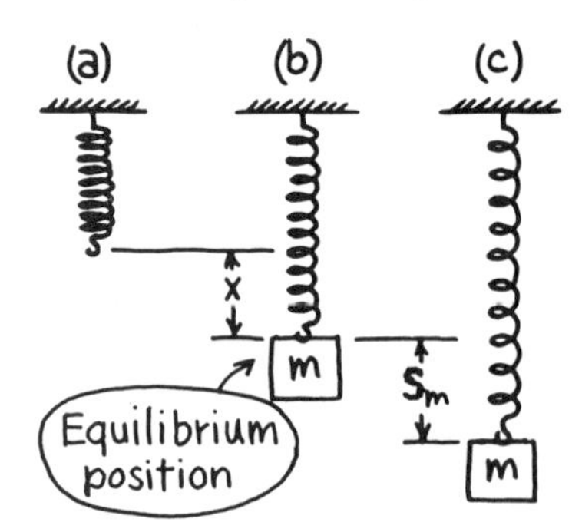

Figure 13-5. A mass suspended from a spring comes to rest in an equilibrium position. Oscillations about this position occur if the mass is further displaced and released.

Example 13-2 A mass of 0.25 kg is hung on a vertical spring, displacing it downward 0.015 m. The mass is pulled downward another 0.010 m from its equilibrium position and released. Friction is negligible and the spring's mass is very small.

(a) What is the frequency of oscillation?

(b) How far above the release position does the mass move before it starts its return?

(c) What is the frequency if the mass is doubled?

Solution to (a)

$\left\{ \begin{array}{l} m = 0.25 \text{ kg} \\ x = 0.015 \text{ m} \end{array} \right\}$ Calculate the spring capacitance and then calculate the frequency.

$$C = \frac{x}{F_C} = \frac{x}{W} = \frac{x}{mg} = \frac{0.015 \text{ m}}{0.25 \text{ kg}(9.81 \text{ m/s}^2)} = 6.1 \times 10^{-3} \text{ m/N}$$

$$f = \frac{1}{2\pi\sqrt{mC}} = \frac{1}{2\pi\sqrt{0.25 \text{ kg}(6.1 \times 10^{-3} \text{ m/N})}} = \boxed{4.1 \text{ Hz}}$$

Solution to (b)

The mass moves a maximum of $2S_m$ from the release position.

$$2S_m = 2(0.010 \text{ m}) = \boxed{0.020 \text{ m}}$$

Solution to (c)

$$\left\{\begin{aligned} m &= 2 \times 0.25\ \text{kg} \\ &= 0.50\ \text{kg} \end{aligned}\right\}$$

$$f = \frac{1}{2\pi\sqrt{0.50\ \text{kg}(6.1 \times 10^{-3}\ \text{m/N})}} = \boxed{2.9\ \text{Hz}}$$

The larger mass oscillates more slowly.

Mechanical Rotational System

Figure 13-6 shows a rotational device called a torsion pendulum. A long thin vertical shaft with rotational capacitance C supports a solid cylinder with moment of inertia I. The cylinder is displaced by a torque τ_C which twists

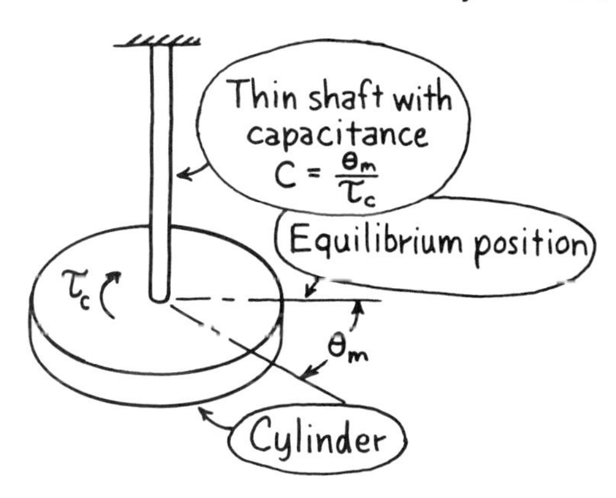

Figure 13-6. A torsion pendulum is an example of a rotational MC system. A torque τ_C twists the vertical shaft through angle θ_m from its equilibrium position. The system oscillates when the torque is suddenly removed.

the shaft through an angle θ_m. Upon release the cylinder rotates with simple harmonic motion at a frequency $f = 1/(2\pi\sqrt{IC})$. Example 13-3 is another simple illustration of a torsional system that you can set up and test in the physics lab.

Example 13-3 A wooden metre stick is clamped to the end of a table and a 0.50-kg mass is taped to the free end 0.90 m from the clamped end. The rotational capacitance of the deflected metre stick is 0.020 rad/(N · m). The end of the stick is pulled down slightly and released. What is the frequency of the resulting oscillation? Assume that the mass of the metre stick is small enough to be neglected and that there is no friction.

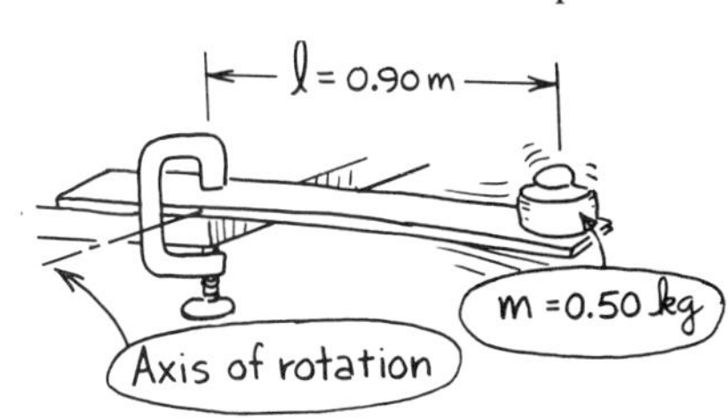

Solution

The moment of inertia of a concentrated mass m that is a distance ℓ from the axis of rotation is (from Appendix B, Table 6)

$$I = m\ell^2 = 0.50\ \text{kg}(0.90\ \text{m})^2 = 0.405\ \text{kg} \cdot \text{m}^2$$

Hence

$$f = \frac{1}{2\pi\sqrt{IC}} = \frac{1}{2\pi\sqrt{(0.405\ \text{kg}\cdot\text{m}^2)0.020\ \text{rad}/(\text{N}\cdot\text{m})}} = \boxed{1.8\ \text{Hz}}$$

Fluid System

Perhaps you are familiar with the noise that can occur in a water pipe when a faucet is suddenly turned on or off. The noise and the oscillatory movement of water that makes the rattling is called *water hammer.* The pipes act as capacitors and the moving water has inertance.

Figure 13-7 illustrates a simple liquid system consisting of water in a symmetrical U-shaped tube. The water levels can be made different by tilting the U-tube, stopping up the left tube, and turning the tube upright again. Oscillations start when you suddenly remove your finger from the left tube. The height X_m is a measure of the amount of liquid above the equilibrium line at the start.

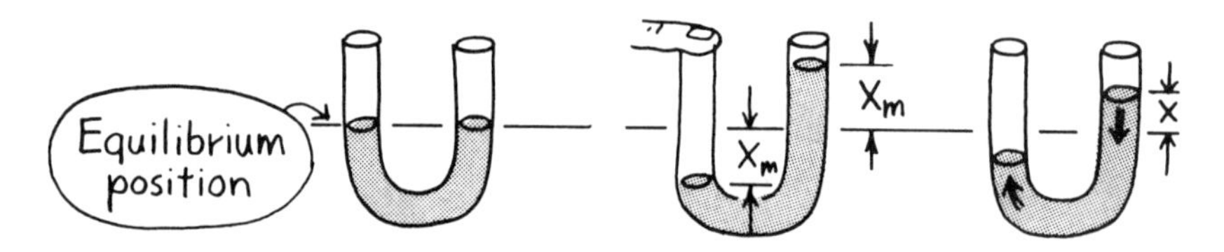

Figure 13-7. An oscillation occurs in a symmetrical U-shaped tube partially filled with water. The water is displaced away from its equilibrium position and released.

Figure 13-8 shows what happens to the water level x in the right tube. One cycle of the water's oscillation is illustrated. A mechanical mass–spring system is superimposed above the fluid system for comparison.

At the start, all the energy is stored in the reservoir (capacitor) as potential energy. When the finger is removed, gravity accelerates the water toward its equilibrium position, and the fluid's inertance causes it to overshoot. The volume flow rate $\dot{V}$ is greatest at time I when the water is at its equilibrium position and x is zero. At this instant all the potential energy stored at the start has been transferred to the water mass as kinetic energy. The water surges up into the left tube because of its inertia. The water loses speed as it moves up, and it comes to rest momentarily at II on the time axis. The height in the left tube at II is the same as the height in the right tube at 0 if no energy has been lost to friction. The water retraces its path back to its starting position at time IV to complete one cycle of the oscillation.

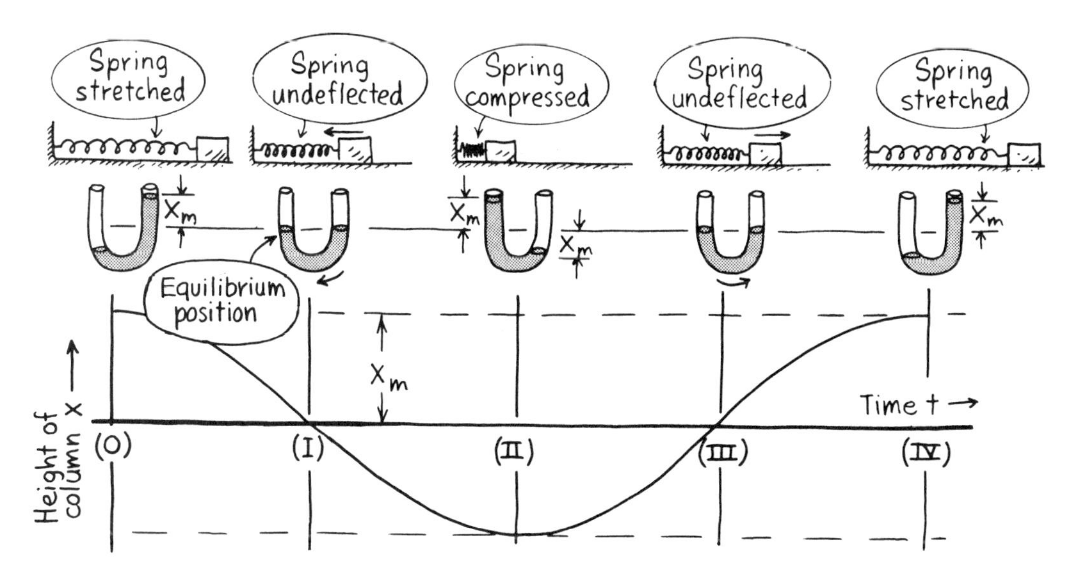

Figure 13-8. Oscillations of water in a symmetrical U-tube are like oscillations of a mass–spring system.

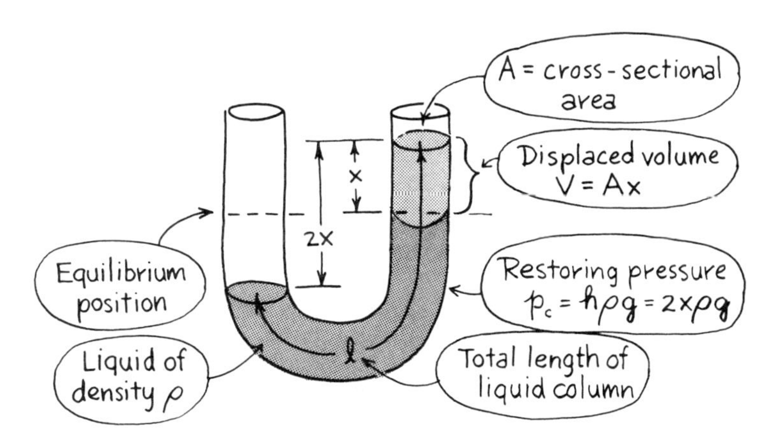

Figure 13-9. U-tube with an oscillating column of liquid.

The capacitance C of the tube (Fig. 13-9) is the ratio of displaced volume to the pressure tending to restore the system to equilibrium:

$$C = \frac{V}{p_C} = \frac{Ax}{2x\rho g}$$

$$= \frac{A}{2\rho g}$$

Fluid inertance M is $\rho\ell/A$. Thus, by Eq. (13-7),

$$f = \frac{1}{2\pi\sqrt{(\rho\ell/A)(A/2\rho g)}}$$

$$= \frac{1}{2\pi\sqrt{\ell/2g}} \qquad (13\text{-}9)$$

Equation (13-9) can also be derived by treating the water as an object in a mechanical translational system and using Newton's second law.

Example 13-4 Suppose that the cross-sectional area of a U-tube is $1.0 \times 10^{-4}\ \text{m}^2$, and the total length of its water column is 1.0 m. The column is displaced from its equilibrium position and released. Calculate the frequency of the resulting oscillation, assuming that no friction is present.

Solution

$\{\ell = 1.0\ \text{m}\}$ Use Eq. (13-9). Area A is not needed.

$$f = \frac{1}{2\pi\sqrt{\ell/2g}} = \frac{1}{2\pi\sqrt{1.0\ \text{m}/[(2)(9.81\ \text{m/s}^2)]}} = \boxed{0.70\ \text{Hz}}$$

Electrical System

An electrical system consisting of a coil and a capacitor is an *MC* system that oscillates if allowed to. Figure 13-10 shows such an *LC* system.

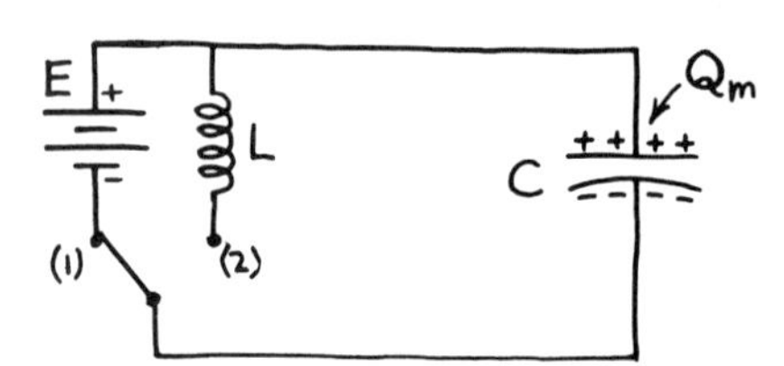

Figure 13-10. The switch in position 1 charges the capacitor to $Q_m = CE$. An oscillatory movement of charge occurs when the switch is thrown to position 2.

With the switch in position 1, the capacitor charges to its maximum value $Q_m = CE$. When the switch is thrown to position 2, the capacitor discharges through the coil, setting up an oscillation that we will describe by comparison with the mechanical mass–spring system. Figure 13-11 shows what happens.

At the start, all the energy is stored in the spring (mechanical system) and the capacitor (electrical system). When it is released, the mass accelerates to the left. Its velocity and kinetic energy increase to a maximum at the center or equilibrium position, where the displacement of the spring is zero. This is point I on the time graph. Meanwhile, in the electric circuit the capacitor discharges, and the resulting current produces a magnetic field around the coil. The

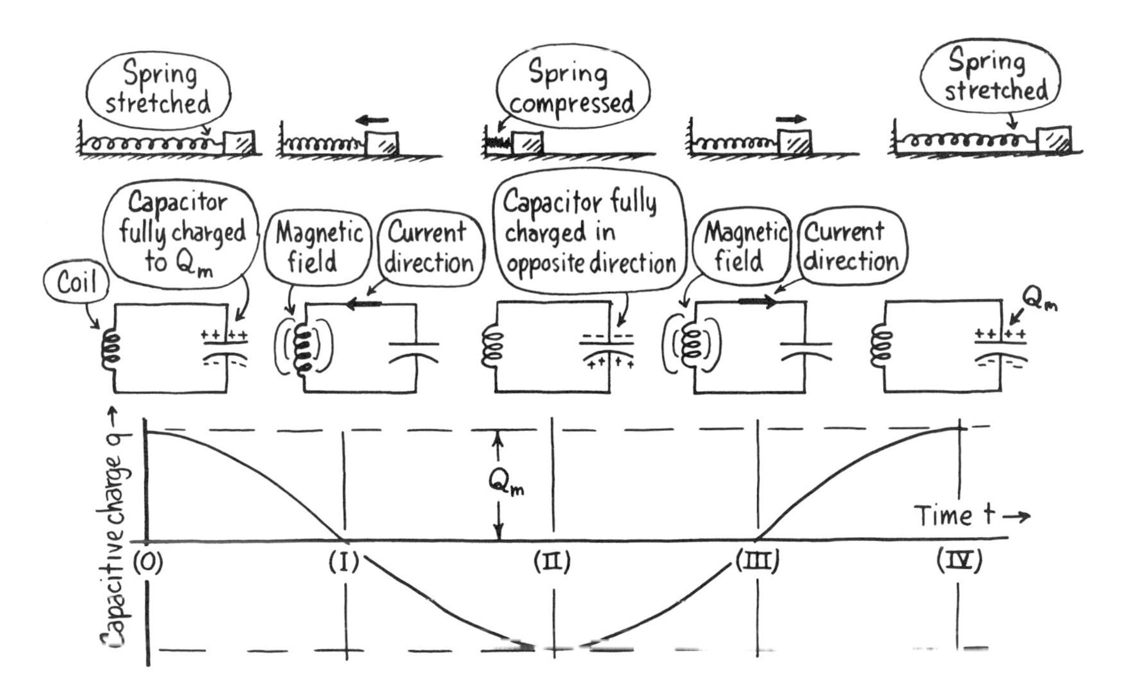

Figure 13-11. Oscillations in an electrical system with coil and capacitor behave like oscillations of a mass–spring system.

amount of "kinetic" energy stored in the magnetic field depends upon the current in the coil. Hence at point I in the time sequence the spring has no deflection, the capacitor has no charge, and all the energy is stored either in the mass as kinetic energy of motion or in the coil as "kinetic" energy of the magnetic field.

The mass passes the center position at time I and then starts to slow up as it compresses the spring. The mass cannot stop instantaneously because of its inertia, but it eventually comes to rest for an instant at time II in the time sequence. Now the energy is stored in the compressed spring. Similarly, the flow of positive charge in the electric circuit does not stop instantaneously. It surges onto the bottom plate of the capacitor. The flow of this charge (current) decreases and stops momentarily at time II in the time sequence. The capacitor is now fully charged (in the opposite direction) and all the energy resides in the capacitor. The current at time II is zero, and there is no magnetic field around the coil.

See if you can follow what happens in both the mechanical and electrical systems in order to complete the cycle. Does the movement of charge or mass ever stop? The answer is theoretically "no," *if* there is no resistance to drain energy away from either system. The mass and the charge continue to slop back and forth forever. Table 13-1 gave the equation for the frequency of the oscillating electrical system:

$$f = \frac{1}{2\pi\sqrt{LC}}$$

Example 13-5 A mass of 0.25 kg attached to a vertical spring oscillates at a frequency of 3.0 Hz. An electrical *LC* series system containing a 10-H coil oscillates at the same frequency.

(a) What is the capacitance of the spring?
(b) What is the capacitance of the capacitor in the electrical system?

Solution to (a)

$\left\{\begin{matrix} m = 0.25\ \text{kg} \\ f = 3.0\ \text{Hz} \end{matrix}\right\}$

$$f = \frac{1}{2\pi\sqrt{mC}} \Rightarrow f^2 = \frac{1}{4\pi^2 mC}$$

$$C = \frac{1}{4\pi^2 mf^2} = \frac{1}{4\pi^2(0.25\ \text{kg})(3.0\ \text{Hz})^2} = \boxed{1.1 \times 10^{-2}\ \text{m/N}}$$

Solution to (b)

$\{L = 10\text{H}\}$

$$C = \frac{1}{4\pi^2 Lf^2} = \frac{1}{4\pi^2(10\ \text{H})(3.0\ \text{Hz})^2} = \boxed{2.8 \times 10^{-4}\ \text{F}}$$

*13-4 SYSTEMS WITH RESISTANCE (*RMC* SYSTEMS)

We now take up the subject of resistance in systems that oscillate. Such oscillations do not continue indefinitely, because resistance continually drains energy away. The oscillations gradually die out and the system comes to rest. The pattern of decay appears to be exponential in nature; it is called *damped harmonic motion.* The mathematics required to develop the equations describing the decay of the oscillations includes a knowledge of calculus and is beyond the scope of our text.

Figure 13-12a shows a frictionless mechanical system displaced and suddenly released. We will work in terms of a general system, where q represents changing parameter (distance, volume, charge, etc.). If no resistance is present, the system oscillates back and forth with a frequency $f = 1/(2\pi\sqrt{MC})$. The period T of this oscillation is $T = 1/f$ or

$$T = 2\pi\sqrt{MC} \tag{13-10}$$

In Fig. 13-12b, a resistor in the form of a dashpot has been added in series to the system. The system exhibits damped harmonic motion if displaced and suddenly released. Two things about this motion are different from the undamped motion:

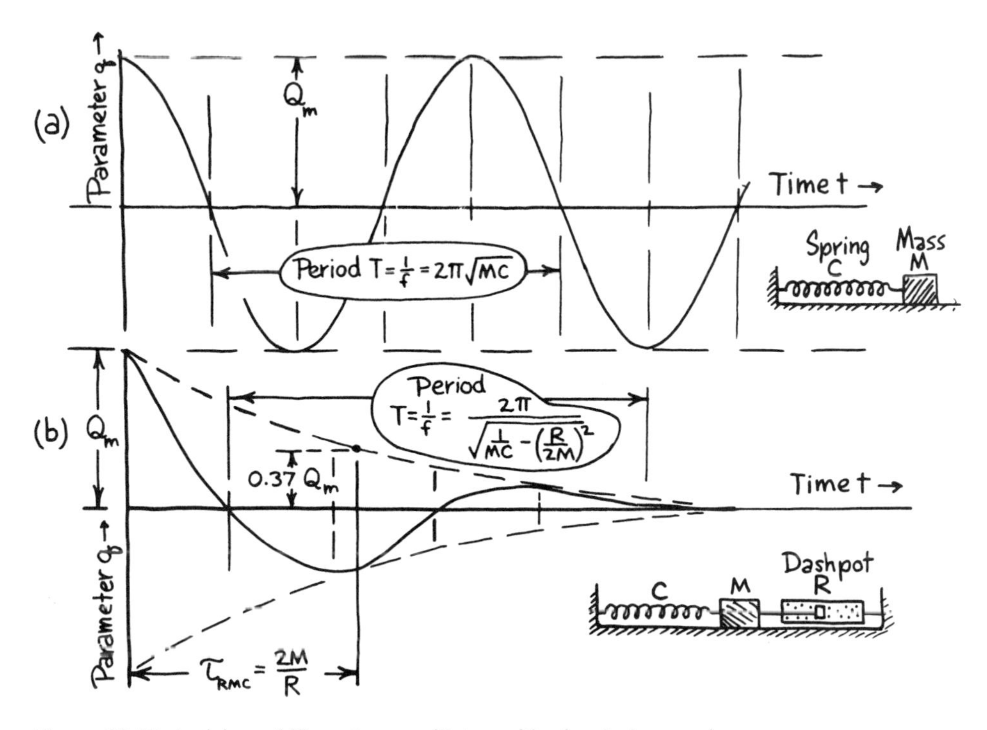

Figure 13-12. In (a) an *MC* system oscillates with simple harmonic motion. In (b) an *RMC* system oscillates with damped harmonic motion. The resistor drains energy from the system, reduces the amplitude, and increases the period of oscillation.

1. The amplitude does not remain constant but decreases in a manner that can be approximated by an exponential equation. The time constant for this exponential decay is twice as large as the time constant for a system with only inertance and resistance.

$$\tau_{RMC} = \frac{2M}{R} \tag{13-11}$$

2. The system oscillates more slowly; that is, the frequency is lower than in the same system with no resistance.

$$f = \frac{\sqrt{(1/MC) - (R/2M)^2}}{2\pi} \tag{13-12}$$

The position of R in the expressions for frequency and for time constant of damped harmonic motion makes sense. Resistance should hasten decay of an oscillation because energy is removed. The time constant τ_{RMC} should become smaller as R increases. The position of R in the denominator of the time constant equation (13-11) agrees with this. We should also expect that resistance makes the oscillations more sluggish, resulting in a lower frequency. An increase in the value of R reduces the numerator of Eq. (13-12), producing a smaller value of f.

Figure 13-13 shows the effect of increasing the resistance in an *RMC* system. The decay occurs more quickly and the frequency of oscillation decreases (period gets larger).

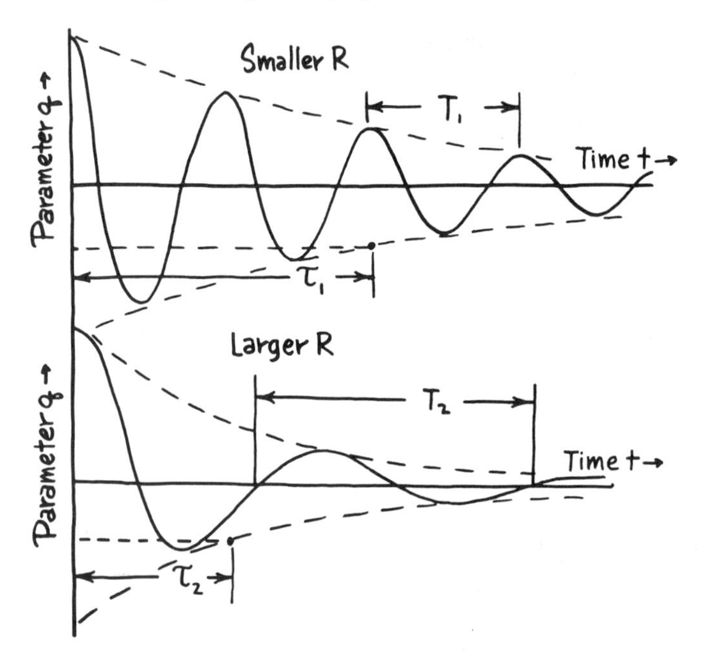

Figure 13-13. Adding resistance to a system reduces the time of decay and increases the period of oscillation (decreases frequency). A system with more resistance has a shorter time constant ($\tau_2 < \tau_1$) and a lower frequency or longer period ($T_2 > T_1$).

Critical Damping

What happens if resistance increases even more? Equation (13-12)

$$f = \frac{\sqrt{(1/MC) - (R/2M)^2}}{2\pi}$$

shows the following:

1. If $R = 0$, Eq. (13-12) boils down to $f = 1/(2\pi\sqrt{MC})$, the frequency of an *MC* system with no resistance (Fig. 13-14a).

2. If $(R/2M)^2$ is *less than* $1/MC$, as in Fig. 13-14b, the top of Eq. (13-12) is smaller, and there is an oscillation with a lower frequency (greater period) than for the undamped system.

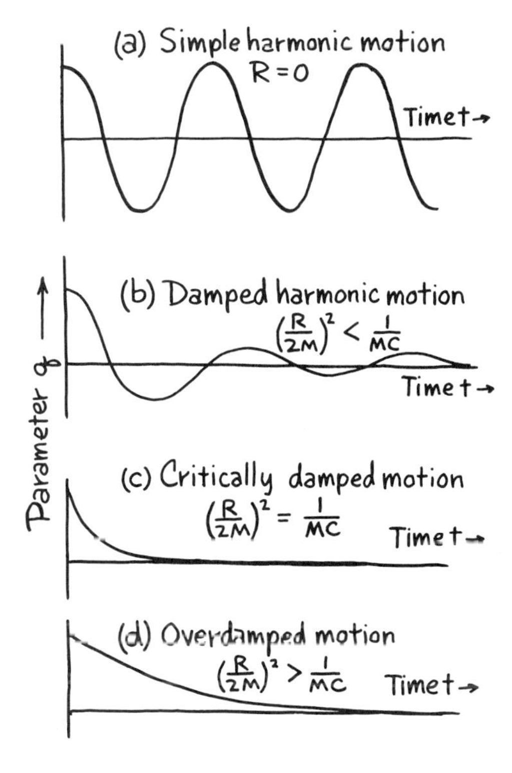

Figure 13-14. Adding resistance eventually stops the oscillation. The system is critically damped when this happens. The addition of more resistance causes the system to become overdamped. Critical damping guarantees the fastest possible return to equilibrium in response to a sudden change of excitation force.

3. If $(R/2M)^2$ is *equal to* $1/MC$ (Fig. 13-14c), the numerator is zero, making the frequency zero and the period infinite. There is no oscillation at all, and the system decays to its equilibrium condition without any jiggling back and forth. The system is said to be *critically damped* when there is just enough resistance present to stop oscillations. The resistance needed to do this is called the *critical resistance* R_{cr}. Since $(R_{cr}/2M)^2 = 1/MC$,

$$R_{cr} = 2\sqrt{M/C} \tag{13-13}$$

The critically damped system does not decay exactly according to a simple exponential, as we had in the previous chapters. However, we can approximate the decay time by saying that it is completed in five or six time constants, where $\tau = 2M/R_{cr}$.

4. If $(R/2M)^2$ is *greater than* $1/MC$, the expression for frequency becomes imaginary (square root of a negative number) and has no meaning. There is no oscillation, and the decay takes even longer than for critical damping (Fig. 13-14d). The system is said to be *overdamped*.

The various effects of adding resistance can be demonstrated using an electrical series *RLC* circuit connected to a square-wave generator. An oscilloscope, connected across the capacitor, shows pictures similar to Fig. 13-14 as resistance is increased in the circuit. Critically damped systems are important because they have the shortest response time. They come to steady state quickest after a sudden change in the excitation force.

Example 13-6 A battery charges the capacitor to 10 V with the switch in position 1.

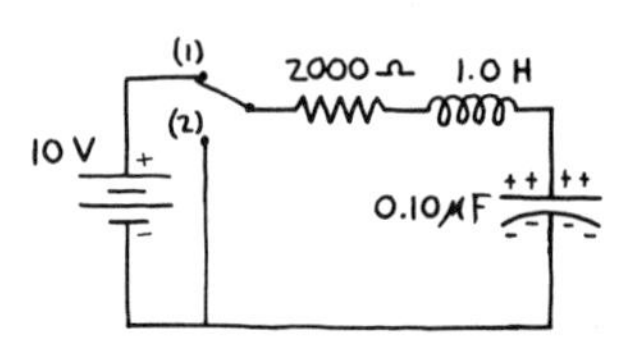

(a) Make a sketch of what happens to the capacitor charge during a 5.0-ms time interval after the switch is thrown to position 2.

(b) What resistance is needed for critical damping?

(c) Estimate the time needed for the capacitor to discharge with the system critically damped.

Solution to (a)

$$\left\{ \begin{array}{l} R = 2000\ \Omega \\ L = 1.0\ \text{H} \\ C = 0.10 \times 10^{-6}\ \text{F} \end{array} \right\}$$

The time constant of the decay is

$$\tau_{RMC} = \frac{2L}{R} = \frac{2(1.0\ \text{H})}{2000\ \Omega} = 1.0 \times 10^{-3}\ \text{s} = \boxed{1.0\ \text{ms}}$$

The period of oscillation is

$$T = \frac{1}{f} = \frac{2\pi}{\sqrt{(1/LC) - (R/2L)^2}}$$

Note that $R/2L = 1/\tau_{RMC}$

$$= \frac{2\pi}{\sqrt{1/[(1.0\ \text{H})(0.10 \times 10^{-6}\ \text{F})] - [(2000\ \Omega/(2 \times 1.0\ \text{H})^2]}}$$

$$= 2.1 \times 10^{-3}\ \text{s} = \boxed{2.1\ \text{ms}}$$

The initial capacitor charge is

$$Q_m = CE = (0.10 \times 10^{-6}\ \text{F})(10\ \text{V}) = \boxed{1.0 \times 10^{-6}\ \text{C}}$$

First sketch in the envelope of the decay on the graph of charge versus time. At $t = \tau_{RMC} = 1.0$ ms, the charge is $0.37(1.0 \times 10^{-6}) = 0.37 \times 10^{-6}$ C. At $t = 5\tau_{RMC} = 5.0$ ms, the charge is almost zero. Next lay off oscillations with period $T \cong 2.0$ ms. About 2.5 cycles fit within the 5-ms time interval before everything stops.

Solution to (b)

$$R_{cr} = 2\sqrt{\frac{L}{C}} = 2\sqrt{\frac{1.0\ \text{H}}{0.10 \times 10^{-6}\ \text{F}}} = \boxed{6.3 \times 10^3\ \Omega}$$

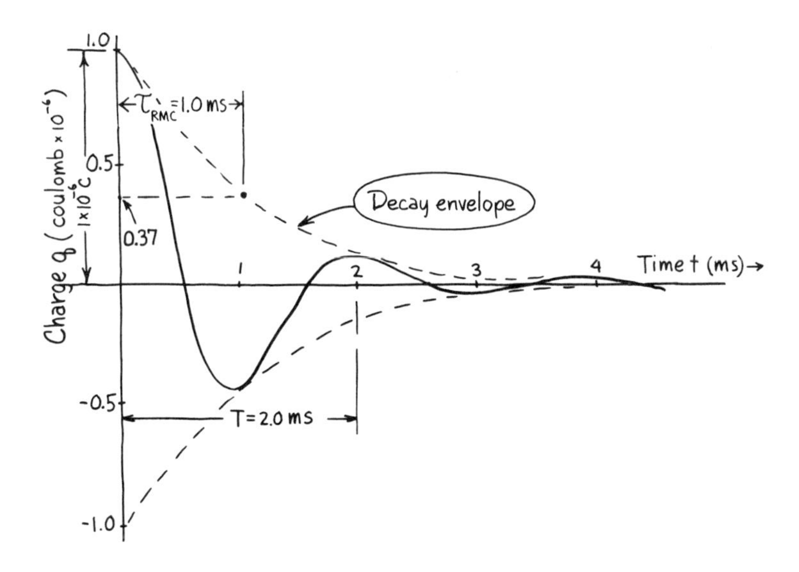

Solution to (c)

$$\tau_{RMC} = \frac{2L}{R_{cr}} = \frac{2(1.0\ \text{H})}{6300\ \Omega} = 0.32 \times 10^{-3}\ \text{s} = 0.32\ \text{ms}$$

$$5\tau_{RMC} = 5(0.32\ \text{ms}) = \boxed{1.6\ \text{ms}}$$

13-5 SUMMARY

1. Systems containing capacitance and inertance (but not resistance) oscillate if disturbed from their equilibrium position. The oscillation follows the equation of a sine or cosine curve and is called *simple harmonic motion.*

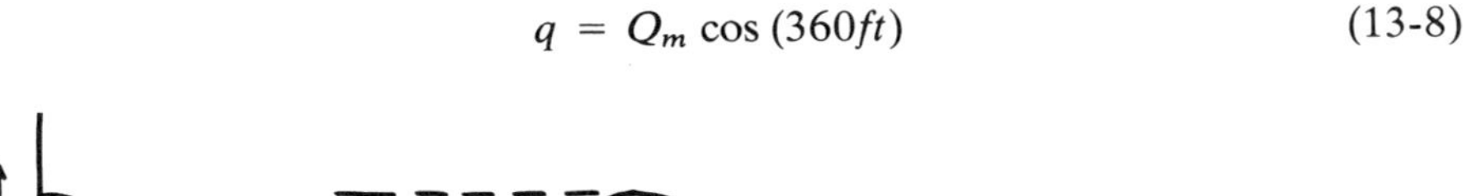

$$q = Q_m \cos(360ft) \tag{13-8}$$

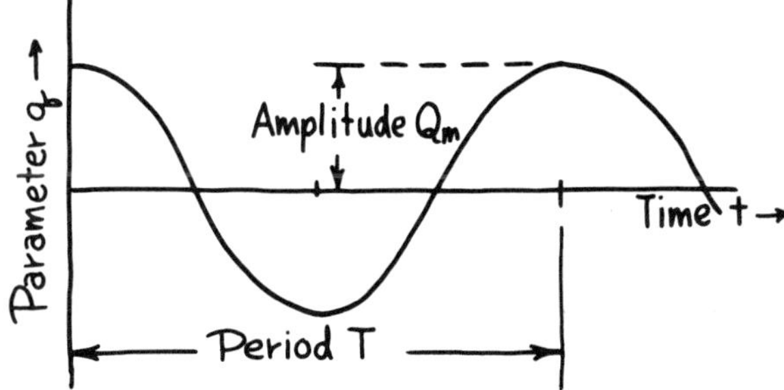

The *amplitude* Q_m of the disturbance is the largest value of the parameter. The *period* T is the time to complete one cycle of motion. The *frequency* f is the number of cycles in 1 s. Frequency is the reciprocal of period:

$$f = \frac{1}{T} \tag{13-1}$$

2. The equation for the frequency f in terms of the system's capacitance C and inertance M is

$$f = \frac{1}{2\pi\sqrt{MC}} \tag{13-7}$$

3. In the various systems this becomes

Mechanical translational: $f = \dfrac{1}{2\pi\sqrt{mC}}$ — m: Mass; C: Linear spring capacitance

Mechanical rotational: $f = \dfrac{1}{2\pi\sqrt{IC}}$ — I: Moment of inertia; C: Rotational spring capacitance

Electrical: $f = \dfrac{1}{2\pi\sqrt{LC}}$ — L: Coil inductance; C: Capacitor capacitance

Fluid: $f = \dfrac{1}{2\pi\sqrt{MC}}$ — M: Fluid inertance; C: Reservoir capacitance

4. The oscillation is *damped* if resistance R is present in the system. Equations for time constant τ_{RMC} of the decay envelope and the frequency f of

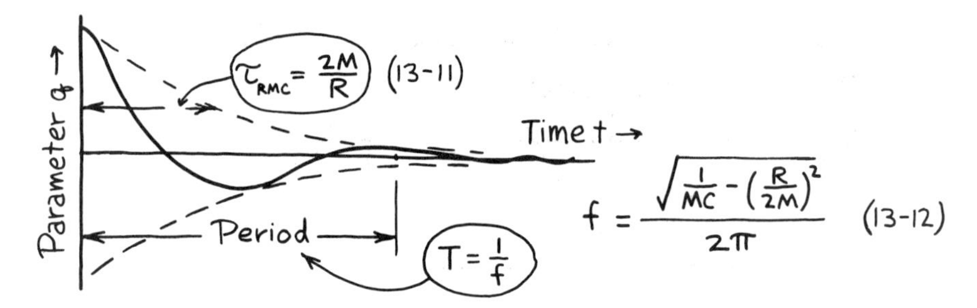

the damped oscillation are shown in the accompanying sketch. If more resistance is added to the system, greater damping occurs, until a point is reached where the system moves to its equilibrium state with no oscillation or overshoot. Such a system is *critically damped* and the resistance that just stops the oscillation is the *critical resistance* R_{cr}.

$$R_{cr} = 2\sqrt{\frac{M}{C}} \tag{13-13}$$

Further increase in system resistance causes *overdamping*. The system moves to its equilibrium position but takes longer to do so.

PROBLEMS

13-1. A mass–spring system with negligible resistance is displaced 0.20 m from its equilibrium position and released. The system makes 20 complete cycles of simple harmonic motion in a 10-s time interval.
(a) What is the frequency of the oscillation?
(b) Where is the mass 0.10 s after it is released?

13-2. A spring stretches 0.10 m when a 0.40-kg mass is hung from the free end. If the mass is displaced downward and released, what is the frequency of the resulting oscillatory motion? Neglect any frictional effects and assume that the mass of the spring is negligible.

13-3. Suppose that the mass of the unsupported part of the metre stick in the figure of Example 13-3 is 180 g. There is, however, *no* concentrated mass at its free end. The unclamped length is 0.90 m and the torsional capacitance is 0.020 rad/(N · m). At what frequency will the stick vibrate if displaced downward and released? (*Hint*: Refer to Appendix B, Table 6, for the moment of inertia of various objects.)

13-4. A torque of 0.050 N · m twists a flywheel through a 30° angle against the torque of a coiled spring. When the system is released, it oscillates with a frequency of 3.0 Hz.
(a) What is the torsional capacitance of the coil spring?
(b) What is the moment of inertia of the flywheel?

13-5. A U-tube with a cross-sectional area of 1.0 cm^2 contains 50 cm^3 of mercury. The liquid is depressed slightly to one side and released so that it sloshes back and forth. Estimate the frequency of the oscillation. Neglect any friction in this system.

13-6. What, if anything, happens to the frequency of oscillation in Problem 13-5 if the volume of mercury is doubled?

13-7. A 0.50-H coil and a 0.010-μF capacitor are connected to a 1.0-V battery as shown. Describe what happens when the switch is suddenly thrown from position 1 to position 2.

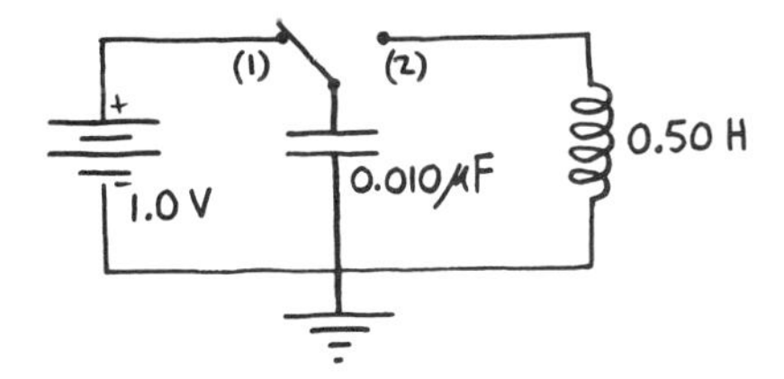

Problem 13-7

13-8. The antenna picks up signals of many different frequencies. In order to listen to just one frequency, the natural frequency of the circuit is adjusted (the circuit is tuned) by changing the value of the capacitance C. When

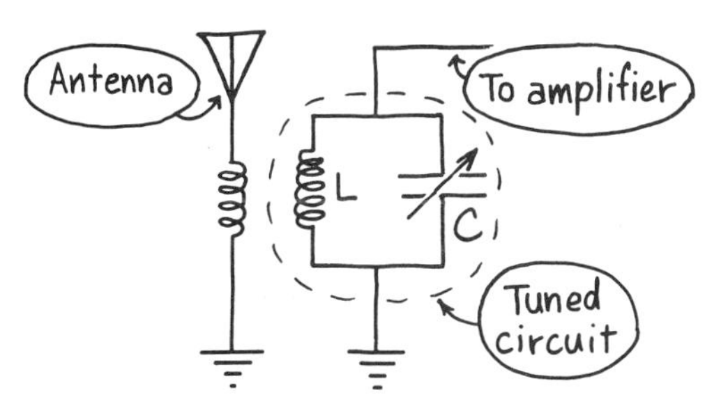

Problem 13-8

the natural circuit frequency matches that of an incoming signal, the others are effectively excluded. If the inductance L is 9.0 mH, what must be the value of C in order to listen to the ball game on 560 kHz?

13-9. The mass in Problem 13-1 is 0.15 kg.
(a) What is the spring capacitance?
(b) What is the largest energy ever stored in the spring?
(c) What is the largest velocity of the mass and where does this occur?

13-10. A solid brass sphere 6.0 inches in diameter is attached to the end of a thin rod that acts as a torsional spring. The sphere, which has a mass of 1.1 slug, is twisted and released. Eight cycles of motion are observed in a 10-s time interval.
(a) What is the moment of inertia of the brass sphere? (*Hint*: Refer to Appendix B, Table 6.)
(b) What is the torsional capacitance of the thin rod in rad/(lb · ft)?

13-11. A mechanical rotational system consists of a torsional spring with a capacitance of 0.40 rad/(lb · ft) and a mass with a moment of inertia of 0.025 slug · ft^2 about its rotational axis. If the system is twisted and released:
(a) What is the frequency of the oscillatory motion?
(b) How long does it take to complete 20 oscillations? Assume that no friction is present in the system.

13-12. The mechanism that controls my wind-up alarm clock is a harmonic oscillator with a frequency of 2.0 Hz. The inertive element is a balance wheel, which rotates on a shaft connected to a capacitive element, the hair spring. The wheel's diameter is 1.4 cm and I calculate its mass to be 1.6 g. Nearly all its mass is around the periphery of the wheel (i.e., the spokes are light, as on a bicycle wheel).
(a) Calculate the rotational spring constant of the hair spring.
(b) What would the frequency be if the balance wheel were made of aluminum instead of brass?

13-13. A 0.050-μF capacitor is charged up by a 6.0-V battery. The battery is removed and the capacitor is suddenly connected in series with a coil of inductance L. What must be the value of L in order to have an oscillation with a frequency of 20,000 Hz? Neglect any effects of resistance on the circuit.

13-14. Refer to problem 13-7 with the switch in position 2.
(a) What is the largest current in the circuit? (*Hint*: Equate the largest energy stored in the capacitor to the largest energy stored in the magnetic field of the coil.)
(b) What resistance must be added to critically damp the oscillation?

13-15. A machine with a mass of 2000 kg is suddenly but gently dropped onto a very light platform. The effective spring constant of the springs acting together is 3.0×10^5 N/m, and the effective resistance of the shock absorbers acting together is 9.8×10^4 N/(m/s).
(a) Is the resulting motion underdamped or overdamped?
(b) What should the shock absorber resistance be for critical damping?
(c) Estimate the time needed for the system to come to rest when critically damped.

Problem 13-12

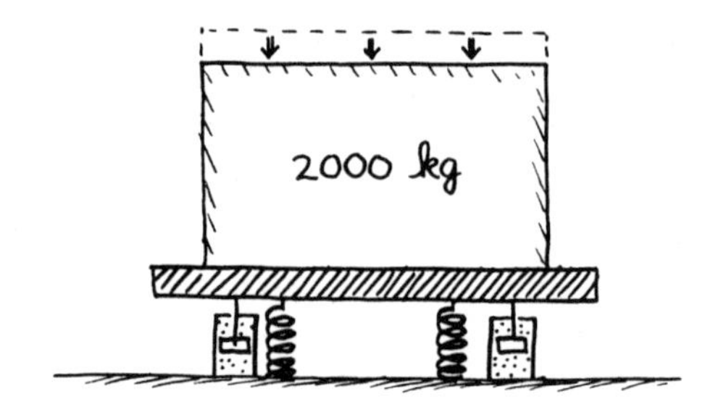

Problem 13-15

13-16. What must be the resistance of a dashpot to critically damp the motion of the mass–spring system in Problem 13-9?

13-17. The brass sphere and rod in Problem 13-10 are immersed in a liquid, twisted, and released. The resulting oscillations are damped out in about 5.0 s.

(a) Estimate the torsional resistance of the liquid acting on the oscillating system.

(b) What is the frequency of the damped oscillation?

13-18. Suggest some ways in which the oscillations of the liquid in the two glass tubes connected by thin rubber tubing can be critically damped.

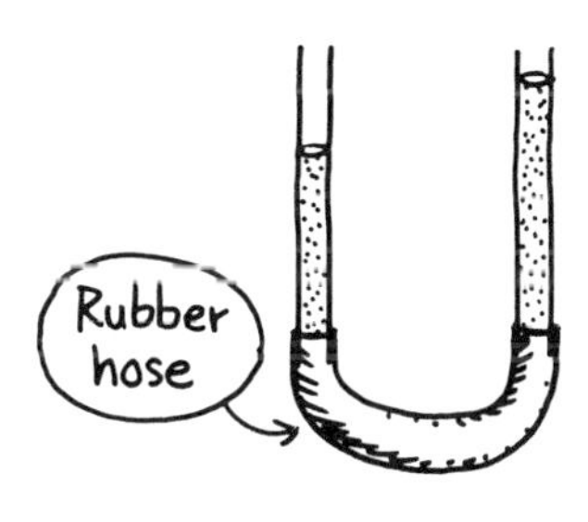

Problem 13-18

13-19. Suppose that a resistance of 5000 Ω is placed in series with the coil and capacitor of Problem 13-7.

(a) How long does it take for the resulting oscillation to die out when the switch is suddenly thrown from position 1 to 2?

(b) What is the period of these oscillations?

13-20. The period of oscillation for a general MC system is $T = 2\pi\sqrt{MC}$. Prove that T has the unit of seconds in the mechanical translational, mechanical rotational, fluid, and electrical systems. Use SI units. (*Hint*: The force unit newton, N, is a kg · m/s^2.)

13-21. A 1000-Ω resistor and a 0.40-H coil are suddenly connected in series to a capacitor previously charged by a 10-V battery.

(a) What is the time constant of the decay envelope and the period of the oscillation if the capacitor is rated at 0.010 μF?

(b) Make a sketch of how the capacitor voltage changes with time. (*Hint*: Capacitor voltage is proportional to capacitor charge.)

13-22. Refer to Problem 13-21.

(a) What should be the capacitance of the capacitor to critically damp any oscillation that results when the capacitor is suddenly connected to the coil and resistor?

(b) Make a sketch of how the capacitor voltage changes with time.

13-23. The oscillations of the water column in Example 13-4 die out to less than 1 percent of the original amplitude in 7.5 s. Calculate the system's fluid resistance. Assume that R is constant.

13-24. A 1000-kg car sags 1.0 cm all the way around when a 100-kg passenger climbs aboard.

(a) Find the natural frequency of the system if the shock absorbers are removed.

(b) Find the resistance R_{cr} of shock absorbers that must be added to produce critical damping.

ANSWERS TO ODD-NUMBERED PROBLEMS

13-1. (a) 2.0 Hz
(b) 0.062 m from equilibrium position

13-3. 5.1 Hz

13-5. 1.0 Hz

13-7. Charge oscillates at a frequency of 2.2×10^3 Hz.

13-9. (a) 4.2×10^{-2} m/N
(b) 0.47 J
(c) 2.5 m/s

13-11. (a) 1.6 Hz
(b) 13 s

13-13. 1.3×10^{-3} H

13-15. (a) Overdamped
(b) 4.9×10^4 N/(m/s)
(c) 0.41 s

13-17. (a) 5.5×10^{-2} lb · ft/(rad/s)
(b) 0.78 Hz

13-19. (a) 1.0×10^{-3} s
(b) 0.47×10^{-3} s

13-21. (a) 0.80×10^{-3} s, 0.40×10^{-3} s

(b)

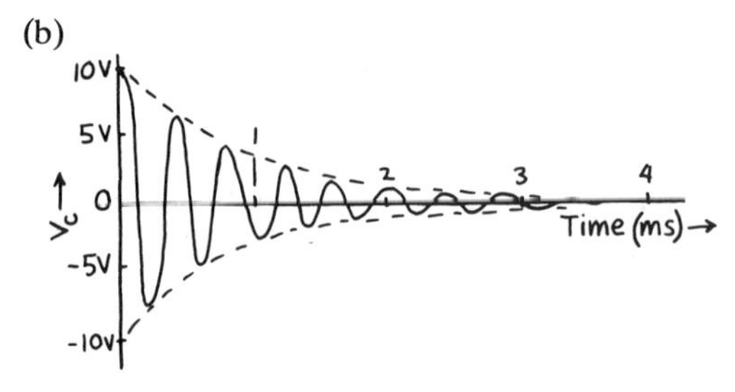

13-23. 8.0×10^{6} Pa/(m^3/s)

14

Forced Oscillations—*R, M,* and *C* Systems

This chapter looks at systems that are forced to oscillate by an excitation force that is neither steady, nor suddenly on, nor suddenly off. The excitation force acts according to the mathematics of a sine curve, pushing first in one direction and pulling next in the other. We will study the system's response to this force, looking in particular at the behavior of the parameter rate (velocity, current, etc.). Observations are to be made only after the excitation force has been acting for some time, so that anything associated with starting up the system has died out according to the time-constant concepts developed in the three preceding chapters. The system is then in the sinusoidal steady-state condition. We will find that the parameter rate also is a sine function, but it may not be in phase with the excitation force. This means that the parameter rate and excitation force do not peak at the same time. Quantities called capacitive reactance and inertive reactance (similar to resistance) describe the relation between excitation force and parameter rate in these oscillating systems.

Only mechanical translational and electrical systems are to be studied in this chapter. Fluid and heat systems will come later. In the mechanical system, a sinusoidal force acts separately on a dashpot, a mass, and a spring. In the electrical system, a sinusoidal emf acts separately on a resistor, a coil, and a capacitor. Usually, this kind of analysis is done on electrical systems, but the mechanical and electrical systems act the same way. Furthermore, you can visualize the dashpot, mass, and spring in motion. This may give you insight into what is happening in sinusoidal systems that you would not get by studying the electrical system alone.

14-1 SINUSOIDAL MOTION

Oscillatory, sinusoidal parameter rates (velocities, volume flow rates, electric current, heat rates), and forces (force, torque, pressure, voltage, temperature) are very common. Generators produce alternating currents and voltages. Rotating machinery does the same for mechanical and fluid systems. Systems are excited by energy that comes from sound, optical, radio, and other wave sources. These are all sinusoidal inputs.

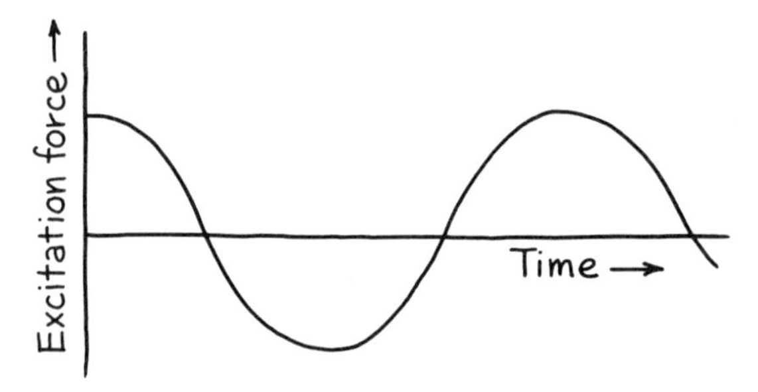

Figure 14-1. The excitation force varies sinusoidally with time.

When the amplitude and frequency of a sinusoidal excitation force and the resulting parameter rate are maintained over long periods of time, a system is said to be in the *sinusoidal steady state.* Note that this does not agree with the previous notion of steady state (Sec. 6-2) for which the parameter rate $\dot{Q}$ remains constant. Here parameter rates and excitation forces continually change not only their magnitudes but also their directions. We consider a system to be in the sinusoidal steady state if amplitudes (maximum value of whatever is oscillating) and frequency (number of oscillations each second) do not change over long periods of time.

The equation for the instantaneous value y of something that varies sinusoidally with time is

Remember that the subscript m is used to indicate the *maximum* value of a changing quantity

$$y = Y_m \sin(360ft) \tag{14-1}$$

y *Instantaneous Value of Whatever Is Changing*	Y_m *Amplitude*	f *Frequency*	t *Time*
Units appropriate to whatever is varying		hertz (Hz)	second (s)

The product $360ft$ in Eq. (14-1) is an angle measured in degrees. The sine curve equation (14-1) is plotted in Fig. 14-2. The *period* T is the reciprocal of frequency f and is the time to complete one cycle of oscillation.

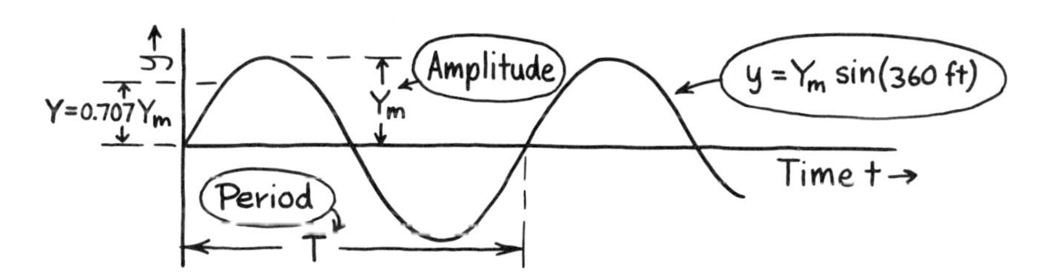

Figure 14-2. Equation (14-1) is a sine curve with period T, amplitude Y_m, frequency f, and an rms value Y. The product 360 ft is an angle measured in degrees.

In keeping with common practice, a lowercase letter such as y will refer to the exact or instantaneous value of a changing function; the uppercase letter with a lowercase m subscript (Y_m) will refer to the peak value.

Another measure of the sine curve's magnitude, denoted by the uppercase letter alone, is the *root mean square* (rms) value. The rms value Y is 0.707 times the amplitude:

$$Y = 0.707Y_m \tag{14-2}$$

The importance of rms values will be discussed later in this chapter.

The horizontal axis of a sine curve can be plotted as angle rather than time. A time of one period is equivalent to 360°. Figure 14-3 shows the time–angle relation for a sine curve. The value of y is zero at 0°, 180°, and 360°. It is positive Y_m at 90° and negative Y_m at 270°.

Phase Angle

The sine curve in Fig. 14-2 starts in such a position that y is zero when t is zero. However, it does not have to be this way. Figure 14-4a shows a sine curve in which y is positive when t is zero, and Fig. 14-4b shows a sine

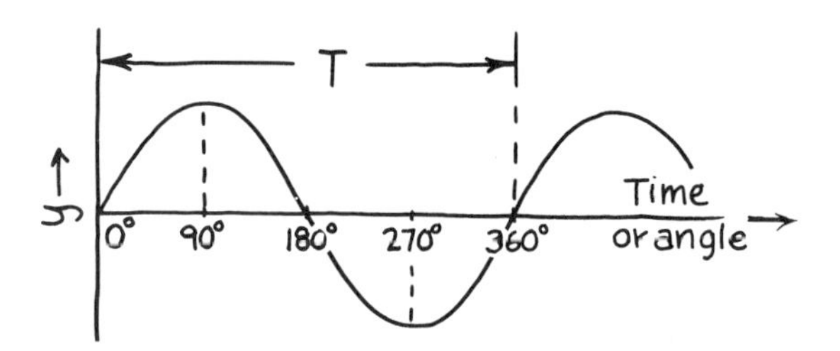

Figure 14-3. The horizontal axis can be plotted in terms of angle as well as time. The period T is equivalent to 360°.

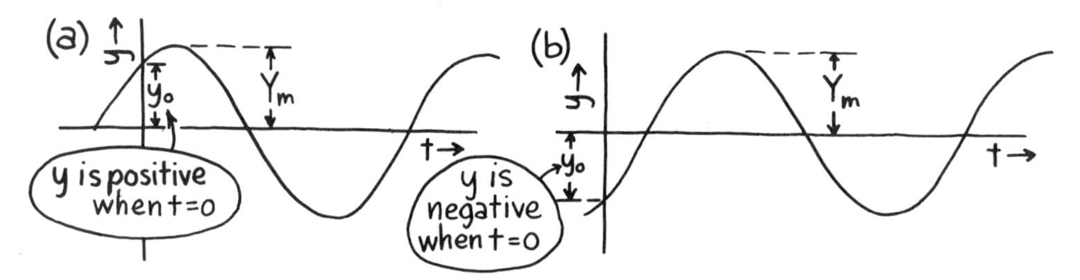

Figure 14-4. The sine curve can be shifted along the time axis so that y is either positive or negative when $t = 0$.

curve with a negative value of y when t is zero. Equation (14-1) can be modified to account for this shift of the sine curve along the axis. The equation becomes

$$y = Y_m \sin(360ft + \phi) \tag{14-3}$$

The angle ϕ (phi) is called the *phase angle* and is measured in degrees. It may be either positive or negative. If ϕ is positive, y is positive when t is zero, and the sine curve is positioned as in Fig. 14-5a. The curve is shifted toward the left and ϕ is plotted in degrees to the left of the origin. Figure 14-5b shows a sine

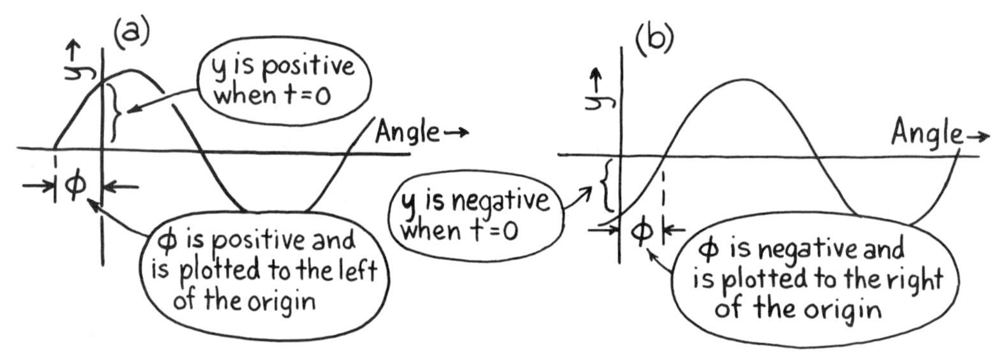

Figure 14-5. *Positive* phase angles plot to the *left* of the origin, and *negative* phase angles to the *right* of the origin.

curve with a negative phase angle. The curve is shifted to the right and the phase angle is plotted to the right of the origin.

Example 14-1 The phase angle, amplitude, and frequency of a sine curve are +30°, 20.0, and 100 Hz, respectively.

(a) What is the value of y when t is zero?
(b) At what time is y a positive maximum?
(c) What is the rms value of y?
(d) Sketch one cycle of the sine curve.

Solution to (a)

$$y = Y_m \sin(360ft + \phi) = 20.0 \sin(0 + 30°) = 20.0(0.50) = \boxed{10}$$

Solution to (b)

y is maximum when $\sin(360ft + \phi)$ is maximum. The sine of an angle is maximum at 90°. Thus

$$y = Y_m \quad \text{when} \quad 360ft + 30° = 90° \Rightarrow t = \frac{90 - 30}{360f} = \frac{60}{360(100\ \text{Hz})}$$

$$= \boxed{1.7 \times 10^{-3}\ \text{s}}$$

Solution to (c)

$$Y = 0.707Y_m = 0.707(20.0) = \boxed{14.1}$$

Solution to (d)

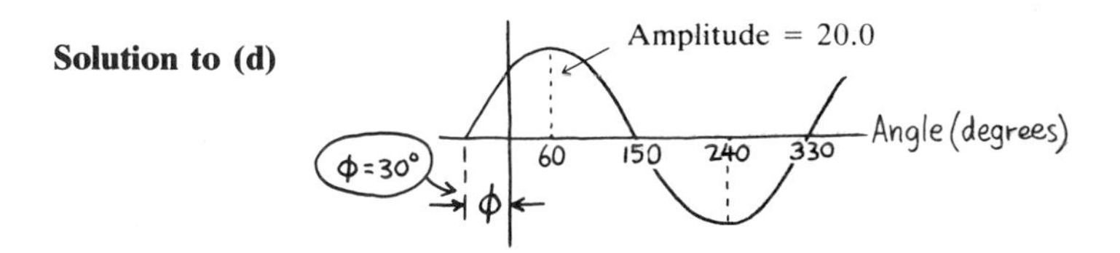

Sine curves like the one just discussed play a prominent part in the remaining sections of this chapter. They will represent sinusoidal forces and voltages, and parameter rates such as velocity and electric current.

14-2 SYSTEMS CONTAINING RESISTANCE

Our first look at the forced oscillations of mechanical and electrical systems involves only resistance. We start with a mechanical system and take some time to study a mechanical device that generates sinusoidal translational velocities and forces.

Figure 14-6 shows such a mechanical generator. A constant-speed motor turns a heavy flywheel clockwise as viewed from the top. An off-center pin is fastened to the flywheel. This fits and slides freely in a slot cut in a plate that

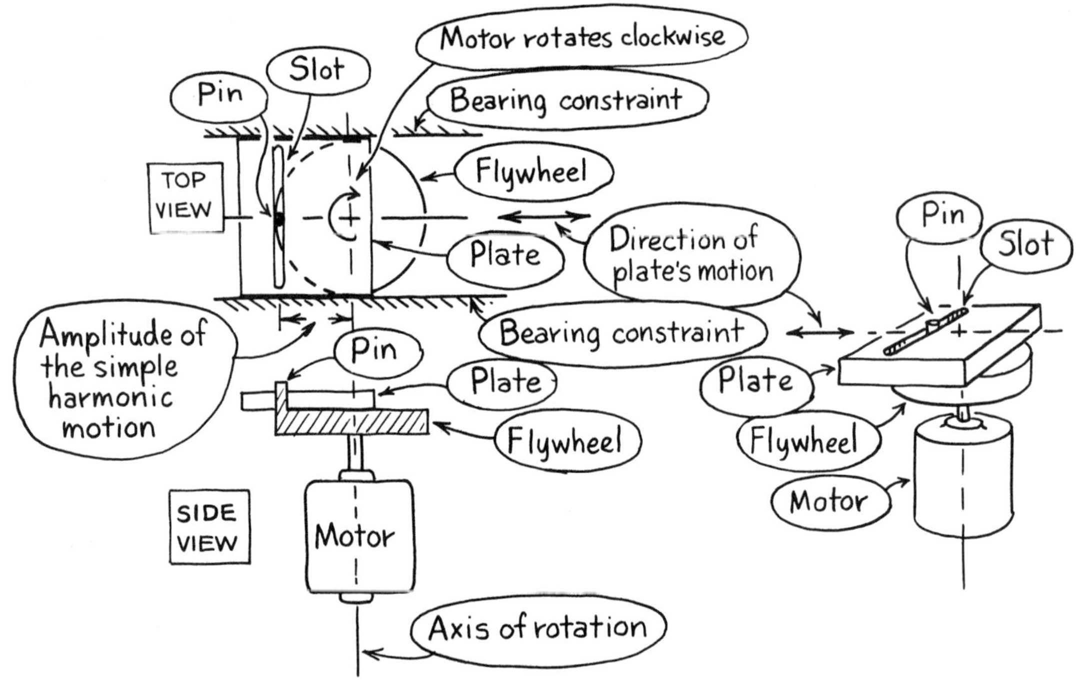

Figure 14-6. Three views of the mechanical generator.

rests on top of the flywheel. The plate is constrained so that it can only move back and forth along a line. Every point on the plate undergoes simple harmonic motion (Sec. 13-1). The frequency of this motion is the motor's angular velocity in revolutions per second. The amplitude of the motion is the distance between the motor's axis and the pin.

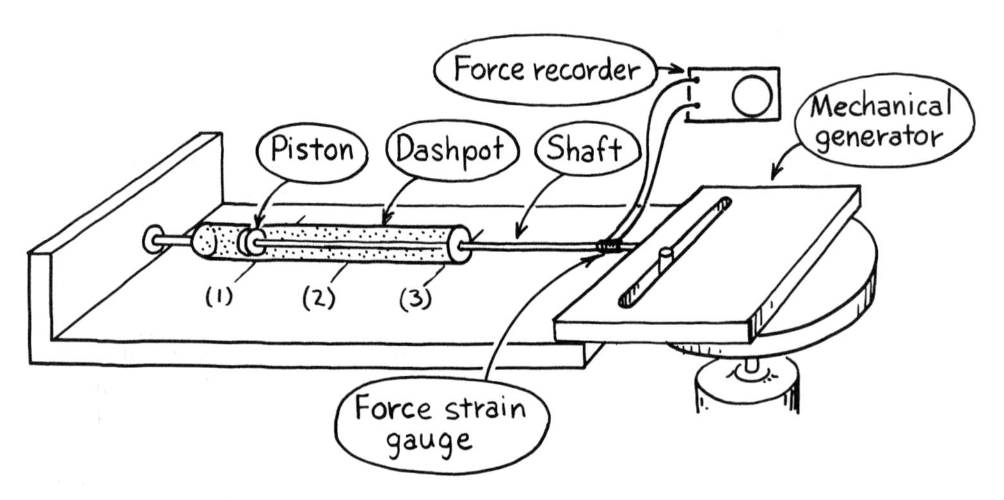

Figure 14-7. A mechanical resistor, the dashpot, is attached to the mechanical sine generator.

Resistance in a Mechanical System

Figure 14-7 shows a dashpot (mechanical translational resistor) attached to the plate of the mechanical generator. A force strain gauge is mounted on the piston's shaft so we can measure the force needed to move the

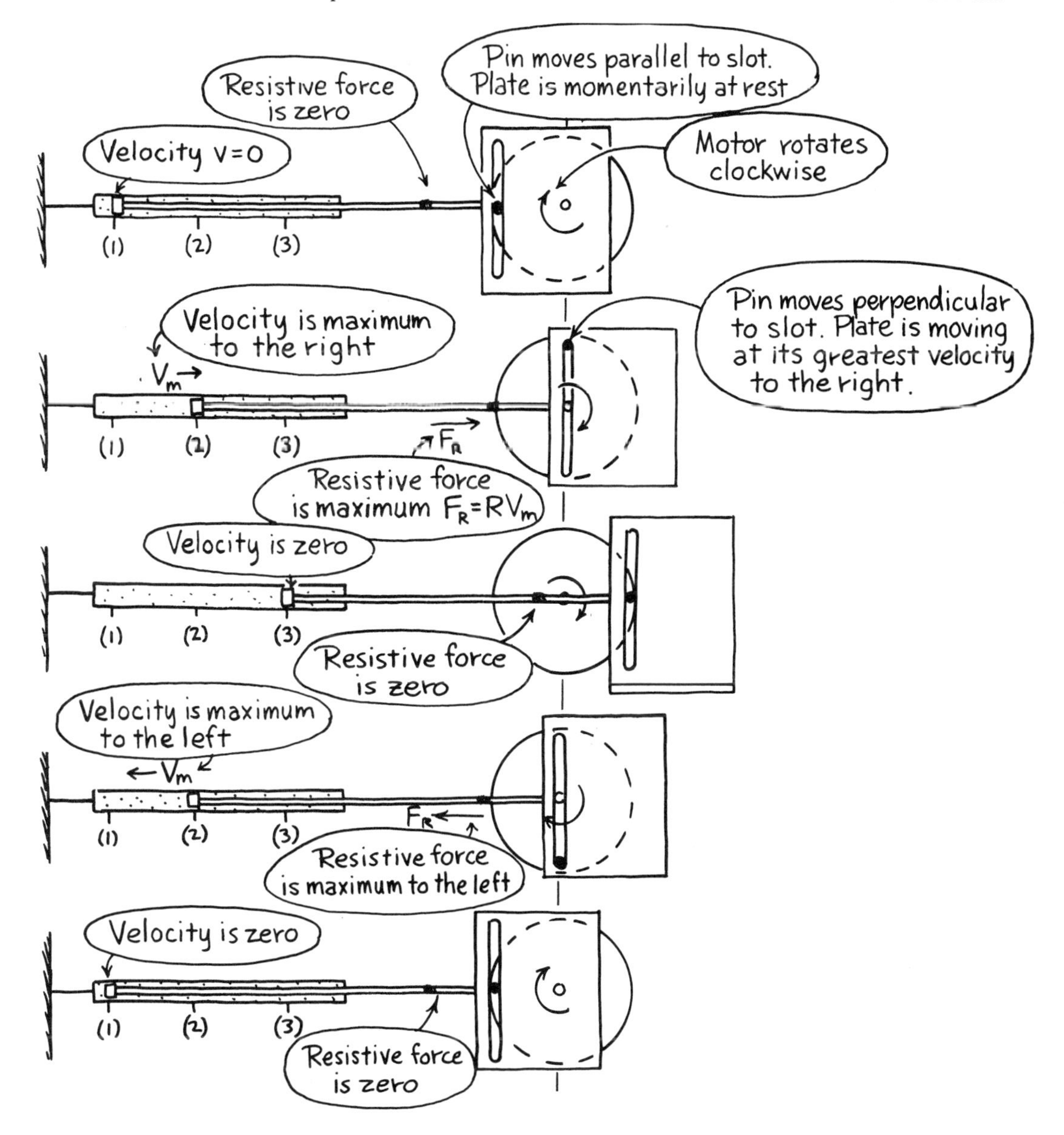

Figure 14-8. The dashpot's velocity changes continually in a sinusoidal manner during one cycle of the mechanical generator's motion.

piston through the fluid-filled cylinder. The masses of the piston and shaft are negligibly small. The motor starts up, and we wait for it to reach constant angular velocity so that the system is in the sinusoidal steady state with the piston oscillating back and forth between positions 1 and 3. Let us now look at the system as it goes through a complete cycle. Figure 14-8 will help.

At position 1 the piston's velocity is zero because the plate comes momentarily to rest. At position 2 the velocity is maximum to the right. The pin moving perpendicular to the slot imparts its full tangential velocity to the plate at this instant. At position 3 the velocity is zero, at position 2 the velocity is maximum to the left, and at position 1 the velocity is again zero to complete one cycle of motion.

If the piston's position varies sinusoidally with time, so does the velocity. However, the velocity is zero at positions 1 and 3 where piston displacement is largest.

Figure 14-9, a graph of velocity versus time, shows the sinusoidal nature of the velocity. Motion to the right is considered positive in this graph. The zero time is picked just as the piston passes position 1.

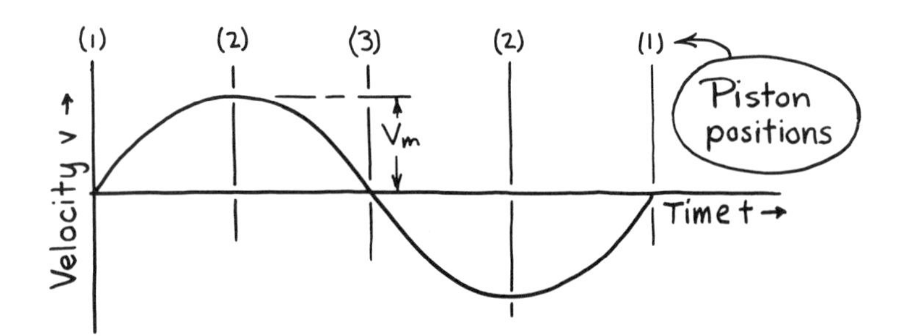

Figure 14-9. The piston's velocity follows the equation of a sine curve. Motion to the right is considered positive.

What happens to the resistive force f_R while all this is going on? The answer is that, if the dashpot is a linear resistor (R constant), f_R acts just like the piston velocity v. Resistance R is the ratio of resistive force to velocity:

$$R = \frac{f_R}{v} \tag{7-6}$$

If the velocity is zero, so is the force, and if the velocity is maximum, so is the force. The force is needed to overcome the fluid's viscous drag, which depends on the piston's velocity. Figure 14-10 is a graph of resistive force versus time showing the sinusoidal nature of the force. The velocity graph is plotted underneath for reference. The force and velocity sine curves are said to be *in phase* because their peaks and troughs coincide.

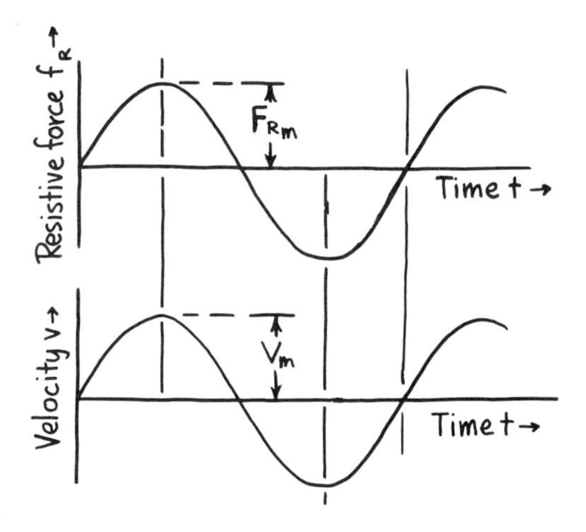

Figure 14-10. Resistive force f_R acts according to the equation of a sine curve. The force and velocity curves are in phase.

The force and velocity sine curves have the same frequency and can be represented by the equations

$$v = V_m \sin(360ft) \tag{14-4}$$

$$f_R = F_{Rm} \sin(360ft) \tag{14-5}$$

v *Instantaneous Velocity*	V_m *Maximum Velocity*	f_R *Instantaneous Resistive Force*	F_{Rm} *Maximum Resistive Force*	f *Frequency*	t *Time*
metre/second (m/s)		newton (N)		hertz (Hz)	second (s)
foot/second (ft/s)		pound (lb)			

Since f_R is everywhere proportional to v, it follows, using Eq. (7-6), that

$$R = \frac{F_{R_m}}{V_m} \tag{14-6}$$

and

$$R = \frac{0.707F_{Rm}}{0.707V_m}$$

or

$$R = \frac{F_R}{V} \tag{14-7}$$

F_{Rm} *Maximum Resistive Force*	F_R *rms Resistive Force*	V_m *Maximum Velocity*	V *rms Velocity*	R *Resistance*
newton (N)		metre/second (m/s)		$\frac{\text{newton}}{\text{metre/second}}$ $\left(\frac{\text{N}}{\text{m/s}}\right)$
pound (lb)		foot/second (ft/s)		$\frac{\text{pound}}{\text{foot/second}}$ $\left(\frac{\text{lb}}{\text{ft/s}}\right)$

The importance of the rms values for force and velocity becomes apparent when we look at energy and power supplied by the mechanical generator.

Energy and Power in the Mechanical Resistance System

The dashpot converts mechanical energy into heat energy because of the frictional opposition of the viscous fluid to any piston movement. This energy is supplied by the mechanical generator. We know from Chapter 6 that power, the time rate of energy change, is the product of force and velocity. In Sec. 6-10 the force and velocity were constant, but here they change continually. Still, the equation for power holds, but this time it is an instantaneous power p.

$$p = f_R v \tag{14-8}$$

p *Instantaneous Power*	f_R *Instantaneous Resistive Force*	v *Instantaneous Velocity*
watt (W)	newton (N)	metre/second (m/s)
$\frac{\text{foot} \cdot \text{pound}}{\text{second}}$ $\left(\frac{\text{ft} \cdot \text{lb}}{\text{s}}\right)$	pound (lb)	foot/second (ft/s)

Figure 14-11 shows how the instantaneous power varies with time. The graph is constructed by finding the product $f_R \cdot v$ for a number of different times using the force and velocity graphs in Fig. 14-10. The generator supplies no

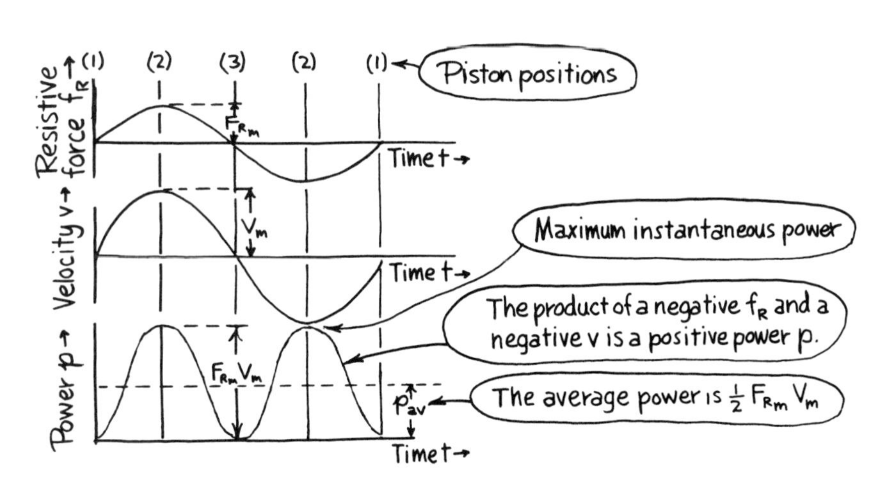

Figure 14-11. Instantaneous power p is the product of resistive force and velocity at any instant of time. Average power is one half of the peak power.

power at positions 1 and 3, because both f_R and v are zero. It supplies maximum power ($F_{Rm} \cdot V_m$) at position 2 when the force and velocity are maximum. Can you see a need for an average value of power? It turns out that the *average power* P_{av} is just half of the maximum instantaneous power:

$$P_{av} = \tfrac{1}{2} F_{Rm} \cdot V_m \tag{14-9}$$

A little mathematical juggling shows that the average power is the product of rms force and velocity:

$$P_{av} = \tfrac{1}{2} F_{Rm} V_m = \frac{F_{Rm}}{\sqrt{2}} \frac{V_m}{\sqrt{2}} = (0.707 F_{Rm})(0.707 V_m)$$

$$P_{av} = F_R \cdot V \tag{14-10}$$

P_{av} *Average Power*	F_{Rm} *Maximum Resistive Force*	F_R *rms Resistive Force*	V_m *Maximum Velocity*	V *rms Velocity*
watt (W)	newton (N)		metre/second (m/s)	
$\frac{\text{foot} \cdot \text{pound}}{\text{second}}$ $\left(\frac{\text{ft} \cdot \text{lb}}{\text{s}}\right)$	pound (lb)		foot/second (ft/s)	

Equation (14-10) looks just like a steady-state power equation. That is why rms forces and parameter rates are so popular.

Example 14-2 The mechanical generator's motor makes one complete revolution every 2.0 s. When the dashpot force is at its maximum value of 15.0 lb, the velocity is 1.5 ft/s.

(a) What is the frequency of the force and velocity sine curves?
(b) What is the dashpot resistance?
(c) What is the rms velocity?
(d) What average power is supplied?

Solution to (a)

$$f = \frac{\text{number of cycles}}{\text{time}} = \frac{1 \text{ cycle}}{2.0 \text{ s}} = \boxed{0.50 \text{ Hz}}$$

Solution to (b)

$$\left\{\begin{matrix} F_{Rm} = 15.0 \text{ lb} \\ V_m = 1.5 \text{ ft/s} \end{matrix}\right\}$$

Since V_{Rm} and F_{Rm} occur at the same instant,

$$R = \frac{F_{Rm}}{V_m} = \frac{15.0 \text{ lb}}{1.5 \text{ ft/s}} = \boxed{10 \frac{\text{lb}}{\text{ft/s}}}$$

Solution to (c)

$$V = 0.707 V_m = 0.707(1.5 \text{ ft/s}) = \boxed{1.1 \text{ ft/s}}$$

Solution to (d)

$$P_{av} = \tfrac{1}{2} F_{Rm} \cdot V_m = \tfrac{1}{2}(15.0 \text{ lb})(1.5 \text{ ft/s}) = 11.25 \rightarrow \boxed{11 \text{ ft} \cdot \text{lb/s}}$$

Resistance in an Electrical System

The discussion of resistance in an electrical system is brief, because we can draw on our knowledge of the mechanical system. According to unified concepts, potential difference is analogous to force and current is analogous to velocity. The sinusoidal voltage generator in Fig. 14-12 pushes charge through the resistor first in one direction and then the other.

From Ohm's law, Eq. (7-3), instantaneous current i will at all times be proportional to the instantaneous potential difference v_R across the resistor. This in turn equals the instantaneous electromotive force e. *For a purely resistive system*:

$$e = v_R = iR \tag{14-11}$$

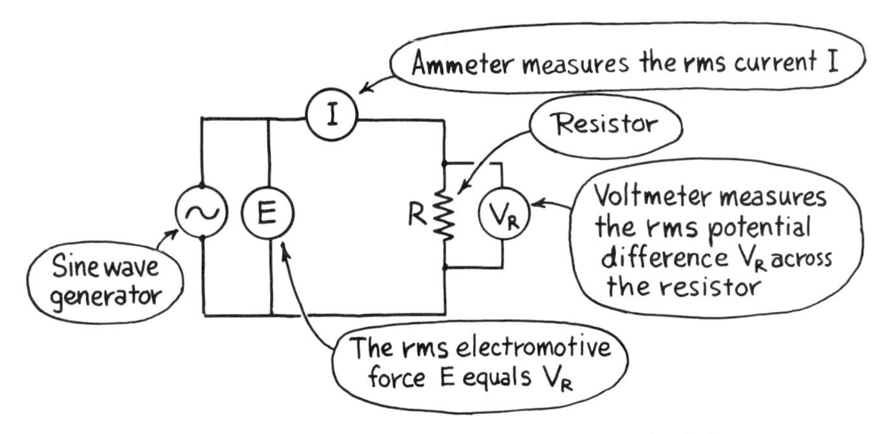

Figure 14-12. An electrical system analogous to the mechanical system of Fig. 14-7.

e *Instantaneous emf*	v_R *Instantaneous Resistor Voltage*	i *Instantaneous Current*	R *Resistance*
volt (V)		ampere (A)	ohm (Ω)

where
$$v_R = V_{Rm} \sin (360ft) \tag{14-12}$$
and
$$i = I_m \sin (360ft) \tag{14-13}$$
Also
$$R = V_{Rm}/I_m \tag{14-14}$$
and
$$R = V_R/I \tag{14-15}$$

V_R *rms Resistor Voltage*	V_{Rm} *Maximum Resistor Voltage*	I_m *Maximum Current*	f *Frequency*	t *Time*	I *rms Current*
volt (V)	volt (V)	ampere (A)	hertz (Hz)	second (s)	ampere (A)

Figure 14-13 shows graphs of v_R and i versus time. Both graphs are sine curves and they are in phase. The voltage and current peaks and troughs coincide. The instantaneous power, the product of v_R and i, is also plotted in Fig. 14-13. It looks just like the result for the mechanical system. The

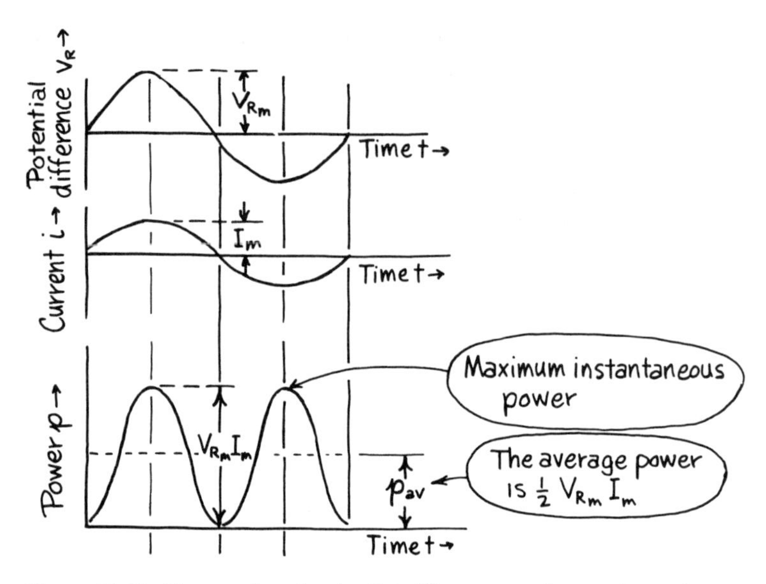

Figure 14-13. The graphs of potential difference and current are sine curves that are in phase. The graph for power looks just like that for a mechanical system.

generator supplies power in bursts to the resistor, which dissipates the energy in the form of heat. There are times when no energy is being supplied and times when the rate of supply is large. The average power equation has the same form as the mechanical system equations [Eqs. (14-9) and (14-10)].

$$P_{av} = \tfrac{1}{2} V_{Rm} \cdot I_m \tag{14-16}$$

$$P_{av} = V_R \cdot I \tag{14-17}$$

Example 14-3 A sinusoidal generator supplies 10.0 V at a frequency of 100 Hz to a single resistor of 48 Ω. (*Note:* All voltages and currents are rms values unless otherwise stated.)

(a) What is the maximum potential difference across the resistor?
(b) What is the rms current and its frequency?
(c) What average power is supplied?

Solution to (a)

$\{V_R = E = 10.0\text{ V}\}$

$$V_{Rm} = \frac{V_R}{0.707} = \boxed{14.1\text{ V}}$$

Solution to (b)

$\{R = 48\ \Omega\}$

$$I = \frac{V_R}{R} = \frac{10.0\text{ V}}{48\ \Omega} = 0.208 \rightarrow \boxed{0.21\text{ A}}$$

Solution to (c)

$$P_{av} = V_R \cdot I = (10.0\text{ V})(0.21\text{ A}) = \boxed{2.1\text{ W}}$$

Summary for Systems Containing Resistance

The following sums things up for a system containing resistance and oscillating in the sinusoidal steady state.

1. The resistive force f_R is sinusoidal and has the same frequency as the parameter rate $\dot{q}$ (velocity, current, etc.).

2. The resistive force sine curve is in phase with the parameter rate sine curve.

3. The resistive force is at all times proportional to the parameter rate ($f_R = R\dot{q}$) if R is constant.

4. The resistor removes energy from the system. The average power needed to supply the resistor is the product of the rms force and the rms parameter rate ($P_{av} = F_R \cdot \dot{Q}$).

14-3 SYSTEMS CONTAINING INERTANCE

We now turn to systems containing inertance—mass in a mechanical system and inductance in an electrical system. We want to find a relation between force and parameter rate, and to study the energy needed to produce oscillations.

Inertance in a Mechanical System

Replace the dashpot with a block of mass m. The block is attached to the oscillating plate of the mechanical generator and moves back and forth on a frictionless surface between positions 1 and 3. Figure 14-14 shows the setup. The force strain gauge measures an inertive force needed to accelerate the block. We start up the motor and wait for the steady state. Figures 14-15

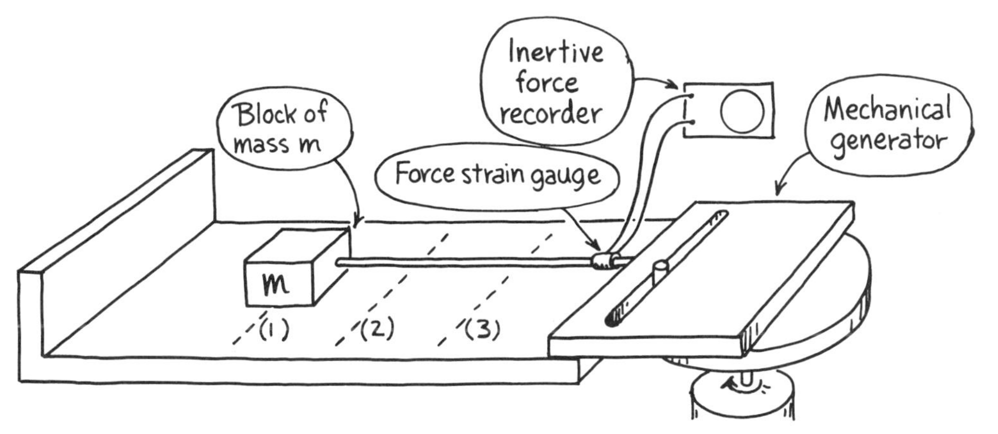

Figure 14-14. A mass is attached to the mechanical generator. It oscillates back and forth between points 1 and 3.

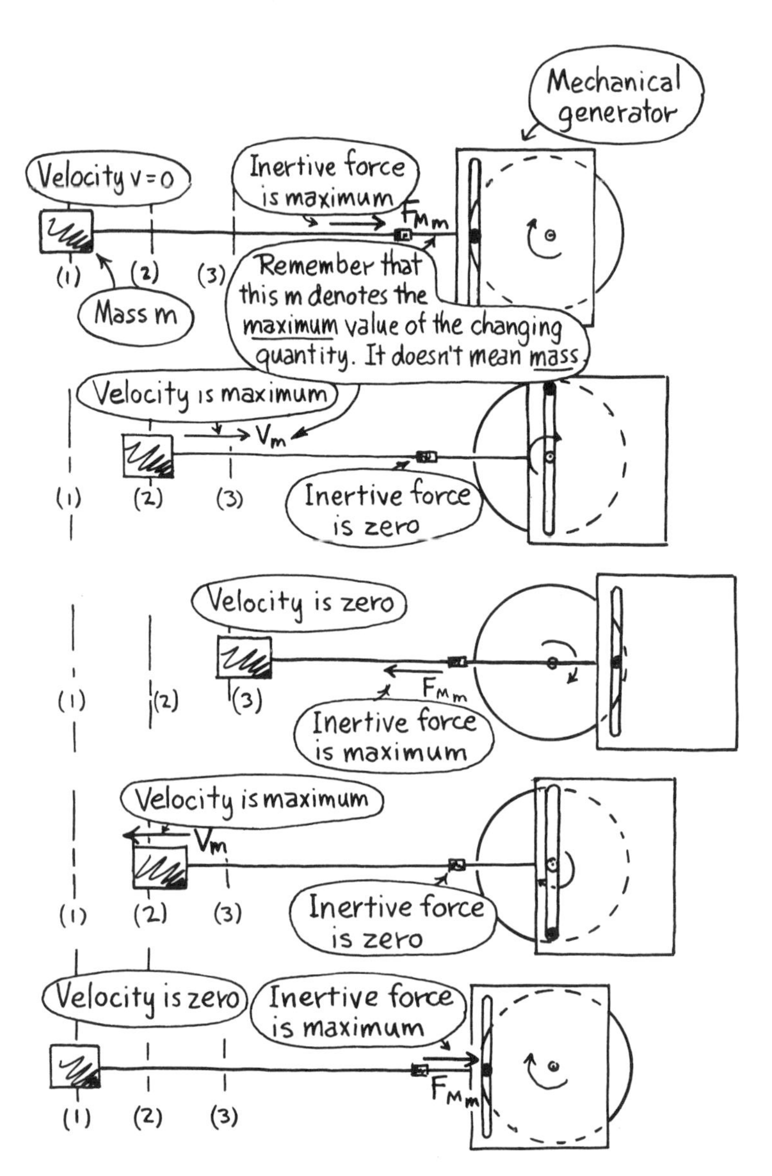

Figure 14-15. Inertive force is maximum when acceleration is greatest. This happens at the points where the velocity is zero.

and 14-16 show what happens. The block's velocity behaves the same as the dashpot piston velocity.

At position 1 the velocity is zero, at position 2 the velocity is maximum to the right, at position 3 it is zero, at position 2 it is maximum to the left, and at position 1 it is zero again to complete one cycle of motion.

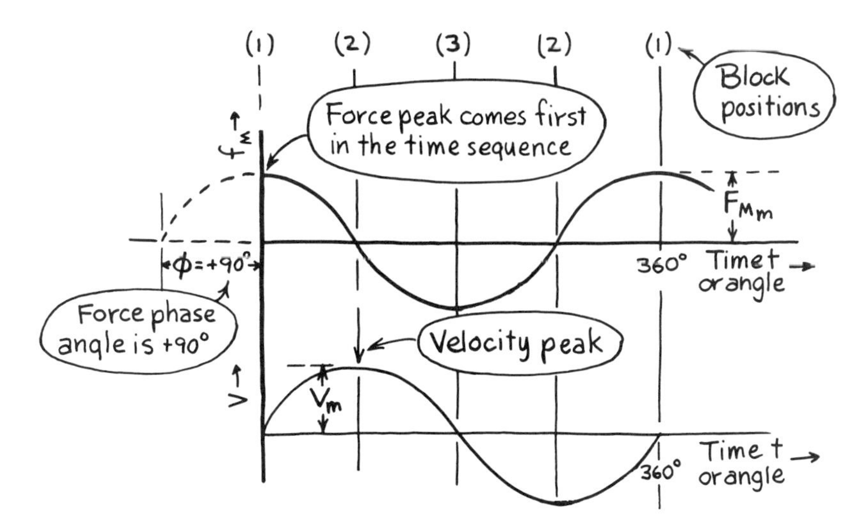

Figure 14-16. The force sine curve has a phase angle of +90° relative to velocity. Its peak comes first in the time sequence. We say that the force leads the velocity by 90°.

The inertive force acts differently than the resistive dashpot force. At position 1 the block, starting from rest, is accelerated to the right by an inertive force supplied by the generator. From Newton's second law, that inertive force is

$$f_M = ma \tag{3-3}$$

The force at position 1 is a maximum because the acceleration is a maximum to the right. At position 2 the block, which has been speeding up, attains a maximum velocity. Since there is no acceleration at this point, there is no force. The block then slows up and stops momentarily at position 3. An inertive force to the left is needed to make the block slow up. After position 3 the force continues to push the block to the left, but the force becomes zero at position 2 when the block's velocity is momentarily constant. Now an inertive force to the right starts to build up and slow the block down to a zero velocity at position 1 to complete the cycle.

Figure 14-16 shows the inertive force and velocity versus time graphs. Position 1 is used as reference for measuring time. A force to the right is considered positive. We know the velocity versus time graph is a sine curve. So is the force versus time graph, except that its peaks and troughs are displaced relative to the velocity curve. The peak of the force sine curve comes first in the time sequence. The force sine curve has a positive phase angle of 90° relative to velocity, and we say that the force *leads* the velocity by 90°. Here is a major difference between resistive and inertive systems. In the resistive system the force and velocity are in phase. In the inertive system, force leads velocity by 90°. Equations for force and velocity are

$$v = V_m \sin(360ft) \tag{14-4}$$

$$f_M = F_{Mm} \sin(360ft + 90°) \tag{14-18}$$

v *Instantaneous Velocity*	V_m *Maximum Velocity*	f_M *Instantaneous Inertive Force*	F_{Mm} *Maximum Inertive Force*	f *Frequency*	t *Time*
metre/second (m/s)		newton (N)		hertz (Hz)	second (s)
foot/second (ft/s)		pound (lb)			

Another difference between the two systems is that the inertive force depends upon the generator frequency. Calculus is needed to show that the relation between the maximum inertive force F_{Mm} and the maximum velocity V_m is

Again note that the two m's stand for different things

$$F_{Mm} = (2\pi fm)V_m \tag{14-19}$$

F_{Mm} *Maximum Inertive Force*	f *Frequency*	m *Mass*	V_m *Maximum Velocity*
newton (N)	hertz (Hz)	kilogram (kg)	metre/second (m/s)
pound (lb)		slug (slug)	foot/second (ft/s)

We should expect that F_{Mm} is proportional to the mass m. A larger mass needs a larger force to accelerate it. But F_{Mm} is also proportional to acceleration ($f_M = ma$), and we would expect that the accelerations of starting and stopping would be greater if the block oscillates back and forth at a greater frequency.

The relation between rms and maximum values of force and velocity is easy to find.

$$F_{Mm} = 2\pi f m V_m \tag{14-19}$$

$$0.707 F_{Mm} = 2\pi f m (0.707 V_m)$$

$$F_M = 2\pi f m V \tag{14-20}$$

If both sides are divided by V, we come out with something that has the same form as resistance (force/parameter rate).

$$X_M = 2\pi f m = \frac{F_M}{V} = \frac{F_{Mm}}{V_m} \tag{14-21}$$

F_M *rms Inertive Force*	V *rms Velocity*	f *Frequency*	m *Mass*	X_M *Inertive Reactance*
newton (N)	metre/second (m/s)	hertz (Hz)	kilogram (kg)	newton / metre/second $\left(\frac{N}{m/s}\right)$
pound (lb)	foot/second (ft/s)		slug (slug)	pound / foot/second $\left(\frac{lb}{ft/s}\right)$

X_M is called *inertive reactance.* It relates force to velocity in an inertive system, just as resistance relates force and velocity in a resistive system. Its units are the same as those of resistance, and it can be thought of in much the same way (i.e., a measure of opposition to flow). The main difference is that no energy is lost in this purely inertive system, as we shall see.

Example 14-4 The mechanical generator's motor makes one complete revolution in 0.50 s. It moves a 0.50-kg block back and forth on a frictionless surface. The maximum inertive force is 3.0 N.

(a) What is the frequency of the force sine curve?
(b) What is the inertive reactance?
(c) What is the block's maximum velocity?
(d) What is the rms value of the inertive force?

Solution to (a)

$\{t = 0.50\text{ s}\}$

$$f = \frac{\text{number of cycles}}{\text{time}} = \frac{1\text{ cycle}}{0.50\text{ s}} = \boxed{2.0\text{ Hz}}$$

Solution to (b)

$\{m = 0.50\ \text{kg}\}$

$$X_M = 2\pi f m = 2\pi(2.0\ \text{Hz})(0.50\ \text{kg}) = 6.28 \rightarrow \boxed{6.3\ \text{N/(m/s)}}$$

Solution to (c)

$\{F_{Mm} = 3.0\ \text{N}\}$

$$X_M = \frac{F_{Mm}}{V_m} \Rightarrow V_m = \frac{F_{Mm}}{X_M} = \frac{3.0\ \text{N}}{6.28\ \text{N/(m/s)}} = \boxed{0.48\ \text{m/s}}$$

Solution to (d)

$$F_M = 0.707 F_{Mm} = 0.707(3.0\ \text{N}) = \boxed{2.1\ \text{N}}$$

Energy and Power in the Mechanical Inertive System

The first thing we want to point out is that the system loses no energy to friction. A mass stores kinetic energy when in motion. The mechanical generator must supply energy to the mass to get it moving in parts of the cycle, but this energy is returned to the generator in other parts of the cycle.

Let's analyze this using the setup of Fig. 14-15 again. Note that the motor must keep the flywheel rotating clockwise. Figure 14-17 shows that energy is needed to accelerate the block from position 1 to position 2. Since this tends to slow up the flywheel, energy from the motor is needed to keep things going. From position 2 to position 3 the block slows up and stops. Its inertia pushes the flywheel in the clockwise direction. The block loses kinetic energy and the flywheel gains it. The process repeats itself on the return to position 1. In

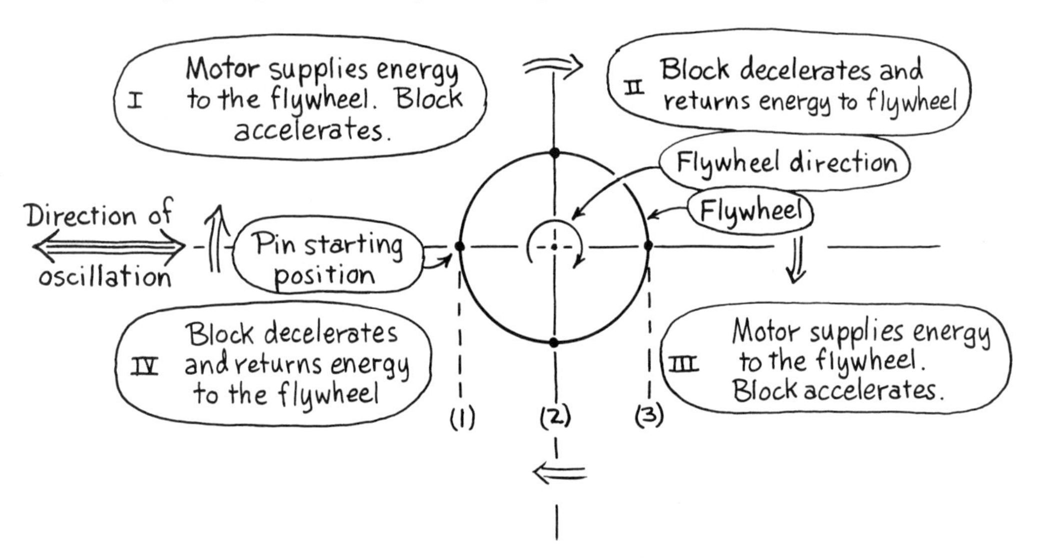

Figure 14-17. The flywheel supplies energy to the block in alternate quarters of the cycle. The block returns this energy to the flywheel in the following quarters of the cycle.

general, the flywheel loses energy to the block when the block accelerates and gains the same energy back when the block decelerates.

Inductance in an Electrical System

Now we will connect our sine-wave generator to a pure inductance, represented by a coil, as shown in Fig. 14-18. Experiments show that the

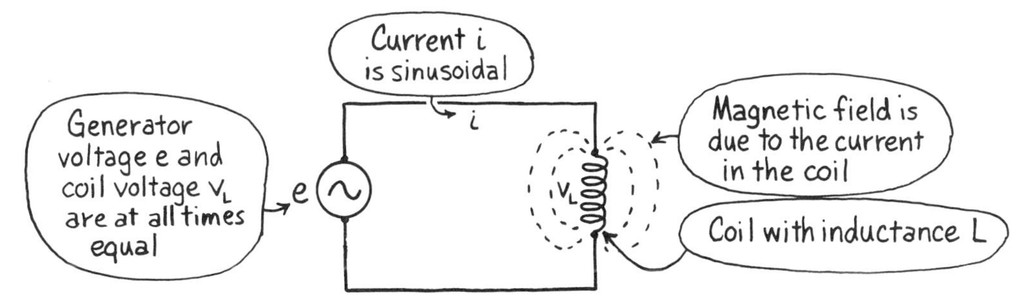

Figure 14-18. The generator pushes current through the coil in a sinusoidal manner. The current is responsible for the magnetic field.

current in the circuit will also be sinusoidal. Any current in the coil produces a magnetic field surrounding the coil. Its strength depends on the current, and it is greatest when the current is greatest. Section 8-2 discusses this in more detail. The magnetic field strength varies sinusoidally, building up in one direction, collapsing, building up in the other direction, etc., always in phase with the current.

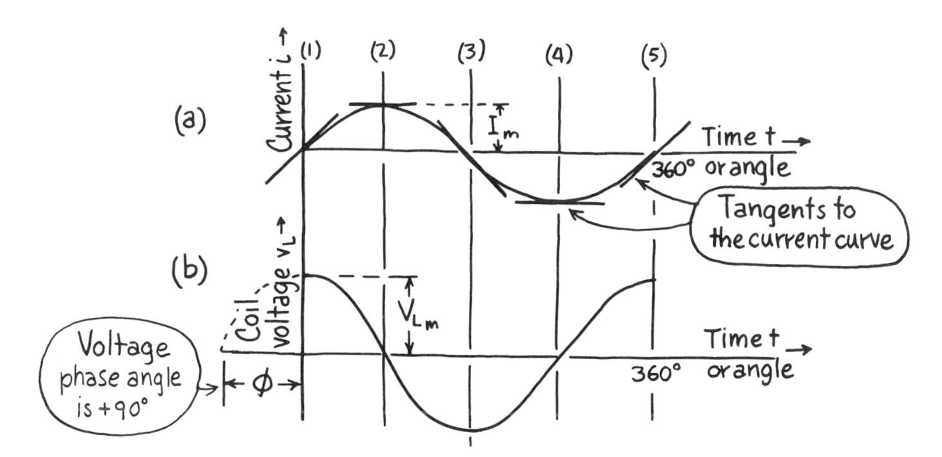

Figure 14-19. At points 1 and 5 tangent slopes are positive. At points 2 and 4 tangent slopes are zero. At point 3 the tangent slope is negative. Coil voltage v_L, proportional to $\Delta i/\Delta t$, is a sine curve with a 90° phase angle.

But a *changing* magnetic field induces a voltage in the coil. This voltage v_L is directly proportional to the magnetic field's rate of change, which in turn depends upon the current's rate of change. Section 10-2 discusses this in more detail.

$$v_L = L\frac{\Delta i}{\Delta t} \qquad (8\text{-}11)$$

Figure 14-19a shows the graph of a sinusoidal current. Little tangents with slope $\Delta i/\Delta t$ are drawn at strategic positions along the current curve. Position 1 is the position where the slope is most positive. The coil voltage v_L is the maximum at this instant, too. It is plotted in Fig. 14-19b at position 1. At position 2 the tangent has zero slope. The current is momentarily constant, $\Delta i/\Delta t$ is zero, and v_L is zero. The pattern continues at positions 3 through 5. See if you can follow this and verify that the coil voltage, which is sinusoidal, leads the current by 90°. The coil in the sinusoidal electrical system acts like the moving block in the sinusoidal mechanical system.

$$i = I_m \sin(360ft) \qquad (14\text{-}13)$$

$$v_L = V_{Lm} \sin(360ft + 90°) \qquad (14\text{-}22)$$

i Instantaneous Current	I_m Maximum Current	v_L Instantaneous Coil Voltage	V_{Lm} Maximum Coil Voltage	f Frequency	t Time
ampere (A)		volt (V)		hertz (Hz)	second (s)

Again, using analogies with the mechanical system, it can be shown mathematically that the relation between maximum current I_m and maximum coil voltage V_{Lm} depends on frequency f and coil inductance L.

$$V_{Lm} = (2\pi fL)I_m \qquad (14\text{-}23)$$

The *inductive reactance* X_L is the ratio of coil voltage to current using either maximum or rms values.

$$X_L = 2\pi fL = \frac{V_{Lm}}{I_m} = \frac{V_L}{I} \qquad (14\text{-}24)$$

V_{Lm} Maximum Coil Voltage	V_L rms Coil Voltage	I_m Maximum Current	I rms Current	f Frequency	L Inductance	X_L Inductive Reactance
volt (V)		ampere (A)		hertz (Hz)	henry (H)	ohm (Ω)

In the absence of resistance, no net power is supplied to a pure inductor connected to a sine-wave generator once the steady state is reached. True, the generator is needed to build up the magnetic field in alternate parts of the cycle and this requires energy. But this energy is returned to the source when the magnetic field collapses. Analogies can again be made to the mechanical system.

Example 14-5 A sine-wave generator supplies 10.0 V rms at 1000 Hz to a 0.150-H coil.

(a) What maximum voltage appears across the coil?
(b) What is the coil's inductive reactance?
(c) What is the rms current?
(d) What is the rms current if the frequency is increased to 2000 Hz?

Solution to (a)

$\{V_L = 10.0\text{ V}\}$

$$V_L = 0.707 V_{Lm} \Rightarrow V_{Lm} = \frac{V_L}{0.707} = \frac{10.0\text{ V}}{0.707} = \boxed{14.1\text{ V}}$$

Solution to (b)

$\left\{\begin{matrix} f = 1000\text{ Hz} \\ L = 0.150\text{ H} \end{matrix}\right\}$

$$X_L = 2\pi f L = 2\pi(1000\text{ Hz})(0.150\text{ H}) = \boxed{942\ \Omega}$$

Solution to (c)

$$I = \frac{V_L}{X_L} = \frac{10.0\text{ V}}{942\ \Omega} = \boxed{10.6 \times 10^{-3}\text{ A}}$$

Solution to (d)

$$X_L = 2\pi f L = 2\pi(2000\text{ Hz})(0.150\text{ H}) = 1885\ \Omega$$

$$I = \frac{V_L}{X_L} = \frac{10.0\text{ V}}{1885\ \Omega} = \boxed{5.31 \times 10^{-3}\text{ A}}$$

Summary for Systems Containing Inertance

The following sums things up for a system containing inertance and oscillating in the sinusoidal steady state.

1. Inertive force f_M is sinusoidal and has the same frequency as the parameter rate $\dot{q}$.

2. The inertive force sine curve leads the parameter rate sine curve by 90°.

3. Inertive reactance X_M, a relation between force and parameter rate, depends on inertance M and frequency f.

$$X_M = \frac{F_M}{\dot{Q}} = 2\pi f M$$

4. No net power is needed to drive a sinusoidal oscillating system containing pure inertance.

14-4 SYSTEMS CONTAINING CAPACITANCE

We will now look at the third kind of energy-handling device, capacitors, all by themselves in oscillating systems. Specifically, we will consider springs in mechanical systems and capacitors in electrical systems. Again we look for a relation between force and parameter rate and at the energy involved in an oscillating capacitive system.

Capacitance in a Mechanical System

The moving mass of the previous section is replaced with a spring, as shown in Fig. 14-20. One end is attached to the mechanical generator through a rigid rod, which has a strain gauge mounted on it to measure spring

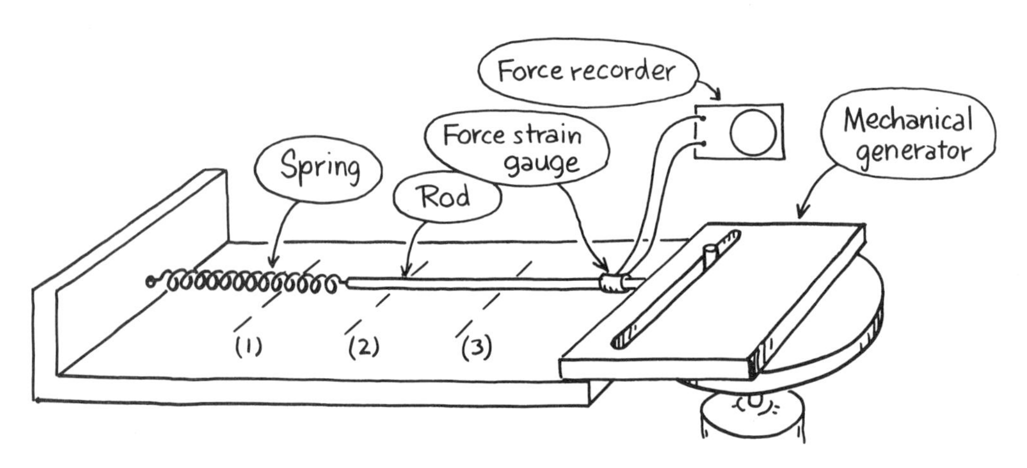

Figure 14-20. A mechanical capacitor (spring) is attached to the mechanical generator. The spring's end oscillates back and forth between points 1 and 3.

forces. Consider the spring as undeflected when the end is at position 2. The spring's end is driven sinusoidally back and forth between positions 1 and 3 once the motor starts up and the steady state is reached.

At position 1 the generator must push to the left with maximum force to fully compress the spring. At position 2 there is no force because the spring is not deflected. At position 3 the generator pulls to the right in order to fully extend the spring. The force pattern is sinusoidal. Figure 14-21 shows how the velocity and the capacitive force change during a cycle, and Fig. 14-22 shows a graph of force and velocity versus time. The force on the spring is positive if

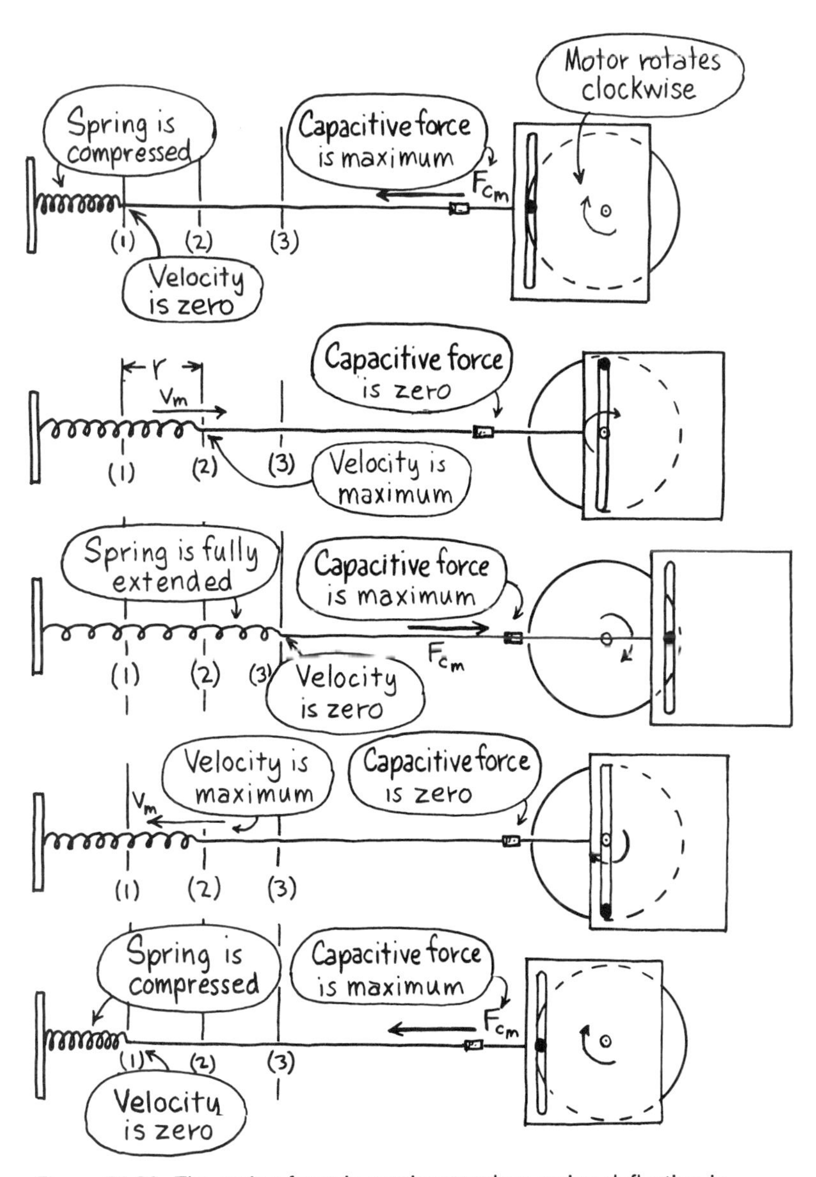

Figure 14-21. The spring force is maximum when spring deflection is greatest. This happens at positions 1 and 3 when velocity is zero.

the generator pulls to the right, and the velocity is positive if the spring's end moves to the right.

The peaks and troughs of the force curve are displaced relative to the velocity sine curve. The force sine curve has a negative phase angle of $-90°$. We say the force *lags* the velocity by 90° when this happens. (Remember that force *leads* velocity by 90° in the inertive system.)

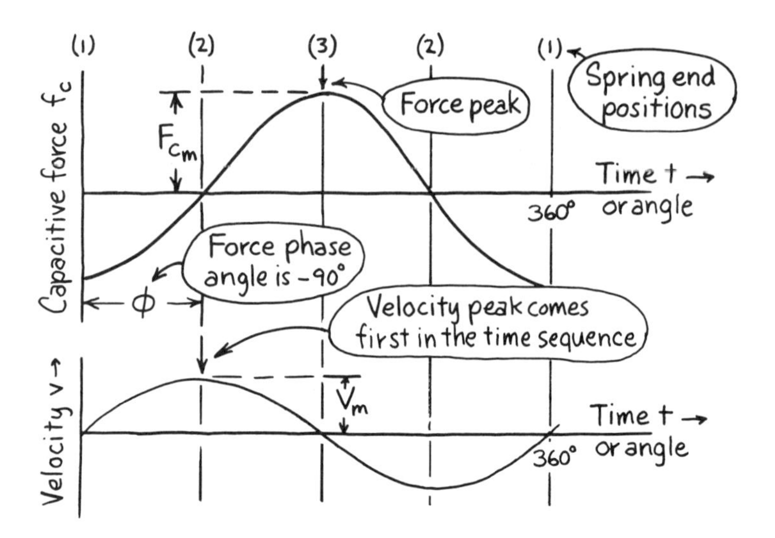

Figure 14-22. The capacitive force sine curve has a phase angle of −90°. Its peak comes later in the time sequence than the velocity peak. We say that the force *lags* velocity by 90°.

Equations for the instantaneous velocity and capacitive force are

$$v = V_m \sin(360ft) \tag{14-4}$$

$$f_C = F_{Cm} \sin(360ft - 90°) \tag{14-25}$$

v Instantaneous Velocity	V_m Maximum Velocity	f_C Instantaneous Capacitive Force	F_{Cm} Maximum Capacitive Force	f Frequency	t Time
metre/second (m/s)		newton (N)		hertz (Hz)	second (s)
foot/second (ft/s)		pound (lb)			

It can be shown using calculus that the relation between maximum capacitive force F_{Cm} and maximum velocity V_m is

$$F_{Cm} = \frac{1}{2\pi fC} \cdot V_m \tag{14-26}$$

The same relation applies to rms values of force and velocity.

$$F_C = \frac{1}{2\pi fC} \cdot V \tag{14-27}$$

The term $1/(2\pi fC)$ is called the *capacitive reactance* X_C of the spring.

$$X_C = \frac{1}{2\pi fC} = \frac{F_{Cm}}{V_m} = \frac{F_C}{V} \tag{14-28}$$

F_{Cm} Maximum Capacitive Force	F_C rms Capacitive Force	V_m Maximum Velocity	V rms Velocity	f Frequency	C Capacitance	X_C Capacitive Reactance
newton (N)		metre/second (m/s)		hertz (Hz	metre/newton (m/n)	newton / metre/second $\left(\frac{N}{m/s}\right)$
pound (lb)		foot/second (ft/s)			foot/pound (ft/lb)	pound / foot/second $\left(\frac{lb}{ft/s}\right)$

Equation 14-26 can be derived without recourse to calculus. The pin makes one revolution (distance $2\pi r$) in one time period ($T = 1/f$). The pin velocity is distance/time $= 2\pi r/T = 2\pi rf$. This is the maximum spring velocity at position 2.

$$V_m = 2\pi rf \quad \text{(velocity equation)}$$

The maximum capacitive force F_{Cm} occurs at either end when the spring's deflection is r.

$$F_{Cm} = \frac{r}{C} \quad \text{(force equation)}$$

Solve for r in the velocity equation, substitute the result in the force equation, and you get

$$r = \frac{V_m}{2\pi f}$$

$$F_{Cm} = \frac{V_M}{2\pi f} \cdot \frac{1}{C} = \frac{1}{2\pi fC} \cdot V_m \tag{14-26}$$

Note that both the spring capacitance C and the frequency f are in the denominator of the equation for capacitive reactance. An increase in either C or f decreases X_C. Compare this with the equation for inertive reactance X_M, where an increase in either M or f increases X_M.

$$X_M = 2\pi f M \tag{14-21}$$

$$X_C = \frac{1}{2\pi f C} \tag{14-28}$$

Example 14-6 The spring capacitance is 0.12 m/N and the maximum spring deflection is 0.20 m. The motor makes two complete revolutions in 1.0 s.

(a) What is the maximum force?
(b) What is the frequency of the force sine curve?
(c) What is the capacitive reactance?
(d) What is the peak velocity?

Solution to (a)

$$\left\{\begin{aligned}\text{maximum}&\\ \text{deflection} &= 0.20\text{ m}\\ C &= 0.12\text{ m/N}\end{aligned}\right\}$$

$$F_{Cm} = \frac{\text{deflection}}{\text{capacitance}} = \frac{0.20\text{ m}}{0.12\text{ m/N}} = 1.67 \rightarrow \boxed{1.7\text{ N}}$$

Solution to (b)

$$f = \frac{\text{number of cycles}}{\text{time}} = \frac{2.0\text{ cycle}}{1.0\text{ s}} = \boxed{2.0\text{ Hz}}$$

Solution to (c)

$$X_C = \frac{1}{2\pi f C} = \frac{1}{2\pi(2.0\text{ Hz})(0.12\text{ m/N})} = \boxed{0.66\,\frac{\text{N}}{\text{m/s}}}$$

Solution to (d)

$$X_C = \frac{F_{Cm}}{V_m} \Rightarrow V_m = \frac{F_{Cm}}{X_C} = \frac{1.67\text{ N}}{0.66\text{ N/(m/s)}} = \boxed{2.5\text{ m/s}}$$

Energy and Power in the Mechanical Capacitive System

In the absence of friction, no net power is needed to oscillate the spring. The mechanical generator need only supply energy during those parts of the cycle when the spring is either being compressed or stretched. This energy is returned to the flywheel as the spring loses its deflection. Figure 14-23 summarizes this.

Only resistors continually drain energy from the source in sinusoidal systems. Those parts of the system that store potential or kinetic energy later return it so that no net power is used.

Capacitance in an Electrical System

Figure 14-24 shows the sequence of steady-state events when a capacitor is connected to a sine-wave generator. In Fig. 14-24a, the capacitor

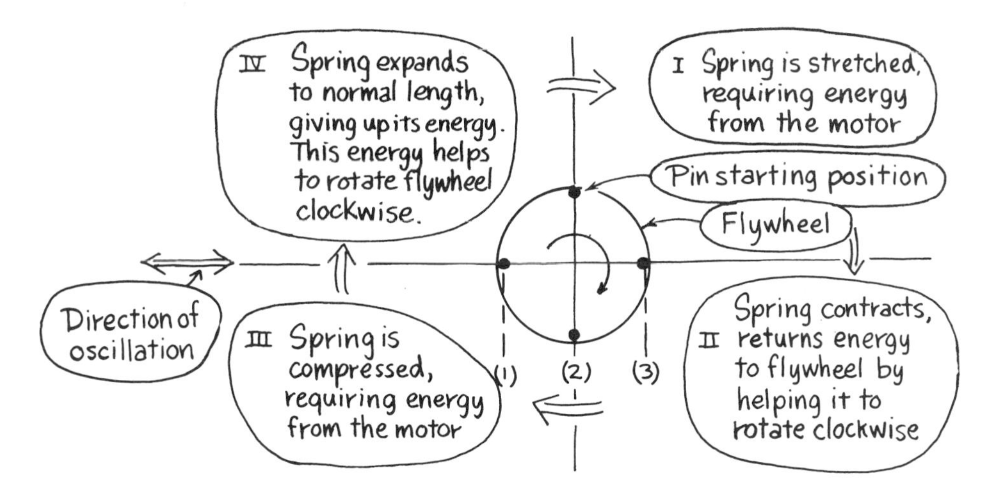

Figure 14-23. The motor supplies energy to the spring through the flywheel in alternate quarters of the cycle. The spring returns this energy to the flywheel in the following quarter-cycles.

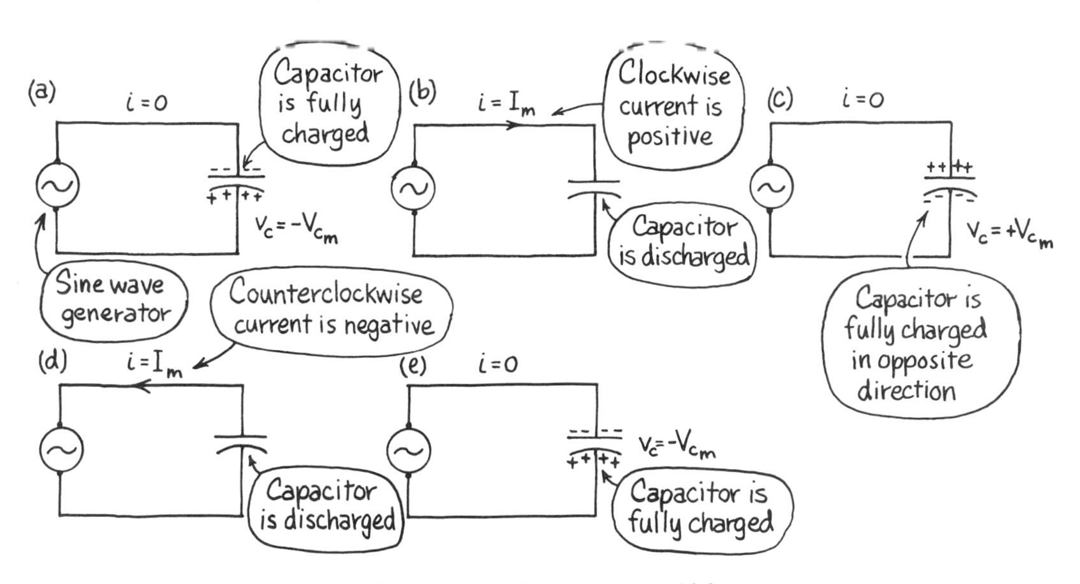

Figure 14-24. Both the current and capacitor voltage are sinusoidal. Whenever one is maximum the other is zero.

voltage v_C is maximum, because the capacitor is fully charged and the sinusoidally changing current is momentarily zero. In Fig. 14-24b, a maximum current surges clockwise toward the upper plate. We take this as the positive direction for the current. The capacitor is discharged and v_C is zero. In Fig. 14-24c, positive charge has surged onto the upper capacitor plate. The

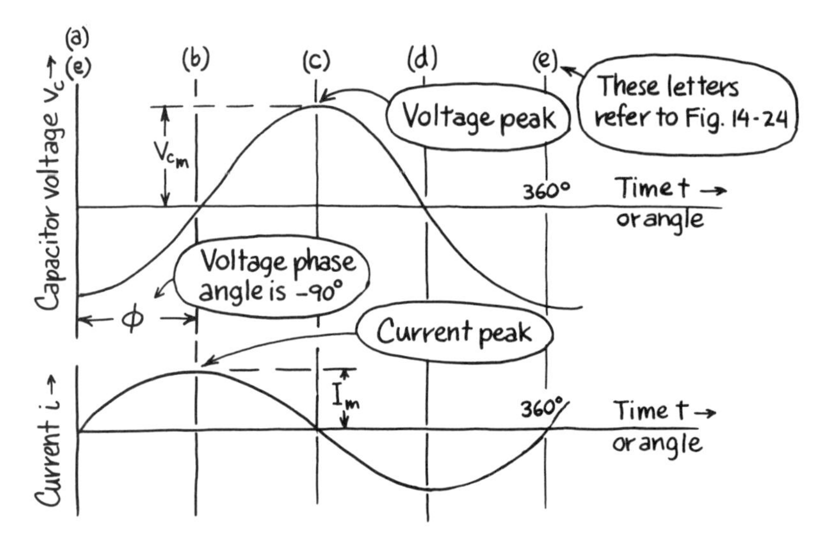

Figure 14-25. The capacitor voltage sine curve lags the current sine curve by 90°. The voltage peak comes later in the time sequence than the current peak.

capacitor is fully charged in the opposite direction and the current is momentarily zero. The pattern repeats itself to complete one full cycle.

Figure 14-25 summarizes the voltage and current patterns for a capacitor. The capacitor voltage lags the current by 90°. The same was true for the capacitive force compared to velocity in the mechanical system.

Here are equations for current and capacitor voltage.

$$i = I_m \sin(360ft) \tag{14-13}$$

$$v_C = V_{Cm} \sin(360ft - 90°) \tag{14-29}$$

$$V_{Cm} = \frac{1}{2\pi fC} \cdot I_m \tag{14-30}$$

The definitions for electrical capacitive reactance X_C are the same as for the mechanical system.

$$X_C = \frac{1}{2\pi fC} = \frac{V_{Cm}}{I_m} = \frac{V_C}{I} \tag{14-31}$$

V_{Cm} *Maximum Capacitor Voltage*	V_C *rms Capacitor Voltage*	I_m *Maximum Current*	I *rms Current*	f *Frequency*	C *Capacitance*	X_C *Capacitive Reactance*
volt (V)		ampere (A)		hertz (Hz)	farad (F)	ohm (Ω)

In the absence of resistance, no net power is supplied to a capacitor connected to a sine-wave generator once the steady state is reached. The generator is needed to build up charge on the capacitor plates, first in one direction and then in the other. This requires energy from the source. But the capacitor returns its stored energy when it discharges. Analogies can again be made to the mechanical system.

Example 14-7 A sine-wave generator supplies 10.0 V rms at 1000 Hz to a 0.100-μF capacitor.

(a) What maximum voltage appears across the capacitor plates?
(b) What is the capacitive reactance?
(c) What is the rms current?
(d) What is the rms current if the frequency is increased to 2000 Hz?

Solution to (a)

$\{V_C = 10.0\text{ V}\}$

$$V_C = 0.707 V_{Cm} \Rightarrow V_{Cm} = \frac{V_C}{0.707} = \frac{10.0\text{ V}}{0.707} = \boxed{14.1\text{ V}}$$

Solution to (b)

$\left\{\begin{matrix} f = 1000\text{ Hz} \\ C = 0.10 \times 10^{-6}\text{ F} \end{matrix}\right\}$

$$X_C = \frac{1}{2\pi fC} = \frac{1}{2\pi(1000\text{ Hz})(0.100 \times 10^{-6}\text{ F})} = 1592 \rightarrow \boxed{1.59 \times 10^3\ \Omega}$$

Solution to (c)

$$I = \frac{V_C}{X_C} = \frac{10.0\text{ V}}{1592\ \Omega} = \boxed{6.28 \times 10^{-3}\text{ A}}$$

Solution to (d)

$\{f = 2000\text{ Hz}\}$

$$X_C = \frac{1}{2\pi fC} = \frac{1}{2\pi(2000\text{ Hz})(0.100 \times 10^{-6}\text{ F})} = 796\ \Omega$$

$$I = \frac{V_C}{X_C} = \frac{10.0\text{ V}}{796\ \Omega} = \boxed{12.6 \times 10^{-3}\text{ A}}$$

Summary for Systems Containing Capacitance

The following sums things up for systems containing capacitance and oscillating in the sinusoidal steady state.

1. The capacitive force f_C is sinusoidal and has the same frequency as the parameter rate $\dot{q}$.

2. The capacitive force lags the parameter rate by 90°.

3. The capacitive reactance X_C, a relation between force and parameter rate, depends on capacitance and frequency.

$$X_C = \frac{F_C}{\dot{Q}} = \frac{1}{2\pi fC}$$

4. No net power is needed to drive a sinusoidal oscillating system containing pure capacitance.

14-5 SUMMARY

Chapter 14 has introduced you to mechanical and electrical systems excited by a sinusoidal generator. Once these systems are in the sinusoidal steady state, some interesting and simple relations develop between force and parameter rate. The relations depend upon whether the system contains a pure resistance, a pure inertance, or a pure capacitance. In general, forces and parameter rates oscillate sinusoidally at the frequency of the generator. There may be phase differences, and the relation between force and parameter rate may depend upon frequency. Energy is consumed only when resistance is present, and no net power is needed for an inertance or a capacitance. Table 14-1 summarizes the findings.

TABLE 14-1 A SUMMARY OF FORCE, PARAMETER RATE, AND POWER RELATIONS IN SINUSOIDAL MECHANICAL AND ELECTRICAL SYSTEMS

System Element	*Phase Relation Between Force and Parameter Rate*	*rms Force and Parameter Rate Relation*	*Power Needed*
Resistance	Force and parameter rate are in phase	$R = \frac{F_R}{\dot{Q}}$	$P_{av} = F_R\dot{Q}$
Inertance	Force leads parameter rate by 90°	$X_M = \frac{F_M}{\dot{Q}} = 2\pi fM$	$P_{av} = 0$
Capacitance	Force lags parameter rate by 90°	$X_C = \frac{F_C}{\dot{Q}} = \frac{1}{2\pi fC}$	$P_{av} = 0$

PROBLEMS

Note: Unless otherwise stated, all forces and parameter rates are given in terms of their rms values.

14-1. A sinusoidal force has an amplitude of 5.0 N and a frequency of 30 Hz.
(a) What is the force's rms value?
(b) What is the period?

14-2. A sine-wave generator delivers a 3.0-V, 400-Hz signal.
(a) What is the peak-to-peak voltage?
(b) What is the period?

14-3. A dashpot experiences a maximum velocity of 4.0 m/s when excited by a 15-N, 2.0-Hz sinusoidal force.
(a) What is the amplitude of the force?

(b) Find the dashpot resistance.

(c) What average power is consumed in the resistor?

14-4. A dashpot with resistance of 2.1×10^{-3} lb/(ft/s) is excited by a 0.050-lb, 2.5-Hz sinusoidal force. What is the maximum velocity of the dashpot piston?

14-5. A 5.0-V, 200-Hz sinusoidal voltage source is connected to a 200-Ω resistor.

(a) What is the rms current?

(b) Is a resistor power rating of 0.25 W sufficient for this situation?

14-6. How many 48-Ω resistors must be connected in series to limit current from a 120-V, 60-Hz source to 0.30 A or smaller?

14-7. A mass of 0.18 slug is excited by a 25-lb, 0.75-Hz sinusoidal force. What is the greatest velocity that the mass attains?

14-8. Which mass has greater inertive reactance, a 64-lb object that makes one oscillation every 3.0 s or a 100-g object oscillating at 100 Hz?

14-9. What is the inductance of a coil that limits the current from a 210-V, 50-Hz source to 1.5 A? Neglect any coil resistance in making the calculation.

14-10. What frequency limits the current through a 10-mH coil to 25 mA when excited by an 8.0-V sinusoidal source? Neglect any coil resistance in making this calculation.

14-11. A spring deflects 10.0 cm when a 500-g mass is hung at one end.

(a) What is the spring's capacitance in metres/newton?

(b) What is the spring's capacitive reactance if excited by a 15-Hz sinusoidal force?

14-12. What rms force is needed to produce a maximum velocity of 88 ft/s at the end of a spring where the force is applied? The spring constant is 120 lb/ft and the frequency is 50 Hz.

14-13. (a) What is the capacitive reactance of a 0.50-μF capacitor when excited by a 12-V, 400-Hz signal?

(b) What is the rms current?

14-14. An ac ammeter reads 8.0 mA when connected into a circuit consisting of 110-V, 60-Hz source and a capacitor. What is the capacitance of the capacitor?

14-15. (a) Write an equation for a sinusoidal voltage that has an amplitude of 30 V, a frequency of 100 Hz, and a value of 15 V when the time is zero seconds. (The phase angle does not exceed 90°.)

(b) Make a sketch of the graph for one complete cycle.

14-16. Sketch a graph of a sinusoidal velocity that has an amplitude of 14 m/s, a period of 0.50 s, and a velocity of 10 m/s to the left when the time is zero seconds. Consider velocity to the right as positive. The magnitude of the phase angle is not greater than 90°.

14-17. For the mechanical generator of Fig. 14-7, the radius of the circle made by the pin is 2.0 in. The generator rotates at 120 rpm.

(a) What is the maximum dashpot velocity in feet/second?

(b) What is the maximum dashpot force if its resistance is 0.15 lb/(ft/s)?

14-18. A steady force of 25 lb moves the piston of a dashpot 6.0 in. during a 1.5-s time interval. What is the amplitude of a sinusoidal force that gives the dashpot piston an rms velocity of 1.8 ft/s?

14-19. (a) Sketch one cycle of the voltage and current graphs in a circuit consisting of a 440-Ω resistor connected to a 220-V, 50-Hz source.

(b) What power is supplied by the source?

14-20. Which limits the current more, a 5000-Ω resistor connected to a 10-V, 1000-Hz source, or a 0.040-μF capacitor connected to the same source?

14-21. A solid cylindrical steel flywheel has a radius of 6.00 in. and a thickness of 2.00 in. What is its inertive reactance if excited by a 60.0-Hz sinusoidal torque? (*Hint:* Refer to Appendix B, Table 6, for the moment of inertia equation.)

14-22. What excitation frequency makes the inertive reactance of a 300-g mass the same as the capacitive reactance of a spring with spring constant of 10 N/m?

14-23. Sketch one complete cycle of the voltage and current graphs in a circuit consisting of a 25-mH coil connected to a 20-V, 500-Hz source. The coil resistance is negligible.

14-24. A coil with negligible resistance is connected to a 10.0-V sinusoidal source. The current is 2.6 mA when the frequency is 1000 Hz. What current would you expect if the source frequency is increased to 1500 Hz?

14-25. A torsional spring rotates 30° about its axis of rotation when subjected to a torque of 10 oz · in. What is the capacitive reactance of the spring in lb · ft/(rad/s) if excited by a 60-Hz sinusoidal torque?

14-26. Sketch one complete cycle of the voltage and current graphs in a circuit consisting of a 1.5-μF capacitor connected to a 25-V, 200-Hz source.

14-27. Which circuit has the smaller current, a 1.5-μF capacitor connected to a 10.0-V, 400-Hz source, or a 12-μF capacitor connected to an 8.0-V, 60-Hz source?

ANSWERS TO ODD-NUMBERED PROBLEMS

14-1. (a) 3.5 N
(b) 0.033 s

14-3. (a) 21 N
(b) 5.3 N/(m/s)
(c) 42 W

14-5. (a) 0.025 A
(b) Yes, power is 0.12 W

14-7. 42 ft/s

14-9. 0.45 H

14-11. (a) 0.020 m/N
(b) 0.52 N/(m/s)

14-13. (a) $8.0 \times 10^2\ \Omega$
(b) 0.015 A

14-15. (a) $e = 30 \sin (36{,}000t + 30°)$
(b)

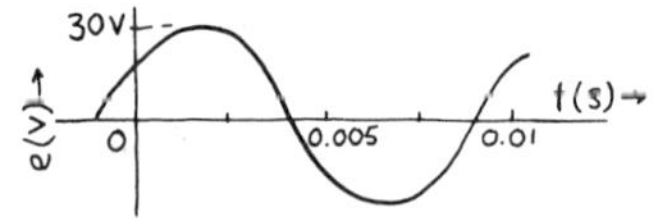

14-17. (a) 2.1 ft/s
(b) 0.31 lb

14-19. (a)

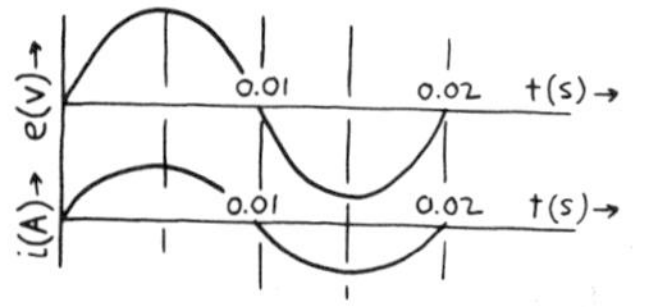

(b) 1.1×10^2 W

14-21. 93.9 lb · ft/(rad/s)

14-23.

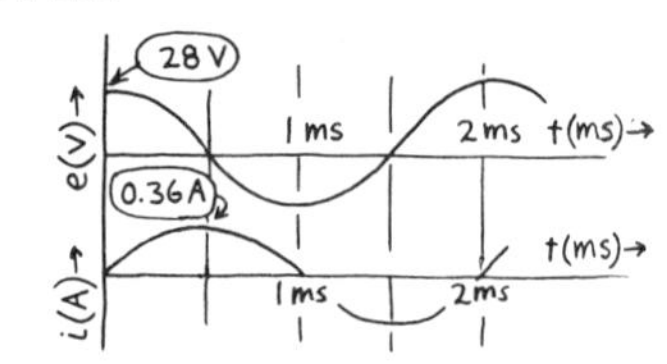

14-25. 2.6×10^{-4} lb · ft/(rad/s)

14-27. $I_1 = 38$ mA
$I_2 = 36$ mA (second one has smaller current)

15

Forced Oscillations— *RC* and *RM* Systems

In this chapter we will put together some of the ideas that arose out of Chapter 14. Instead of R, M, and C all alone, we will put R and C together and then R and M together in sinusoidally excited systems. Several new ideas will be introduced to explain these more complicated systems. One idea is a ***phasor,*** *which, represented by an arrow, gives information about a sine curve's amplitude and phase angle.*

Impedance *is another new idea. Impedance gives information about how a system reacts to a sinusoidal force in much the same way that resistance gives information about how a system reacts to a constant force. The adding of sine curves of force and of parameter is a technique used to explain how systems react to sinusoidal forces. These additions are made simple using phasor diagrams, which treat sine curve phasors as if they were vectors.*

The chapter is primarily concerned with making comparisons between mechanical and electrical systems, as was done in the last chapter, although we refer briefly to heat and fluid systems. Both series and parallel systems are discussed, although series systems are treated in greater detail. The chapter opens up possibilities for branching into a vast new area of knowledge, but we have tried to limit this exploration so that you won't get lost in detail.

15-1 PHASORS AND IMPEDANCE

To fully describe a sine curve, we must know the amplitude and the phase angle, as well as frequency. The value of y at any time t can be calculated using

$$y = Y_m \sin(360ft + \phi) \tag{14-3}$$

There is another way—a simple and useful method of representing the value of a sinusoidally varying quantity. The method uses an arrow called a *phasor.* The arrow's length represents the sine curve's amplitude, and the arrow's direction is the phase angle. The phasor, represented by an arrow, behaves very much like a vector, as we shall see.

Figure 15-1 shows some force sine curves and their corresponding phasors. These force sine curves come from the $R, M,$ and C systems in Chapter 14. The velocity sine curve (Fig. 15-1a) is used as a reference for measuring phase angle in all the others.

In Chapter 14 the ratio of the amplitudes of the force and parameter rate was shown to be either resistance R, inertive reactance X_M, or capacitive

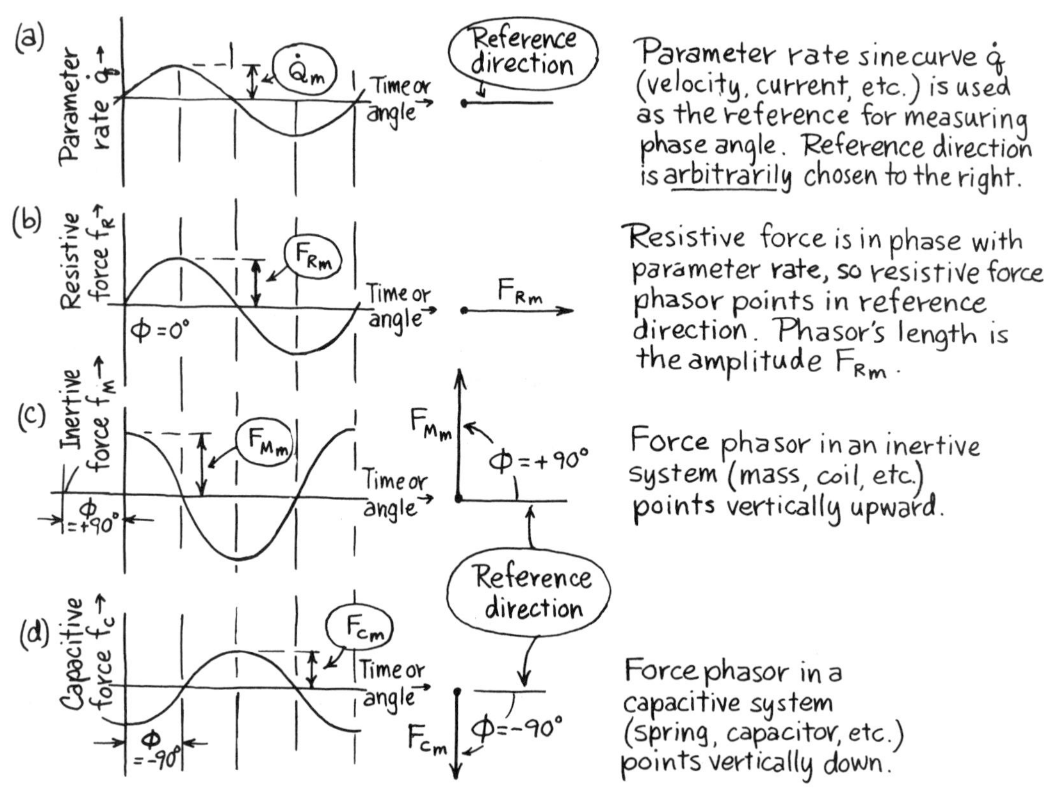

Figure 15-1. Force sine curves and their corresponding force phasors. The phasor gives two pieces of information. Phasor length is the amplitude and the direction is phase angle of the force sine curve.

reactance X_C. The force sine curves are either in phase or 90° out of phase with parameter rate. The effects of R, X_M, and X_C can all be put together into a single descriptive term called *impedance*. Impedance Z is one of those quantities that involves two pieces of information. Its size or magnitude Z is the ratio of the amplitudes (maximum values) of excitation force and parameter rate. It also involves the phase angle ϕ of the excitation force relative to parameter rate. The definition of Z and the commonly used notation for phase angle are embodied in Eq. (15-1):

$$Z = \frac{F_{Em}}{\dot{Q}_m} \angle\phi \tag{15-1}$$

Impedance (points to Z); Phase angle of excitation force relative to parameter rate (points to $\angle\phi$); Ratio of amplitudes of excitation force and parameter rate (points to $F_{Em}/\dot{Q}_m$)

Since the ratio of sine curve amplitudes is the same as the ratio of their rms values

$$Z = \frac{F_E}{\dot{Q}} \angle\phi \tag{15-2}$$

F_E and $\dot{Q}$ are rms values

The units of impedance are any appropriate units of force to parameter rate. They are just the same as for resistance. Examples are newton/(metre/second) and ohm. Think of impedance as describing the overall opposition that a system offers when it is excited by a force. Large impedances mean small parameter rates, and small impedances mean large parameter rates for a given excitation force.

Table 15-1 summarizes impedance for the three elements (R, M, and C) treated individually in Chapter 14. In these special cases, impedance becomes either R, X_M, or X_C, with the phase angle tagged on at the end for additional information.

Here are a couple of simple examples that illustrate phasors and impedance.

Example 15-1 A mass, excited by a sinusoidal force with an amplitude of 12.0 N, oscillates with a peak velocity of 2.0 m/s.

(a) Draw the excitation force phasor.
(b) What is the impedance of the system?

Solution to (a)

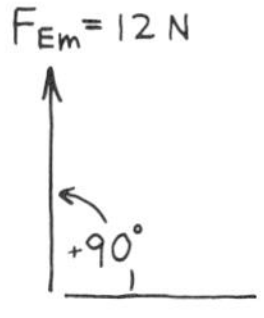

$F_{Em} = 12$ N

+90°

Since the excitation force leads the velocity by 90° in inertive systems, the force phasor points vertically upward, 90° from the horizontal reference direction.

TABLE 15-1 IMPEDANCE IN *R*, M, AND *C* SYSTEMS

System	*Force and parameter rate sine curves*	*Impedance equation*
R	$\dot{Q}_m$ $F_{Rm} = F_{Em}$	$Z = R\underline{/0^\circ}$ (15-3) where $R = \left(\frac{F_{Rm}}{\dot{Q}_m}\right) = \left(\frac{F_R}{\dot{Q}}\right)$ ← rms values amplitudes
M	$\dot{Q}_m$ $F_{Mm} = F_{Em}$	$Z = X_M\underline{/+90^\circ}$ (15-4) where $X_M = \left(\frac{F_{Mm}}{\dot{Q}_m}\right) = \left(\frac{F_M}{\dot{Q}}\right) = 2\pi fM$ amplitudes rms values
C	$\dot{Q}_m$ $F_{Cm} = F_{Em}$	$Z = X_C\underline{/-90^\circ}$ (15-5) where $X_c = \left(\frac{F_{Cm}}{\dot{Q}_m}\right) = \left(\frac{F_C}{\dot{Q}}\right) = \frac{1}{2\pi fC}$ amplitudes rms values

Solution to (b)

$$\left\{\begin{matrix} F_{Em} = 12.0\text{ N} \\ V_m = 2.0\text{ m/s} \end{matrix}\right\}$$

$$Z = \frac{F_{Em}}{V_m}\underline{/+90^\circ} = \frac{12.0\text{ N}}{2.0\text{ m/s}}\underline{/+90^\circ} = \boxed{6.0\,\underline{/+90^\circ}\,\frac{\text{N}}{\text{m/s}}}$$

Here V_m stands for maximum *velocity*. Don't confuse it with *volume* or *voltage*.

Example 15-2 A 10.0-V, 5000-Hz sinusoidal emf excites a 0.20-μF capacitor.

(a) Draw the emf phasor for the circuit.
(b) What is the impedance of the capacitor?
(c) What is the current? (Unless otherwise specified, we mean rms value.)

Solution to (a)

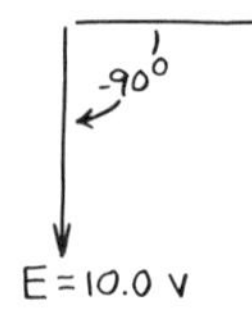

Since the excitation emf lags the current by 90° in a purely capacitive circuit, the voltage phasor points vertically downward, 90° in the negative direction, from the horizontal reference.

Solution to (b)

$$\left\{\begin{array}{l} f = 5000\text{ Hz} \\ C = 0.20\ \mu\text{F} \\ \quad = 0.20 \times 10^{-6}\text{ F} \end{array}\right\}$$

$$X_C = \frac{1}{2\pi fC} = \frac{1}{2\pi(5000\text{ Hz})(0.20 \times 10^{-6}\text{ F})}$$

$$= 159 \rightarrow 1.6 \times 10^2\ \Omega$$

$$Z = X_C = \boxed{1.6 \times 10^2 \angle -90^\circ\ \Omega}$$

Solution to (c)

$$I = \frac{V_C}{X_C} = \frac{E}{X_C} = \frac{10.0\text{ V}}{159\ \Omega} = \boxed{6.3 \times 10^{-2}\text{ A}}$$

15-2 *RM* SERIES SYSTEMS

Phasors and impedance are important because they help to describe the behavior of more complicated systems excited by sinusoidal forces. One such mechanical system, shown in Fig. 15-2, is made up of a dashpot and mass

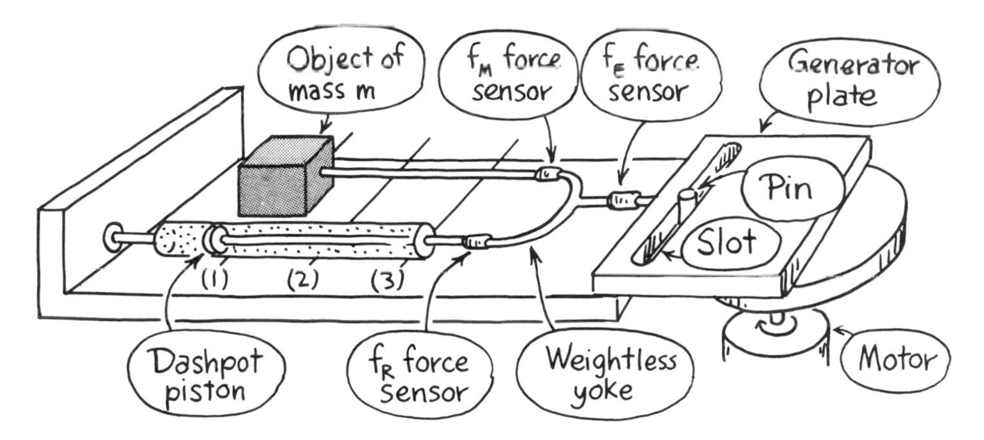

Figure 15-2. A mechanical *RM* series system. A dashpot and a mass are attached through a weightless yoke to the mechanical generator. The mass and dashpot are in a series arrangement.

connected to the mechanical sinusoidal generator of Chapter 14. A weightless yoke has been attached to the generator plate to tie the elements together. Three strain gauges mounted on the yoke measure resistive force f_R, inertive force f_M, and excitation force f_E. The yoke and the dashpot piston have essentially no mass, so f_E must at all times equal the sum of f_R and f_M:

$$f_E = f_R + f_M \tag{15-6}$$

The mechanical generator supplies the same sinusoidal motion to the mass and the dashpot piston. They are forced to move with a common sinusoidal velocity. The forces are additive ($f_E = f_R + f_M$), and the parameter rates are the same. This setup thereby satisfies the requirements for a series system as defined in Sec. 7-6. The mass and dashpot are in series although they appear to be in parallel.

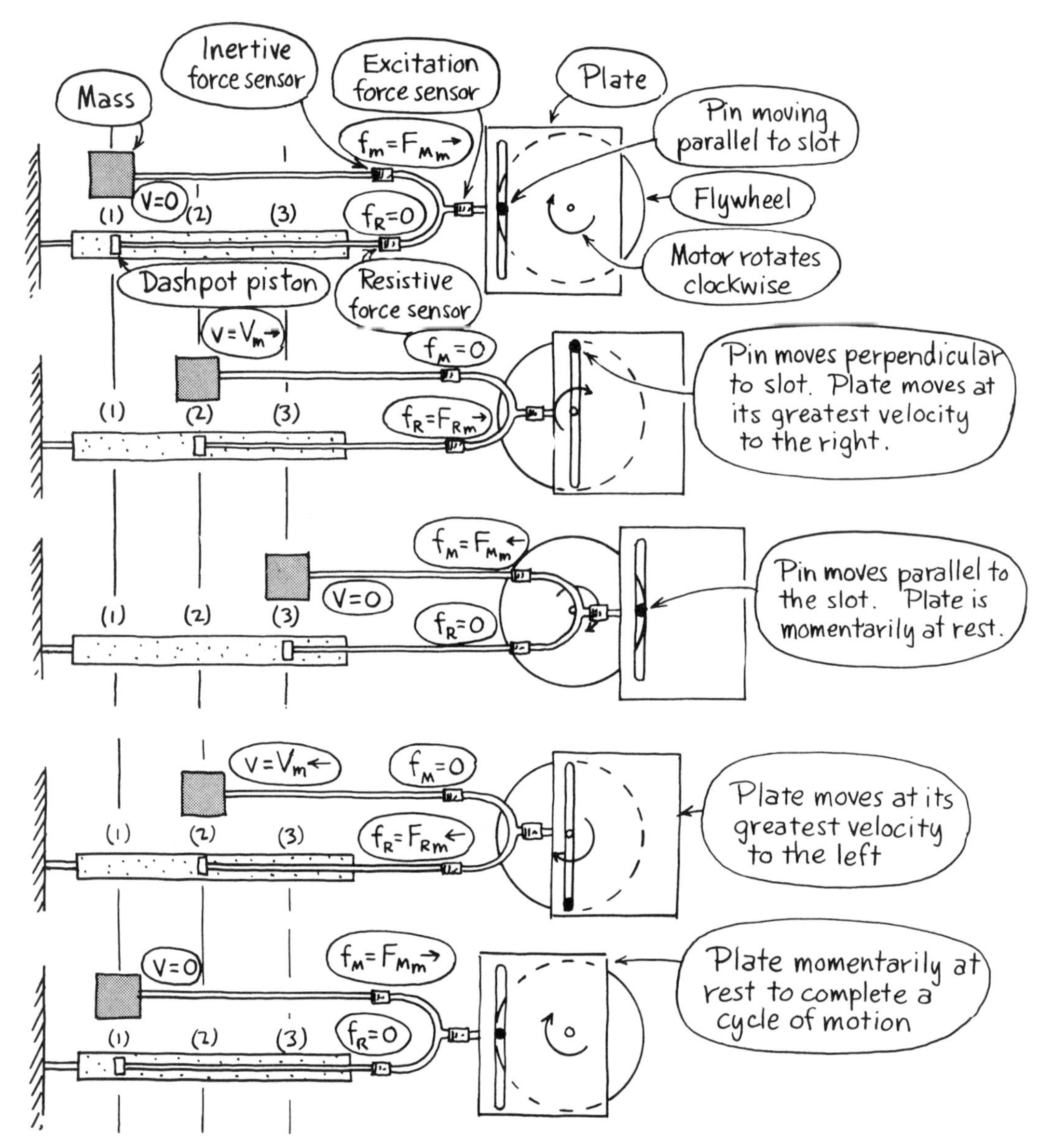

Figure 15-3. Forces and velocities change sinusoidally during one cycle of motion.

The motor starts up and we wait for the system to reach the sinusoidal steady state. Both piston and mass move back and forth between positions 1 and 3, as shown in Fig. 15-3, which analyzes a complete cycle of motion. The information in the figure is taken from Fig. 14-8 for the dashpot and Fig. 14-15 for the mass.

Figure 15-4 includes plots of the common velocity v, the resistive force f_R, and the inertive force f_M against time for one cycle of motion. The three quantities plot as sine curves. Resistive force is in phase with velocity, and

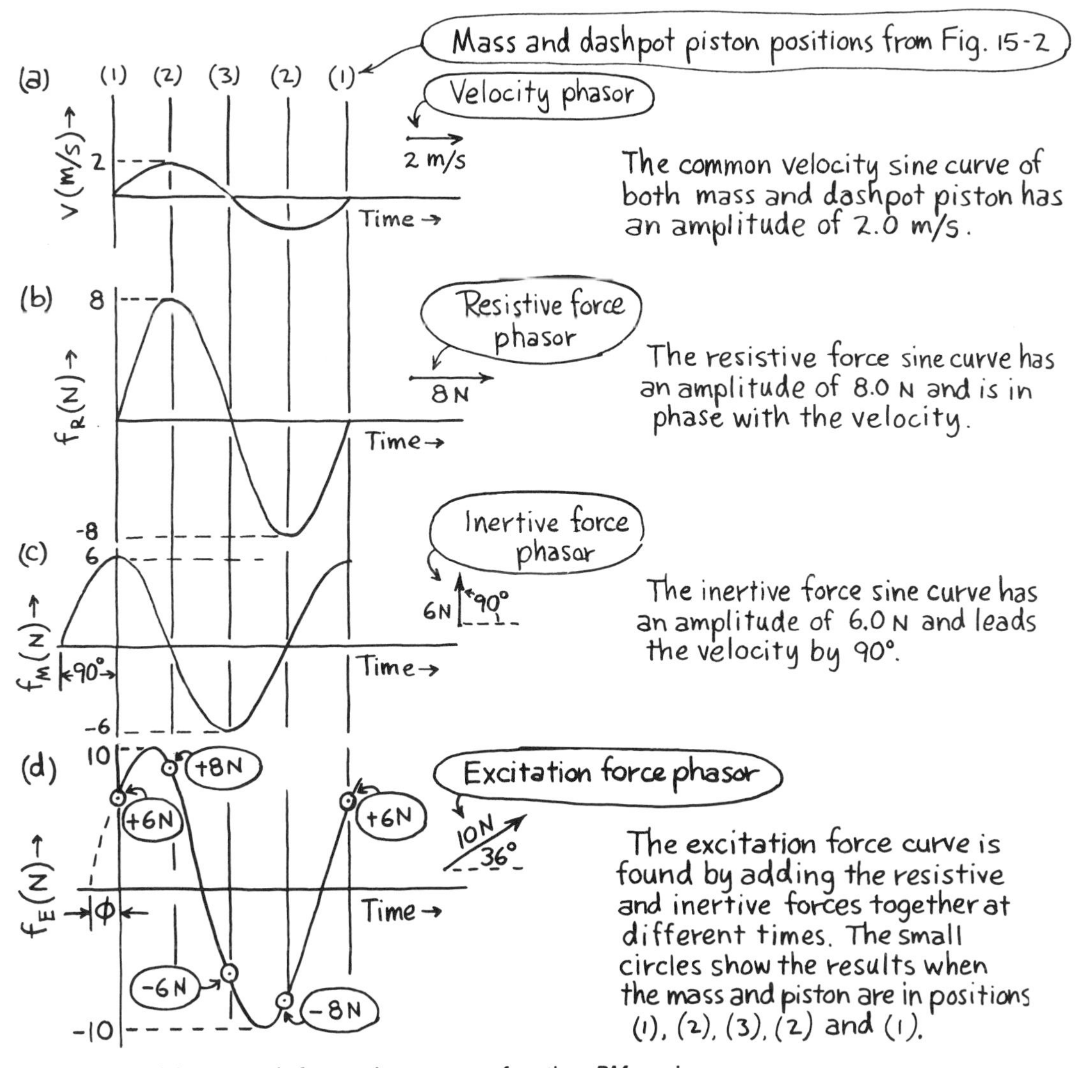

Figure 15-4. Velocity and force sine curves for the *RM* series mechanical system.

inertive force leads the velocity by 90°. Velocities and forces to the right are considered positive. We will arbitrarily assign values to v, f_R, and f_M for this example. Assume that their amplitudes are 2.0 m/s, 8.0 N, and 6.0 N, respectively. All this is shown in Fig. 15-4.

The sine curves of Fig. 15-4a, b, and c show how v, f_R, and f_M change with time. But what about f_E? According to Eq. (15-6), at any time instant f_E must be equal to the sum of f_R and f_M. Let's calculate f_E for some strategic times when we know the values of forces f_R and f_M. These times occur when both mass and dashpot piston are at positions 1, 2, and 3 in Fig. 15-3. The calculated

Position of Dashpot and Mass	f_R (N)	f_M (N)	$f_R + f_M = f_E$ (N)
1	0	6.0	0 + 6.0 = 6.0
2	8.0	0	8.0 + 0 = 8.0
3	0	−6.0	0 − 6.0 = −6.0
2	−8.0	0	−8.0 + 0 = −8.0
1	0	6.0	0 + 6.0 = 6.0

excitation forces f_E have been plotted in Fig. 15-4d. Their locations are shown by small circles. If additional times are used to calculate f_E as the sum of f_R and f_M, it becomes clear that the plot of f_E versus time is a sine curve. This technique of adding curves is called *superposition*. From Fig. 15-4d, f_E has an amplitude of about 10 N and a positive phase angle of between 35 and 40° with respect to the velocity sine curve. The phase angle can be estimated by looking at the position of the sine curve along the time axis. The peak is displaced about 0.4 of the way from position 2 toward position 1. The peak occurs earlier (f_E leads v) by about (0.4)(90°) = 36°. Figure 15-5 shows all this for the excitation force sine curve, together with its phasor representation, which for the first time is at an angle other than 0 or 90°. Note also that the amplitude of the excitation force F_{Em} (10 N) is *not* the algebraic sum of the amplitudes of the resistive force F_{Rm} (8.0 N) and the inertive force F_{Mm} (6.0 N).

Force and Impedance Phasor Diagrams

Is it possible to find the amplitude and phase angle of the excitation force without resorting every time to the painful process of sine curve addition? The answer fortunately is yes; phasors, treated as vectors, provide the method. Look at what happens when the resistive and inertive force phasors are added as vectors in Fig. 15-6. Answers for the amplitude and phase angle are very

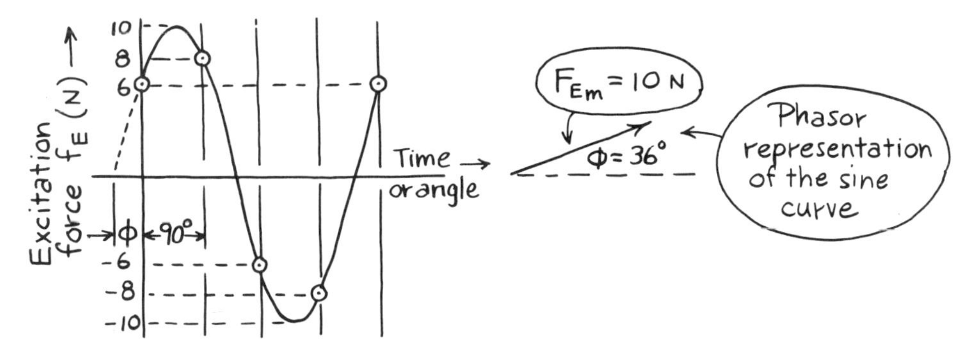

Figure 15-5. The excitation force is a sine curve with an amplitude of 10 N. It leads the velocity sine curve by 36°.

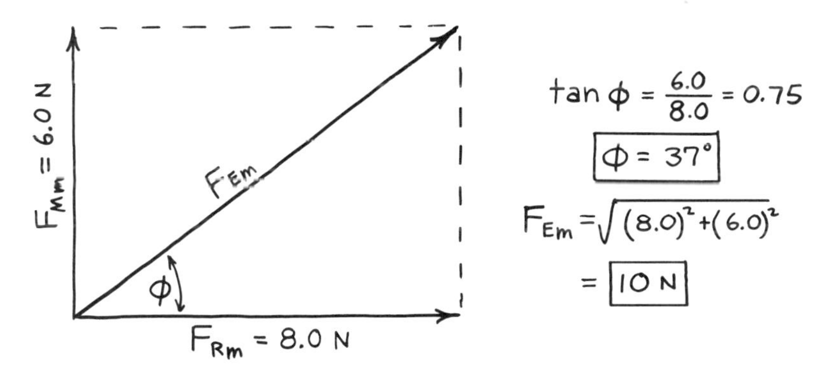

Figure 15-6. The vector addition of F_{Rm} and F_{Mm} yields answers for both F_{Em} and its phase angle with respect to the common velocity.

close to what was obtained using sine curve addition. In fact, theory shows that phasor vector addition gives exact answers, whereas sine curve addition depends on the accuracy of the sketch.

Suppose that the force phasors in Fig. 15-6 are each divided by the velocity amplitude of 2.0 m/s. The result gives values of R, X_M, and Z.

$$R = \frac{F_{Rm}}{V_m} \quad (14\text{-}6) \qquad X_M = \frac{F_{Mm}}{V_m} \quad (14\text{-}21) \qquad Z = \frac{F_{Em}}{V_m} \quad (15\text{-}1)$$

$$= \frac{8.0\ \text{N}}{2.0\ \text{m/s}} \qquad = \frac{6.0\ \text{N}}{2.0\ \text{m/s}} \qquad = \frac{10.0\ \text{N}}{2.0\ \text{m/s}}$$

$$= 4.0\frac{\text{N}}{\text{m/s}} \qquad = 3.0\frac{\text{N}}{\text{m/s}} \qquad = 5.0\frac{\text{N}}{\text{m/s}}$$

Look what happens if another vector triangle is formed using R and X_M as sides and Z as the diagonal, as in Fig. 15-7. We get a 3, 4, 5 impedance triangle that is similar to the 6, 8, 10 force triangle. This illustrates that R, X_M, and Z can be treated as phasor quantities. The vector sum of R and X_M equals the impedance Z.

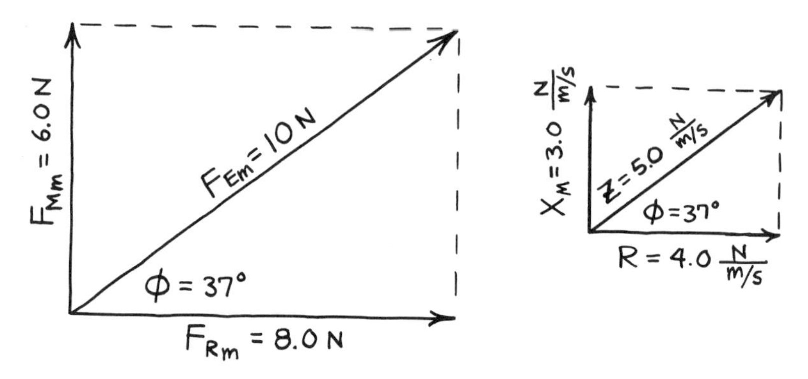

Figure 15-7. The triangle formed by force vectors is similar to the triangle formed by *R*, X_M, and *Z*.

Figure 15-8 shows a general force phasor diagram and a general impedance phasor diagram for series *RM* systems. Equations for calculating forces and impedances are included with each figure. Use these diagrams and equations when making calculations on *RM* series systems.

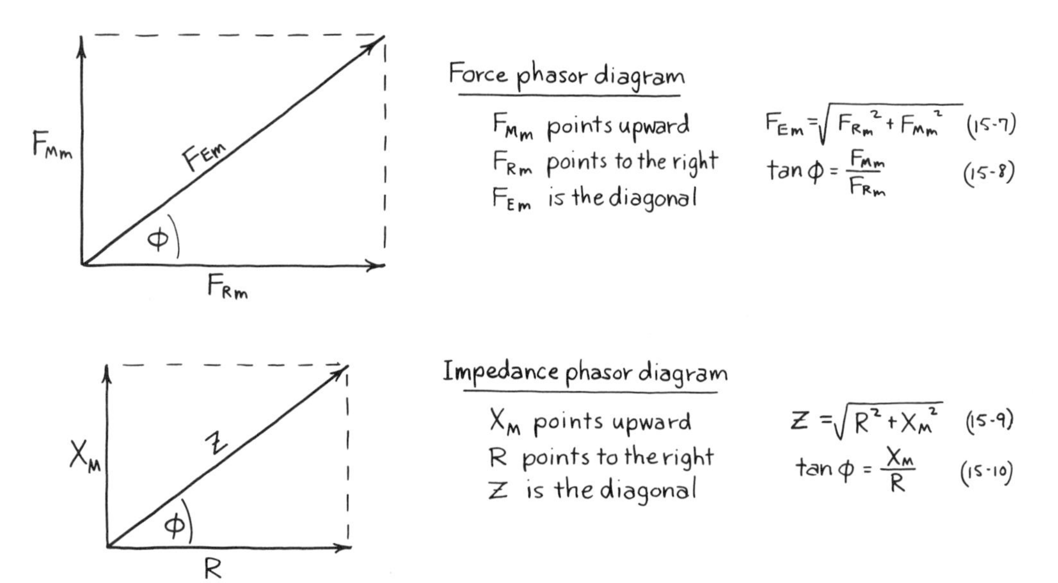

Figure 15-8. Force and impedance phasor diagrams for *RM* series systems.

Power in an *RM* System

The resistive dashpot continually takes energy from the generator. The mass needs energy too, but any energy it gets is returned to the source (Secs. 14-2 and 14-3). Therefore, it is easy to calculate average power P_{av} consumed in an *RM* system. It is just the average power supplied to the dashpot.

$$P_{av} = F_R \cdot V \tag{14-10}$$

The average power for an *RM* system can be calculated using the excitation force F_E from the force phasor diagram of Fig. 15-8,

$$\cos \phi = \frac{F_{Rm}}{F_{Em}} = \frac{F_R}{F_E}$$

$$F_R = F_E \cos \phi$$

and substituting the result into Eq. (15-10):

$$P_{av} = F_R \cdot V = (F_E \cos \phi)V$$

$$P_{av} = F_E V \cos \phi \tag{15-11}$$

The $\cos \phi$ part of Eq. (15-11) is called the *power factor.* The power factor tells what fraction of the product of input quantities (excitation force times velocity) is needed to supply power to the system.

An Electrical Analogy: The *RL* Series System

Now let us try to apply our knowledge about the mechanical *RM* system directly to an electrical *RL* system. Figure 15-9 shows a coil and a resistor connected to a sine-wave generator. The figure also shows current and voltage sine curves, the voltage phasor diagram, and the impedance phasor diagram associated with the circuit.

Everything that was said about the mechanical system applies to this electrical system when the dashpot is replaced by the resistor, the mass by the coil, and the mechanical generator by the electrical sine-wave generator. Example 15-3 shows how to make some calculations on an *RL* series circuit.

Example 15-3 A coil with negligible resistance is connected in series to a 2000-Ω resistor and a sine-wave generator whose output is 10.0 V (rms) at 5000 Hz. The measured voltage across the coil is 6.0 V.

(a) What is the resistor voltage?
(b) What is the common current?
(c) What is the circuit impedance and the phase angle?

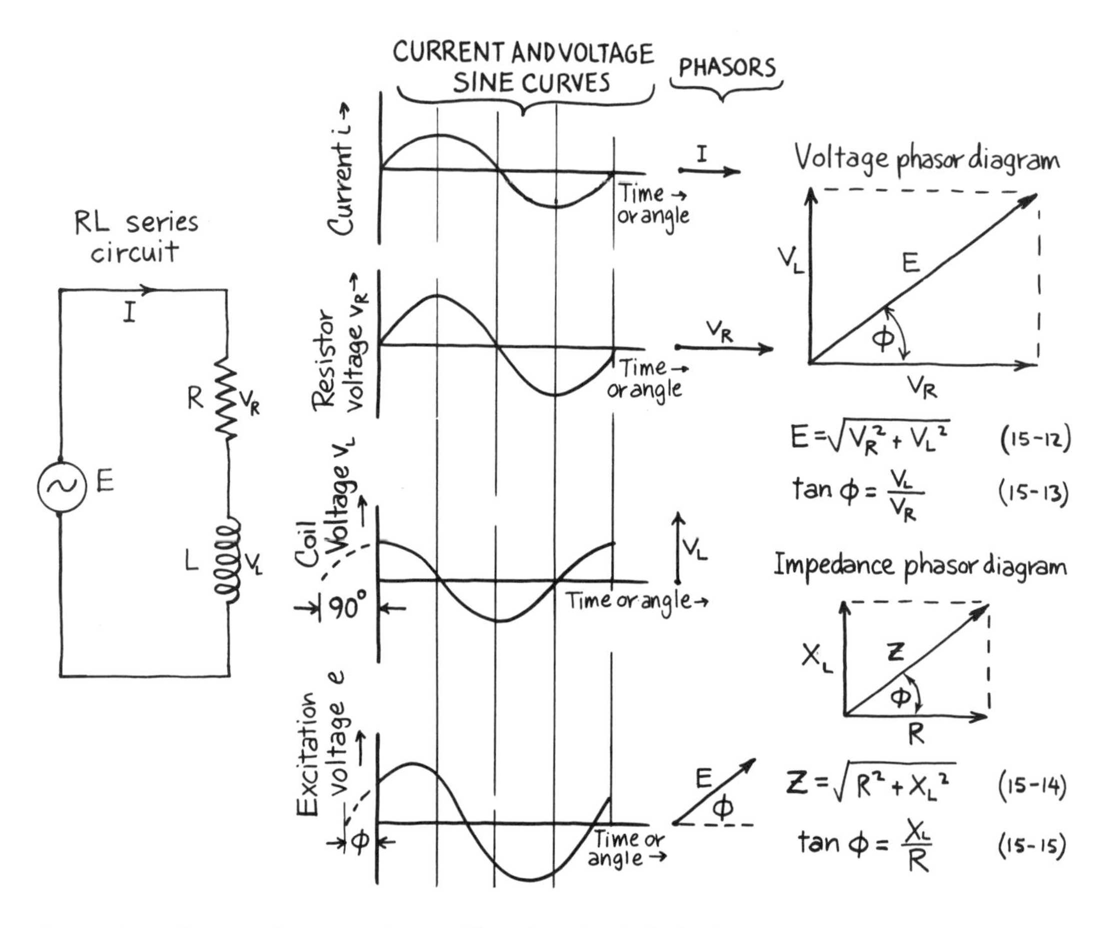

Figure 15-9. The complete story for an *RL* series electrical circuit unfolds by making analogies to a mechanical *RM* system.

Solution to (a)

$\left\{\begin{matrix} E = 10.0 \text{ V} \\ V_L = 6.0 \text{ V} \end{matrix}\right\}$

Use the voltage phasor diagram:

V_L= 6.0 V, E=10.0 V, φ, V_R

$$V_R = \sqrt{E^2 - V_L^2} = \sqrt{(10.0\ V)^2 - (6.0\ V)^2} = \boxed{8.0\ V}$$

Solution to (b)

Ohm's law works nicely:

$\{R = 2000\ \Omega\}$

$$I = \frac{V_R}{R} = \frac{8.0\text{ V}}{2000\ \Omega} = \boxed{4.0 \times 10^{-3}\text{ A}}$$

Solution to (c)

Use the definition of impedance and the impedance phasor diagram:

$$Z = \frac{E}{I} = \frac{10.0 \text{ V}}{4.0 \times 10^{-3} \text{ A}} = 2.5 \times 10^{3} \ \Omega$$

$$\cos \phi = \frac{R}{Z} = \frac{2000 \ \Omega}{2500 \ \Omega} = 0.80 \Rightarrow \phi = 37°$$

$$Z = \boxed{2.5 \times 10^{3} \underline{/37°} \ \Omega}$$

15-3 *RC* SERIES SYSTEMS

Much of the background needed to study more complicated sinusoidal systems has been laid out. We know about impedance, sine curve addition, and phasor diagrams. It should not be too difficult to understand a different system, the *RC* series system, using this body of knowledge. Let's try it.

In Fig. 15-10 the mass has been replaced by a spring that is undeflected when one end is at position 2. This end is attached to the weightless yoke. The

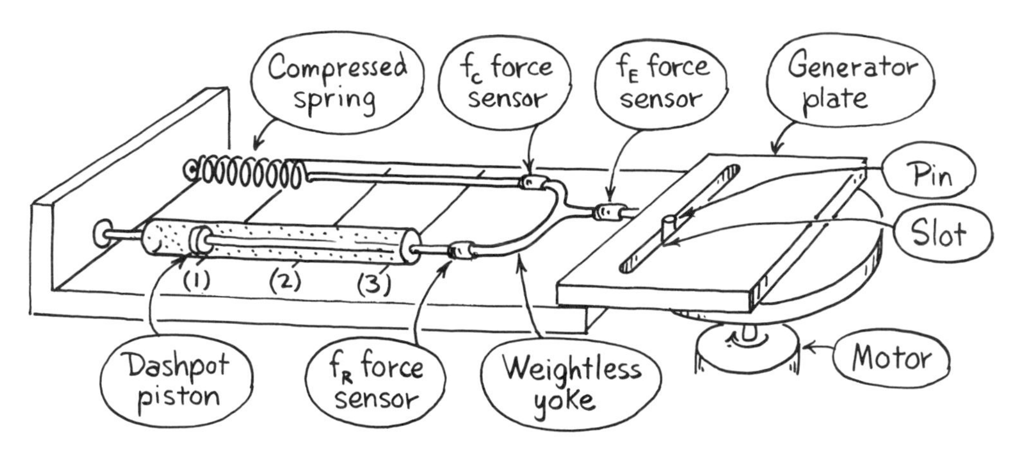

Figure 15-10. A dashpot and a spring are attached through a weightless yoke to the mechanical generator. The spring and dashpot form a series *RC* mechanical system.

sum of the spring capacitive force f_C and the dashpot resistive force f_R must at all times equal the generator excitation force f_E:

$$f_E = f_R + f_C \tag{15-16}$$

Again we rev up the generator and wait for the sinusoidal steady state. Figure 15-11 analyzes a complete cycle of motion. The information for this figure is taken from Fig. 14-8 for the dashpot and Fig. 14-21 for the spring. In

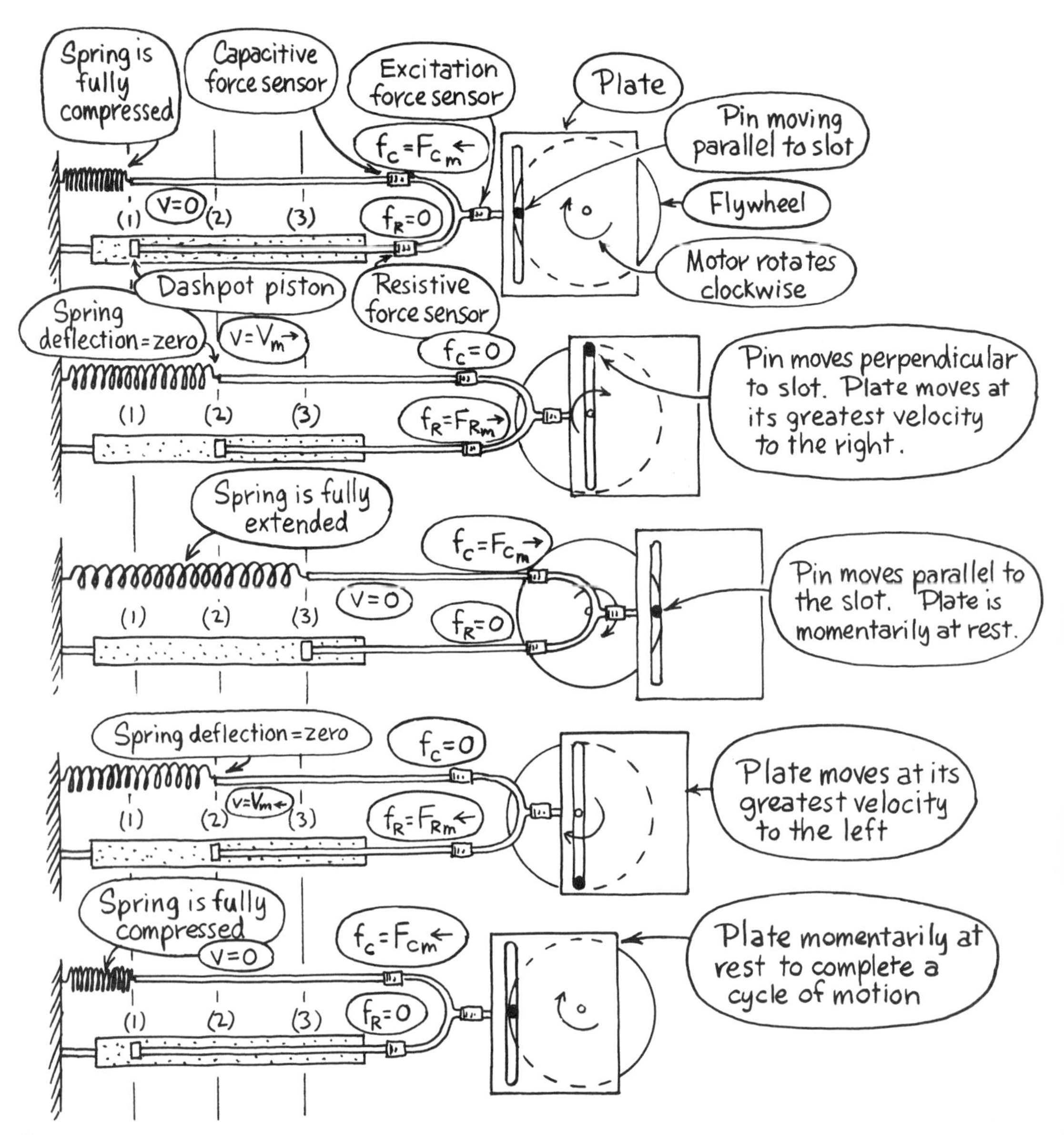

Figure 15-11. Forces and velocities change sinusoidally during one cycle of the *RC* series system. Compare this with the *RM* system shown in Fig. 15-3.

this example we use the same amplitudes that were used in the series *RM* discussion: velocity $V_m = 2.0\,\text{m/s}$; resistive force $F_{Rm} = 8.0\,\text{N}$; capacitive force $F_{Cm} = 6.0\,\text{N}$.

Figure 15-12 then shows the sine curves of the common velocity v, the resistive force f_R, and the capacitive force f_C for one cycle of motion. The

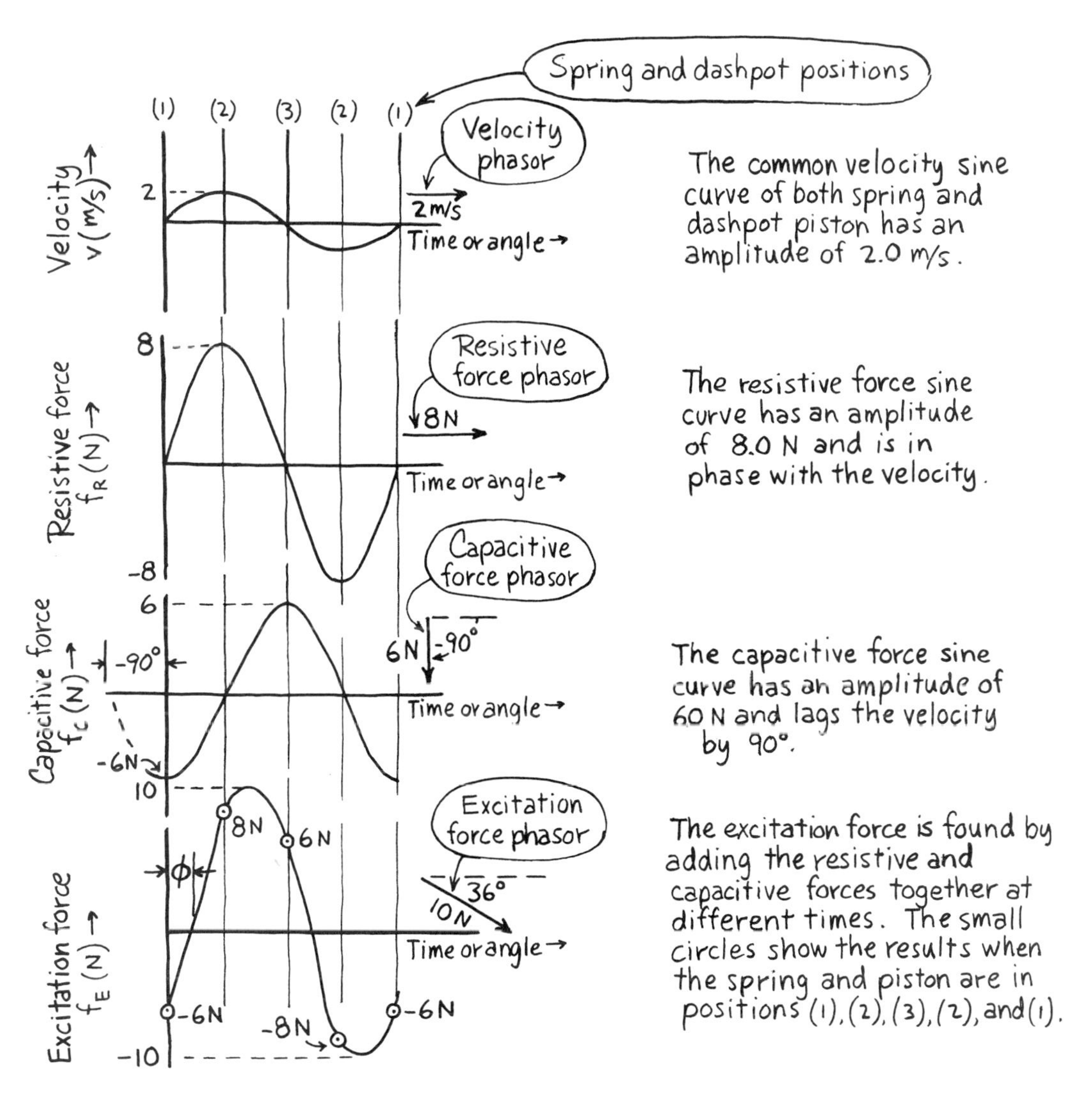

Figure 15-12. Velocity and force sine curves for the *RC* series mechanical system.

graphs are almost the same as Fig. 15-4 for the *RM* series system except that the capacitive force *lags* rather than leads the common velocity. The excitation force is found by adding f_R to f_C at different times. It has an amplitude of 10 N and a *negative* phase angle of about 36°. The negative phase angle is the only thing different from the *RM* system.

The excitation force amplitude and phase angle can be easily found using a force phasor diagram (Fig. 15-13). Note that the capacitive force phasor points downward because its phase angle is −90°.

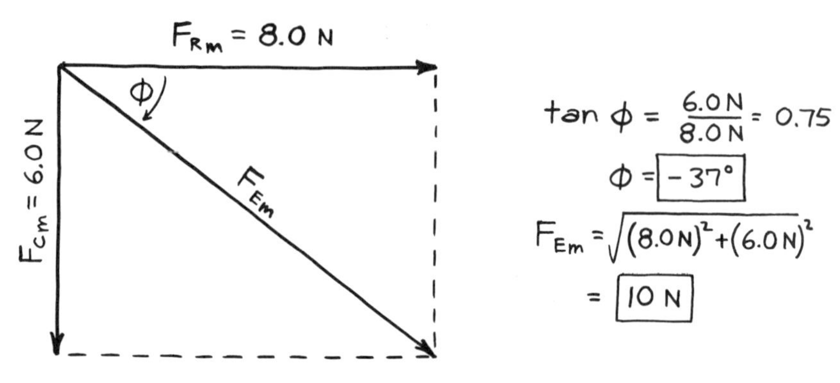

Figure 15-13. The vector addition of F_{Rm} and F_{Cm} yields answers for both F_{Em} and its phase angle ϕ with respect to the common velocity.

An impedance phasor diagram can be constructed for this system by dividing each of the force phasors by the common velocity amplitude. Figure 15-14 shows the results. The figure includes the necessary simple equations for making calculations and the results from the *RM* series system for comparison.

In summary, the patterns for *RM* and *RC* series systems are similar. The only real difference is between the phase angles of inertive and capacitive force

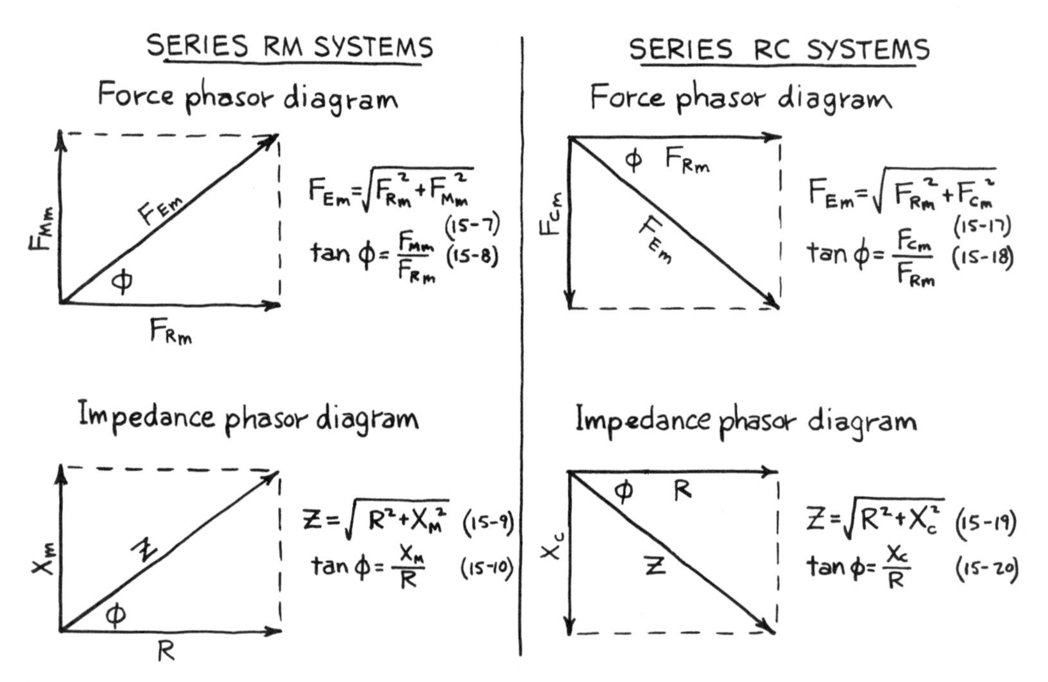

Figure 15-14. Phasor diagrams and equations for *RM* and *RC* series systems.

sine curves. The inertive force leads and the capacitive force lags the common velocity, each by 90°. Thus the phasors for inertive force and reactance point upward, and those for capacitive force and reactance point downward.

Power needed to drive the *RC* series system can be found using the same equations as for the *RM* system.

$$P_{av} = F_R \cdot V \tag{14-10}$$

$$P_{av} = F_E V \cos \phi \tag{15-11}$$

Only the dashpot requires power (Sec. 14-2). The spring receives energy but returns it to the generator in other parts of the cycle (Sec. 14-4).

An Electrical Analogy: The *RC* Series System

Our knowledge gained from the mechanical system can be used to describe the series *RC* electrical system shown in Fig. 15-15. The resistor replaces the dashpot, the capacitor does the job of the spring, and the electrical sine-wave generator replaces the mechanical generator. Figure 15-15 also

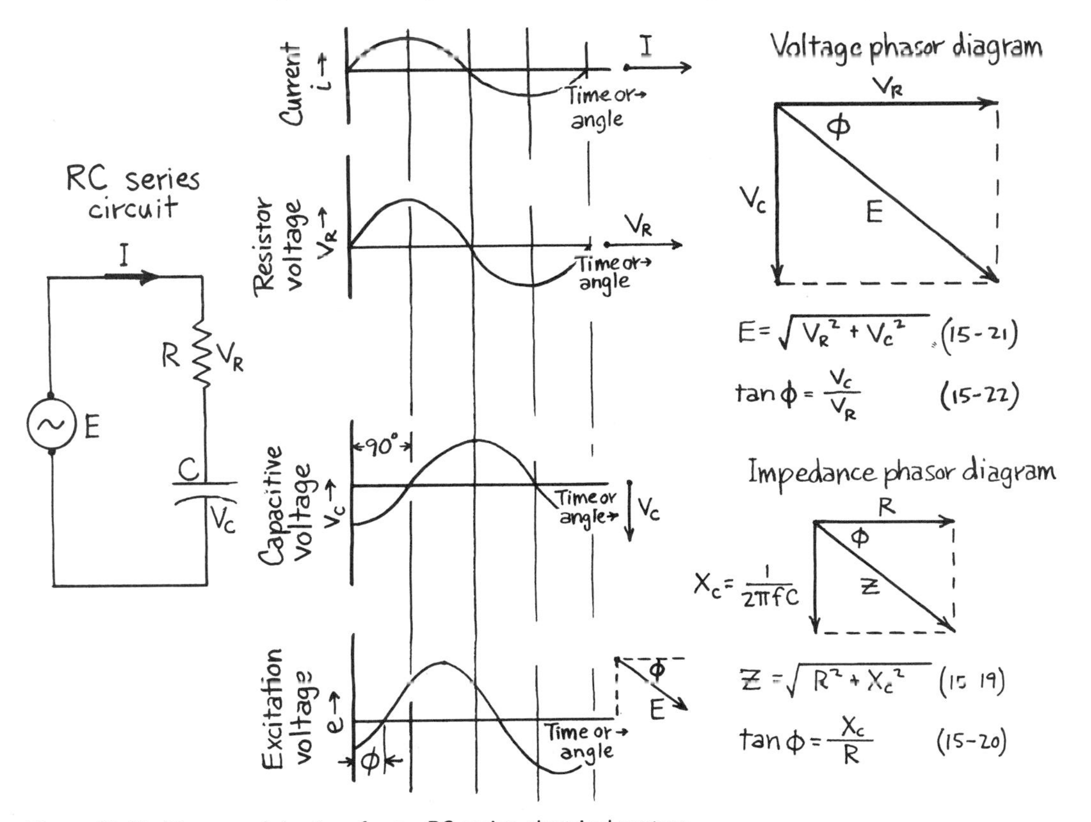

Figure 15-15. The complete story for an *RC* series electrical system.

shows current and voltage sine curves, phasor diagrams, and simple equations for solving numerical problems. Now to apply this knowledge to Example 15-4.

Example 15-4 In Fig. 15-15 the resistance is 2000 Ω, the capacitance is 0.010 μF, and the generator excitation voltage is 10.0 V (rms).

(a) Calculate the capacitor voltage V_C, the resistor voltage V_R, and the phase angle ϕ when the generator frequency is 16,000 Hz.

(b) What generator frequency makes the phase angle −45°?

Solution to (a)

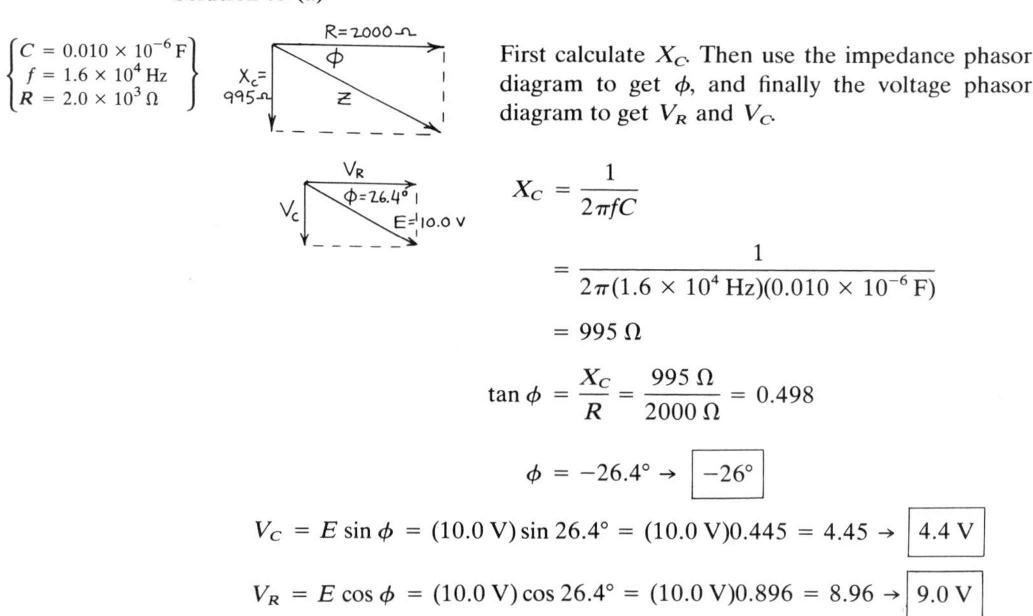

$$\left\{\begin{array}{l} C = 0.010 \times 10^{-6}\,\text{F} \\ f = 1.6 \times 10^{4}\,\text{Hz} \\ R = 2.0 \times 10^{3}\,\Omega \end{array}\right\}$$

First calculate X_C. Then use the impedance phasor diagram to get ϕ, and finally the voltage phasor diagram to get V_R and V_C.

$$X_C = \frac{1}{2\pi f C}$$

$$= \frac{1}{2\pi(1.6 \times 10^{4}\,\text{Hz})(0.010 \times 10^{-6}\,\text{F})}$$

$$= 995\,\Omega$$

$$\tan\phi = \frac{X_C}{R} = \frac{995\,\Omega}{2000\,\Omega} = 0.498$$

$$\phi = -26.4° \rightarrow \boxed{-26°}$$

$$V_C = E\sin\phi = (10.0\,\text{V})\sin 26.4° = (10.0\,\text{V})0.445 = 4.45 \rightarrow \boxed{4.4\,\text{V}}$$

$$V_R = E\cos\phi = (10.0\,\text{V})\cos 26.4° = (10.0\,\text{V})0.896 = 8.96 \rightarrow \boxed{9.0\,\text{V}}$$

Solution to (b)

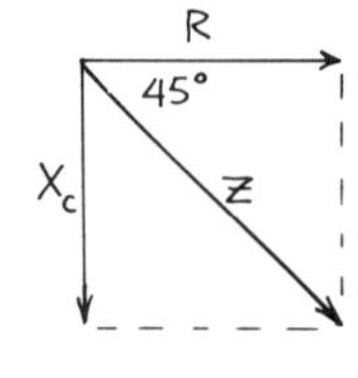

Use the impedance phasor diagram. The phase angle is −45° when $X_C = R$.

$$R = X_C = \frac{1}{2\pi f C}$$

$$f = \frac{1}{2\pi RC} = \frac{1}{2\pi(2000\,\Omega)(0.010 \times 10^{-6}\,\text{F})}$$

$$= 7.96 \times 10^{3} \rightarrow \boxed{8.0 \times 10^{3}\,\text{Hz}}$$

*A Heat System Analogy: Thermal Lag

Electrical RC circuits can be used to analyze heat systems using unified concepts. The analogous relations are laid out in the following table.

Unified Concept	*Heat System*	*Electrical System*	*Chapter Reference*
Force	Temperature difference	Potential difference	4
Resistance	Thermal resistance	Electrical resistance	7
Capacitance	Thermal capacitance	Electrical capacitance	9
Parameter rate	Heat rate	Electric current	6

We can use an electrical analogy to study the behavior of a mercury thermometer as it measures a sinusoidally changing temperature. The thermometer has both a thermal resistance and a thermal capacitance. Figure 15-16a shows how the thermometer reacts to a sudden "step" change of temperature (i.e., to a suddenly applied excitation force), and to a sinusoidal temperature change. The excitation force is the difference between temperatures of the thermometer and of its surroundings. The reaction to the step change in temperature is the familiar exponential curve described by a time constant τ_{RC} (Sec. 11-4). In the sinusoidal case of Fig. 15-16b, the measured temperature follows a sine curve that lags the excitation temperature and has an amplitude smaller than the excitation temperature.

The thermometer's behavior can be better understood by making an analogy to a simple *RC* electrical circuit (Fig. 15-17). The voltage phasor

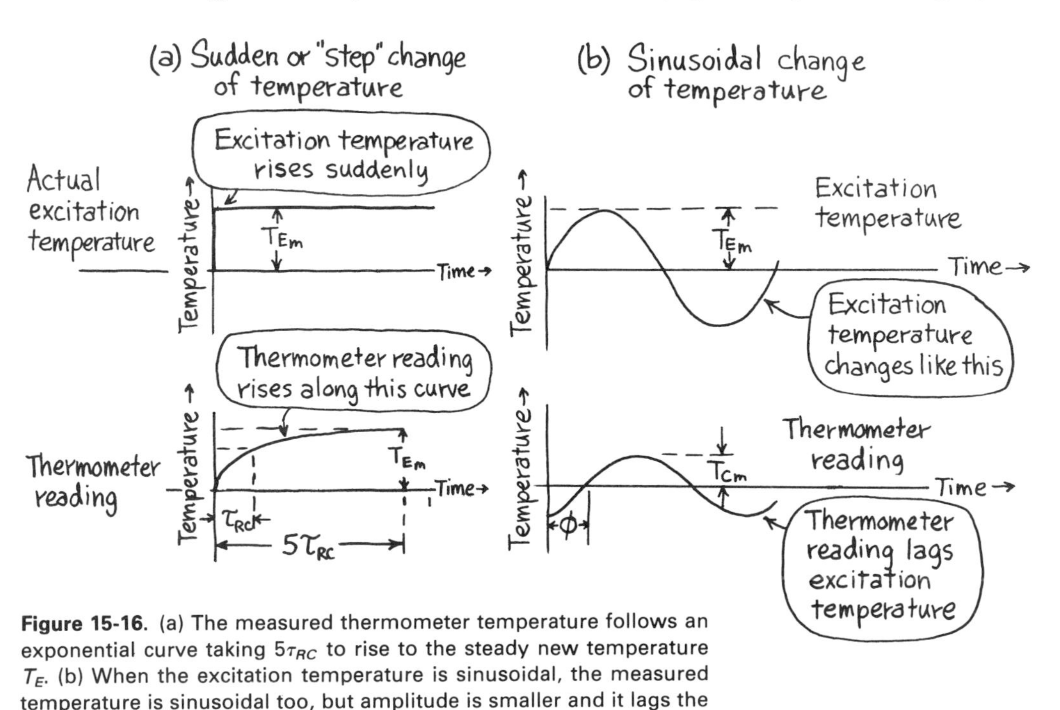

Figure 15-16. (a) The measured thermometer temperature follows an exponential curve taking $5\tau_{RC}$ to rise to the steady new temperature T_E. (b) When the excitation temperature is sinusoidal, the measured temperature is sinusoidal too, but amplitude is smaller and it lags the excitation temperature.

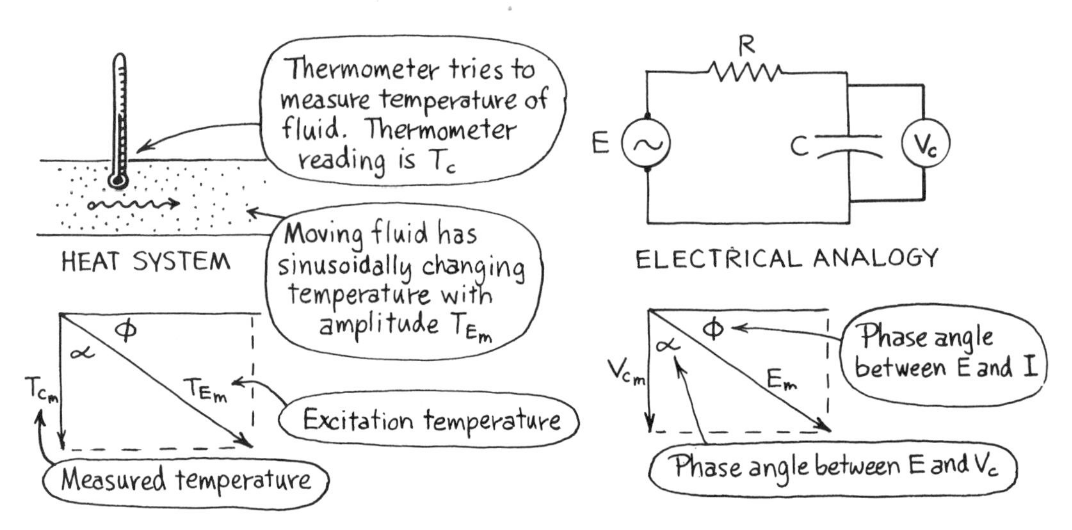

Figure 15-17. The heat system is analogous to a simple *RC* electrical circuit. The voltage and temperature phasor diagrams help to describe the relation between the measured temperature T_C and excitation temperature T_E.

diagram shows that the peak capacitor voltage V_{Cm} is smaller than the peak excitation voltage E_m, and also lags the excitation voltage by the angle α. In the temperature phasor diagram, the peak thermometer reading T_{Cm} (analogous to V_{Cm}) is smaller than the peak excitation temperature T_{Em} (analogous to E_m). The thermometer reading also lags the excitation temperature by the angle α.

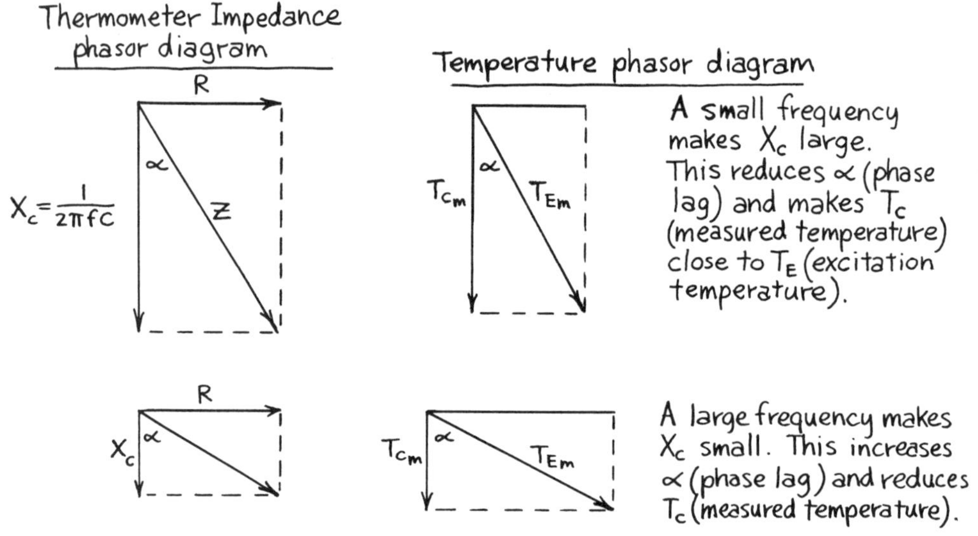

Figure 15-18. Impedance phasor diagrams help to describe how frequency affects the thermometer-measured temperature.

Impedance phasor diagrams give insights as to how the thermometer reacts when the excitation temperature frequency changes. Remember that X_C decreases with increased frequency, since $X_C = 1/2\pi fC$ [Eq. (14-28)]. If the frequency is small (slowly changing sinusoidal temperature), X_C is large, α is small, and T_{Cm} almost equals T_{Em} (Fig. 15-18). This means that the thermometer faithfully measures a slowly changing temperature. If the frequency is large (rapidly changing temperature), X_C is small, the phase lag α is large, and T_{Cm} (thermometer reading) is small. The thermometer may not measure any change at all for a rapidly changing temperature. These ideas are important in systems that control energy.

The situation in which the phase angle is $-45°$ (T_C lags T_E by 45°) is especially interesting. The frequency that causes this makes $X_C = R$, and can be easily calculated as

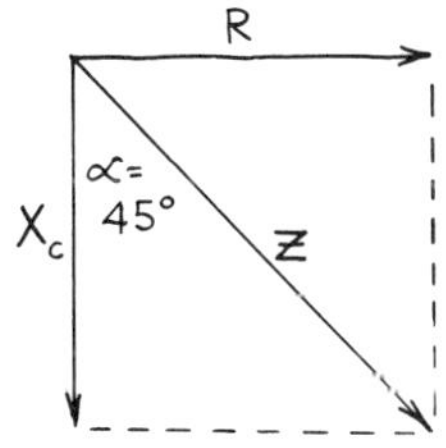

$$X_C = R$$

$$\frac{1}{2\pi fC} = R$$

$$f = \frac{1}{2\pi(RC)}$$

But

$$\tau_{RC} = RC \tag{11-5}$$

$$f = \frac{1}{2\pi\tau_{RC}} \tag{15-23}$$

The time constant comes back again, this time to help us answer questions about sinusoidal systems. See if you can prove that in this situation the thermometer reading T_{Cm} is 0.707 times the excitation temperature T_{Em}. Here is an example.

Example 15-5 The time constant of a mercury thermometer is 50 s. Would you recommend that it be used to measure a sinusoidally changing temperature that has a period of 1.0 min?

Solution

The frequency f_1 of the sinusoidal temperature is

$$f_1 = \frac{1}{T} = \frac{1}{60\text{ s}} = 0.017\text{ Hz}$$

The frequency f_2 at which we can expect a 45° phase lag is

$$f_2 = \frac{1}{2\pi\tau_{RC}} = \frac{1}{2\pi \cdot 50\text{ s}} = 0.0032\text{ Hz}$$

Since f_1 is $0.017/0.0032 = 5$ times larger than f_2, we can expect a poor performance as the thermometer tries to follow the changing excitation temperature. The measured temperature will vary sinusoidally with a frequency f_1 of 0.017 Hz. The measured temperature amplitude will be small compared to the excitation temperature amplitude and will lag it by an angle of between 45 and 90°.

15-4 PARALLEL SYSTEMS

Up to this point we have looked only at series-connected systems. But many practical situations may include elements connected in parallel. In a series system the parameter rate (current, velocity, etc.) is common to all elements, and the sum of the forces across each element at all times equals the excitation force. In a parallel system the same excitation force is applied to each element. The parameter rate of the source splits up into two or more branches, but at all times it is equal to the sum of parameter rates in the individual branches. These ideas were discussed in Sec. 7-6, and they are summarized in Fig. 15-19.

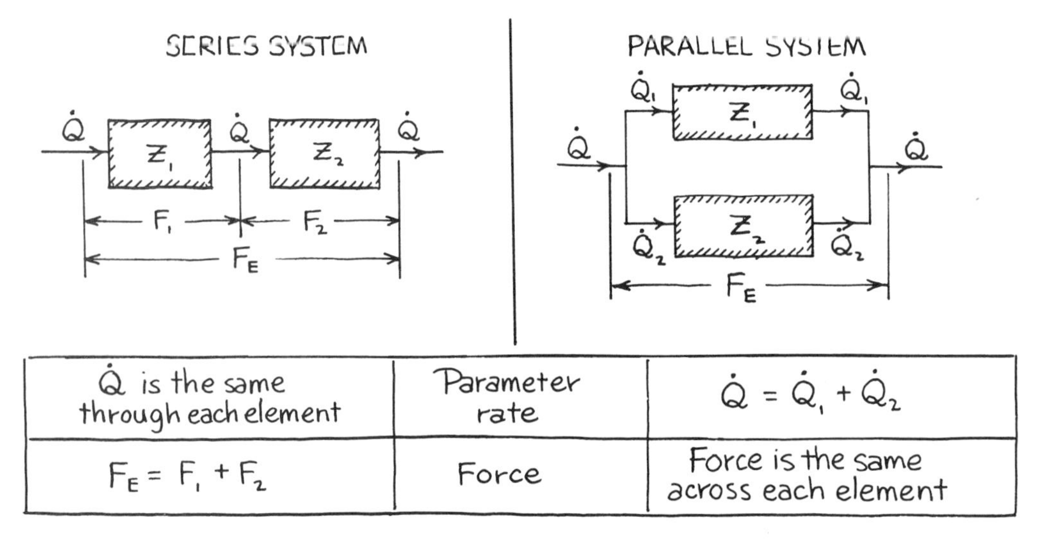

$\dot{Q}$ is the same through each element	Parameter rate	$\dot{Q} = \dot{Q}_1 + \dot{Q}_2$
$F_E = F_1 + F_2$	Force	Force is the same across each element

Figure 15-19. A comparison of parameter rate and force for series and parallel systems.

Series and parallel arrangements are fairly easy to spot in electrical systems, but they may not be so apparent in other systems. Furthermore, we now have three kinds of elements to contend with—resistors, capacitors, and inertive devices—rather than just resistors. Let us look at some systems and try to decide whether the elements that make them up are in series or parallel arrangements.

Figure 15-20 shows two electrical circuits. In Fig. 15-20a the resistor and coil are obviously in a series arrangement. There is a common current, and the excitation voltage at all times equals the sum of the potential differences across the resistor and the coil. In Fig. 15-20b the resistor and capacitor are in a

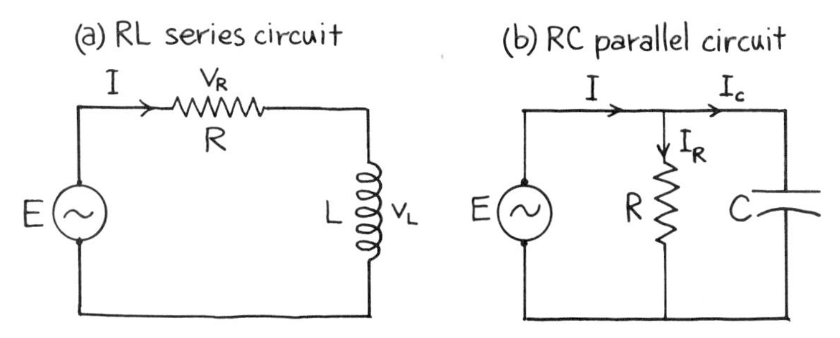

Figure 15-20. Series and parallel electrical circuits.

parallel arrangement. The excitation voltage E is common to both elements, and the source current I is at all times the sum of the branch currents I_R and I_C.

Figure 15-21 shows two heat systems. In Fig. 15-21a the metal fins are arranged to conduct heat away from a hot, moving fluid. This is a parallel arrangement, since each fin conducts a portion of the total heat being transferred and the temperature difference between hot and cold surfaces is the same for each fin. In Fig. 15-21b the wall of a furnace is made up of an insulation layer and an outer steel plate. This is a series arrangement, because the same heat passes through both materials, and the total temperature difference is the sum of the temperature differences across each material.

Figure 15-22 shows a fluid system in which a mass of water is accelerated back and forth by a piston inside a rough pipe. This is a series RM system.

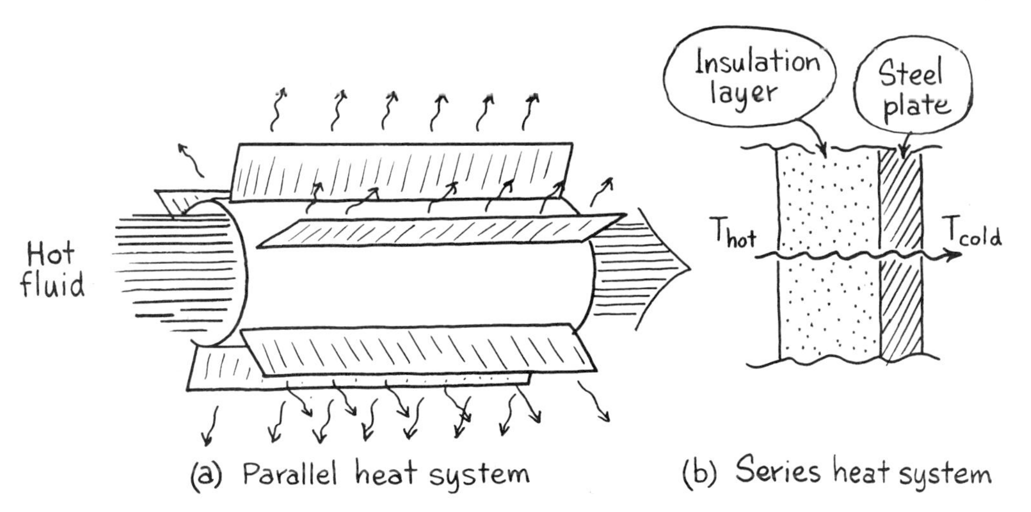

Figure 15-21. Series and parallel heat systems. (a) Each fin conducts a portion of the heat. All fins share a common temperature difference. This is a parallel system. (b) The same heat flows through both wall materials. This is a series arrangment.

There is only one volume flow rate, and the piston excitation pressure must at all times provide pressure to accelerate the water and to overcome fluid friction.

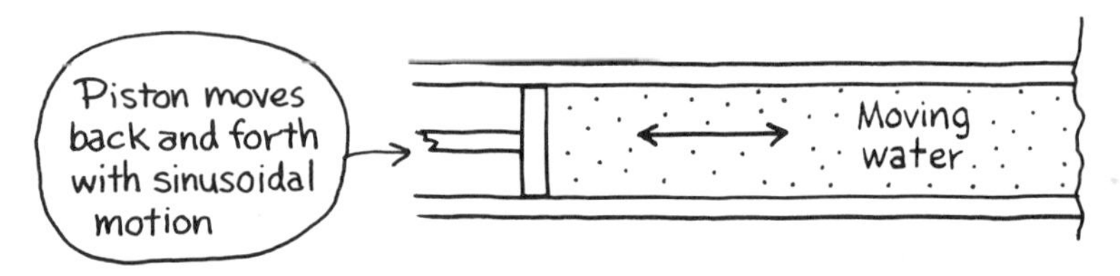

Figure 15-22. The piston excitation pressure must accelerate the water and overcome friction forces. There is a common volume flow rate. Both facts establish this as a series *RM* system.

Figure 15-23 shows a mechanical mass–dashpot system. In Fig. 15-23a the generator, mass, and dashpot piston are all rigidly connected. This means that the generator deflection Δs_E at all times equals the mass and dashpot deflections Δs_M and Δs_R. Hence their velocities ($v_E = \Delta s_E/\Delta t$, $v_M = \Delta s_M/\Delta t$, and $v_R = \Delta s_R/\Delta t$) are equal. Furthermore, the net force f_M needed to accelerate the mass is $f_M = f_E - f_R$ or $f_E = f_M + f_R$. The excitation force f_E at all times equals the sum of the inertive force f_M and the resistive force f_R. Both statements (same velocity and additive forces) satisfy the conditions for a series arrangement of elements.

You wouldn't think that reversing the mass and dashpot would make any difference, but it does. Look at Fig. 15-23b. Here the generator is connected

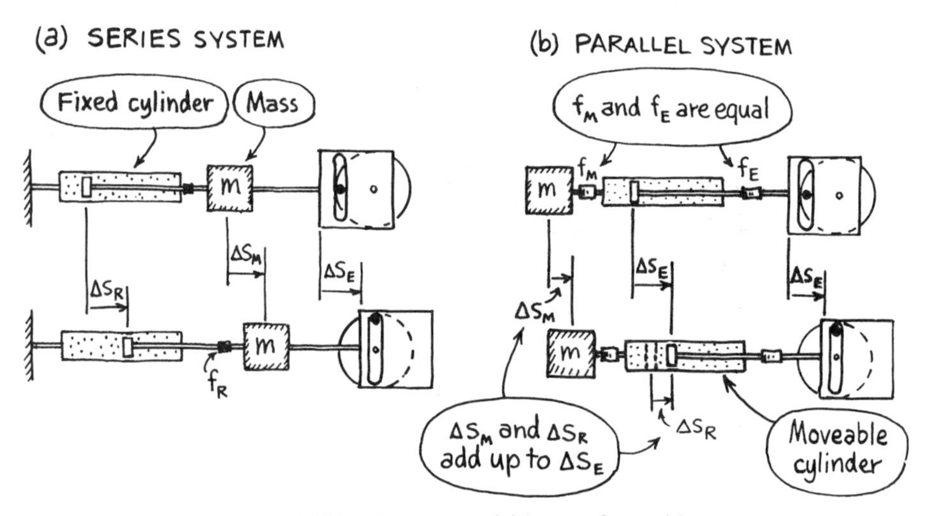

Figure 15-23. A mechanical *RM* series system (a) is transformed into an *RM* parallel system (b) merely by interchanging the mass and dashpot positions with respect to the mechanical sinusoidal generator.

directly to the dashpot piston. Frictional forces cause the dashpot cylinder to move, but it does not move as much as the piston. The mass, connected rigidly to the cylinder, has the same motion as the cylinder. The movement Δs_R of the piston relative to its cylinder is the difference between the generator movement Δs_E and the mass movement Δs_M. $\Delta s_R = \Delta s_E - \Delta s_M$ or $\Delta s_E = \Delta s_R + \Delta s_M$. Furthermore, any force from the generator is transmitted directly through the weightless dashpot to the mass. All forces are equal; that is, $f_E = f_R = f_M$. Both statements (additive motions and same forces) are in agreement with the definition for a parallel arrangement of elements. The mass and dashpot appear physically to be in series, but they are in parallel according to our definition of a parallel system.

Figure 15-24 shows two mechanical *RC* systems. We will leave it up to you to decide which arrangement is series connected and which is parallel connected.

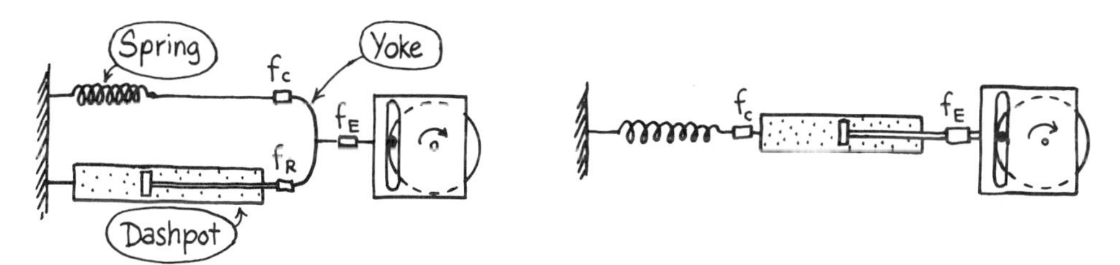

Figure 15-24. Which of the mechanical *RC* systems is in series and which is in parallel?

*Calculations in Parallel Systems

Calculations can be made on parallel systems. However, don't use the force and impedance phasor diagrams developed so far in this chapter. They only work for series systems. Something that we will call a *parameter rate phasor diagram* (for velocity, current, etc.) is used in solving parallel systems. It is based on the fact that the sum of the branch sinusoidal parameter rates at all times equals the total parameter rate from the generator. Figure 15-25 shows a series and a parallel electrical circuit together with their corresponding phasor diagrams. Note that the current phasor I_C points *up* and the voltage phasor V_C points *down*. This is because the reference for measuring phase angle is different. In the series diagram the common current is used as reference. In the parallel diagram the common voltage is used as reference. In either case, capacitor voltage lags capacitor current by 90°.

The strategy in calculating the impedance of the series system is to find the excitation voltage using the voltage phasor diagram and divide by the common current. To calculate the impedance of the parallel system, find the total current using the current phasor diagram and divide it into the common excitation voltage. The series and parallel impedances do not have the same value. The parallel impedance is smaller. (See Problems 15-25 and 15-26.)

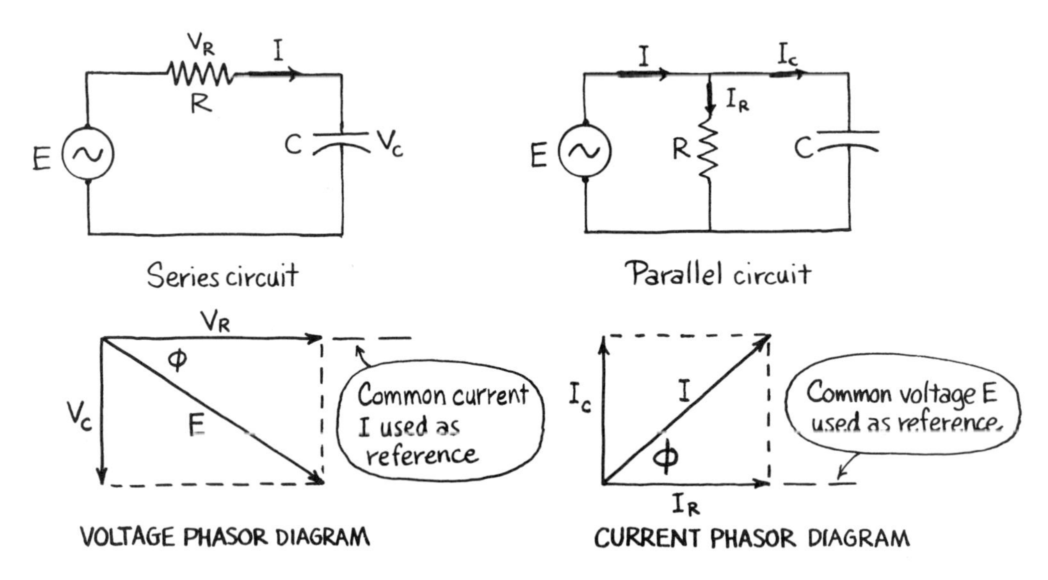

Figure 15-25. A voltage phasor diagram is used to add sinusoidal voltages in the series circuit, and the common current is used as reference for measuring phase angle. In a parallel circuit, a current phasor diagram is used to add sinusoidal currents. The common source voltage *E* serves as reference for measuring phase angle.

15-5 SUMMARY

This chapter has brought out a lot of new information about the more complicated *RM* and *RC* sinusoidal systems. We will try to summarize it for you.

1. The amplitude and phase angle of a sine curve can be represented by an arrow called a phasor. The length of the arrow is the amplitude and the direction is the phase angle with respect to another reference sine curve.

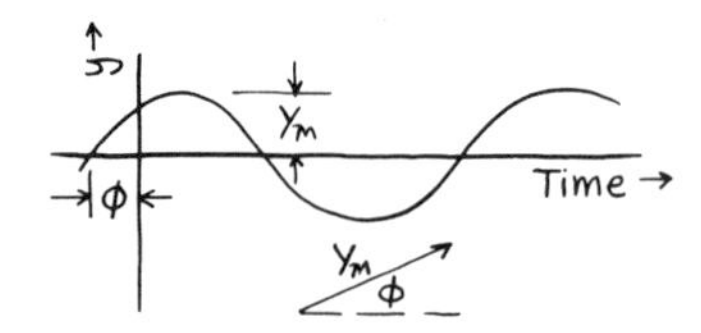

2. Impedance includes two pieces of information about a system. One is the ratio of excitation force to input parameter rate amplitudes. The other

piece of information is the phase angle of the excitation force with respect to the common parameter rate.

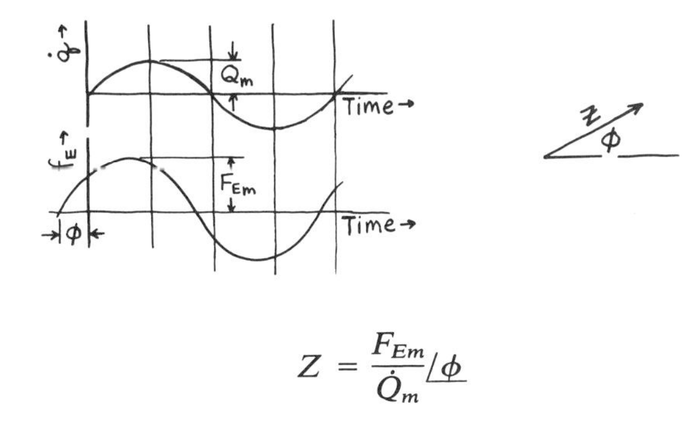

$$Z = \frac{F_{Em}}{\dot{Q}_m}\underline{/\phi}$$

Think of impedance as the total opposition that a system offers when excited by a constant or changing force.

3. In series systems the elements share a common parameter rate. Forces across the elements are additive, and their sum equals the excitation force. The addition is done by treating force phasors as vectors and adding them together. Impedance phasors also add as vectors in series systems.

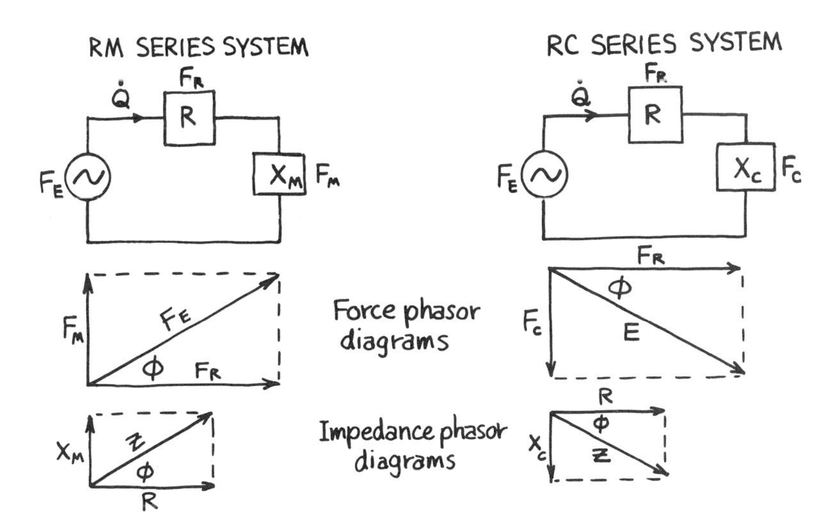

4. In parallel systems the elements share a common excitation force. Parameter rates through each branch are additive. Their sum equals the total parameter rate coming from the generator. Parameter rate phasors treated as

vectors are used to perform the addition. The impedance of a parallel system is found by dividing the total parameter rate into the common excitation force. Impedance phasor diagrams for series systems should not be used for parallel systems.

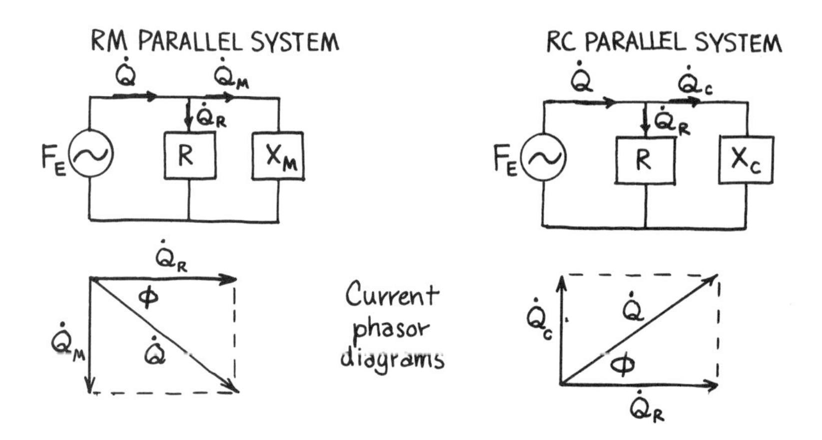

The next chapter puts inertance and capacitance elements together in a system. We did this in Chapter 13 and discovered that the system oscillated at a definite resonant frequency. What happens if we sinusoidally excite such a system at this same frequency? We should expect some kind of distinctive system reaction.

PROBLEMS

15-1. A mass of 900 kg vibrates at a frequency of 30 Hz. What is the impedance (magnitude and angle) of the mass?

15-2. An open door with a moment of inertia of 30 kg · m^2 about its axis of rotation is subjected to an angular vibration of 2.0 Hz. What is the impedance (magnitude and angle) of the door?

15-3. What is the impedance (magnitude and angle) of a 0.20-μF capacitor if excited by a 1.5-V, 4000-Hz signal?

15-4. A coil has an inductance of 0.30 H.

(a) What frequency must a source sine-wave generator have in order for the coil to have an impedance of 3000 Ω? Neglect any resistance the coil may have.

(b) Make a sketch of the coil's impedance phasor diagram.

15-5. A spring deflects 0.25 ft when stretched by a 15-lb force. What is the spring impedance if it vibrates at a frequency of 5.0 Hz? (*Hint*: Refer to Sec. 9-2 for a definition of spring capacitance.)

15-6. Velocity and exicitation force graphs for a

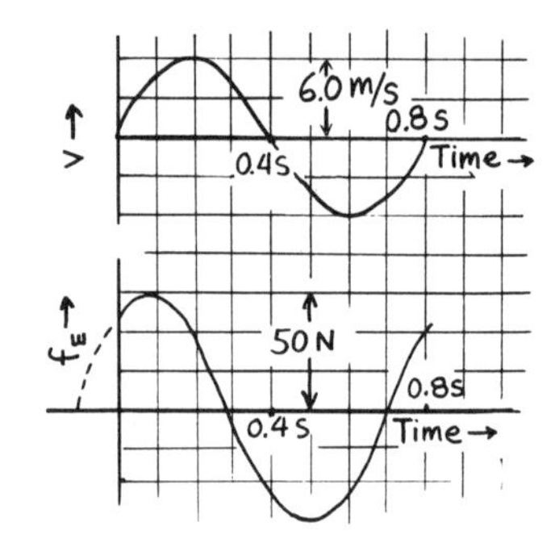

Problem 15-6

sinusoidal mechanical system are shown in the sketches.

(a) What is the impedance of the system? Give both magnitude and phase angle.

(b) What is the frequency of the excitation?

15-7. Graphs of current and excitation voltage from a sine-wave generator are shown in the sketch.

(a) What is the impedance of the circuit to which the generator is connected? Give both magnitude and phase angle.

(b) What is the generator's frequency?

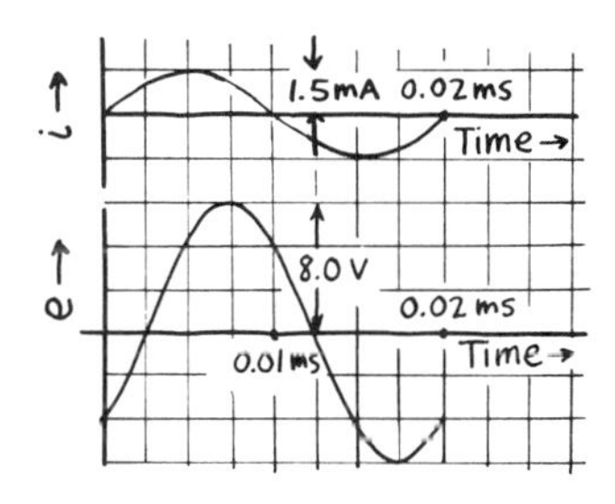

Problem 15-7

15-8. The impedance of a sinusoidal electrical circuit is 6000 Ω. Voltage leads current by 30°. The measured current has an amplitude of 4.0 mA.

(a) What is the excitation voltage amplitude?

(b) Make a sketch of the excitation voltage graph using the current graph as reference.

15-9. The impedance of a sinusoidal mechanical rotational system is 500 lb·ft/(rad/s). The phase angle of the system is −60°. The measured angular velocity has an amplitude of 0.020 rad/s.

(a) What is the excitation torque amplitude?

(b) Make a sketch of the excitation torque graph using the angular velocity graph as reference.

15-10. Refer to the sketches in Problem 15-6.

(a) Does the excitation force lead or lag the velocity?

(b) Calculate the time interval between the occurrence of the peak force and the peak velocity.

15-11. Refer to the sketches in Problem 15-7.

(a) Does the excitation voltage lead or lag the current?

(b) What is the time interval between the occurrence of the peak current and the peak excitation voltage?

15-12. The sketch shows the impedance phasor of a sinusoidal series electrical circuit. The excitation voltage is 7.0 V rms.

(a) What is the rms current?

(b) What average power is supplied to the circuit?

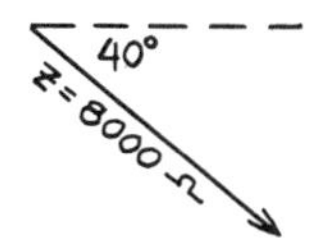

Problem 15-12

15-13. The sketch shows the impedance phasor of a sinusoidal series mechanical system. The measured velocity amplitude is 0.050 ft/s.

(a) What is the amplitude of the sinusoidal excitation force?

(b) What average power in ft·lb/s is supplied to the system?

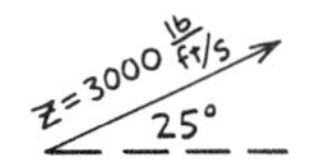

Problem 15-13

15-14. A force sine curve has an amplitude of 10 lb. Another force sine curve with the same frequency has an amplitude of 15 lb and leads the first curve by 90°.

(a) Graphically add the two force sine curves together.

(b) What is the phase angle of the combined force curve relative to the 10-lb force curve?

15-15. Find the amplitude and phase angle of the combined force in Problem 15-14 using a force phasor diagram.

15-16. The current from a sine-wave generator has an amplitude of 2.0 mA when connected to an electronic circuit. The generator voltage has an amplitude of 6.0 V and leads the current by 30°.

(a) What is the circuit impedance? Give both magnitude and angle.

(b) Make sine curve sketches of the generator

current and voltage. Show the amplitudes and the phase relation between the two curves on your sketches.

15-17. A coil has a resistance of 150 Ω and an inductance of 0.95 H. What is its impedance (magnitude and phase angle) if connected to a 60-Hz sinusoidal source?

15-18. A temperature-measuring device has a time constant of 30 s. Can it be used to accurately measure temperature in a system where temperature varies sinusoidally with a period of 15 min?

15-19. A pressure sine curve has an amplitude of 5.0 psi. Another pressure sine curve with the same frequency has an amplitude of 10.0 psi and lags the first pressure curve by 90°. Graphically add the two pressure sine curves together. What is the amplitude and phase angle of the combined pressures relative to the first curve?

15-20. Find the amplitude and phase angle of the combined pressure in Problem 15-19 using a pressure phasor diagram.

15-21. A 15.0-V, 400-Hz source is connected across a 600-Ω resistor.

(a) What power is supplied by the source?

(b) What size capacitor must be added in series to make the phase angle between source voltage and current exactly 45°?

(c) What power is supplied to the circuit with the capacitor present? (This problem illustrates that the source supplies only half as much power when the phase angle changes from 0 to 45°.)

15-22. A temperature-sensing device has a mass of 2.0 g, a specific heat of 0.10 cal/(g · °C), and a time constant of 70 s.

(a) What frequency causes a 45° lag between input and measured sinusoidal temperatures?

(b) What is the capacitance of the sensor in J/°C? (*Hint*: Refer to Sec. 9-2 to calculate the capacitance.)

(c) Estimate the thermal resistance of the sensor in °C/W. (*Hint*: Approximate the behavior of the sensor by comparing it to a series *RC* electrical circuit.)

15-23. A 0.100-H coil and a 500-Ω resistor are connected in series to a 10.0-V, 1000-Hz source.

(a) What is the coil reactance?

(b) What is the circuit impedance (magnitude and phase angle)?

(c) What is the current?

15-24. A 0.200-μF capacitor and a 500-Ω resistor are connected in series to a 10.0-V, 1000-Hz source.

(a) What is the capacitor's reactance?

(b) What is the circuit impedance (magnitude and phase angle)?

(c) What is the current?

15-25. The 0.100-H coil and 500-Ω resistor in Problem 15-23 are connected in parallel rather than series to the 10.0-V, 1000-Hz source.

(a) What is the total current? Does it lead or lag the source voltage?

(b) What is the circuit impedance (magnitude and phase angle)?

(c) Why would you expect the parallel circuit impedance to be smaller than the series impedance?

15-26. The 0.200-μF capacitor and 500-Ω resistor in Problem 15-24 are connected in parallel rather than in series to the 10.0-V, 1000-Hz source.

(a) What is the total current? Does it lead or lag the source voltage?

(b) What is the circuit impedance (magnitude and phase angle)?

(c) Why would you expect the parallel circuit impedance to be smaller than the series impedance?

ANSWERS TO ODD-NUMBERED PROBLEMS

15-1. $1.7 \times 10^5 \underline{/+90^\circ}$ N/(m/s)

15-3. $2.0 \times 10^2 \underline{/-90^\circ}$ Ω

15-5. $1.9 \underline{/-90^\circ}$ lb/(ft/s)

15-7. (a) $5.3 \times 10^3 \underline{/-45^\circ}$ Ω
(b) 50×10^3 Hz

15-9. (a) 10 lb · ft
(b)

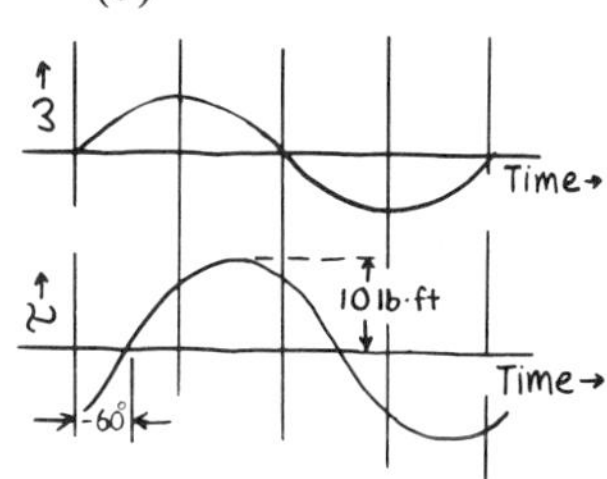

15-11. (a) Lags because voltage peak comes last in the time sequence
(b) 2.5×10^{-3} s

15-13. (a) 1.5×10^2 lb
(b) 3.4 ft · lb/s

15-15. 18 lb; 56°

15-17. $3.9 \times 10^2 \underline{/+67.3^\circ}$ Ω

15-19. 11 psi; −65°

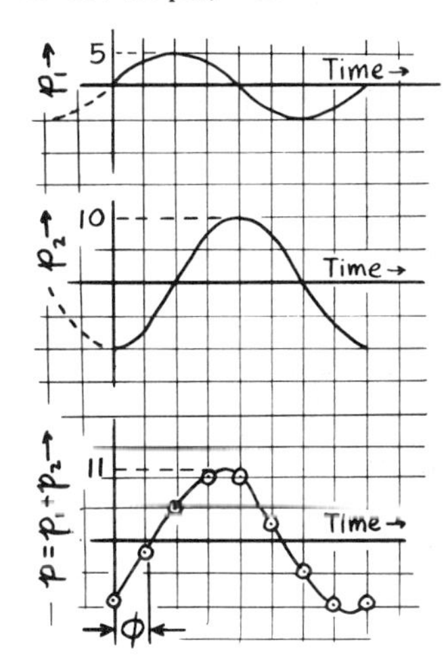

15-21. (a) 0.375 W
(b) 6.63×10^{-7} F
(c) 0.187 W

15-23. (a) 628 Ω
(b) $803 \underline{/+51.5^\circ}$ Ω
(c) 0.0125A

15-25. (a) 2.56×10^{-2} A
(b) $391 \underline{/+38.5^\circ}$ Ω
(c) More paths in parallel system provide less overall opposition to flow of charge.

16

Forced Oscillations and Resonance—*RMC* Systems

Resonance is a large sinusoidal oscillation in a system driven by a relatively small periodic excitation force. There are many examples of resonance in nature, ranging from the very large to the very small. Extreme tidal ranges in parts of the earth's ocean basins are the result of resonance brought on by gravitational forces among earth, moon, and sun—a very large system. The ocean's water is blue because of resonance in the very small system of an atom in the atmosphere reacting to frequencies in the blue color region of our vision. Pushing a child on a swing is an example of resonance if the push is applied at the right time. Tuning a radio to pick up certain stations takes advantage of resonance in an electronic circuit. Selecting the right motor mounts to avoid excessive vibration is a problem in resonance for a mechanical system. Designing or selecting valves to avoid natural vibrations in hydraulic systems is an interesting problem in resonance. The list is potentially endless.

Resonance occurs in systems that store both potential and kinetic energy. We saw in Chapter 13 that MC systems have a natural frequency at which the storage devices just exchange their potential and kinetic energies without any help from the source.

We first study resonance in this chapter in a series-connected mechanical system (Section 16-1) in which we can see what happens to the mass, the spring, and the dashpot, and how the energy moves back and forth. The results of these studies are generalized for other systems in Section 16-2. A series-connected electrical circuit is analyzed in Section 16-3. Calculations are made that allow us to construct curves of current and impedance versus frequency. Circuit quality Q is introduced to describe the steepness of these curves. Section 16-4 looks at resonance in parallel-connected systems, and Section 16-5 shows a number of worked examples in both series- and parallel-connected systems.

16-1 RESONANCE IN A SERIES MECHANICAL SYSTEM

Resonance requires that both potential and kinetic energy storage devices be present in a system. So let's attach a spring and a mass to the mechanical sinusoidal generator of Chapters 14 and 15. A dashpot is also added to provide resistance in the system (Fig. 16-1).

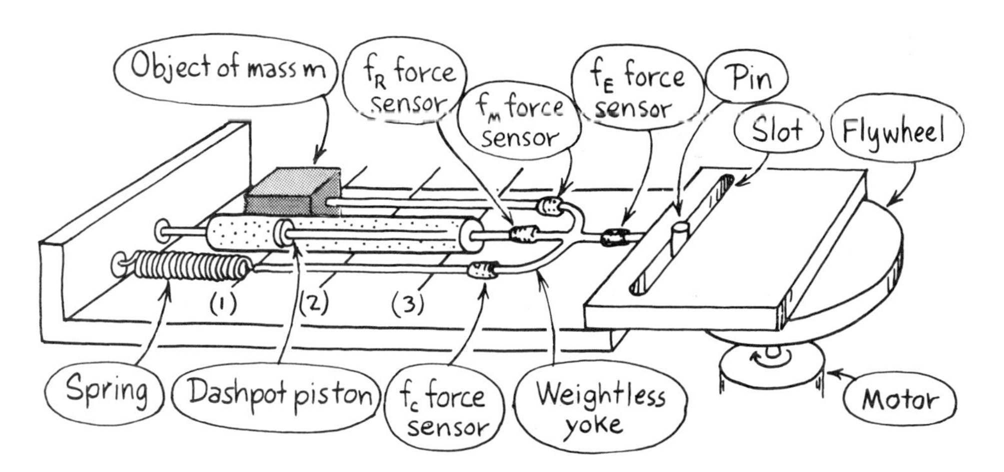

Figure 16-1. A mechanical *RMC* series system. The dashpot, mass, and spring are attached through a weightless yoke to the sinusoidal mechanical generator.

The yoke is fitted with force sensors to measure dashpot resistive force f_R, mass inertive force f_M, spring capacitive force f_C, and the excitation force f_E. The yoke and dashpot piston have negligible mass, so f_E at all times equals the sum of f_R, f_M, and f_C.

$$f_E = f_R + f_M + f_C \tag{16-1}$$

The mechanical generator supplies the same sinusoidal motion to all three elements. All move with a common sinusoidal velocity. This is a system connected in series.

The motor starts up and we wait for the system to reach the sinusoidal steady state. The oscillating system is a combination of the *RM* and *RC* series systems discussed in Chapter 15. Figure 15-3 describes the mass–dashpot (*RM*) system, and Fig. 15-11 describes the spring–dashpot (*RC*) system. All the information needed from these sections can be summarized as follows:

1. The resistive force f_R varies sinusoidally *in phase* with the common velocity.

2. The inertive force f_M describes a sine curve that *leads* the common velocity by 90°.

3. The capacitive force f_C describes a sine curve that *lags* the common velocity by 90°.

For the sake of discussion here, we will assign the same arbitrary values to v, f_R, f_M, and f_C as was done in Chapter 15. Assume that their amplitudes are 2.0 m/s, 8.0 N, 6.0 N, and 6.0 N, respectively. The amplitudes of f_M and f_C

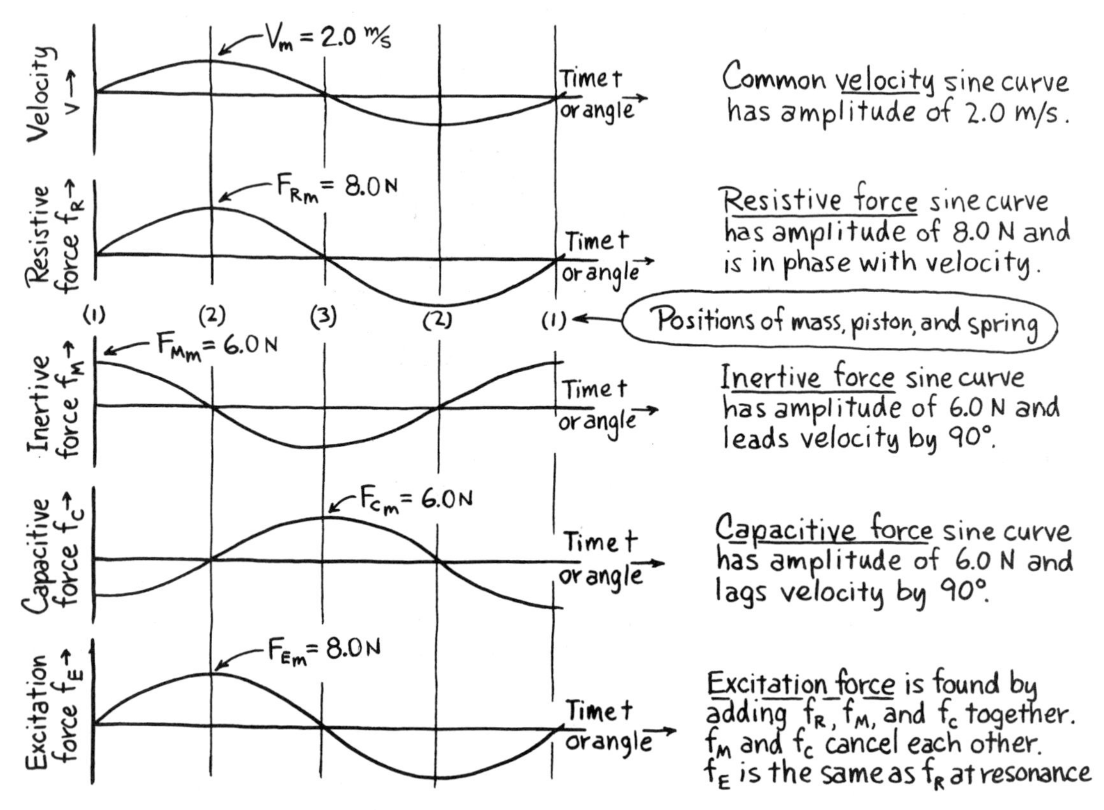

Figure 16-2. Velocity and force curves for *RMC* mechanical system at resonance.

have purposely been made equal. Graphs of all these are shown in Fig. 16-2. Velocities and forces to the right are taken as positive.

The excitation force f_E is found by adding f_R, f_M, and f_C together at different times using the superposition principle. Note that f_M and f_C cancel each other out, because they are exactly out of phase with each other. The mechanical generator does not "see" either the spring or the mass. It only supplies a force to move the dashpot ($f_E = f_R$). This situation is called *series resonance.*

Conditions for Series Resonance

Series resonance occurs when the inertive and capacitive forces have the same amplitude. This makes inertive reactance $X_M = F_{M_m}/V_m$ and capacitive reactance $X_C = F_{Cm}/V_m$ equal, since mass and spring have a common velocity.

$$F_{Mm} = F_{Cm}$$

$$X_M = \frac{F_{Mm}}{V_m} = \frac{F_{Cm}}{V_m} = X_C$$

and

$$X_M = X_C \quad \text{(resonance)} \tag{16-2}$$

There must be a single frequency f_r that makes $X_M = X_C$ possible for a given mass m and a spring capacitance C. It is easy to find an equation for f_r, the *resonant frequency.*

$$X_M = 2\pi f_r m \tag{14-21}$$

Don't confuse this with changing resistive force f_R (arrow pointing to f_r)

$$X_C = \frac{1}{2\pi f_r C} \tag{14-28}$$

$$X_M = X_C$$

$$2\pi f_r m = \frac{1}{2\pi f_r C}$$

$$f_r^2 = \frac{1}{4\pi^2 mC}$$

$$f_r = \frac{1}{2\pi\sqrt{mC}} \tag{16-3}$$

f_r Resonant Frequency	m Mass	C Spring Capacitance
hertz (Hz)	kilogram (kg)	$\frac{\text{metre}}{\text{newton}}$ (m/N)
	slug (slug)	$\frac{\text{foot}}{\text{pound}}$ (ft/lb)

Equation (16-3) is identical to Eq. (13-3) for the natural frequency of an *MC* system. You may wonder why an *RMC* system in this chapter oscillates with the frequency of an *MC* system in Chapter 13. The difference is that now a source continually feeds energy into the system to compensate for the energy lost to resistance. The resistor tends to slow down the vibration rate, but the source makes up for it.

Equal inertive and capacitive force amplitudes at resonance lead to equal reactances and the equation for resonant frequency. The fact that resistive and excitation force amplitudes are equal at resonance ($F_{Rm} = F_{Em}$) leads to a couple of other interesting conclusions. One is that the magnitude of the system's impedance $Z = F_{Em}/V_m$ just equals the dashpot resistance $R = F_{Rm}/V_m$. Also, the excitation force is in phase with velocity. The phase angle of the impedance phasor is zero.

$$Z = R\angle 0° \qquad \text{(resonance)} \tag{16-4}$$

TABLE 16-1 CONDITIONS OF SERIES RESONANCE

Amplitudes of inertive and capacitive force are equal	$F_{Mm} = F_{Cm}$	
Amplitudes of resistive and excitation force are equal	$F_{Em} = F_{Rm}$	
Inertive and capacitive reactances are equal	$X_M = X_C$	(16-2)
System impedance equals system resistance; the phase angle is zero	$Z = R\angle 0°$	(16-4)
The expression for resonant frequency is identical to the natural frequency of an *MC* system	$f_r = \dfrac{1}{2\pi\sqrt{MC}}$	(16-5)

Table 16-1 summarizes the various conditions of resonance. These conditions apply equally well to the other systems, too. Bear in mind that you must use the correct analogies. For example, inertance M is moment of inertia, fluid inertance, and inductance, as well as mass. The resonant frequency equation $f_r = 1/(2\pi\sqrt{MC})$ in the table [Eq. (16-5)] is applicable to other resonating systems.

Example 16-1 The mass in Fig. 16-1 is 0.239 kg and the frequency is 2.0 Hz. The various forces are as shown in Fig. 16-2.

(a) What is the impedance of the system at resonance?

(b) What must be the value of the spring capacitance in order to have the resonant situation?

Solution to (a)

$$\left\{\begin{matrix} F_{Rm} = 8.0\text{ N} \\ V_m = 2.0\text{ m/s} \end{matrix}\right\}$$

$$R = \frac{F_{Rm}}{V_m} = \frac{8.0\text{ N}}{2.0\text{ m/s}} = 4.0\,\frac{\text{N}}{\text{m/s}}$$

$$Z = R\underline{/0^\circ} = \boxed{4.0\underline{/0^\circ}\,\frac{\text{N}}{(\text{m/s})}}$$

Solution to (b)

$$\left\{\begin{matrix} m = 0.239\text{ kg} \\ f_r = 2.0\text{ Hz} \end{matrix}\right\}$$

$$X_M = 2\pi f_r m = 2\pi(2.0\text{ Hz})(0.239\text{ kg}) = 3.00\text{ N/(m/s)}$$

But

$$X_M = X_C = \frac{1}{2\pi f_r C}$$

$$C = \frac{1}{2\pi f_r X_C} = \frac{1}{2\pi(2.0\text{ Hz})[3.00\text{ N/(m/s)}]} = \boxed{2.7\times10^{-2}\text{ m/N}}$$

Energy Flow at Resonance

Another interesting way of looking at resonance is to study energy storage and transfer between spring and mass once the system is in the sinusoidal steady state. Each energy storage device was analyzed separately in Chapter 14 (Sec. 14-3 for inertance and Sec. 14-4 for capacitance). These studies show that each energy storage device receives energy from the mechanical generator during portions of a cycle, but returns the same energy to the mechanical generator during the remaining portions of the cycle. Figures 14-17 and 14-23 show the energy flow for the separately connected mass and spring. What happens if both mass and spring are together in the same system at resonance? Figure 16-3 and Table 16-2 answer this question. Study the figure and the contents of the table carefully.

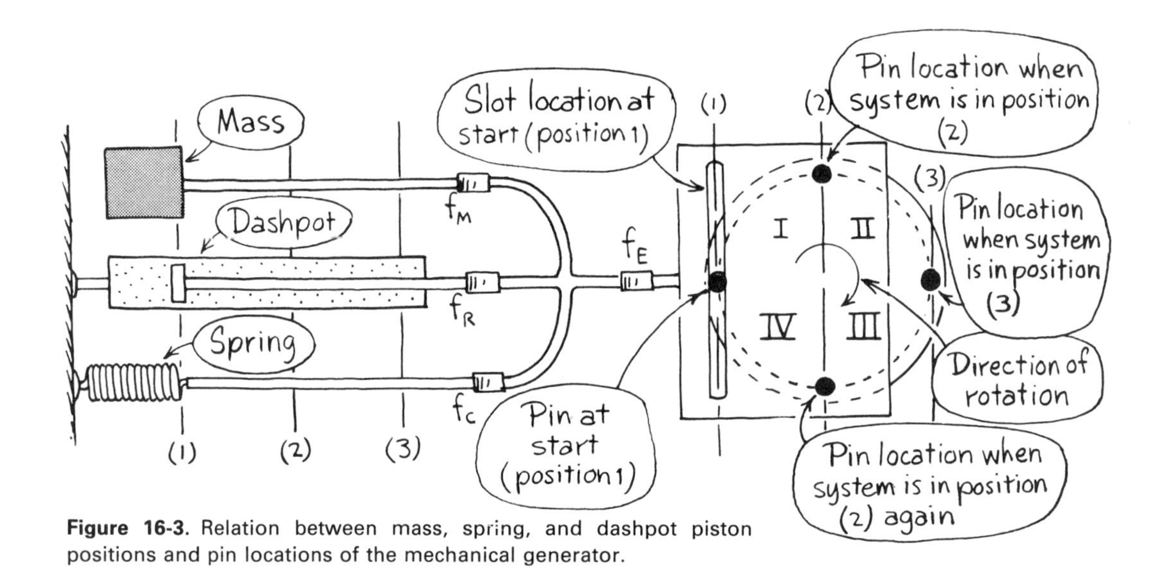

Figure 16-3. Relation between mass, spring, and dashpot piston positions and pin locations of the mechanical generator.

When one storage device requires energy the other is giving it up. At resonance, with $F_{Mm} = F_{Cm}$, there is no need for the mechanical generator to supply any energy at all to either storage device. It need only supply energy to the dashpot. Resonance, therefore, is a situation in which energy storage devices absorb and release energies to each other without any outside help other than what is needed to overcome dissipative resistance. This is a nice way to look at resonance.

TABLE 16-2 KINETIC AND POTENTIAL ENERGY FLOW AT RESONANCE

Quarter of Cycle	*Pin Movement*	*What Happens to the Spring (Potential Energy Change)*	*What Happens to the Mass (Kinetic Energy Change)*
I	Position 1 to position 2	Spring returns to unstretched length. Potential energy decreases.	Mass speeds up. Kinetic energy increases.
II	Position 2 to position 3	Spring stretches. Potential energy increases.	Mass slows up and stops. Kinetic energy decreases.
III	Position 3 to position 2	Spring returns to unstretched length. Potential energy decreases.	Mass speeds up. Kinetic energy increases.
IV	Position 2 to position 1	Spring is compressed. Potential energy increases.	Mass slows up and stops. Kinetic energy decreases.

16-2 GENERALIZED BEHAVIOR OF A SERIES *RMC* SYSTEM

The information from Table 16-1 about series resonance in a mechanical system applies equally well to other systems using unified concept analogies. But suppose that these systems are not sinusoidally excited at the resonant frequency? Can the common parameter rate (velocity, current, volume flow rate, etc.), the impedance, and the various force amplitudes be calculated? The answers to these questions can be found by expanding the force and impedance phasor diagrams of Chapter 15 to include three phasors rather than two.

Figure 16-4 shows generalized force and impedance phasor diagrams for a series *RMC* system. The phasor diagrams are based on the fact that the

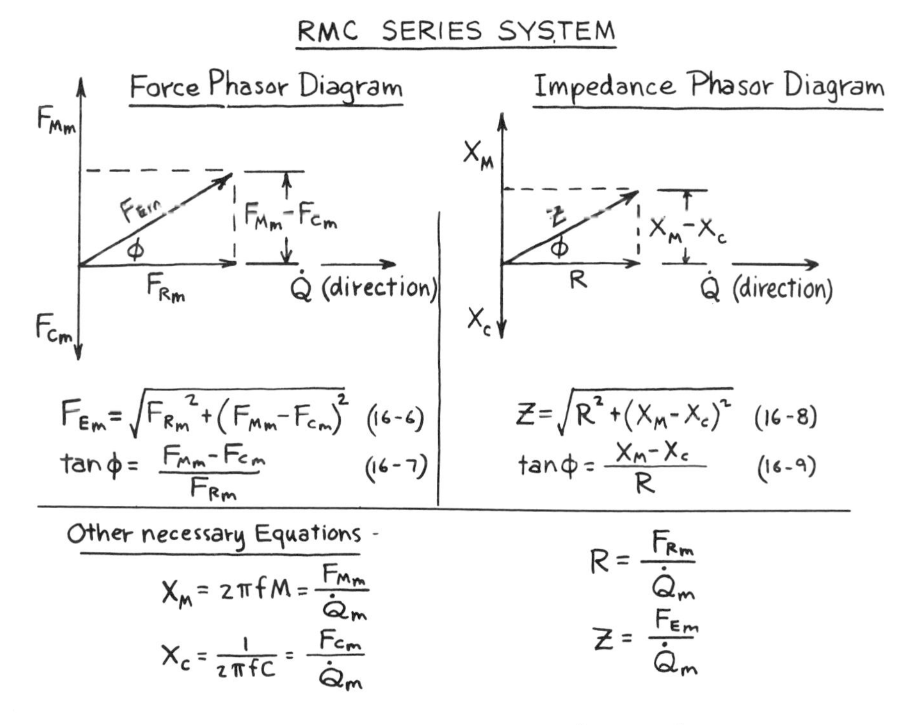

Figure 16-4. Force phasor and impedance phasor diagrams for a generalized *RMC* series system. Equations for F_{Em}, $\tan\phi$ and Z come directly from the phasor diagrams. Other equations needed to complete the picture are included.

excitation force is the sum of the resistive, inertive, and capacitive forces at any instant. All these forces vary sinusoidally, and their sum can be found by treating their phasors as vectors. This is all explained in detail in Sec. 15-2 and Sec. 15-3. Equations (16-6) through (16-9) for the excitation force amplitude

F_{Em}, the phase angle ϕ, and the impedance Z come directly from the diagrams, and are listed together with other equations necessary to complete the picture. The frequency is chosen such that inertive force is larger than capacitive force.

Study the phasor diagrams carefully. Note that what we know about resonance agrees with this expanded treatment. At resonance, $F_{Mm} = F_{Cm}$ and the phase angle is zero. The F_{Em} phasor falls on top of the F_{Rm} phasor and is equal in size. Also, $X_M = X_C$. The impedance phasor Z falls on top of R and the magnitudes are equal.

We are now in a position to study a system's response to any excitation frequency, including the resonant frequency. This is done in the next section for an electrical system.

16-3 RESONANCE IN A SERIES ELECTRICAL SYSTEM

Figure 16-5 shows a coil (L = 1.00 H), resistor (R = 1000 Ω), and capacitor (C = 0.100 μF) connected in series to a sine-wave generator adjusted so that

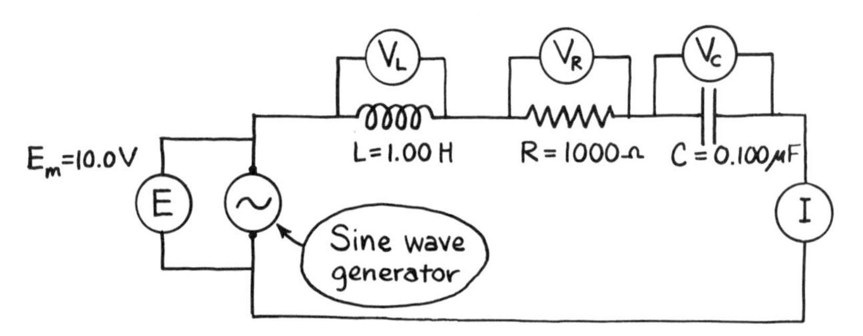

Figure 16-5. A series connected *RLC* circuit.

the amplitude of the excitation emf is always 10.0 V. It is easy to calculate the resonant frequency of this circuit. Just use Eq. (16-5):

$$f_r = \frac{1}{2\pi\sqrt{MC}} = \frac{1}{2\pi\sqrt{LC}} = \frac{1}{2\pi\sqrt{(1.00\text{ H})(0.100 \times 10^{-6}\text{ F})}} = 503\text{ Hz}$$

The impedance of the circuit at resonance is just the resistance R:

$$Z = R = 1000\ \Omega$$

The current at resonance is

$$I_m = \frac{E_m}{Z} = \frac{E_m}{R} = \frac{10.0\text{ V}}{1000\ \Omega} = 1.00 \times 10^{-2}\text{ A} \quad \text{or} \quad 10.0\text{ mA}$$

and the phase angle of course is zero. Coil and capacitor reactances and voltages are easy to calculate.

$$X_L = 2\pi fL = 2\pi(503\text{ Hz})(1.00\text{ H}) = 3160\ \Omega \rightarrow 3.16 \times 10^3\ \Omega$$

$$V_{Lm} = I_m X_L = (0.0100\text{ A})(3160\ \Omega) = 31.6\text{ V}$$

$$X_C = \frac{1}{2\pi fC} = \frac{1}{2\pi(503\text{ Hz})(0.100 \times 10^{-6}\text{ F})} = 3164 \rightarrow 3.16 \times 10^3\ \Omega$$

$$V_{Cm} = I_m X_C = (0.0100\text{ A})(3164\ \Omega) = 31.64 \rightarrow 31.6\text{ V}$$

The coil and capacitor voltages are equal (31.6 V). They are considerably larger than the excitation voltage of 10.0 V. The calculations have been made using voltage and current amplitudes (peak values). Usually, rms values are used in electrical measurements. It really makes no difference because rms values are just 0.707 times the peak values (see p. 368).

The next set of calculations involves frequencies of 400 and 600 Hz, above and below the resonant frequency of 503 Hz. Values of X_L, X_C, Z, ϕ, I_m, V_{Lm}, and V_{Cm} are found using the information in Fig. 16-4. Study these calculations carefully.

Figure 16-6 shows the relation between excitation voltage E_m and the current I_m for frequencies of 400, 503, and 600 Hz. The current is greatest at resonance when the phase angle is zero. The phase angle is positive or negative for frequencies other than resonance.

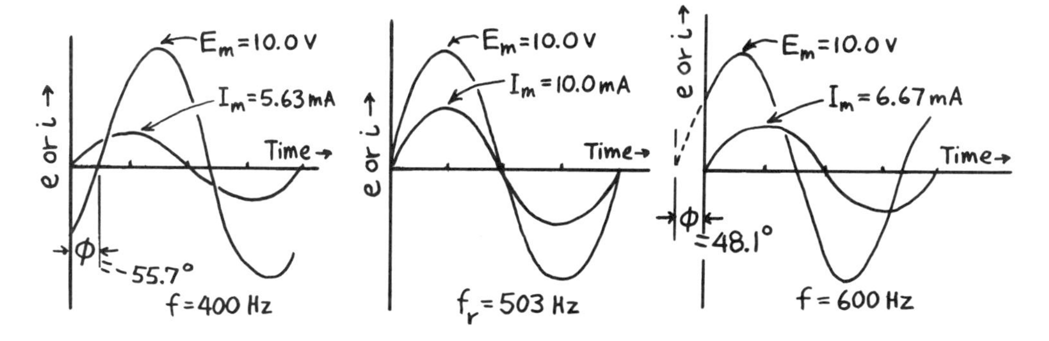

Figure 16-6. The current is greatest at resonance. The phase angle is positive or negative for frequencies other than resonance.

Figure 16-7 presents graphs of the results of these and similar calculations for frequencies on either side of resonance. These graphs show the reactance and voltage for both storage devices, the phase angle, the impedance, and the current, all as functions of the frequency.

	$f = 600\ \text{Hz}$	$f = 400\ \text{Hz}$
Reactance	$X_L = 2\pi fL = 2\pi(600\ \text{Hz})(1.00\ \text{H}) = 3770 \rightarrow \boxed{3.77 \times 10^3\ \Omega}$	$X_L = 2\pi fL = 2\pi(400\ \text{Hz})(1.00\ \text{H}) = 2513 \rightarrow \boxed{2.51 \times 10^3\ \Omega}$
	$X_C = \dfrac{1}{2\pi fC} = \dfrac{1}{2\pi(600\ \text{Hz})(0.100 \times 10^{-6}\ \text{F})} = 2653 \rightarrow \boxed{2.65 \times 10^3\ \Omega}$	$X_C = \dfrac{1}{2\pi fC} = \dfrac{1}{2\pi(400\ \text{Hz})(0.100 \times 10^{-6}\ \text{F})} = 3979 \rightarrow \boxed{3.98 \times 10^3\ \Omega}$
Impedance and phase angle	$\tan\phi = \dfrac{X_L - X_C}{R} = \dfrac{1117\ \Omega}{1000\ \Omega} = 1.117 \Rightarrow \phi = \boxed{48.2°}$	$\tan\phi = \dfrac{X_L - X_C}{R} = \dfrac{-1466\ \Omega}{1000\ \Omega} = -1.466 \Rightarrow \phi = \boxed{-55.7°}$
	$Z = \sqrt{R^2 + (X_L - X_C)^2} = \sqrt{(1000\ \Omega)^2 + (1117\ \Omega)^2} = 1499 \rightarrow \boxed{1.50 \times 10^3\ \Omega}$	$Z = \sqrt{R^2 + (X_L - X_C)^2} = \sqrt{(1000\ \Omega)^2 + (-1466\ \Omega)^2} = 1775 \rightarrow \boxed{1.78 \times 10^3\ \Omega}$
Current	$I_m = \dfrac{E_m}{Z} = \dfrac{10.0\ \text{V}}{1499\ \Omega} = \boxed{6.67 \times 10^{-3}\ \text{A}}$	$I_m = \dfrac{E_m}{Z} = \dfrac{10.0\ \text{V}}{1775\ \Omega} = \boxed{5.63 \times 10^{-3}\ \text{A}}$
Voltage	$V_{Lm} = I_m X_L = (6.67 \times 10^{-3}\ \text{A})(3770\ \Omega) = \boxed{25.1\ \text{V}}$	$V_{Lm} = I_m X_L = (5.63 \times 10^{-3}\ \text{A})(2513\ \Omega) = \boxed{14.2\ \text{V}}$
	$V_{Cm} = I_m X_C = (6.67 \times 10^{-3}\ \text{A})(2653\ \Omega) = \boxed{17.7\ \text{V}}$	$V_{Cm} = I_m X_C = (5.63 \times 10^{-3}\ \text{A})(3979\ \Omega) = \boxed{22.4\ \text{V}}$
Impedance phasor diagrams	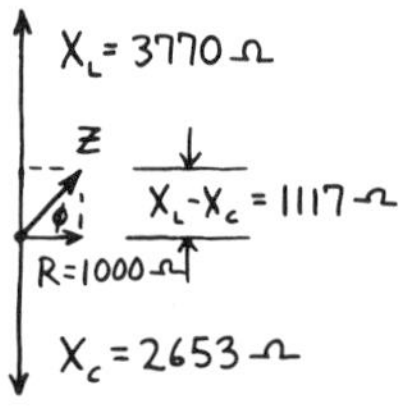 $X_L = 3770\ \Omega$; Z; ϕ; $X_L - X_C = 1117\ \Omega$; $R = 1000\ \Omega$; $X_C = 2653\ \Omega$	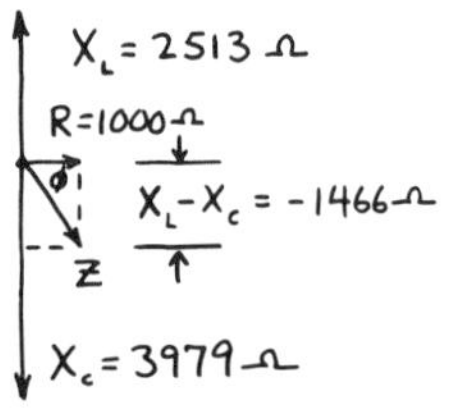 $X_L = 2513\ \Omega$; $R = 1000\ \Omega$; ϕ; $X_L - X_C = -1466\ \Omega$; Z; $X_C = 3979\ \Omega$

The graphs of Fig. 16-7 are worth a really close look. Take time to study them. Notice the capacitor and coil *voltages* (Fig. 16-7a, b). They both change with frequency and are greatest at resonance. In fact, they are several times greater than the 10-V excitation voltage. The *phase angle* (Fig. 16-7c) starts off negative and near 90° for low frequencies. It becomes zero at resonance and goes positive at frequencies greater than f_r. *Impedance* (Fig. 16-7d) is as small as it can get at resonance, because the capacitive and inertive (inductive)

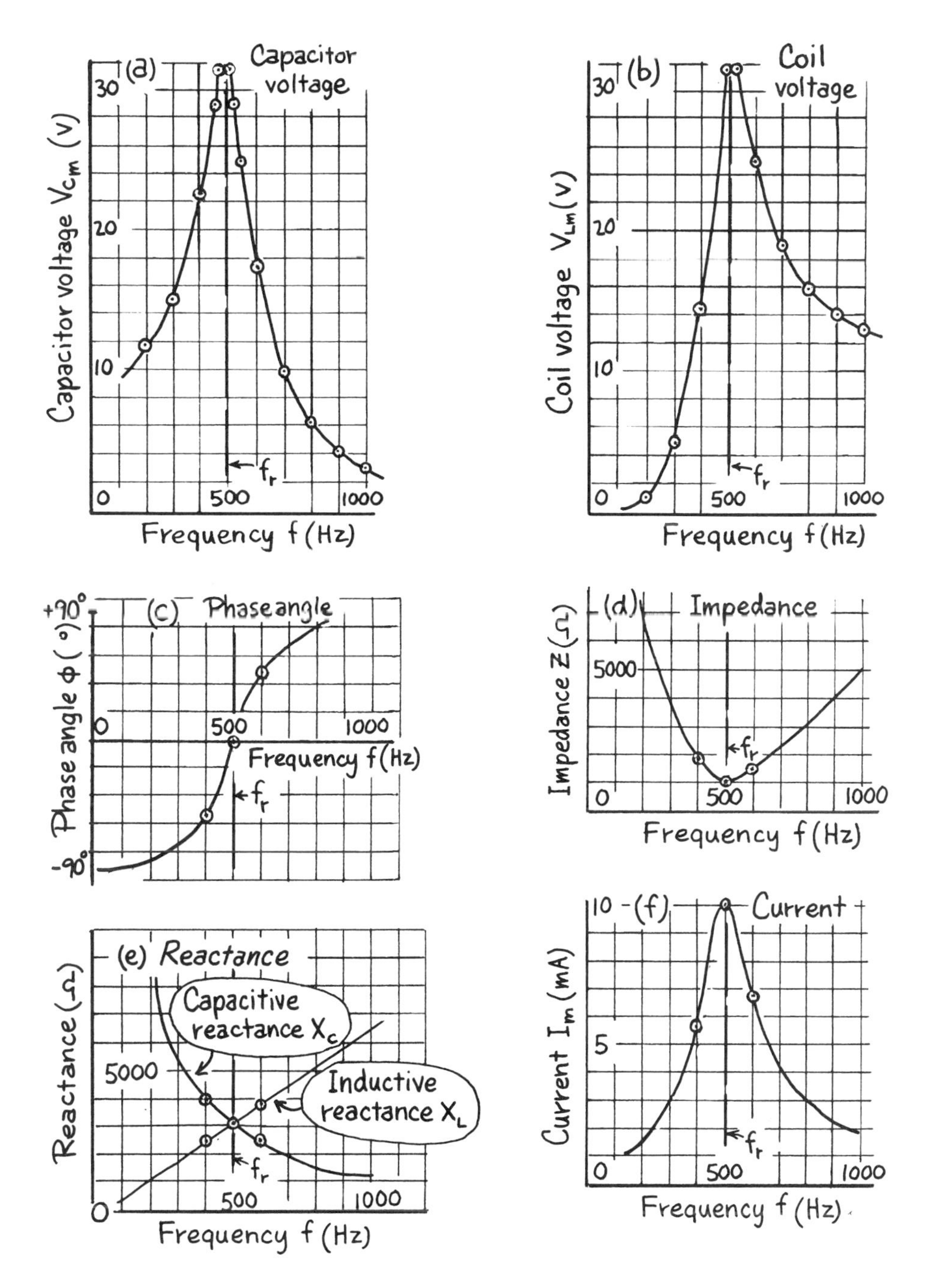

Figure 16-7. A picture story of series resonance showing graphs of capacitor and coil voltage, phase angle, impedance, inertive and capacitive reactance, and current, all plotted against frequency.

reactances (Fig. 16-7e) are equal at that frequency. Thus the *current* (Fig. 16-7f) has its greatest value at resonance.

The Shape of the Resonance Curve

The shape of the graph of current versus frequency (Fig. 16-7f) can be described using a dimensionless ratio called the *quality* Q of the circuit. It is a measure of the steepness of the current versus frequency graph near resonance. (Please, by the way, do not confuse the symbol Q used for quality with the symbol Q used for parameter. They are the same symbol but they stand for different things.)

Quality is defined as

$$\underset{\text{Quality}}{Q} = \frac{\overbrace{2\pi f_r L}^{X_L \text{ at resonance}}}{\underset{\text{Resistance}}{R}} \tag{16-10}$$

Quality can be thought of as a measure of the spread or steepness of the skirts of the current peak near resonance. The spread of the current peak is expressed as a bandwidth of frequencies Δf.

$$\Delta f = f_2 - f_1 \tag{16-11}$$

where frequencies f_2 and f_1 are those that produce currents 0.707 times the current at resonance. Figure 16-8 shows the bandwidth Δf and the two frequencies f_2 and f_1.

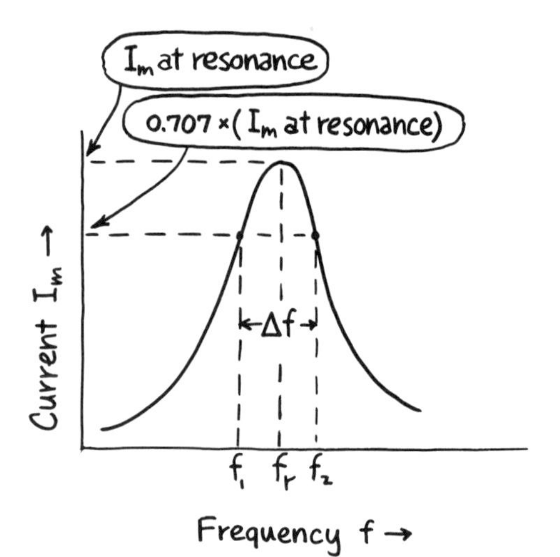

Figure 16-8. The bandwidth $\Delta f = f_2 - f_1$ is a measure of the steepness of the current versus frequency curve near resonance.

The bandwidth Δf can be approximated if the Q of the circuit is known using

$$\Delta f = \frac{f_r}{Q} \tag{16-12}$$

Note that Q is simply the ratio of reactance to resistance in a system at resonance. If a circuit contains only resistance, the current does not change with frequency, and a graph like Fig. 16-7f would be a straight horizontal line. Increasing the reactance makes the system more frequency dependent. The current versus frequency graph becomes steeper, and the value of Q is larger.

The quality of our electrical circuit under study is

$$Q = \frac{2\pi f_r L}{R} = \frac{2\pi(503\text{ Hz})(1.00\text{ H})}{1000\ \Omega} = 3.16$$

and the bandwidth is

$$\Delta f = \frac{f_r}{Q} = \frac{503\text{ Hz}}{3.16} = 159\text{ Hz}$$

This value for Δf agrees nicely with the bandwidth found by measuring the frequency spread on the actual current versus frequency curve (Fig. 16-7f). Go back and verify that the spread of frequencies between the two points where the current is 0.707 times the resonant current is about 160 Hz.

The points where the currents are 0.707 times the resonant current (Fig. 16-8) are called *half-power points.* At the frequencies of these two points, the phase angle is 45°. At resonance the phase angle ϕ is zero and the impedance Z is equal to the resistance R. Hence at resonance the average power P_{av} being used is

$$P_{av} = EI \cos \phi \tag{15-10}$$

But

$$I = \frac{E}{Z} \tag{15-2}$$

Substituting

$$P_{av} = E\frac{E}{Z}\cos 0°$$

Cos 0° = 1 and $Z = R$, so $P_{av} = E^2/R$ at resonance. An impedance phasor diagram for a phase angle of 45° shows that $R = Z \cos 45°$ or $Z = R/\cos 45° = R/0.707$. But, again, $I = E/Z = E/(R/0.707) = 0.707E/R$. Putting this into the equation for average power,

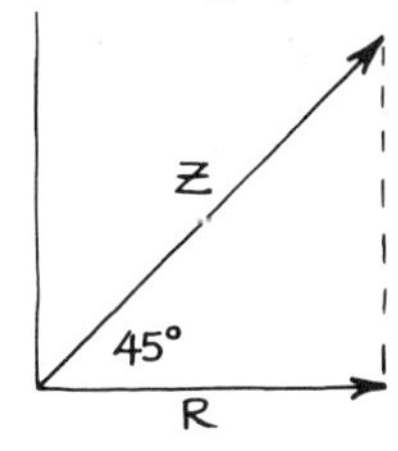

$$P_{av} = EI \cos 45° = E\frac{0.707E}{R}(0.707) = 0.5\ E^2/R$$

The power when ϕ is 45° is just one half the power delivered to the circuit at resonance. That's why the two points are called half-power points.

Quality is also related to the potential difference at resonance across either coil or capacitor. For example, the potential difference across the coil at resonance is

$$V_{Lm} = X_L I_m = (2\pi f_r L) I_m$$

But

$$I_m = \frac{E_m}{R} \quad \text{and} \quad Q = \frac{2\pi f_r L}{R}$$

Hence

$$V_{Lm} = 2\pi f_r L \frac{E_m}{R} = \frac{2\pi f_r L}{R} E_m = Q E_m$$

$$V_{Lm} = Q E_m \tag{16-13}$$

The potential difference QE_m is the largest the circuit experiences. It appears alternately across both the coil and the capacitor. Looked at in this way, Q is a measure of the height of the resonance peak.

We can now make quick estimates of how a current versus frequency curve should look in series resonance. Example 16-2 does this for a circuit like the one we have been dealing with except that resistance R is changed.

Example 16-2 An *RLC* series circuit has $R = 500\ \Omega$, $L = 1.00$ H, and $C = 0.100\ \mu$F.

(a) Calculate the resonant current, the coil voltage at resonance, and the bandwidth.

(b) Sketch a graph of current versus frequency.

Solution to (a)

The resonant frequency is the same as for the circuit we have been looking at because L and C are unchanged.

$$f_r = 503 \text{ Hz}$$

$\left\{ \begin{matrix} E_m = 10.0 \text{ V} \\ R = 500\ \Omega \end{matrix} \right\}$ At resonance the impedance is all resistance:

$$I_m = \frac{E_m}{Z} = \frac{E_m}{R} = \frac{10.0 \text{ V}}{500\ \Omega} = 0.020 \text{ A} = \boxed{20 \text{ mA}}$$

$\{L = 1.00\ H\}$ The quality Q of the circuit is

$$Q = \frac{2\pi f_r L}{R} = \frac{2\pi(503 \text{ Hz})(1.00 \text{ H})}{500\ \Omega} = \boxed{6.32}$$

The coil voltage at resonance is

$$V_{Lm} = QE_m = 6.32(10.0\ \text{V}) = \boxed{63.2\ \text{V}}$$

And the bandwidth is

$$\Delta f = \frac{f_r}{Q} = \frac{503\ \text{Hz}}{6.32} = \boxed{80\ \text{Hz}}$$

Solution to (b)

The half-power points are at 0.707(20 mA) = 14.1 mA. A sketch of the current versus frequency curve can be based on the resonance point and the half-power points. We get a pretty good idea of the curve using just these three points.

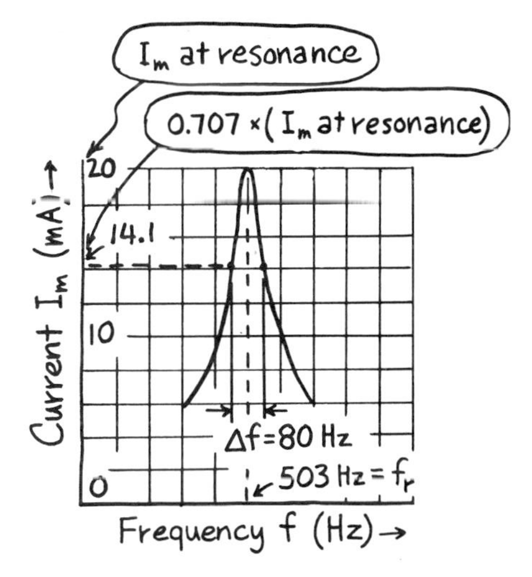

Notice that a higher value of Q corresponds to increased resonant current and decreased bandwidth. Three curves of the same circuit with different resistances have been plotted on the same axes in Fig. 16-9 to show this effect. The graph of current versus frequency for the circuit with 500-Ω resistance peaks very steeply. A circuit like this is very *sensitive*, as indicated by the height of the peak, and very *selective*, because there is only a small band of frequencies associated with large currents. The circuit with the 2000-Ω resistor has a much flatter peak. This circuit is less sensitive and less selective.

The expression for bandwidth $\Delta f = f_r/Q$ is not very accurate for small values of Q. It should not be used for Q values less than 5, although we got satisfactory results with $Q = 3.16$.

Electronics engineers sometimes want highly sensitive resonant circuits. They can do this by decreasing the resistance in the circuit, as implied by

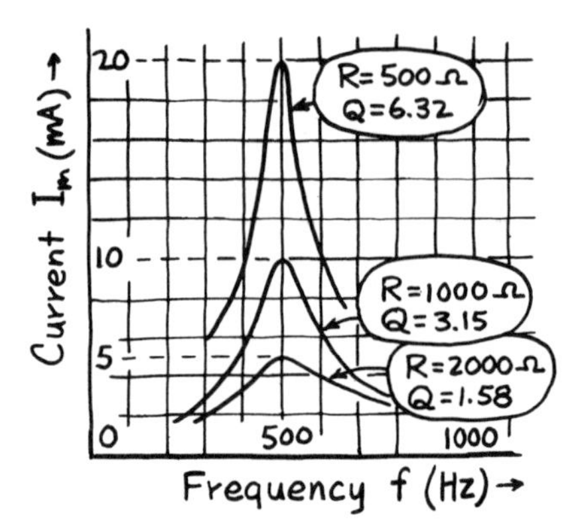

Figure 16-9. Adding resistance cuts down and flattens the resonance peak.

Fig. 16-9. Similar effects can be accomplished by decreasing the capacitance in the system. However, the inductance must then be increased so that the product LC stays the same. Otherwise, the resonant frequency $f_r = 1/(2\pi\sqrt{LC})$ would be changed.

An electronics engineer may not want to see a decrease in bandwidth for a particular application, because this limits the amount of energy that can be carried by oscillations near the resonant frequency. Mechanical engineers may want to flatten out the resonance peak if they are stuck with a certain frequency of vibration. This can be done by introducing more resistance in the form of shock absorbers or "viscous dampers."

16-4 PARALLEL RESONANCE

Resonance in series-connected systems is characterized by a frequency at which the parameter rate (velocity, current, etc.) is maximum and the impedance is minimum. Resonance exists in parallel-connected systems, too. However, just the opposite happens. At the resonant frequency of a parallel system, the parameter rate (the *excitation* parameter rate) is minimum and impedance is maximum. Parallel resonance is sometimes called antiresonance because of this. Again, let us look at an electrical example with the same coil and capacitor as in the last section (Fig. 16-5), but connected in parallel rather than in series. We will use a larger resistance to dramatize the parallel resonance characteristics. Figure 16-10 shows the circuit. The coil's inductance $L = 1.00\,\text{H}$, resistance $R = 10{,}000\,\Omega$, and capacitance $C = 0.100\,\mu\text{F}$. They are connected to a sine-wave generator adjusted so that the amplitude E_m of the excitation emf is always 10.0 V.

Since we are dealing with a parallel circuit, the excitation emf is at all times the same across the elements R, L, and C. This is a characteristic of parallel circuits. But the currents can be different, both in magnitude and phase angle. Remember that current phase angle is given relative to the common voltage in parallel electrical systems (Sec. 15-4). In a resistor the current and excitation force (emf) are in phase. In a coil the excitation emf leads the

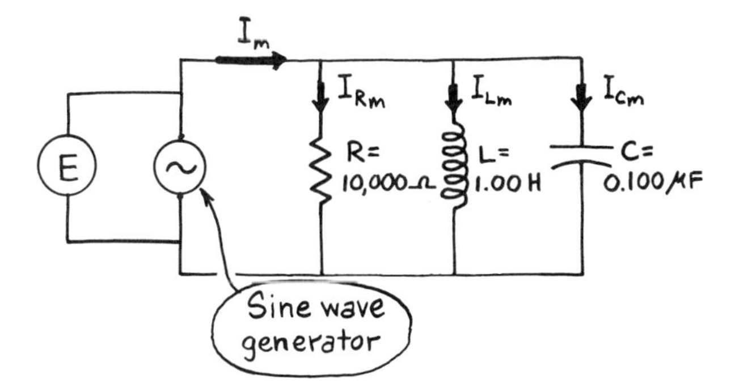

Figure 16-10. An *RLC* parallel-connected circuit.

current by 90°, or, said in another way, the current lags the emf by 90°. In a capacitor the excitation emf lags the current by 90°, or, said another way, current leads emf by 90°. Figure 16-11 shows this. Current phasors using the common excitation emf as reference are included in the figure.

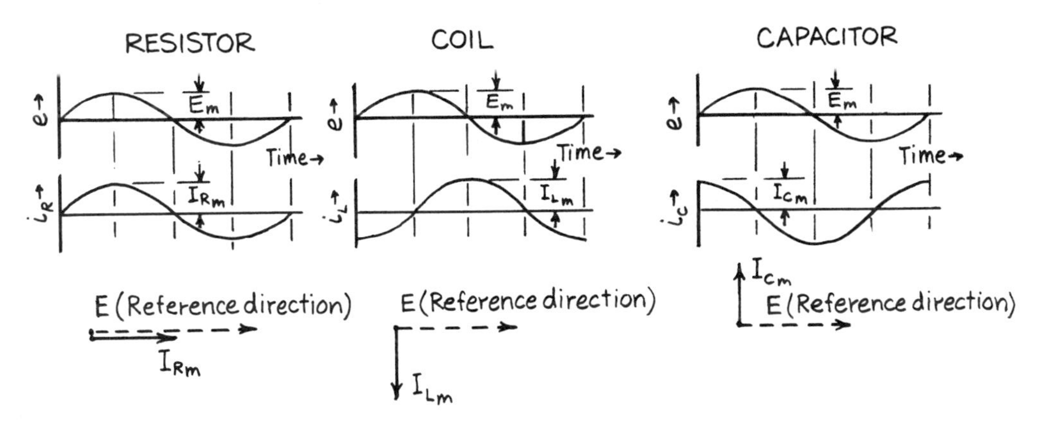

Figure 16-11. Current and potential difference are in phase in a resistor, 90° out of phase (current lagging) in a coil, and 90° out of phase (current leading) in a capacitor.

The source current I_m from the sine-wave generator is the sum of I_{Rm}, I_{Lm}, and I_{Cm}. Sinusoidal currents can be added by vector addition of the phasors, which represent the currents. The current phasor vector addition is shown in Fig. 16-12. By inspection, the equation for I_m is

$$I_m = \sqrt{I_{Rm}^2 + (I_{Cm} - I_{Lm})^2} \tag{16-14}$$

Resonance occurs in the parallel circuit when I_{Lm} and I_{Cm} are equal. Hence

$$X_L = \frac{E_m}{I_{Lm}} = \frac{E_m}{I_{Cm}} = X_C$$

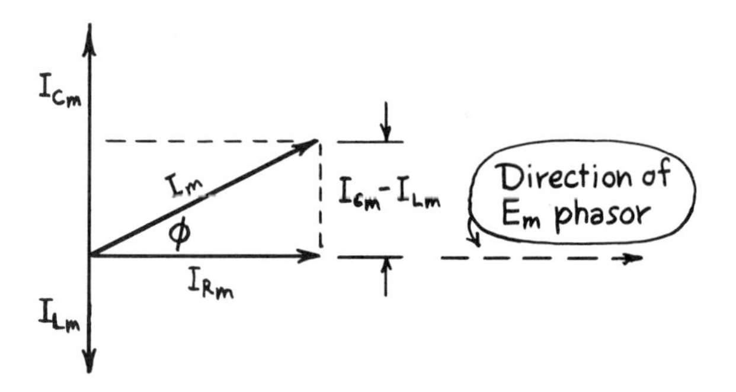

Figure 16-12. Current phasor diagram for a parallel circuit. The source current I_m is the vector sum of the branch currents I_{Rm}, I_{Lm} and I_{Cm}. The common excitation voltage E_m is used as reference for measuring phase angles.

and X_L equals X_C. But this condition ($X_L = X_C$) is true for series resonance, too. Hence the equation for the resonant frequency of a parallel circuit must be the same as for a series circuit. Equation (16-5), stated in electrical terms is

$$f_r = \frac{1}{2\pi\sqrt{LC}}$$

At resonance the source current amplitude I_m just equals the resistor current amplitude I_{Rm}. The phase angle is zero and the impedance Z is just the resistance ($Z = E_m/I_m = E/I_R = R$). These also are the conditions of series resonance. Summarizing all this for the particular circuit under discussion,

$$f_r = \frac{1}{2\pi\sqrt{LC}} = \frac{1}{2\pi\sqrt{(1.00\text{ Hz})(0.100 \times 10^{-6}\text{ F})}} = 503\text{ Hz}$$

$$I_m = I_{Rm} = \frac{E_m}{R} = \frac{10.0\text{ V}}{10{,}000\ \Omega} = 1.00 \times 10^{-3}\text{ A} \quad \text{or} \quad 1.00\text{ mA}$$

$$\phi = 0°$$

$$Z = R = 10{,}000\ \Omega$$

It seems strange that the source need not supply any current to either the coil or the capacitor at resonance. Is there any current at all in these storage devices? There is indeed, and it can be easily calculated. First find the coil reactance at resonance:

$$X_L = 2\pi fL = 2\pi(503\text{ Hz})(1.00\text{ H}) = 3160\ \Omega$$

Then calculate the current:

$$I_{Lm} = I_{Cm} = \frac{E_m}{X_L} = \frac{10.0\text{ V}}{3160\ \Omega} = 3.16 \times 10^{-3}\text{ A} = 3.16\text{ mA}$$

The current of 3.16 mA is called a *tank current.* The coil and capacitor together comprise a *tank circuit,* so-called because the energy is alternately stored in the two elements. Charge circulates back and forth as the energy surges between the capacitor's electric field and the coil's magnetic field. Thus the source does not have to supply current to the coil or capacitor at resonance, because they alternately supply current to each other.

The next set of calculations involves frequencies other than resonance. These frequencies are 400 and 600 Hz, just as for the series circuit calculations in Sec. 16-3. Values of X_L, X_C, I_{Lm}, I_{Cm}, I_m, and Z are calculated. The key to understanding what follows is the current phasor diagram (Fig. 16-12) and the corresponding equation (16-14).

Figure 16-13 graphs the results of these and similar calculations for frequencies on either side of resonance. These graphs are just for current and impedance. The graphs for the series circuit are included for comparison.

f = 600 Hz	f = 400 Hz
Reactance	
$X_L = 2\pi fL = 2\pi(600\text{ Hz})(1.00\text{ H})$ $= 3770 \rightarrow \boxed{3.77 \times 10^3\ \Omega}$	$X_L = 2\pi fL = 2\pi(400\text{ Hz})(1.00\text{ H})$ $= 2513 \rightarrow \boxed{2.51 \times 10^3\ \Omega}$
$X_C = \dfrac{1}{2\pi fC} = \dfrac{1}{2\pi(600\text{ Hz})(0.100 \times 10^{-6}\text{ F})}$ $= 2653 \rightarrow \boxed{2.65 \times 10^3\ \Omega}$	$X_C = \dfrac{1}{2\pi fC} = \dfrac{1}{2\pi(400\text{ Hz})(0.100 \times 10^{-6}\text{ F})}$ $= 3979 \rightarrow \boxed{3.98 \times 10^3\ \Omega}$
Current	
$I_{Lm} = \dfrac{E_m}{X_L} = \dfrac{10.0\text{ V}}{3770\ \Omega} = \boxed{2.65 \times 10^{-3}\text{ A}}$	$I_{Lm} = \dfrac{E_m}{X_L} = \dfrac{10.0\text{ V}}{2513\ \Omega} = \boxed{3.98 \times 10^{-3}\text{ A}}$
$I_{Cm} = \dfrac{E_m}{X_C} = \dfrac{10.0\text{ V}}{2653\ \Omega} = \boxed{3.77 \times 10^{-3}\text{ A}}$	$I_{Cm} = \dfrac{E_m}{X_C} = \dfrac{10.0\text{ V}}{3979\ \Omega} = \boxed{2.51 \times 10^{-3}\text{ A}}$
$I_{Rm} = \dfrac{E_m}{R} = \dfrac{10.0\text{ V}}{10{,}000\ \Omega} = \boxed{1.00 \times 10^{-3}\text{ A}}$	$I_{Rm} = \dfrac{E_m}{R} = \dfrac{10.0\text{ V}}{10{,}000\ \Omega} = \boxed{1.00 \times 10^{-3}\text{ A}}$
$I_m = \sqrt{I_{Rm}^2 + (I_{Cm} - I_{Lm})^2}$ $= \sqrt{(1.00\text{ mA})^2 + (1.12\text{ mA})^2} = \boxed{1.50\text{ mA}}$	$I_m = \sqrt{I_{Rm}^2 + (I_{Cm} - I_{Lm})^2}$ $= \sqrt{(1.00\text{ mA})^2 + (-1.47\text{ mA})^2} = \boxed{1.78\text{ mA}}$
Impedance	
$Z = \dfrac{E_m}{I_m} = \dfrac{10.0\text{ V}}{1.50 \times 10^{-3}\text{ A}}$ $= 6667 \rightarrow \boxed{6.67 \times 10^3\ \Omega}$	$Z = \dfrac{E}{I_m} = \dfrac{10.0\text{ V}}{1.78 \times 10^{-3}\text{ A}}$ $= 5618 \rightarrow \boxed{5.62 \times 10^3\ \Omega}$

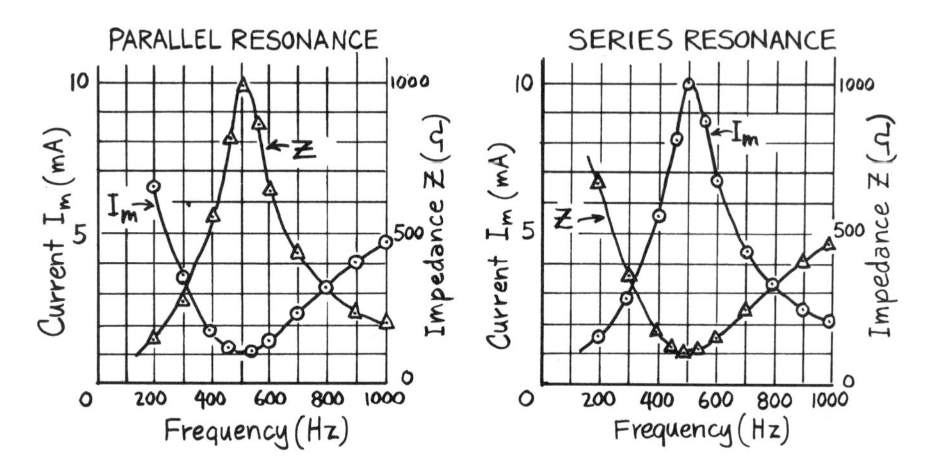

Figure 16-13. Parallel resonance is characterized by a small excitation current and a large impedance. Series resonance is characterized by large current and small impedance.

Practical parallel circuits have resistance usually in the coil branch of the circuit (Fig. 16-14). Such a circuit is more difficult to analyze, but its behavior is similar to our simple *RLC* parallel circuit of Fig. 16-10 if the resistance is small.

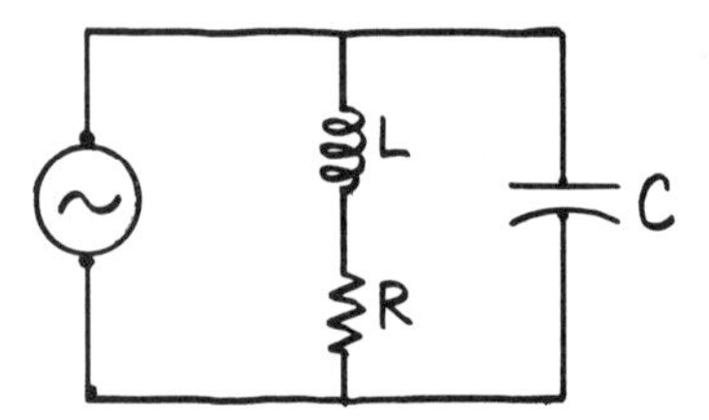

Figure 16-14. A practical parallel circuit usually has some resistance in the coil branch of the circuit.

16-5 EXAMPLES OF RESONANCE

This section presents a number of worked examples that illustrate resonance in various physical systems excited by a sinusoidal source. Resonance occurs if the natural frequency f_r of the system is close to the frequency of the disturbance.

Example 16-3 A turbine rotor blade vibrates at a natural frequency of 43 Hz. What is one angular turbine speed that can introduce resonant vibrations in the blade? Give your answer in rpm.

Solution

The period T of the vibration in the blade is

$$T = \frac{1}{f} = \frac{1}{43\text{ Hz}} = \frac{1}{43}\text{ s}$$

Resonance is possible if one complete revolution of the turbine takes place in a time interval equal to T. Hence

$$\omega = \frac{1 \text{ revolution}}{T \text{ second}} = \frac{1 \text{ rev}}{\frac{1}{43} \text{ s}} = 43 \frac{\text{rev}}{\text{s}} \times \frac{60 \text{ s}}{1 \text{ min}} = 2580 \rightarrow \boxed{2.6 \times 10^3 \text{ rpm}}$$

Example 16-4 A concentrated mass of 0.30 slug is firmly attached to the end of a 1.00-ft rod. A 40-lb horizontal force acting on the mass bends the rod so that the mass moves 0.010 ft. What mechanical vibration frequency causes the mass and rod to vibrate with large amplitude (i.e., to resonate)?

Solution

$\left\{\begin{matrix} s = 0.010 \text{ ft} \\ F_C = 40 \text{ lb} \end{matrix}\right\}$ First calculate the capacitance C of the rod:

$$C = \frac{s}{F_C} = \frac{0.010 \text{ ft}}{40 \text{ lb}} = 2.5 \times 10^{-4} \text{ ft/lb}$$

$\{m = 0.30 \text{ slug}\}$ And then the resonant frequency f_r:

$$f_r = \frac{1}{2\pi\sqrt{mC}} = \frac{1}{2\pi\sqrt{(0.30 \text{ slug})(2.5 \times 10^{-4} \text{ ft/lb})}} = \boxed{18 \text{ Hz}}$$

Example 16-5 What must be the inductance of a coil used with a 2.0-pF capacitor in a circuit that will resonate when excited by a 100×10^6 Hz signal?

Solution

$\left\{\begin{matrix} C = 2.0 \text{ pF} \\ = 2.0 \times 10^{-12} \text{ F} \\ f_r = 100 \times 10^6 \text{ Hz} \end{matrix}\right\}$ The equation for resonant frequency f_r can be used directly.

$$f_r = \frac{1}{2\pi\sqrt{LC}} \Rightarrow f_r^2 = \frac{1}{4\pi^2 LC} \Rightarrow L = \frac{1}{4\pi^2 f_r^2 C}$$

$$L = \frac{1}{4\pi^2(100 \times 10^6 \text{ Hz})^2(2.0 \times 10^{-12} \text{ F})} = \boxed{1.3 \times 10^{-6} \text{ H}}$$

Example 16-6 A motor weighs 600 lb and causes the springs that support it to compress 0.042 ft when at rest. Is there a possible resonance problem with the system if the motor rotates at 1800 rpm?

Solution

$\left\{\begin{matrix} s = 0.042 \text{ ft} \\ F_C = 600 \text{ lb} \end{matrix}\right\}$ This is a series-connected mechanical system because the motor mass and springs are all rigidly connected. They share a common displacement and velocity if vibrated. Vibrations at the resonant frequency f_r will cause large motor movements. First calculate the spring capacitance C.

$$C = \frac{s}{F_C} = \frac{0.042 \text{ ft}}{600 \text{ lb}} = 7.0 \times 10^{-5} \text{ ft/lb}$$

The motor's mass is

$$m = \frac{W}{g} = \frac{600\ \text{lb}}{32.2\ \text{ft/s}^2} = 18.6\ \text{slug}$$

Next calculate the resonant frequency f_r:

$$f_r = \frac{1}{2\pi\sqrt{mC}} = \frac{1}{2\pi\sqrt{(18.6\ \text{slug})(7.0 \times 10^{-5}\ \text{ft/lb})}} = \boxed{4.4\ \text{Hz}}$$

Compare f_r to the forced frequency from the 1800-rpm rotation.

$$\left(1800\frac{\text{rev}}{\text{min}}\right)\left(\frac{1\ \text{min}}{60\ \text{s}}\right) = 30\frac{\text{rev}}{\text{s}} \quad \text{or} \quad 30\ \text{Hz}$$

The natural frequency of 4.4 Hz is considerably lower than the forced frequency of 30 Hz, so no problem should be encountered. Some resonance effects may be observed as the motor, starting from rest, comes up past the resonant frequency to its rated angular speed.

Example 16-7 A car weighs 3220 lb. Its suspension springs are compressed 0.50 ft when the system is at rest. What is the frequency of a sinusoidal disturbance from a bumpy, corrugated dirt road that causes excessive vibration of the body of the car? Assume that the car's shock absorbers are so worn that their resistance can be neglected.

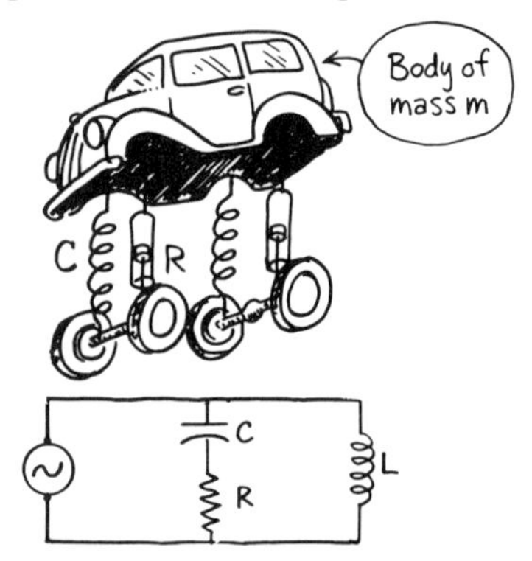

Solution

$\left\{ W = F_C = 3220\ \text{lb},\ s = 0.50\ \text{ft} \right\}$

Some study reveals that the car's suspension system is a parallel-connected mechanical system. The excitation force is applied to the wheels by the bumps in the road. The wheels have negligible mass relative to the car's body. An equivalent electrical circuit is shown. The difference between the wheel displacement and the body displacement is the displacement of the spring and shock absorbers. At resonance a small wheel movement can produce violent shaking of the car's mass, limited only by shock absorber (dashpot) resistance. First calculate the car's mass and the spring's capacitance.

$$m = \frac{W}{g} = \frac{3220\ \text{lb}}{32.2\ \text{ft/s}^2} = 100\ \text{slug}$$

$$C = \frac{s}{F_C} = \frac{0.50\ \text{ft}}{3220\ \text{lb}} = 1.55 \times 10^{-4}\ \text{ft/lb}$$

Then calculate the resonant frequency f_r:

$$f_r = \frac{1}{2\pi\sqrt{mC}} = \frac{1}{2\pi\sqrt{(100 \text{ slug})(1.55 \times 10^{-4} \text{ ft/lb})}} = \boxed{1.3 \text{ Hz}}$$

Example 16-8 The vertical surge pipe (capacitor) in the illustrated hydraulic system acts to reduce the sudden large pressure that would arise from rapidly decelerating water when a valve suddenly closes. The system is essentially one of constant flow, but the pump can introduce a small periodic ripple pressure superimposed on top of any constant pressures. What frequency can make the system resonate? All pipes have cross-sectional areas of 0.016 ft^2 ($\frac{1}{2}$-in. pipe diameters). Neglect any frictional effects.

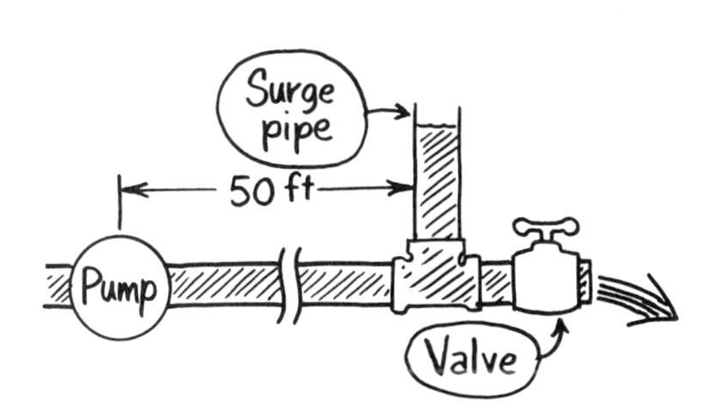

Solution

$\ell = 50$ ft
$A = 0.016$ ft^2
$\rho g = 62.4$ lb/ft^3
(App. B, Table 1)

Calculate the inertance $M = \rho\ell/A$ of the 50-ft pipe, and the capacitance $C = V/p_C = Ah/\rho gh = A/\rho g$ of the surge pipe.

$$M = \frac{\rho\ell}{A} = \left(\frac{62.4 \text{ lb/ft}^3}{32.2 \text{ ft/s}^2}\right)\left(\frac{50 \text{ ft}}{0.016 \text{ ft}^2}\right) = 6.06 \times 10^3 \frac{\text{lb/ft}^2}{\text{ft}^3/\text{s}^2}$$

$$C = \frac{A}{\rho g} = \frac{0.016 \text{ ft}^2}{62.4 \text{ lb/ft}^3} = 2.56 \times 10^{-4} \frac{\text{ft}^3}{\text{lb/ft}^2}$$

Then calculate the resonant frequency f_r:

$$f_r = \frac{1}{2\pi\sqrt{MC}} = \frac{1}{2\pi\sqrt{\left(6.06 \times 10^3 \frac{\text{lb/ft}^2}{\text{ft}^3/\text{s}}\right)\left(2.56 \times 10^{-4} \frac{\text{ft}^3}{\text{lb/ft}^2}\right)}} = \boxed{0.13 \text{ Hz}}$$

These are just a few examples that illustrate resonance in various engineering systems. Other examples are described in problems at the end of the chapter.

16-6 SUMMARY

Resonance occurs in systems that store both potential and kinetic energy (MC and RMC systems). A system resonates when the storage devices just interchange their potential and kinetic energies without any help from the source.

Resonance occurs in both series- and parallel-connected systems. In series-connected systems the equation for the resonant frequency f_r is

$$f_r = \frac{1}{2\pi\sqrt{MC}}$$

The equation holds for parallel-connected systems if the resistance in the tank portion of the system is small. Series resonance is characterized by large parameter rates (velocities, currents, etc.) and small impedance. Parallel resonance is characterized by a large impedance and small excitation parameter rate. However, large parameter rates within the tank portion of the system are possible and in fact characteristic of parallel resonance. Table 16-3 summarizes the various conditions of resonance in systems that are forced to oscillate.

TABLE 16-3 RESONANCE IN SERIES- AND PARALLEL-CONNECTED SYSTEMS

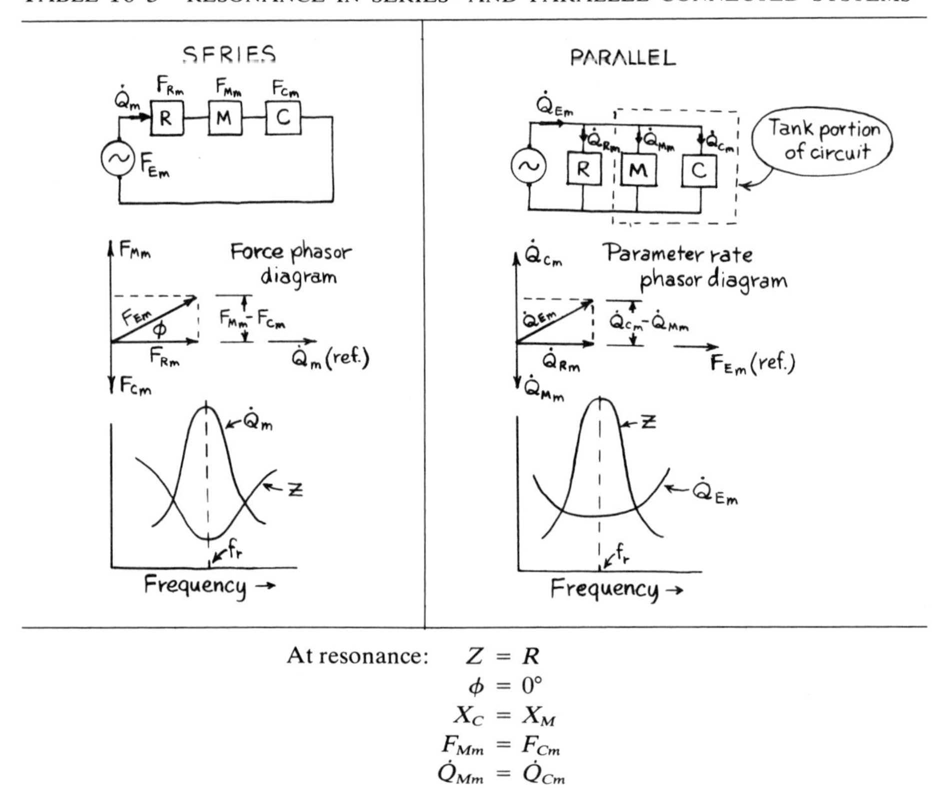

At resonance:

$$Z = R$$
$$\phi = 0°$$
$$X_C = X_M$$
$$F_{Mm} = F_{Cm}$$
$$\dot{Q}_{Mm} = \dot{Q}_{Cm}$$

In series-connected systems the quality Q is a measure of the steepness of the resonance curve (Fig. 16-15). The term quality is also used to describe parallel-connected systems, but is defined in a slightly different way. You can read about this in textbooks on electronic circuits.

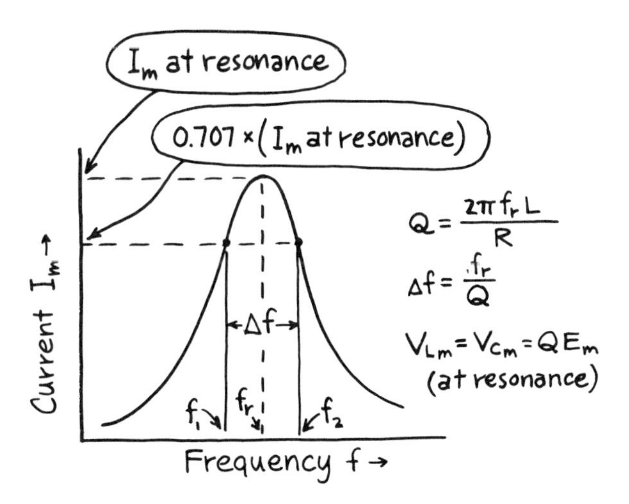

Figure 16-15. Quality Q is a measure of the steepness of the resonance curve in a series circuit.

We now know something about the basic behavior of oscillating systems. In our examples, energy has been confined to systems that store it, interchange it between elements, or dissipate it through a resistor. In the final two chapters we will find that energy can be dispersed or carried away as a wave through an extended system of storage elements. Chapter 17 investigates the mechanisms of wave motion, and Chapter 18 looks at an important type of energy transfer called electromagnetic radiation.

PROBLEMS

Note: Unless otherwise stated, all forces and parameter rates are given in terms of their rms values.

16-1. A spring stretches 5.0 cm when a 200-g object is hung from one end. What is the frequency of an excitation vibration that causes the mass–spring system to exhibit resonant behavior?

16-2. Large vibrational amplitudes occur in one part of an engine at a frequency of 40 Hz. What is one engine angular velocity in rpm that can cause these vibrations?

16-3. Calculate the resonant frequency and the bandwidth of frequencies for a 0.10-H coil and a 0.10-μF capacitor connected in series to a variable frequency sinusoidal source. The coil has a resistance of 50 Ω.

16-4. (a) The circuit of Problem 16-3 is excited at a frequency of 1000 Hz. Which is greater, the coil voltage or the capacitor voltage?
(b) At a frequency of 2000 Hz, does the source emf lead or lag the current? (*Hint*: No calculation is necessary. Just examine the curves in Fig. 16-7.)

16-5. (a) What is the current in Problem 16-3 if the input excitation emf is 5.0 V?
(b) What is the capacitor voltage at resonance?
(c) What average power is drawn from the source at resonance?

16-6. Suppose that a resistance of 50 Ω is added in series with the coil and capacitor of Problem 16-3. What effect does this have on the resonant frequency and the bandwidth of frequencies?

16-7. A glass U-tube contains a water column of 30 cm total length. What frequency of oscillation causes the water in the U-tube to exhibit resonant behavior?

16-8. One end of a flexible rod is clamped so that the rod is held firmly in a horizontal position. The free end deflects 0.050 m when a mass of 0.27 kg is attached to it. What vertical excitation frequency causes the system to resonate?

16-9. A 0.40-kg mass is hung from the bottom end of a spring with a capacitance of 0.030 m/N. The top end of the spring is excited by a small sinusoidal vibration of 5.0 Hz.

(a) Is the mechanical system series or parallel connected?

(b) Does the 5.0-Hz excitation cause any resonance effects in the system?

16-10. Suppose that the mass–spring system of Problem 16-9 is forced to oscillate at its natural resonant frequency. Will the mass exhibit large or small displacements? Why?

16-11. Use the *RLC* series circuit of Fig. 16-5, where $E_m = 10.0$ V, $L = 1.00$ H, $R = 1000\ \Omega$, and $C = 0.100\ \mu$F. Calculate the impedance, the phase angle, and the current amplitude when the frequency is 700 Hz. Use the sample calculations in Sec. 16-3 as a guide, and compare your answers with the curves in Fig. 16-7.

16-12. Repeat Problem 16-11 with a frequency of 300 rather than 700 Hz.

16-13. Use the *RLC* parallel circuit of Figure 16-10, where $E_m = 10.0$ V, $L = 1.00$ H, $R = 10{,}000\ \Omega$, and $C = 0.100\ \mu$F. Calculate the current amplitude, phase angle, and impedance when the frequency is 300 Hz. Use the sample calculations in Sec. 16-4 as a guide, and compare your answers with the curves in Figure 16-13.

16-14. Repeat Problem 16-13 with a frequency of 700 rather than 300 Hz.

16-15. The prongs of a tuning fork are metal bars fixed at one end and free at the other. The moment of inertia and torsional capacitance of each bar are 1.30×10^{-4} kg · m^2 and 2.80×10^{-3} rad/(N · m), respectively. What is the frequency of a sinusoidal disturbance from the air that causes the bars in the fork to vibrate?

16-16. Values of the rms velocity of a vibrating mass are plotted against frequency for a certain series-connected mechanical suspension system.

(a) What is the resonant frequency?

(b) What is the bandwidth of frequencies?

(c) Suggest a method for flattening out the resonance peak.

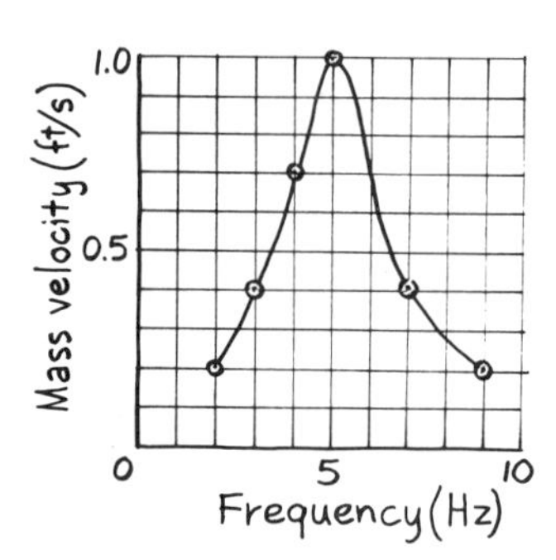

Problem 16-16

16-17. What is the quality of an *RLC* series circuit that includes a 0.75-H coil with a resistance of 220 Ω if the circuit resonates at a frequency of 1000 Hz?

16-18. The illustrated mechanical system is driven by an electric motor whose output motion is sinusoidal.

(a) Is the mechanical system connected in series or in parallel?

(b) Why is it possible for the system to exhibit resonant behavior?

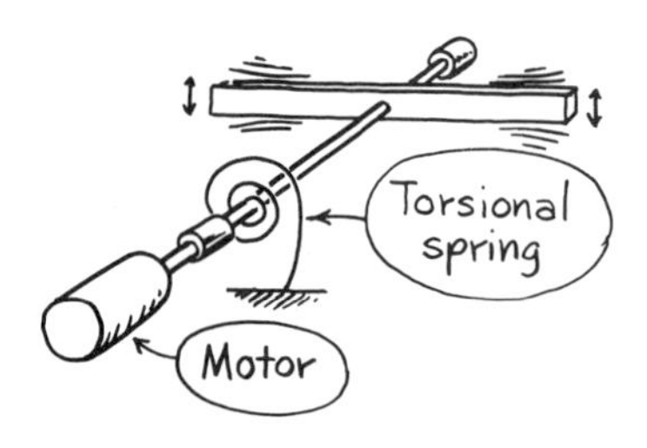

Problem 16-18

16-19. Values of sinusoidal angular velocity for the system in Problem 16-18 are plotted against frequency in the accompanying graph.

(a) What is the resonant frequency?

(b) What is the bandwidth of frequencies?

(c) Suggest a way of flattening out the resonance effects.

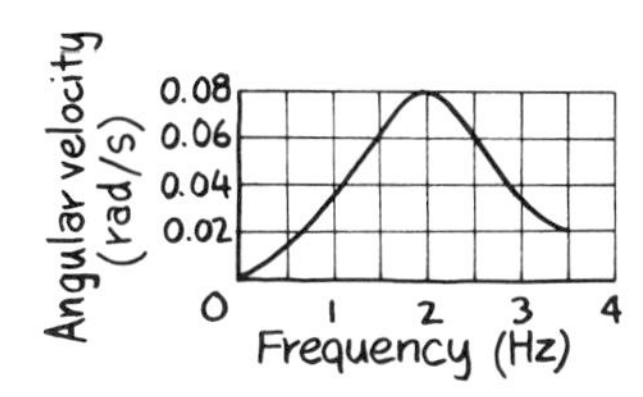

Problem 16-19

16-20. An *RCL* series circuit passes frequencies between 19,000 and 21,000 Hz. A frequency "passes" if it is between the system's half-power points.
(a) What is the resonant frequency?
(b) What is the quality of the circuit?
Assume that all the circuit resistance resides in the coil.

16-21. One end of a metre stick is clamped horizontally to a table. The length of the unclamped portion of the metre stick is 90 cm and the mass is 180 g. A force of 2.0 N causes the free end to deflect 0.040 m.
(a) What is the torsional capacitance of the metre stick?
(b) What is the moment of inertia of the metre stick about the clamped end? (Refer to Appendix B, Table 6, for moment of inertia of various objects.)
(c) What is the frequency of a sinusoidal disturbance applied to the fixed end that causes the system to exhibit resonant behavior?

16-22. A coil has an inductance of 10.0 mH and a resistance of 82.0 Ω.
(a) What capacitance must be added in series so that the circuit resonates at a frequency of 1.00×10^4 Hz?
(b) What range of frequencies are passed by the circuit? A frequency qualifies as passing if it falls between the half-power points.

16-23. A motor and its platform have a combined mass of 500 lb. It is supported at the corners by four springs. The motor shaft rotates at a constant angular speed of 2100 rpm.
(a) What is the frequency generated by the motor's rotation?
(b) What must be the equivalent capacitance of the four support springs to produce a resonant situation with the frequency calculated for the motor in part (a)?
(c) What must be the capacitance of each of the four supporting springs for the resonance effects to occur? (*Hint*: Although the springs appear to be in parallel, they actually are series connected. Make an analogy to series-connected electrical capacitors to answer this question.)

16-24. The springs proposed for the motor mounts in Problem 16-23 have a spring constant of 250 lb/in. Would you recommend that they be used if the intent is to keep the motor from shaking violently on its supports?

16-25. (a) What capacitor must be used to keep the same resonant frequency as in Problem 16-3 if the coil is replaced by a 0.050-H coil? The resistance of the new coil is still 50 Ω.
(b) What is the bandwidth of frequencies with this new combination of *L* and *C*?

16-26. Refer to the mechanical system in Fig. 16-1 and the values used in Fig. 16-2. Suppose that the system is operating at its resonant frequency of 2.0 Hz.
(a) What is the angular velocity of the motor in rpm?
(b) What is the amplitude of the displacement of the series-connected elements? (*Hint*: V_m is the tangential velocity of the pin. The pin travels a distance equal to the circumference of the circle it moves in during a time interval of one period.)

16-27. An engineer wants to block out signals with a frequency near 50,000 Hz by using a 10-mH coil. Suggest a circuit and calculate the value of capacitance needed. Neglect any resistance that the coil may have.

16-28. Refer again to the mechanical system in Fig. 16-1 and the conditions of resonance illustrated in Fig. 16-2. Suppose that the resonant frequency is 2.0 Hz.
(a) What average power is required from the motor in watts?
(b) What is the largest kinetic energy that the mass of 0.239 kg attains?
(c) Can you show that, given the conditions in Fig. 16-2, the mass must be 0.239 kg?

16-29. (a) What is the largest energy stored in the magnetic field of the coil in the circuit of Fig. 16-5 at resonance? Give your answer in joules.
(b) What is the largest energy stored in the capacitor at resonance?

ANSWERS TO ODD-NUMBERED PROBLEMS

16-1. 2.2 Hz
16-3. 1.6×10^3 Hz; 80 Hz
16-5. (a) 0.10 A
(b) 1.0×10^2 V
(c) 0.50 W
16-7. 1.3 Hz
16-9. (a) Parallel
(b) No problem, because $f_r = 1.5$ Hz
16-11. $2.35 \times 10^3\ \Omega$; +64.8°; 426 mA
16-13. 3.57 mA; +73.7°; $2.80 \times 10^3\ \Omega$
16-15. 264 Hz
16-17. 21
16-19. (a) 2.0 Hz
(b) 1.0 Hz
(c) Add some resistance to system
16-21. (a) 0.025 rad/(N · m)
(b) 0.049 kg · m^2
(c) 4.6 Hz
16-23. (a) 35 Hz
(b) 1.3×10^{-5} ft/lb
(c) 5.2×10^{-5} ft/lb
16-25. (a) 0.20×10^{-6} F
(b) 1.6×10^2 Hz
16-27. 1.0×10^{-9} F
16-29. (a) 5.0×10^{-5} J
(b) 5.0×10^{-5} J

17

Waves

Throughout this book we have been dealing with energy transfer by direct means. A force acts directly upon an object. An engine is coupled to a load by a drive shaft and gears. A centrifugal pump pushes directly against a fluid to move it through pipes. Sources of emf have been connected to loads by wires, and energy is given up as charge flows through a resistor. Energy has moved along with fluids, solid objects, and flowing charge.

Direct contact or transfer of material, however, is not a requirement for energy transfer. A magnet lifts an object without touching it. The sun exerts enough force on the earth to keep it orbiting at a distance of 93×10^6 *miles. At the same time, the sun supplies enough energy to support all life on our planet. The armature of an electric motor rotates without any direct mechanical connection to a driving member. Sensitive detection devices called seismographs receive energy from earthquakes that occur many thousands of miles away. Radio and television programs are transmitted great distances without the use of connecting wires.*

In many such situations the energy transfer is accomplished in the form of a ***wave****. In this chapter we will attempt to tell you enough about waves to give you a chance of understanding a subject worthy of many volumes the size of this text. The discussion begins with mechanical systems. Later, electrical systems are described, and in Chapter 18 the complex topic of electromagnetic waves is developed.*

17-1 PROPAGATION OF AN ENERGY PULSE

Energy transfer can be accomplished in more than one way. The energy may be stored as potential or kinetic energy in an object that moves bodily from one point to another. The energy moves along with the object and is transferred only when there is physical contact with another object. For example, a cannon ball fired through the wall of your house imparts enough energy to give it a good shake or even to knock it down, depending upon its state of repair.

When the next earthquake comes, your house will also shake (or fall down), but not as a result of something flying through the air and striking it. The energy imparted during the quake will come through the earth and will not be attached to any particular piece of matter. There will be no net redistribution of objects, except perhaps for the dishes that fall off the shelf and the occasional roof that collapses. Earthquake energy travels as a wave. The energy is transferred with no net movement of the medium through which it travels.

A wave is sent out, or *propagates*, when energy at one point in a medium affects adjacent points. The simplest sort of example to start with is propagation in one direction along a line. You may have witnessed a good example of one-dimensional wave propagation if you have ever lived near a railroad track. A train at rest on a level track sends a pulse (single burst) of energy from engine to caboose as it starts. In Fig. 17-1 the engine jerks forward, and slack in the first coupling (engine to tender) is taken up, producing a resounding clank. The tender jerks forward taking up slack in the

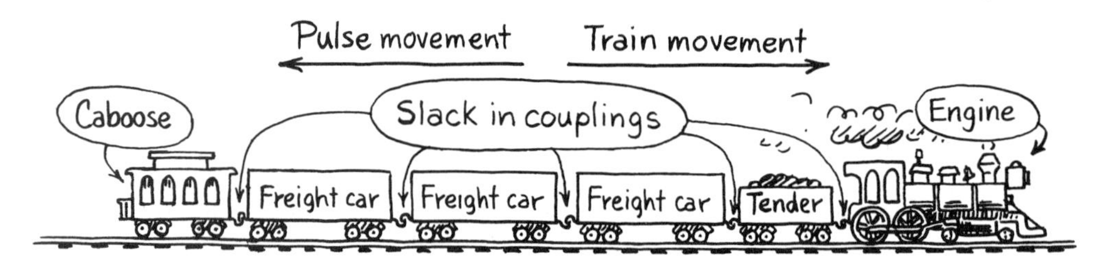

Figure 17-1. An energy pulse is transmitted backward from car to car as the train begins to move forward.

second coupling, the first freight car moves, more slack is taken up, and so it goes, creating a series of clanks that travels back along the line of cars. Ultimately, the pulse of energy (the original engine jerk) is transmitted to the caboose. But the train has hardly begun to move. In fact, the energy pulse in this case travels in a direction opposite to the movement of the train itself. In addition, the speed of the pulse has little relation to the speed of the train. The train as a whole may be essentially standing still while the energy pulse travels backward at a rate of two or three car lengths/second.

Another commonly misunderstood example is the pulse produced in the circulatory system each time the heart contracts. The pulse is normally taken at an artery in the wrist of a patient. A bit of analysis shows that the pulse occurs only very shortly after contraction (thump) of the heart. It is not blood rushing from heart to wrist in that short time, but a pressure pulse that travels through the circulatory system far faster than the blood actually flows. This pulse is propagated in much the same way as the railroad cars banging into one another. The "cars" in this case are fluid molecules. They are much smaller, more numerous, and the collisions are less abrupt—more like one coil of a spring tugging on the next.

Energy pulses propagate in electrical systems, too. When the telegraph key is pushed in Fig. 17-2, the effect is detected almost immediately at the

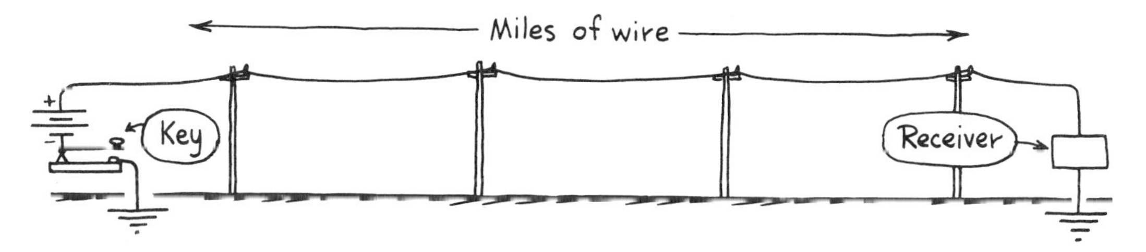

Figure 17-2. A telegraph signal travels rapidly over great distances while the electrons hardly move at all.

receiving end. The charge-carrying electrons, however, move at only a few millimetres/second. It is the pulse or "signal" that travels along the wire at nearly the speed of light through the slowly migrating electrons.

Blood, railroad tracks, trees, air, and in fact all matter consists of particles (atoms, molecules) joined by forces that are elastic to some degree. There is springiness to even the most brittle substance. Thus any material has the ability to store elastic potential energy, and that ability can be described by capacitance in the sense that we have used it in this book.

Matter is also characterized by its mass or the unified concept of inertance, which is related to kinetic energy storage. If friction can be neglected (and it can in many actual situations) we have no resistance to worry about. Thus it turns out that we are looking basically at an *MC* system—the same sort of thing as discussed in Chapter 13. The difference is that this system is extended. It is not just a single *M* hooked to a single *C*; it is a whole string of *MCMCMC* . . . .

Consider the model illustrated in Fig. 17-3. The particles of matter are represented by blocks, each of inertance *M*. The forces that hold atoms and molecules together or apart can fairly well be represented by springs, each of capacitance *C*. *C* and *M* are not necessarily constant in the real situation, but that does not matter for this qualitative discussion.

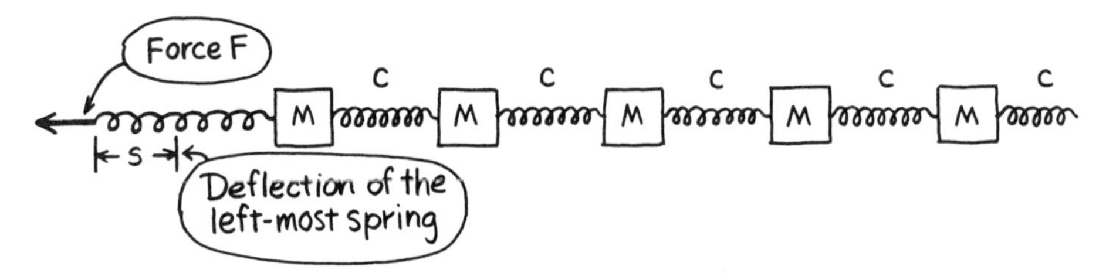

Figure 17-3. Molecular matter can be thought of as a line of blocks joined by springs.

If suddenly a force F is applied to the first spring on the left, it quickly stretches by an amount s to a new length. Force F does work on the spring, and that work is stored in the spring as elastic potential energy. This situation obviously doesn't last long. As soon as tension appears in this spring, it tugs on the first block. The block begins to move, and the potential energy stored in the spring starts pouring into the block as kinetic energy. But as soon as this block moves a bit, the next spring stretches, and the first block's kinetic energy gets transferred as potential energy stored in the next spring.

So it goes. The energy put in at the left end is stored alternately as potential and kinetic energy in elements of the system farther and farther from where the initial work was done. The energy travels off to the right and out of sight while springs and blocks remain pretty much where they started.

17-2 SPEED OF AN ENERGY PULSE

Evidently, it takes time for all the above to happen. Even the electrical impulse does not travel infinitely fast. The question is, what determines how fast?

The Effect of Stiffness or Capacitance

A spring's capacitance C, as we have seen (Chapter 9), is the reciprocal of stiffness as represented by the spring constant k. If the springs of Fig. 17-3 are replaced by stiffer springs (greater k, smaller C), the initial tug on the first spring is felt sooner by the first block. The next block in turn need not move as far to produce a given force on the next spring, and so on. The result is that the energy pulse takes less time to move from element to element within the system. The pulse travels faster when the springs are stiffer.

If the stiff springs are replaced by weak ones (small k, large C), a given displacement (amount of stretch) in the spring produces less force. Thus the acceleration of the first block is slower. The next spring takes a longer time to stretch, and the acceleration of the next block is reduced just like the first, and so on. The whole sequence of events proceeds more slowly. Thus the speed of the energy pulse depends on the stiffness of the springs. Wave speed is faster in stiffer materials with low capacitance.

The Effect of Inertance

Looking at the setup illustrated by Fig. 17-3 and using the same arguments, you can see that heavier blocks with greater inertance accelerate more slowly for a given force, and lighter blocks accelerate faster. Imagine the difference between a string of baseballs and a string of cannonballs of the same size. The less dense baseballs are easier to accelerate, move faster, and transmit the pulse faster. Wave speed is faster in materials with low inertance.

The Effect of Amplitude

It may seem to you that the speed of a pulse should also depend on how far a spring is stretched and how far a block moves. It may seem so, but it doesn't. Experiments show that the size of the displacement (the amplitude) does not affect the speed of the pulse. If a spring is stretched farther (Fig. 17-4), it exerts more force on the next block. The larger force produces a greater acceleration, so the block moves faster on the average and travels a greater distance in the same time. It takes the same amount of time for the block to accept energy from a spring and to pass it on to the next spring. Pulse speed does not depend upon amplitude.

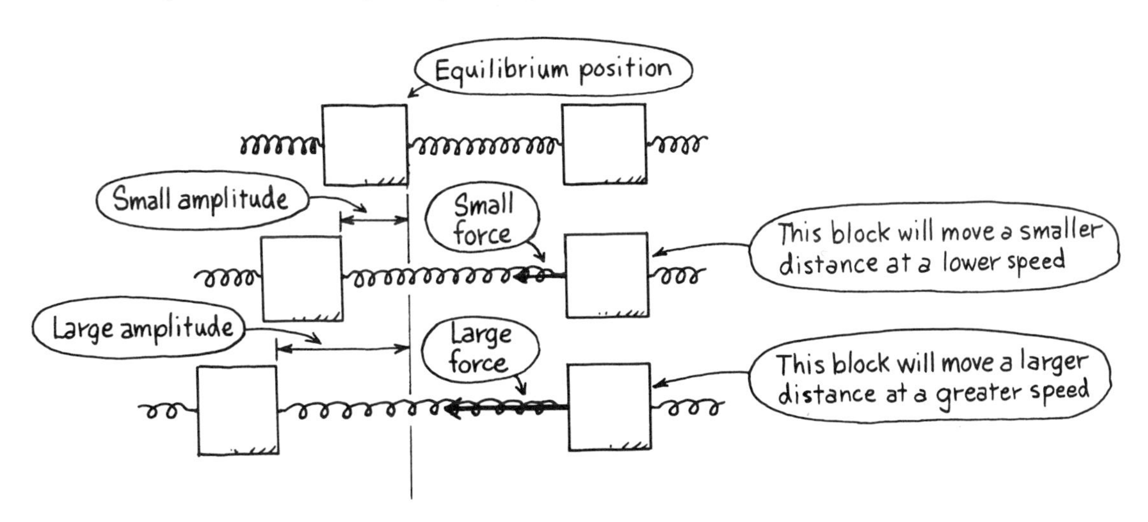

Figure 17-4. Pulse speed does not depend on how far a given element stretches or moves. In each situation the block will receive and transmit energy in the same amount of time.

A General Expression for Wave Speed

All the above, worked out intuitively, is quite valid but not the whole story. You would expect pulse or wave speed to be proportional to an expression containing stiffness divided by inertance, or an expression containing both capacitance and inertance in the denominator. Invariably, careful mathematical analysis and experimental results show this to be true, except that the pulse speed is the square root of such factors.

$$\text{pulse (wave) speed } v = \sqrt{\frac{\text{stiffness factor}}{\text{inertance factor}}}$$

$$v = \frac{1}{\sqrt{\begin{pmatrix}\text{inertance}\\ \text{factor}\end{pmatrix} \times \begin{pmatrix}\text{capacitance}\\ \text{factor}\end{pmatrix}}} \tag{17-1}$$

The *stiffness factor* is a measure of the tendency of a material to snap back to its equilibrium shape. It is usually not just the spring constant $k = 1/C$, as we shall see. Equation (17-1) is about as specific as one equation can be about pulse speed, because there are so many kinds of physical systems through which a pulse can travel. This equation does not solve all our pulse speed problems, but if you understand what it says, the rest of the chapter will make a lot more sense.

17-3 LONGITUDINAL AND TRANSVERSE PULSES

The wave pulse discussed so far involves a particular kind of deformation of the elastic medium. The line of blocks and springs moved along the same line as the propagating pulse. Such a pulse or wave, carried by particles moving parallel to the direction of propagation, is called a *longitudinal* wave (Fig. 17-5a).

It is also perfectly reasonable to deform the medium perpendicular to the line of blocks and springs. In this case the force F suddenly applied to the first

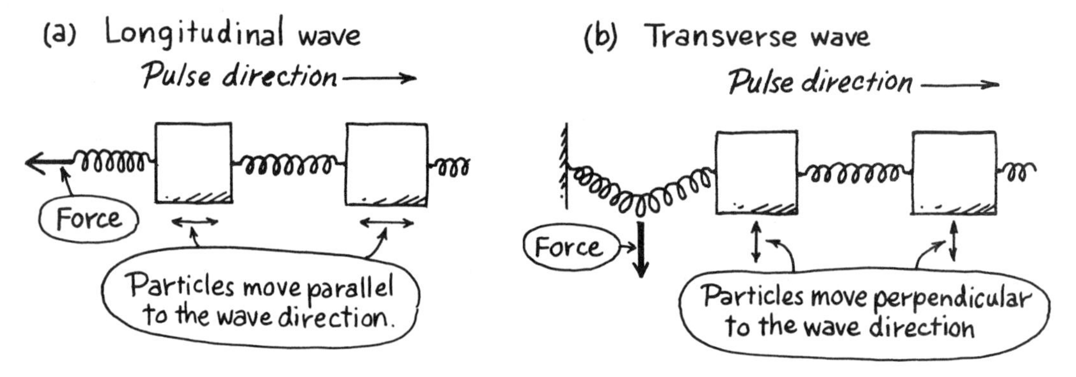

Figure 17-5. Two kinds of wave motion involve forces and motions parallel (longitudinal) and perpendicular (transverse) to the direction that the energy moves.

spring pulls it off to one side (Fig. 17-5b). Again the effect is propagated when the work done is stored alternately as potential and kinetic energy. Energy put in at the left travels down the line to the right while the blocks and springs undergo only a slight perpendicular offset. A wave propagated by this lateral offset of the medium is called a *transverse* wave.

Transverse and longitudinal waves act differently. One difference is that the restoring force in a transverse pulse (tending to make it snap back) is not the same as in a longitudinal pulse. Although the offset produced by force F stretches the spring, there is not a direct relation between F and the amount of stretch. Without going into the details, you can probably sense that it takes more force to stretch a spring a given distance s than to just deflect it the same distance sideways (Fig. 17-6). The stiffness factor is smaller for a transverse wave than for a longitudinal wave in the same medium. This makes longitudinal waves travel faster than transverse waves, all else being equal.

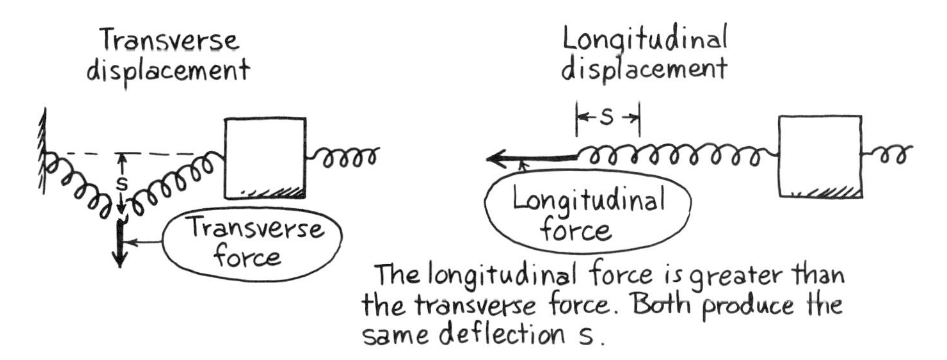

Figure 17-6. A spring deflected longitudinally acts stiffer than the same spring deflected transversely. This accounts for the different speeds of pulses produced by the two kinds of disturbances.

The velocity difference between longitudinal and transverse waves is put to good use by geophysicists (people who study the physical nature of the earth). Most of what we know, or at least strongly suspect, about the earth's interior comes from analyzing waves that are produced when earthquakes occur. Earthquakes send out longitudinal waves (called P waves for "pressure") and transverse waves (called S waves for "shear"). These waves travel through the earth at rates depending upon the stiffness and density of the rock material. P waves travel faster than S waves. The farther the waves come, the greater the time lag between the arrival of P and S waves at a recording station. It is no great problem to calculate the distance traveled if the time lag and wave speeds are known. Earthquakes are located using such information from three recording stations.

17-4 WAVE SPEED IN VARIOUS SYSTEMS

The generalization about wave speeds embodied in Eq. (17-1) is good for all sorts of waves.

$$\text{wave speed } v = \sqrt{\frac{\text{stiffness factor}}{\text{inertance factor}}}$$

$$v = \frac{1}{\sqrt{\begin{pmatrix}\text{inertance}\\ \text{factor}\end{pmatrix} \times \begin{pmatrix}\text{capacitance}\\ \text{factor}\end{pmatrix}}} \tag{17-1}$$

The stiffness and inertance factors vary somewhat, depending on the situation. Table 17-1 lists some wave speed formulas. Following the table are four examples that use formulas from the table.

TABLE 17-1 MISCELLANEOUS WAVE SPEED FORMULAS

	Medium	*Longitudinal*	*Transverse*
Propagation in *one* dimension	Stretched string (e.g., guitar string)	$v = \sqrt{\frac{Y}{\rho}}$ (Y ← Young's modulus; ρ ← Density)	$v = \sqrt{\frac{F}{m/\ell}}$ (F ← Tension; m/ℓ ← mass/length)
	Rigid rod (e.g., steel bar)		
Propagation in *three* dimensions	Fluid (e.g., sound waves in air)	$v = \sqrt{\frac{B}{\rho}}$ (B ← Bulk modulus; ρ ← Density)	No transverse wave in fluids because shear strength is zero ($G = 0$)
	Solid (e.g., earthquake waves in rock)	*P* wave: $v = \sqrt{\frac{B + \frac{4}{3}G}{\rho}}$ (B ← Bulk modulus; G ← Shear modulus; ρ ← Density)	*S* wave: $v = \sqrt{\frac{G}{\rho}}$ (G ← Shear modulus; ρ ← Density)
	Electromagnetic waves (e.g., light, X rays, radio waves in free space)	No such thing; these waves seem to be strictly transverse	$v = \sqrt{\frac{1}{\mu\varepsilon}}$ (μ ← Permeability; ε ← Permittivity)

Note: Young's modulus *Y*, bulk modulus *B*, and shear modulus *G* are discussed in Sec. 9-4. Their values for various substances appear in App. B, Table 7. Values for permeability and permittivity of a vacuum are given in the table of physical constants located in Appendix A.

Example 17-1 A guitar string 50 cm long has a mass of 5.0 g. It is stretched to a tension of 650 N. What is the speed of a transverse wave in the string?

Solution

$$\left\{\begin{aligned} \ell &= 50\text{ cm} \times \left(\frac{1\text{ m}}{100\text{ cm}}\right) \\ &= 0.50\text{ m} \\ m &= 5.0\text{ g} \times \left(\frac{1\text{ kg}}{1000\text{ g}}\right) \\ &= 5.0 \times 10^{-3}\text{ kg} \\ F &= 650\text{ N} \end{aligned}\right\}$$

From Table 17-1,

$$v = \sqrt{\frac{F}{m/\ell}} = \sqrt{\frac{650\text{ N}}{5.0 \times 10^{-3}\text{ kg}/0.50\text{ m}}}$$

$$= \boxed{2.5 \times 10^2\text{ m/s}}$$

Example 17-2 Calculate the speed of sound in a steel rod such as a railroad track.

Solution

App. B, Table 7

$$\left\{\begin{aligned} Y &= 21 \times 10^{10}\text{ N/m}^2 \\ \rho &= 7.8 \times 10^3\text{ kg/m}^3 \end{aligned}\right\}$$

App. B, Table 1

From Table 17-1,

$$v = \sqrt{\frac{Y}{\rho}} = \sqrt{\frac{21 \times 10^{10}\text{ N/m}^2}{7.8 \times 10^3\text{ kg/m}^3}}$$

$$= \boxed{5.2 \times 10^3\text{ m/s}}$$

Example 17-3 The speed of sound in air at 1 atm and 0°C is about 330 m/s. Calculate the effective bulk modulus of air under these conditions.

Solution

App. B, Table 1

$$\left\{\begin{aligned} \rho &= 1.29\text{ kg/m}^3 \\ v &= 330\text{ m/s} \end{aligned}\right\}$$

From Table 17-1,

$$v = \sqrt{\frac{B}{\rho}}$$

$$B = v^2\rho = (330\text{ m/s})^2(1.29\text{ kg/m}^3) = \boxed{1.4 \times 10^5\text{ N/m}^2}$$

Example 17-4

(a) Granite has a density ρ of about $2.7 \times 10^3\text{ kg/m}^3$ and a shear modulus G of $3.0 \times 10^{10}\text{ N/m}^2$. What is the speed of a transverse wave in granite?

(b) Geophysicists find that transverse earthquake waves (S waves) travel a distance of 9100 km through the earth in 20 min. Is the earth made of pure granite?

Solution to (a)

From Table 17-1,

$$\left\{\begin{aligned} \rho &= 2.7 \times 10^3\text{ kg/m}^3 \\ G &= 3.0 \times 10^{10}\text{ N/m}^2 \end{aligned}\right\}$$

$$v = \sqrt{\frac{G}{\rho}} = \sqrt{\frac{3.0 \times 10^{10}\text{ N/m}^2}{2.7 \times 10^3\text{ kg/m}^3}} = \boxed{3.3 \times 10^3\text{ m/s}}$$

Solution to (b)

$$\left\{\begin{aligned} s &= 9100\ \text{km}\left(\frac{1000\ \text{m}}{1\ \text{km}}\right) \\ &= 9.1 \times 10^6\ \text{m} \\ t &= 20\ \text{min}\left(\frac{60\ \text{s}}{1\ \text{min}}\right) \\ &= 1.2 \times 10^3\ \text{s} \end{aligned}\right\}$$

$$v = \frac{s}{t} = \frac{9.1 \times 10^6\ \text{m}}{1.2 \times 10^3\ \text{s}} = \boxed{7.6 \times 10^3\ \text{m/s}}$$

The average S wave velocity of 7.6×10^3 m/s calculated in solution (b) is more than twice the speed of an S wave in granite, 3.3×10^3 m/s, calculated in solution (a). Therefore, the earth is not pure granite.

It appears from looking over Table 17-1 that the speed v_L of longitudinal waves has the form

$$v_L = \sqrt{\frac{\text{elastic modulus}}{\text{density}}} \tag{17-2}$$

An elastic modulus, described in Sec. 9-4, is the ratio of stress to strain. In solids, Young's modulus Y for a rigid rod of cross-sectional area A and length ℓ is defined by Eq. (9-15):

$$Y = \frac{\text{stress}}{\text{strain}} = \frac{F/A}{\Delta\ell/\ell} \tag{9-15}$$

where F is the force that stretches the rod by an amount $\Delta\ell$. This can be rearranged to

$$Y = \frac{F}{\Delta\ell}\frac{\ell}{A}$$

If the rod is thought of as a very stiff spring with capacitance $C = \Delta\ell/F$, then

$$Y = \frac{1}{C}\frac{\ell}{A}$$

Also, the rod's density ρ is

$$\rho = \frac{\text{mass}}{\text{volume}} = \frac{m}{A\ell}$$

where m is the rod's mass. Substituting these expressions for Y and ρ in the velocity equation, Eq. (17-2),

$$v_L = \sqrt{\frac{\text{elastic modulus}}{\text{density}}} = \sqrt{\frac{Y}{\rho}} = \sqrt{\left(\frac{1}{C}\frac{\ell}{A}\right)\left(\frac{A\ell}{m}\right)}$$

or

$$v_L = \frac{1}{\sqrt{(m/\ell)\cdot(C/\ell)}} \tag{17-3}$$

Equation (17-3) tells us that the speed of a longitudinal wave depends upon the capacitive and inertive properties of the solid elastic medium.

By the same sort of manipulation, Eq. (17-2) can be changed for fluids to

$$v_L = \sqrt{\frac{\text{elastic modulus}}{\text{density}}} = \sqrt{\frac{B}{\rho}} = \frac{1}{\sqrt{(m/V)\cdot(C/V)}} \tag{17-4}$$

where m is the mass and C the capacitance of a fluid volume V.

Note the similarity between the equations for v_L in a solid [Eq. (17-3)] and a fluid [Eq. (17-4)]. They both look very much like Eq. (17-1).

$$\text{wave speed } v = \frac{1}{\sqrt{\begin{pmatrix}\text{inertance}\\ \text{factor}\end{pmatrix} \times \begin{pmatrix}\text{capacitance}\\ \text{factor}\end{pmatrix}}} \tag{17-1}$$

17-5 WAVE TRAINS

So far in this chapter we have looked at waves as a one-shot proposition, watching a single burst of energy (a pulse) as it passes through an elastic medium. What would happen if another pulse is sent down the line right after the first one? It is easy to find out. Grab the end of a long spring or rope and shake it. You will find that the second, third, and ensuing pulses do just what the first one does.

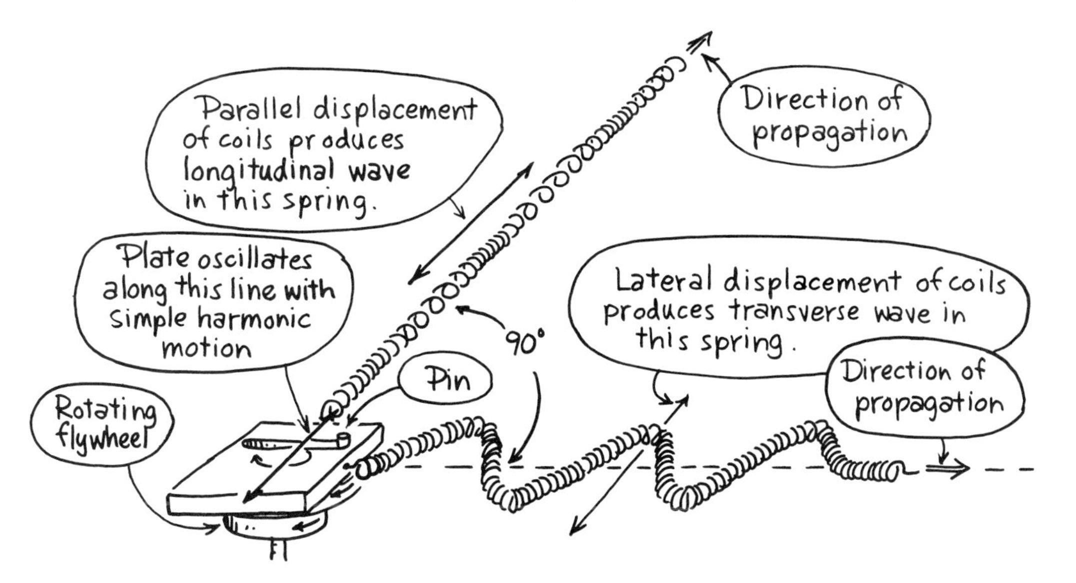

Figure 17-7. Demonstration wave machine produces transverse and longitudinal waves of the same frequency. Springs (Slinkys) are lying on a smooth horizontal surface.

Our next step is to use a sinusoidal energy input to produce a whole string of pulses, one following the other. This can be done nicely using the flywheel and slotted plate drive described in Sec. 14-2. For demonstration purposes, we want to use a medium in which pulses travel slowly enough so that we can see what goes on. The perfect low-stiffness, high-inertia medium is the popular Slinky, a large, floppy spring available at your local toy store. Figure 17-7 shows a machine that will produce both longitudinal and transverse waves that are in phase and of the same frequency. The source moves the attached ends of the Slinkys with sinusoidal motion, and the disturbances produced travel down each Slinky as a wave with constant velocity. The spring receiving the transverse displacement takes the form of a migrating sine curve, as explained in Sec. 14-2. The spring receiving the longitudinal displacement exhibits alternate *compressions* where coils are close together and *rarefactions* where coils are spread apart.

Figure 17-8 illustrates the actual shape of the springs at the particular instant when the generator has made $2\frac{1}{2}$ complete cycles of motion. Compare this with Figs. 17-9 and 17-10, which are records (graphs) of where the single

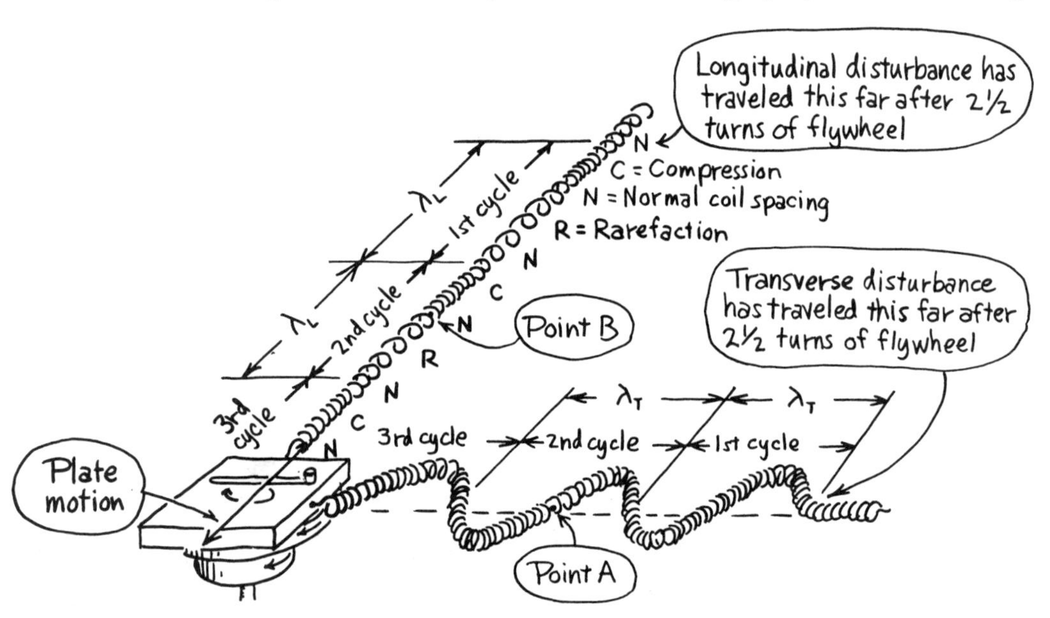

Figure 17-8. Snapshot of wave machine after $2\frac{1}{2}$ turns of the flywheel. Longitudinal wavelength λ_L is greater than transverse wavelength λ_T because the longitudinal wave travels faster.

points A and B on the springs have been at various times. It is easy to confuse such graphs with actual wave shapes and you should guard against it.

Figures 17-9 and 17-10 are time graphs of points A and B on the springs subjected to transverse and longitudinal excitation. These graphs show the

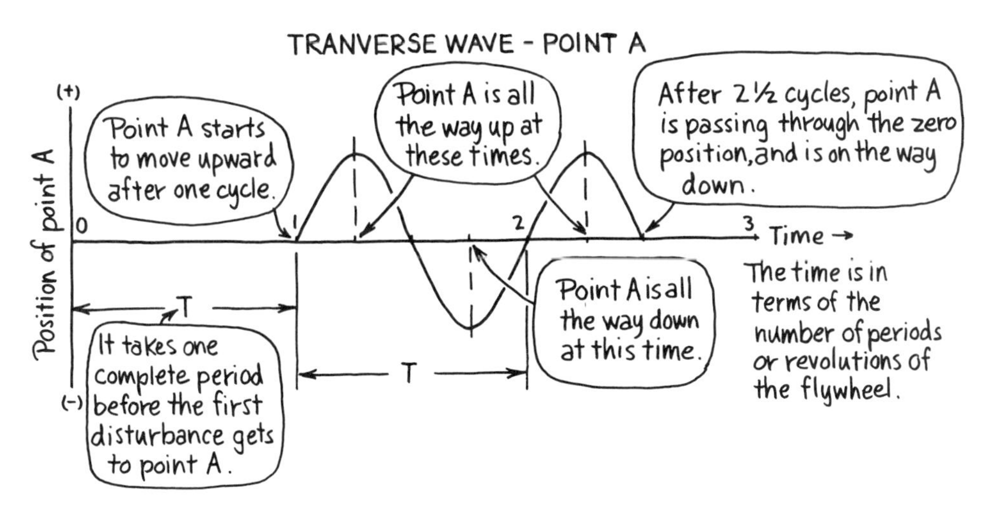

Figure 17-9. Time graph of point *A*'s position in the transverse wave of Fig. 17-8. Time can be represented by the number of flywheel revolutions. This is *not* the shape of the actual spring.

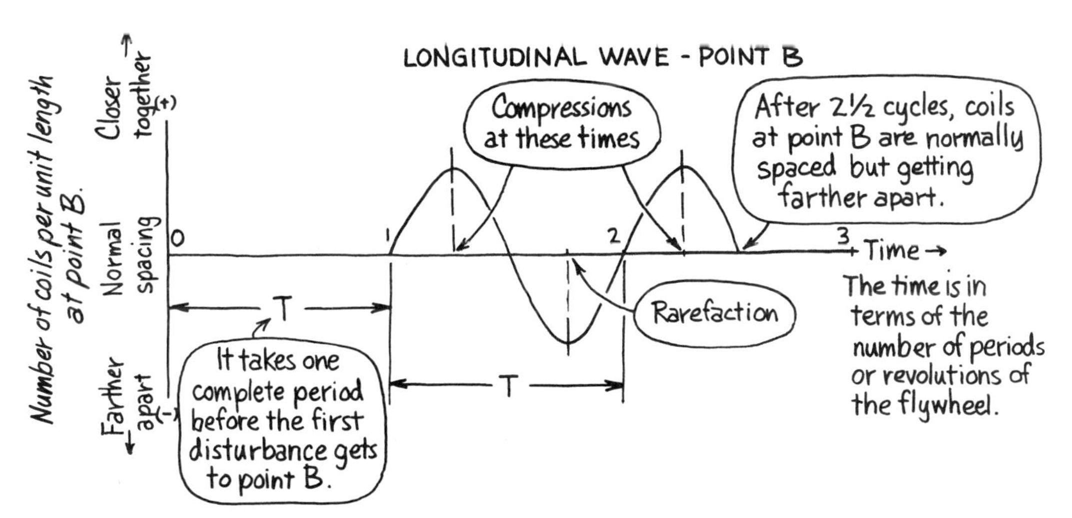

Figure 17-10. Time graph of the closeness of coils ("pressure") at point *B* in the longitudinal wave of Fig. 17-8. Obviously this is not a diagram of the spring itself.

activity of points *A* and *B* (Fig. 17-8) during the first $2\frac{1}{2}$ turns of the flywheel. Figure 17-9 is a record of the lateral displacement of point *A* from its original position due to the transverse waves. Figure 17-10 graphs the "closeness" of the coils (number of coils per unit length) at point *B* as time passes. If this were a sound wave, closeness would be directly related to pressure and could actually be measured with a pressure transducer.

The peak-to-peak length of one cycle in Fig. 17-8 is the distance that the wave travels during that cycle. This distance is called the *wavelength* λ (Greek letter lambda). The time to complete one cycle is the *period* T. The period is shown on the transverse and longitudinal wave time graphs (Figs. 17-9 and 17-10). *Frequency* f is the reciprocal of period T. Points A and B were selected so that the first disturbance reaches them just as one cycle (i.e., one turn of the flywheel) is completed. Study these figures carefully and be sure that you understand them before you continue.

Each wave of Fig. 17-8 travels at its own constant speed v and moves a distance λ in one time period T. Thus

$$v = \frac{\lambda}{T} = \left(\frac{1}{T}\right)\lambda = f\lambda$$

$$v = f\lambda \tag{17-5}$$

v Wave Speed	f Frequency	λ Wavelength
metre/second (m/s)	hertz* (Hz)	metre (m)
foot/second (ft/s)		foot (ft)

* Remember that 1 hertz (Hz) = 1 cycle/second. The "cycle" is dimensionless.

Equation (17-5) is not difficult to understand and is a very useful thing to keep in mind. Notice that it is strictly a kinematic (distance and time) equation based on definitions of wavelength and period. It doesn't say anything about stiffness or inertia or amplitude or why the velocity is what it is. Example 17-5 will illustrate its use.

Example 17-5 One end of a rope is shaken up and down with a frequency of 4.0 Hz. This produces a series of pulses (a wave train) with adjacent crests separated by 0.50 m.

(a) At what speed do the pulses move along the rope?
(b) What frequency would produce waves 3.0 m long?

Solution to (a)

$$\left\{\begin{matrix} f = 4.0\ \text{Hz} \\ \lambda = 0.50\ \text{m} \end{matrix}\right\} \qquad v = f\lambda = (4.0\ \text{Hz})(0.50\ \text{m}) = \boxed{2.0\ \text{m/s}}$$

Solution to (b)

$$\left\{\begin{matrix} \lambda = 3.0\ \text{m} \\ v = 2.0\ \text{m/s (part a)} \end{matrix}\right\} \qquad f = \frac{v}{\lambda} = \frac{2.0\ \text{m/s}}{3.0\ \text{m}} = \boxed{0.67\ \text{Hz}}$$

17-6 REFLECTION AND TRANSMISSION

So far we have not worried about what ultimately becomes of wave energy after it passes our observation point. There are three possible choices. It can be stored as kinetic energy or potential energy or it can be lost through a resistor. Any material system contains some resistance, and the energy lost appears as heat.

But what happens to the energy that is not lost—the energy that makes it all the way to the end of the Slinky? The answer depends upon the conditions at that end. The Slinky is an energy *source*, and whatever lies beyond is a *load*, the element to which the energy is delivered. Transfer of energy from the Slinky to the load depends upon the relative impedances (Sec. 15-1) of source and load. One ideal situation is to connect another identical Slinky to the end of the first one. Source and load have the same properties and thus the same impedance in this situation. A wave will pass smoothly through the junction between the Slinkys. It is as though there were no junction at all—just one continuous uniform medium. The energy in this case is efficiently *transmitted* from source to load (Fig. 17-11).

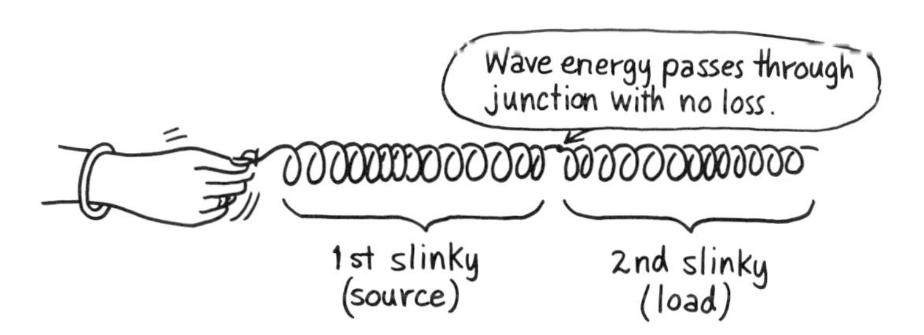

Figure 17-11. If properties (impedances) are matched, energy passes from one medium to another with no problem.

Suppose, however, that the first Slinky is attached at its far end to something that has a much different capacitance or inertance (Fig. 17-12). A very stiff, heavy spring such as a coil spring from an automobile suspension

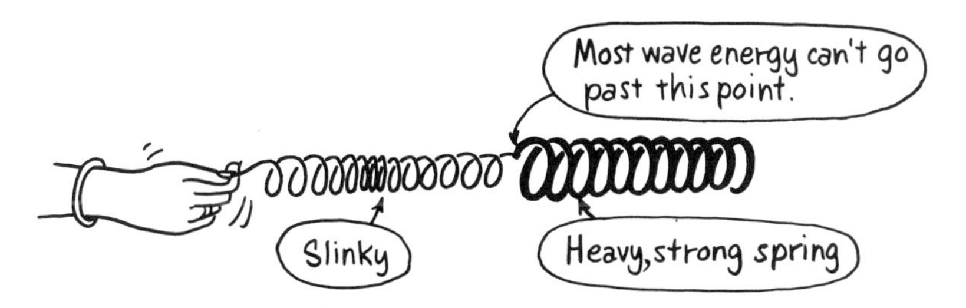

Figure 17-12. Energy does not pass easily to the heavy spring because force applied by the Slinky has little effect.

would not accept energy efficiently. Transferring energy from a Slinky to such a spring would be like transferring energy from a tack hammer to a railroad spike. The impedance of the load is effectively very large, and the small force exerted by the Slinky is hardly enough to make the heavier spring quiver. The Slinky may as well be tied to a brick wall for all the energy that gets through.

Let us go to the other extreme and join the Slinky to a very lightweight, low-stiffness spring, such as might be used in a delicate spring balance or the movement of a galvanometer (Fig. 17-13). In this case the load is so light and weak (small impedance) that energy transfer is difficult. A given displacement of the Slinky's end displaces the light spring by the same amount. Less potential energy and kinetic energy can be stored and transmitted by the light spring, because its capacitance is high and its inertance is low. Here the analogy is the inability of a user, the source, to put enough energy into the tack hammer, the load, to do the job of spike driving. The Slinky in effect is "punching air."

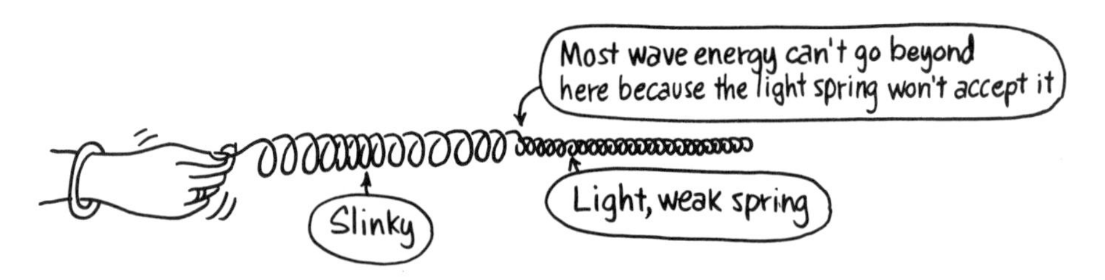

Figure 17-13. Energy does not transfer efficiently to the light spring.

As the light spring is made lighter and weaker, we approach the condition of having the far end of the Slinky attached to nothing at all. In this extreme case, no energy at all is propagated beyond the end of the Slinky. In general, then, the more nearly matched the media, the better the energy transfer. The question now is, what happens to the energy that does not get past the junction?

There are two possibilities. The energy can either be lost or stored. Let us assume that no resistance is present at the junction. No energy is lost, so it must be stored somewhere. Energy that does not get past the junction must remain in the first Slinky; but in what form? Does it pile up at the far end like tumbleweeds blowing against a fence? Obviously not, for energy in such a medium must be on the move.

Thus energy that cannot pass out of the far end has no choice but to turn right around and come back along the same path. This phenomenon is called *reflection*. Reflected waves are propagated with the same frequency and at the same speed as the incident waves arriving at the junction. The energy of the reflected wave, proportional to the square of the amplitude, depends upon the relative amount of incident energy transmitted and reflected at the junction.

The *phase* of a reflected wave relative to the incident wave also depends on conditions at the junction. Let us look at the Slinky under two extreme end conditions. One condition is when the end is securely attached to something, and the other is when the end hangs free.

A pulse arriving at a fixed end (Fig. 17-14) has no place to go. The coils pull against the floor and are themselves pulled back by the floor's reaction force, resulting in a reflected pulse on the side opposite the incident pulse. The reflected pulse has been shifted to the other side. This is thought of as a 180° phase shift, because the pulse looks like the next half of the cycle that would follow it in a wave train.

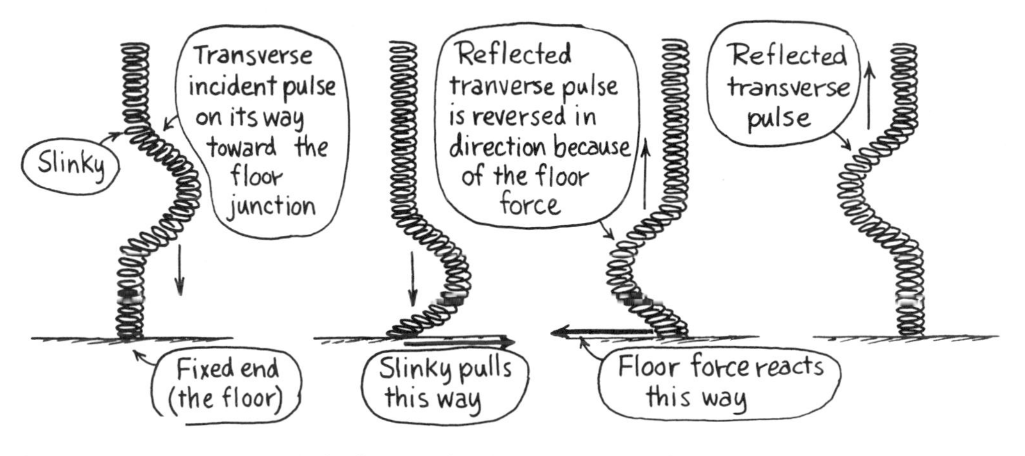

Figure 17-14. When one end of a Slinky is fixed, a 180° phase shift occurs on reflection.

When a pulse arrives at a free end (Fig. 17-15), the unconstrained end coils of the Slinky move to the same side where the incident pulse is located. The Slinky's inertia causes the end to overshoot and thereby produce a reflected pulse on the same side. The reflected pulse is effectively in phase with the incident pulse. There is no phase shift at a free end.

Figure 17-15. At a free end the reflected pulse is on the same side and is in phase with the incident pulse.

Unless the end of the Slinky is completely anchored or perfectly free, some of the incident wave's energy must be transmitted and some reflected. The transmitted part is always exactly in phase with the incident wave, whereas the reflected part may or may not be in phase with the incident wave. The relative amounts of energy reflected and transmitted depend on how different the media are. The more nearly matched the media, the better the transfer of energy from source to load. This will be discussed again in Sec. 17-9.

17-7 INTERFERENCE

A number of interesting things result when a wave is reflected. Pulses from two sources (incident and reflected) now travel along the same Slinky at the same

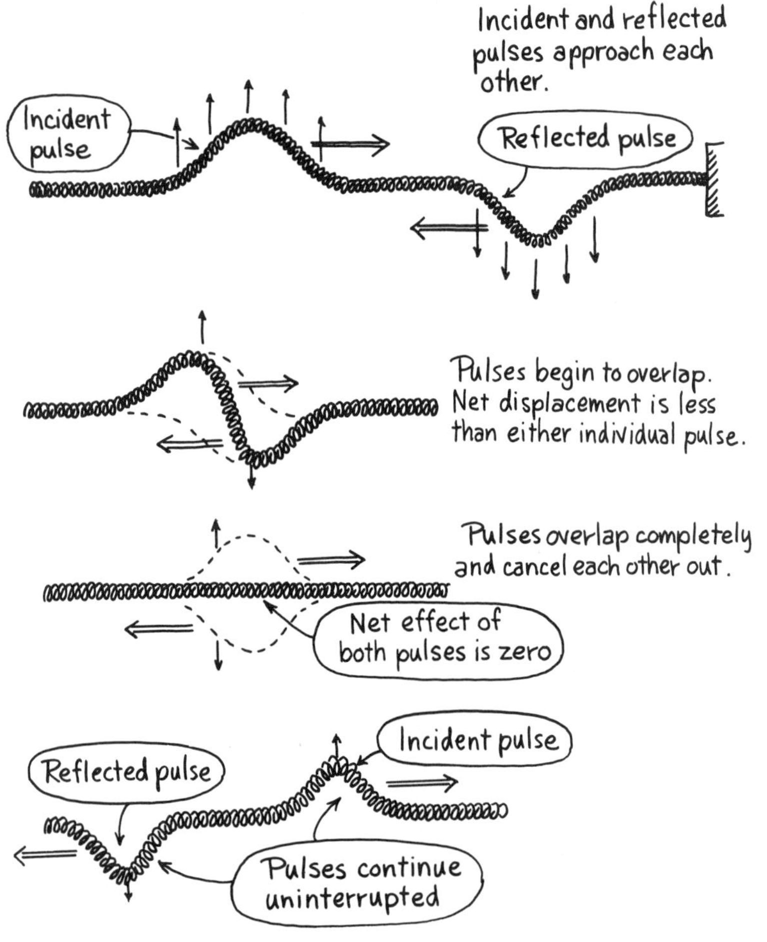

Figure 17-16. Destructive interference between two overlapping pulses. After passage the pulses continue as if nothing had happened.

time, but in opposite directions. What happens when the two pulses meet? Will they collide and rebound like a couple of billiard balls? The answer is no. The pulses pass right through each other almost as though the other one was not there. Their combined effect while passing, however, is different than if only one were present. Think of a transverse pulse as being produced by a line of little men pulling sideways on the Slinky. The net transverse offset at a given point is the algebraic sum of the displacements of the two pulses. For example, when two transverse pulses on opposite sides of the Slinky overlap (Fig. 17-16), the force of one little man is canceled by that of another pulling in the opposite direction. When two pulses overlap this way, the total effect is destroyed, and thus the term *destructive interference* is used to describe this. After passing, the pulses continue as though nothing had happened.

If the pulses are on the same side (Fig. 17-17), the offsetting forces are in the same direction and reinforce each other, making the displacement greater than if only one pulse were present. Thus the term *constructive interference.* After passing, the waves continue as if nothing had happened.

The above discussion involves only transverse pulses simply because they are easier to illustrate. The same arguments apply to longitudinal pulses. Compressions and rarefactions cancel or augment one another as they move back and forth through the medium. You may recognize the methods used here as the principle of superposition applied to waves. The same principle was introduced in Sec. 15-2 to combine sinusoidal graphs of voltages and currents.

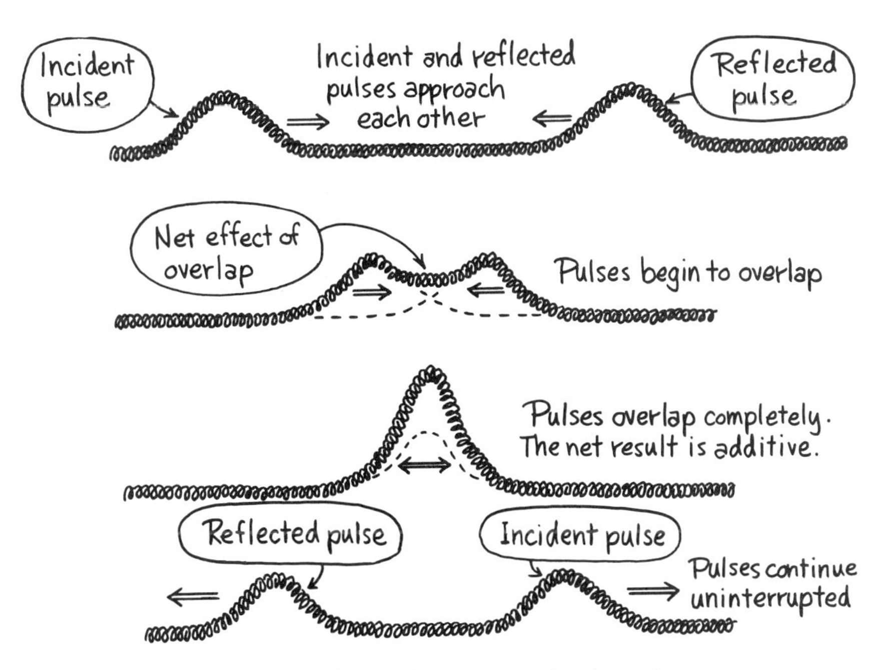

Figure 17-17. Constructive interference between overlapping pulses.

*17-8 STANDING WAVES

Some interesting and useful interference effects can be produced if the timing of incident and reflected pulses is made just right. Figure 17-18 illustrates the details of one possible situation. The free end of the rope is moved up and down with simple harmonic motion, as described in Chapter 13.

In Fig. 17-18a the rope is at rest. Energy is put into the rope by raising the free end sinusoidally (Fig. 17-18b). This upward disturbance travels as an incident wave down the rope. The start of the upward incident disturbance bounces off the fixed end just as the free end rises to its maximum height (Fig. 17-18c). After this, things get complicated by the reflected wave. The reflected wave begins to travel back as a downward disturbance; at the same time the free end is forced downward. Figure 17-18d shows this. In Fig. 17-18e, incident and reflected waves exactly cancel each other out. The rope is straight.

The free end passes through the starting position on its way downward (Fig. 17-18e). It is urged downward by the reflected disturbance, which now turns around at the free end and heads back in such a way as to augment or add to the incident wave from the sinusoidal source (Fig. 17-18f). After $\frac{3}{4}$ cycle the free end has twice the displacement that it was originally given (Fig. 17-18g). From this time on the rope oscillates with this double amplitude at the free end, provided that enough energy is added to compensate for resistive loss. The amplitude (maximum displacement) at a given point along the rope is constant. The amplitude varies with position along the rope. The rope is said to display a *standing wave.*

During the first half-cycle, the free end had to be forced to move as it did. Work was done on the rope; that is, energy was put into the rope by a source at the free end. After one half-cycle, that energy began to arrive back at the free end as a downward-reflected disturbance. During the second half-cycle, the free end was pulled along by the energy already in the rope. Effectively, the energy was returned to the source for reuse during the next half-cycle.

Does this all sound familiar? We are talking about a specific example of resonance—large sinusoidal oscillations produced by a relatively small excitation force—as discussed in Sec. 16-1.

Modes of Vibration

While the free end in Fig. 17-18 moves from the maximum positive to the maximum negative position (elapsed time = $\frac{1}{2}T$, where T is the period), the pulse and its reflection travel down the rope and back (distance = 2ℓ where ℓ is the rope's length). Wave velocity v = distance/elapsed time = $2\ell/\frac{1}{2}T = 4\ell/T$. But $v = \lambda/T$, and therefore $\lambda = 4\ell$. In this case the wavelength is four times the length of the string, and the frequency is $f = v/\lambda = v/4\ell$. The standing wave is said to describe half of a "loop" (Fig. 17-19a). The points that remain stationary are *nodes*, and the points with maximum amplitude are *antinodes.*

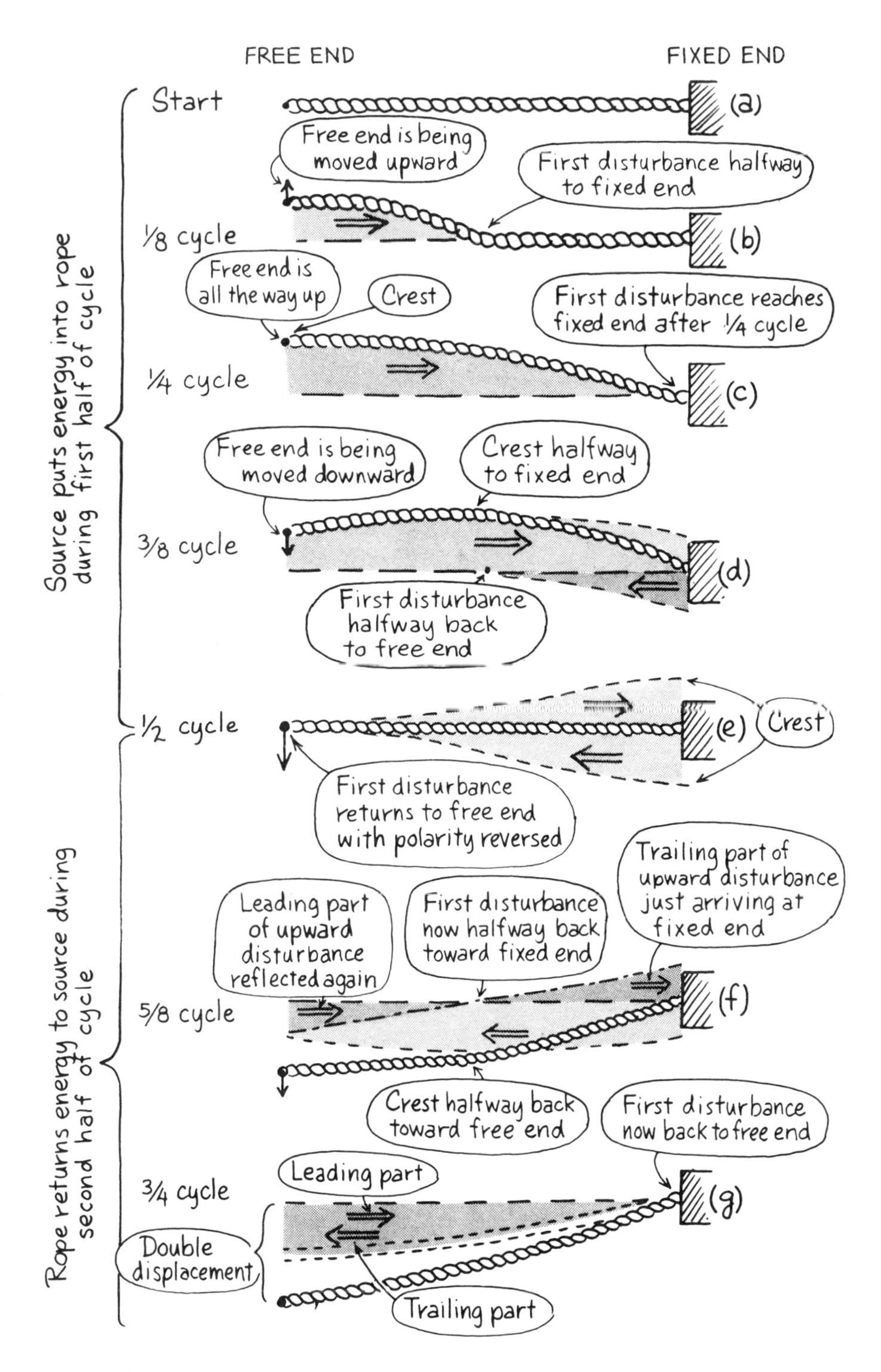

Figure 17-18. Sequence showing the buildup of a standing wave in a rope. The rope has a resonant frequency like any other *RMC* system. Large-amplitude vibrations build up quickly when a source puts energy in at that frequency.

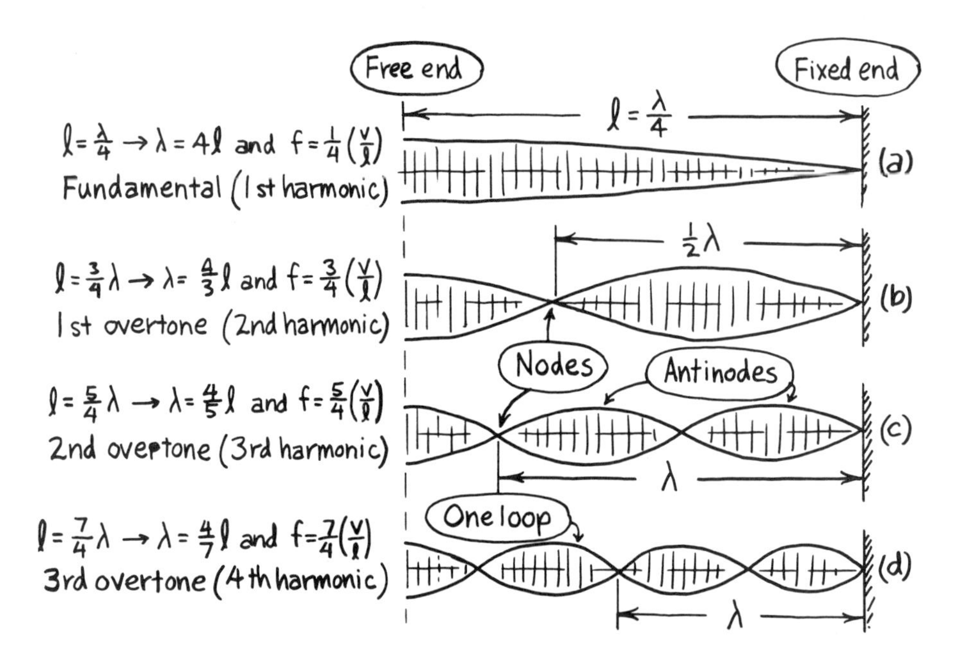

Figure 17-19. Standing wave patterns in a string fixed at one end and free at the other.

Figures 17-19b, c, and d illustrate some of the other ways in which a rope can vibrate with one end fixed and the other end free. Note that for each of these modes or styles of vibration there is a specific frequency. A little experimentation shows that these (and higher) frequencies are the only ones that produce regular vibrations in the string. Any attempt to shake the free end at another frequency produces a mess, with incident and reflected waves meeting and interfering in a disorderly way. The frequencies corresponding to standing waves are the *natural* or the *characteristic* or the *resonant* frequencies of the system. The frequency corresponding to half a loop is called the *fundamental* or first *harmonic.* The succeeding higher frequencies are referred to as *overtones* or additional harmonics. Figure 17-19 shows the standing wave patterns and their description in terms of overtones and harmonics. Note that the frequencies of the overtones are multiples of the fundamental frequency.

A string fixed at both ends displays a different set of standing wave patterns and thus different characteristic frequencies. A guitar or piano string produces only these frequencies when excited. Others very quickly die out as the result of disorderly interference. Figure 17-20 shows the fundamental mode and the first three overtones of a string fixed at both ends.

The *pitch* of a note (highness or lowness) is related to frequency. The greater the frequency, the higher the note. The *quality* (not the same as the quality discussed in Sec. 16-3) of a note is determined by the combination of

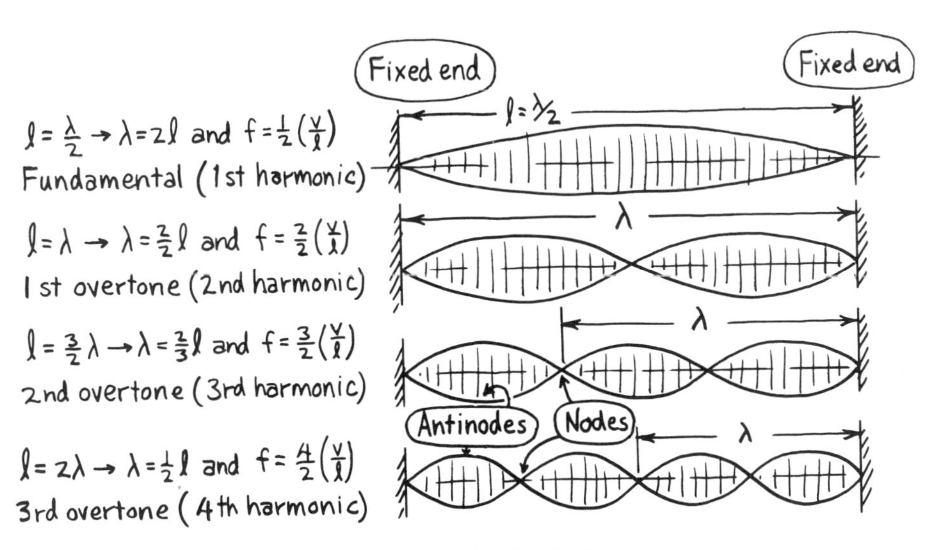

Figure 17-20. Standing wave patterns in a string fixed at both ends.

overtones produced along with the fundamental. Quality is what distinguishes the sound of a guitar from the sound of a piano.

There is one more set of possible standing wave patterns—those in a medium with both ends free. A string must be hung vertically and the upper end shaken back and forth to accomplish this (Fig. 17-21).

You may at this point be saying to yourself, "sure but who ever hangs strings and shakes them like that?" The answer is almost nobody. But all three

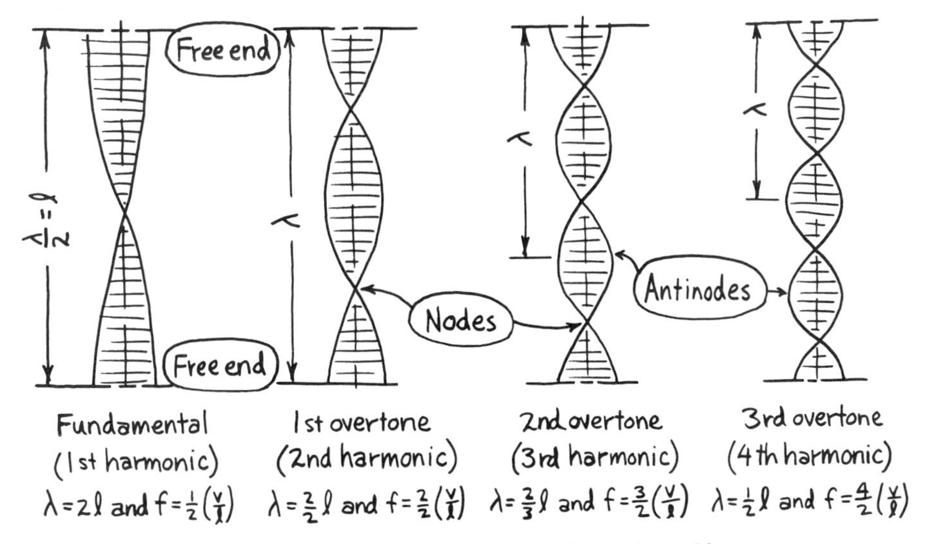

Figure 17-21. Fundamental and first three overtones in a string with both ends free.

ways of vibrating strings have very close analogs in other physical systems. You should have guessed—right?

Standing waves can be produced in a fluid medium. For example, standing longitudinal waves (sound waves) are established in the air column of a wind instrument such as a flute, horn, or pipe organ. As with strings, the fundamental frequency of a wind instrument depends upon the length of the air column and whether the ends are closed (fixed end) or open (free end). A node forms at a closed end and an antinode at an open end. It is part of everyone's experience that larger instruments (bass horns, bull fiddles) make lower notes, whereas higher notes come from short things (trumpets, violins). Now you know why.

Let's look at a couple of problems dealing with standing waves.

Example 17-6 What is the wave speed in a guitar string 0.50 m long whose fundamental frequency is 660 Hz?

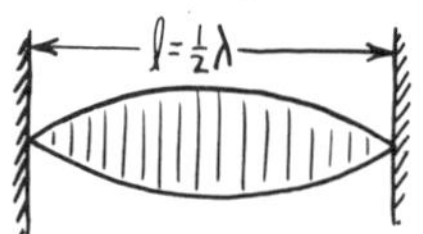

Solution

$\{f = 660\ \text{Hz}\}$

Since the guitar string is fixed at both ends $\ell = \frac{1}{2}\lambda$ and $\lambda = 2\ell = 2(0.50\ \text{m}) = 1.0\ \text{m}$:

$$v = f\lambda = (600\ \text{Hz})(1.0\ \text{m}) = \boxed{660\ \text{m/s}}$$

Example 17-7 A pennywhistle is essentially a vibrating air column closed at one end and open at the other. How long must the whistle be to sound the musical note of A, which has a fundamental frequency of 440 Hz? The speed of sound in air at sea level and 0°C is 331 m/s.

Solution

$\left\{\begin{array}{l} f = 440\ \text{Hz} \\ v = 331\ \text{m/s} \end{array}\right\}$

A node forms at the closed end and an antinode at the open end of the whistle. Thus $\ell = \lambda/4$ and $v = f\lambda$:

$$\lambda = \frac{v}{f} = \frac{331\ \text{m/s}}{440\ \text{Hz}} = 0.75\ \text{m}$$

$$\ell = \frac{\lambda}{4} = \frac{0.75\ \text{m}}{4} = \boxed{0.19\ \text{m}}$$

17-9 ELECTRICAL PULSES AND WAVES

Electrical signals are transmitted at great speed through the components of an electric circuit. It is commonly stated that electrical pulses travel at the speed of light, but this is not necessarily true. Electrical waves travel from one charged particle to another with relatively little net displacement of the particles themselves. The wave's velocity depends on the characteristics of the medium through which it travels, just as in the mechanical systems we have been looking at.

In Sec. 17-1 we found that, as a wave propagates, its energy is stored alternately as potential and kinetic energy in the capacitive and inertive elements of the conducting medium. In Sec. 17-2 a general expression for wave velocity v was related to the inertance and capacitance of the conducting medium:

$$\text{wave speed } v = \frac{1}{\sqrt{\begin{pmatrix}\text{inertance}\\ \text{factor}\end{pmatrix} \times \begin{pmatrix}\text{capacitance}\\ \text{factor}\end{pmatrix}}} \tag{17-1}$$

Section 17-4 showed that equations for the velocity v_L of longitudinal waves in solids and fluids looked very much like Eq. (17-1).

$$\text{Solid:} \quad v_L = \frac{1}{\sqrt{(m/l)(C/\ell)}} \tag{17-3}$$

$$\text{Fluid:} \quad v_L = \frac{1}{\sqrt{(m/V)(C/V)}} \tag{17-4}$$

It turns out that there are real, useful electrical systems that are perfectly analogous to these mechanical systems with respect to wave velocities.

Electromagnetic Delay Lines

An electromagnetic delay line is an electrical conducting medium designed to delay a signal for a specific length of time by lowering the speed of a propagating electrical wave. Delay lines have important application in radar and other devices that depend upon critical timing of electrical pulses.

Figure 17-22 illustrates a *lumped parameter* delay line that is conceptually very close to the line of blocks and springs of Fig. 17-3 discussed earlier in the

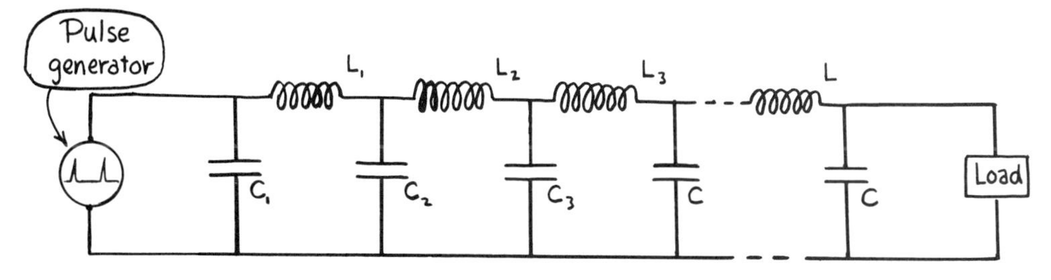

Figure 17-22. An electromagnetic delay line with lumped parameters.

chapter. If suddenly the pulse generator supplies some energy by imposing an emf across the delay line terminals, the first effect is to charge the closest capacitor C_1, which momentarily stores the energy in the electric field between its plates. This is analogous to stored potential energy in a spring. When C_1 discharges through the inductor L_1, the energy changes from stored energy in an electric field to stored energy in the magnetic field about L_1. This is analogous to stored kinetic energy in a mass. As that field collapses, it charges the next capacitor C_2, storing the energy once again in the newly forming electric field. So it goes. The energy is handed from C_2 to L_2 to C_3 and on down the line to the load.

Just as in the mechanical analogue, the time delay is the sum of the small but finite times involved in transferring energy from one storage device to the next. The delay time for a given pair of L and C turns out to be

$$t_d = \sqrt{LC} \tag{17-6}$$

t_d *Delay Time for an LC Pair*	L *Inductance*	C *Capacitance*
second (s)	henry (H)	farad (F)

In many actual situations, such as two-wire transmission lines (coaxial cable, TV antenna "twin lead," etc.), inductance and capacitance are distributed uniformly along the length of the line instead of being lumped as individual inductors and capacitors (Fig. 17-23). Here it is more reasonable to speak of L and C per unit length of the line.

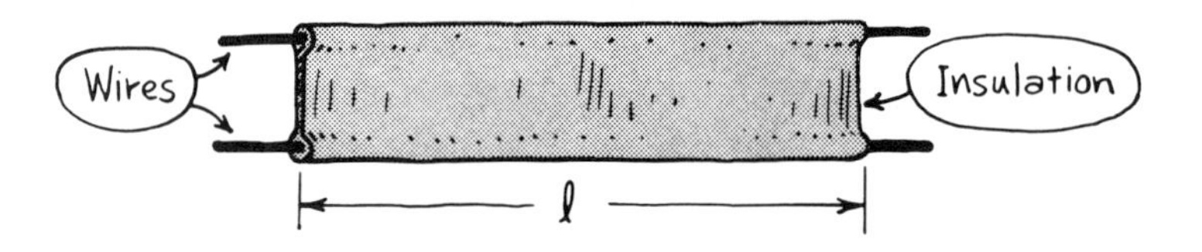

Figure 17-23. This length of TV cable has a distributed capacitance C and a distributed inductance L. It takes time t_d for an electrical signal to travel a distance ℓ from one end to the other.

Divide both sides of Eq. (17-6) by the line length ℓ to get

$$\frac{t_d}{\ell} = \sqrt{\frac{L}{\ell}\frac{C}{\ell}}$$

The quantity t_d/ℓ is elapsed time divided by distance traveled, which is the reciprocal of the wave velocity v. Thus $v = \ell/t_d$ and

$$v = \frac{1}{\sqrt{(L/\ell)(C/\ell)}} \tag{17-7}$$

v Wave Velocity	L/ℓ Inductance per Unit Length	C/ℓ Capacitance per Unit Length
metre/second (m/s)	henry/metre (H/m)	farad/metre (F/m)

which is precisely the same form as Eq. (17-3)! Let's try a couple of example problems.

Example 17-8 How long does it take for an electrical pulse to travel a lumped parameter delay line containing four pairs of L and C, with $L = 20$ mH and $C = 0.10\ \mu$F?

Solution

$\left\{\begin{array}{l} L = 20 \times 10^{-3}\ \text{H} \\ C = 0.10 \times 10^{-6}\ \text{F} \end{array}\right\}$ For each pair,

$$t_d = \sqrt{LC} = \sqrt{(20 \times 10^{-3}\ \text{H})(0.10 \times 10^{-6}\ \text{F})} = 4.47 \times 10^{-5}\ \text{s}$$

and for four pairs

$$4t_d = 4(4.47 \times 10^{-5}\ \text{s}) = 18 \times 10^{-5}\ \text{s} \rightarrow \boxed{0.18\ \text{ms}}$$

Example 17-9 What is the speed of a pulse in a two-wire transmission cable that has 1.3×10^{-10} F and 1.6×10^{-4} H per metre of length?

Solution

$\left\{\begin{array}{l} L/\ell = 1.6 \times 10^{-4}\ \text{H/m} \\ C/\ell = 1.3 \times 10^{-10}\ \text{F/m} \end{array}\right\}$

$$v = \frac{1}{\sqrt{(L/\ell)(C/\ell)}} = \frac{1}{\sqrt{(1.6 \times 10^{-4}\ \text{H/m})(1.3 \times 10^{-10}\ \text{F/m})}}$$

$$= \frac{1}{1.44 \times 10^{-7}} = \boxed{6.9 \times 10^{6}\ \text{m/s}}$$

Note: The speed of light is 3.0×10^{8} m/s $= 300 \times 10^{6}$ m/s.

Reflection of Electrical Pulses

Electromagnetic delay lines and transmission lines display reflection and transmission characteristics just like their mechanical analogues. Unless the end of the line is provided with a load matched to the line (i.e., same impedance as the line), the energy will not transfer efficiently and some must reflect back into the line (Fig. 17-24a).

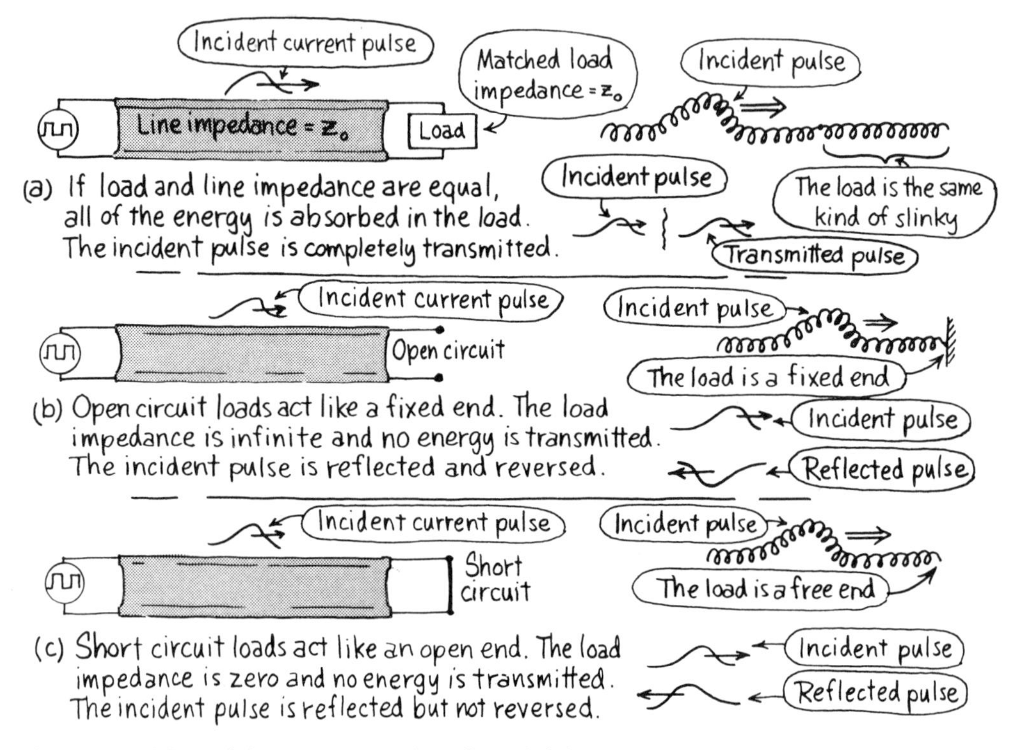

Figure 17-24. Part of the energy must be reflected if the terminating impedance (load) is not the same as the line impedance (source). Open-circuit and short-circuit ends reflect all the energy.

Figure 17-24b illustrates the case in which the conductors of a delay line are not connected to anything. The open-circuit delay line has an infinite impedance load. This corresponds to a fixed end in the mechanical system. No motion is allowed at a fixed end, so the velocity must always be zero. Likewise, there can be no electrical $\dot{Q}$ or current at the open end of a delay line. As in the mechanical system, a positive incident current pulse must always be canceled by a negative reflected pulse of current. The reflected $\dot{Q}$ pulse is 180° out of phase with the incident $\dot{Q}$ pulse.

Figure 17-24c shows a delay line in which the conductors are connected by a thick wire. This short-circuit delay line has a "load" of zero impedance and allows maximum current. The reflected $\dot{Q}$ pulse is in phase with the incident $\dot{Q}$ pulse. This is comparable to the free-end behavior of the Slinky in the mechanical system.

As you might expect, standing waves can be set up in electrical systems in a fashion analogous to the mechanical systems of Sec. 17-8. Electrical standing waves are tremendously important in the design of electronic communication equipment.

Coefficient of Reflection

Thus, depending upon the degree of matching between line (source) and load, wave energy can be completely reflected or totally accepted by the load. In nearly every real situation, some combination of reflection and transmission occurs. The relative amount of wave energy reflected can be expressed as a *coefficient of reflection* ρ, defined as

$$\rho = \frac{Z_1 - Z_2}{Z_1 + Z_2} \tag{17-8}$$

ρ *Coefficient of Reflection*	Z_1 *Impedance of Source (line)*	Z_2 *Impedance of Load*
Units cancel out; dimensionless ratio	Units of impedance appropriate to system; ohm (Ω), $\frac{\text{newton}}{\text{metre/second}}$ $\left(\frac{\text{N}}{\text{m/s}}\right)$, etc.	

The coefficient of reflection ρ is a dimensionless ratio, and the units of Z are appropriate to the particular system under study (force/parameter rate).

We want to look at three situations involving reflection.

1. If the line and load impedances are matched, $Z_2 = Z_1$. The numerator of Eq. (17-8) becomes zero and $\rho = 0$, indicating that no energy is reflected.

2. If $Z_2 = 0$ (equivalent to a free end), $\rho = +1$, indicating complete reflection with no change of sign for $\dot{Q}$ (no phase shift in $\dot{Q}$).

3. If Z_2 is infinitely large (equivalent to a fixed end), $\rho = -1$, indicating complete reflection of energy and a change of sign (180° phase shift) for $\dot{Q}$. The following examples illustrate situations in between the extremes of total reflection and total transmission.

Example 17-10 A delay line of 1000-Ω impedance is terminated by a 500-Ω load impedance.

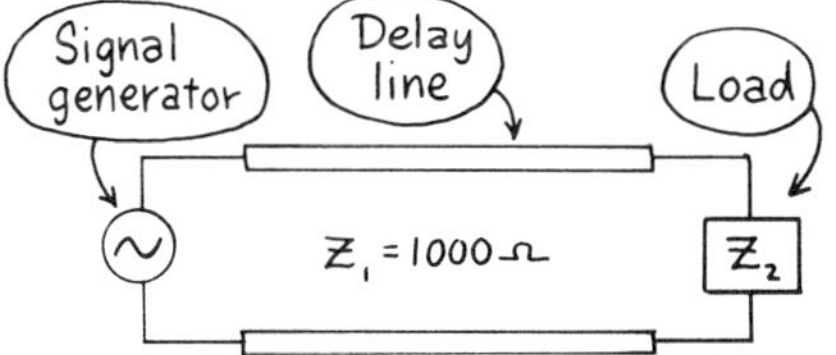

(a) What fraction of the incident energy is reflected from the load?

(b) What is the phase relation between incident and reflected current pulses?

Solution to (a)

$$\begin{Bmatrix} Z_1 = 1000\ \Omega \\ Z_2 = 500\ \Omega \end{Bmatrix} \quad \rho = \frac{Z_1 - Z_2}{Z_1 + Z_2} = \frac{(1000 - 500)\ \Omega}{(1000 + 500)\ \Omega} = \boxed{+0.33}$$

← One-third of the energy is reflected.

Solution to (b)

The positive sign for ρ means that the reflected current pulse is in phase with the incident pulse.

Example 17-11 A 300-Ω two-wire transmission line (a distributed parameter delay line) is terminated by a 4000-Ω load.

(a) What fraction of the incident energy is reflected from the load?
(b) What is the phase relation between incident and reflected pulses?

Solution to (a)

$$\begin{Bmatrix} Z_1 = 300\ \Omega \\ Z_2 = 4000\ \Omega \end{Bmatrix} \quad \rho = \frac{Z_1 - Z_2}{Z_1 + Z_2} = \frac{(300 - 4000)\ \Omega}{(300 + 4000)\ \Omega} = \boxed{-0.86}$$

Solution to (b)

The negative sign for ρ means that the reflected pulse is 180° out of phase with the incident pulse.

17-10 ELECTROMAGNETIC WAVES

Energy that comes to the earth from the sun is transmitted through a medium that is quite different from what we have discussed to this point. This energy flows through almost perfectly empty space. There are no blocks and springs to appeal to for energy-storage duties along the way. *Electromagnetic radiation* includes the sun's energy that we receive, as well as energy from countless other natural and artificial sources—light bulbs, radio transmitters, X-ray machines, hot rocks, and flashes of lightning. It includes light, heat, radio, ultraviolet, and gamma radiation.

This apparently great diversity of energy types is actually just one sort of thing. Experiments show that a complete description of electromagnetic radiation must include a wave-type motion of electric and magnetic fields. This is a big topic, so we will delay the discussion of electromagnetic waves until Chapter 18.

17-11 SUMMARY

We have seen that energy need not be attached to a particular piece of matter in order to travel. When energy moves as a wave, it is handed from one storage element to another down the line like a bucket brigade.

1. The speed v at which a wave propagates depends upon the energy storage characteristics (inertance and capacitance) of the medium through which it travels.

$$\text{wave speed } v = \frac{1}{\sqrt{\begin{pmatrix}\text{inertance}\\ \text{factor}\end{pmatrix} \times \begin{pmatrix}\text{capacitance}\\ \text{factor}\end{pmatrix}}} \tag{17-1}$$

2. Wave pulses can travel either in a longitudinal or transverse form. In a longitudinal pulse the particles making up the medium vibrate parallel to the direction of propagation. In a transverse pulse the particles making up the medium vibrate perpendicular to the direction of propagation. Table 17-1 gives a number of velocity equations for longitudinal and transverse waves in different mediums.

3. Transverse and longitudinal wave trains are produced by disturbing the medium sinusoidally. The distance between corresponding peaks is called the

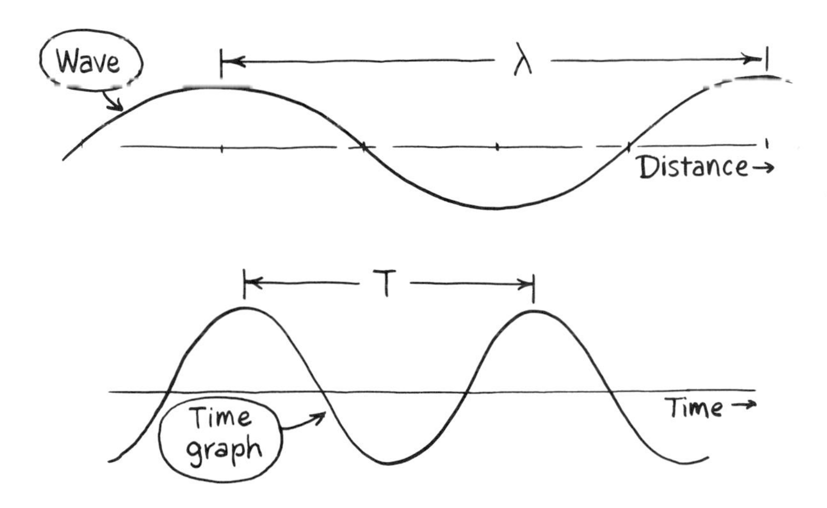

wavelength λ, and the time it takes for successive peaks to pass a given point is the period T. Frequency f is the reciprocal of period. The wave speed v is related to wavelength λ and frequency f.

$$v = f\lambda \tag{17-5}$$

4. Two wave pulses passing a given point in a medium interfere as they overlap to give either destructive or constructive interference. Such interference effects lead to standing waves. Three situations are shown in Table 17-2. Equations for frequency of the fundamental and the overtones are included in the table.

TABLE 17-2 STANDING WAVE FREQUENCIES OF THE FUNDAMENTAL MODE AND ITS OVERTONES

Diagram of the fundamental mode of vibration	$\ell = \lambda/4$ Free end — Fixed end	$\ell = \lambda/2$ Fixed end — Fixed end	$\ell = \lambda/2$ Free end — Free end
Fundamental (1st harmonic)	$f = \frac{1}{4}\frac{v}{\ell}$	$f = \frac{1}{2}\frac{v}{\ell}$	$f = \frac{1}{2}\frac{v}{\ell}$
First overtone (2nd harmonic)	$f = \frac{3}{4}\frac{v}{\ell}$	$f = \frac{2}{2}\frac{v}{\ell}$	$f = \frac{2}{2}\frac{v}{\ell}$
Second overtone (3rd harmonic)	$f = \frac{5}{4}\frac{v}{\ell}$	$f = \frac{3}{2}\frac{v}{\ell}$	$f = \frac{3}{2}\frac{v}{\ell}$
Third overtone (4th harmonic)	$f = \frac{7}{4}\frac{v}{\ell}$	$f = \frac{4}{2}\frac{v}{\ell}$	$f = \frac{4}{2}\frac{v}{\ell}$

5. Delay lines contain inductance and capacitance that is either lumped or distributed. Delay times or pulse velocities can be calculated for these lines.

Lumped parameter:

$$t_d = \sqrt{LC} \tag{17-6}$$

where t_d is the delay time for a given pair of L and C.

Distributed parameter:

$$v = \frac{1}{\sqrt{\left(\frac{L}{\ell}\right) \times \left(\frac{C}{\ell}\right)}} \tag{17-7}$$

where v is the wave velocity.

6. Energy arriving at a junction between two media may be absorbed if a resistive load is present. Wave energy that is not absorbed is transmitted or reflected in amounts depending on the impedances of source and load. Three situations were discussed in some detail.

(a) $Z_1 = Z_2$. The source impedance Z_1 (transmitting medium) and the load impedance Z_2 are equal. All the wave energy is transmitted to the load. There is no reflected wave.

(b) $Z_2 = \infty$. The source is coupled to an infinitely large impedance. No energy is transmitted. The wave is completely reflected out of phase with the incident wave. Examples of this are a Slinky fixed at one end and current in an electrical transmission line that is open circuited.

(c) $Z = 0$. The source is coupled to a load with no impedance at all. No energy is transmitted. The wave is completely reflected in phase with the incident wave. Examples of this are a Slinky free at one end, and current in an electrical transmission line that is short circuited. For situations in between these extremes, use the coefficient of reflection ρ to calculate the fraction of energy reflected.

$$\rho = \frac{Z_1 - Z_2}{Z_1 + Z_2} \tag{17-8}$$

7. Useful wave velocities to remember are:

(a) The speed of sound in air at sea level and 0°C is $v = 331$ m/s = 1086 ft/s.

(b) The speed of light in a vacuum is $c = 3.00 \times 10^8$ m/s = 186,000 mi/s.

PROBLEMS

17-1. What is the speed of a longitudinal wave in aluminum?

17-2. A guitar string is stretched by a force of 500 N. The length of the string is 80 cm and its mass is 10 g. What is the speed of a transverse wave in the string?

17-3. How fast does sound travel in water?

17-4. What is the speed of sound in mercury?

17-5. A piano string (wire) 1.5 m long has a mass of 5.0 g.

(a) What must be the tension in the wire to produce a transverse wave speed of 440 m/s?

(b) What will be the speed if the tension is doubled? Will it be twice as much?

17-6. How long does it take for a shear wave to propagate through 20 km of granite? Granite has a density ρ of 2.7×10^3 kg/m^3 and a shear modulus G of 3.0×10^{10} N/m^2.

17-7. The Golden Gate Bridge is about 6000 ft long. How soon after a car drives onto one end of the bridge can the sound of the car be heard at the other end (a) by listening in the normal fashion and (b) by putting your ear to the steel frame of the bridge?

17-8. Calculate the velocity of an electromagnetic wave in a vacuum using the equation listed in Table 17-1 and values for permeability and permittivity found in the Table of Physical Constants in Appendix A. How does the calculated value compare with the measured speed of 3.00×10^8 m/s?

17-9. If ocean waves approach the shore at a rate of one every 10 s and if the crests are separated by 50 ft, what is their speed?

17-10. What is the wavelength of electromagnetic (radio) waves of frequency 560 kHz? Electromagnetic waves travel at the speed of light.

17-11. What is the frequency of the "30-metre" short-wave broadcast band?

17-12. An X-ray machine produces waves of length 1.54×10^{-10} m. What is the frequency of these waves?

17-13. An organ pipe open at both ends is 0.70 m long.

(a) What is the wavelength of the fundamental mode of vibration?

(b) What is the fundamental frequency?

17-14. (a) What is the wavelength of the third overtone of the pipe in Problem 17-13?

(b) What is its frequency?

17-15. An air column, closed at one end, is

1.00 m long.

(a) What is the wavelength of the fundamental mode of vibration?

(b) What is the fundamental frequency?

17-16. A tuning fork vibrates at a natural frequency of 264 Hz (middle C). The fork is struck and held over the open end of a glass cylinder 100 cm long partially filled with water. How high (depth h) must the water column be so that the air remaining in the cylinder reinforces the sound from the tuning fork?

17-17. The speed of a transverse wave in a piano string is 2000 ft/s. The string, fixed at both ends, has a length of 34.0 in. What is the frequency of the fundamental mode if this string is plucked?

17-18. What is the frequency of the second harmonic in Problem 17-17? Sketch the standing wave pattern for this mode of vibration and show where the nodes and antinodes are located.

17-19. What must be the length of a glass tube open at both ends for the air inside the tube to reinforce the sound from a vibrating 330 Hz tuning fork? Report your answer in metres.

17-20. An electromagnetic delay line consists of two capacitors each of 0.20 μF and two coils each of 5.0 mH. How long does it take a signal to get through the line?

17-21. How many coil–capacitor pairs are needed to delay the transmission of an electrical signal by at least 170 μs? Each coil is 25 mH and each capacitor is 0.10 μF.

17-22. A delay line contains 0.50 μF of distributed capacitance. What is the value of the line's distributed inductance if the line delays a signal by 4.0 μs?

17-23. What is the total capacitance of 100 km of transmission line that has 0.050 mH of inductance per metre of length if an electrical signal travels at 0.90 times the speed of light in the line?

17-24. A transmission line with an impedance of 4000 Ω is connected to a load of 2000 Ω.

(a) What fraction of the energy is reflected?

(b) What is the phase relation between incident and reflected current pulses?

17-25. The impedance of a transmission line is 20×10^3 Ω. What impedance must a load have to absorb 75 percent of the energy from the transmission line?

17-26. The mass of a Slinky is 250 g and a force of 1.0 N will stretch it to a length of 2.0 m.

(a) Calculate the speed of a transverse pulse in the Slinky when it is in this condition.

(b) What is the pulse speed if the Slinky is stretched to a length of 4.0 m? (*Hint*: Consider the Slinky as acting like a stretched string.)

ANSWERS TO ODD-NUMBERED PROBLEMS

17-1. 5.1×10^3 m/s

17-3. 1.5×10^3 m/s

17-5. (a) 6.5×10^2 N
(b) 6.2×10^2 m/s; no, it will increase by $\sqrt{2}$

17-7. (a) 5.5 s
(b) 0.35 s

17-9. 5.0 ft/s

17-11. 1.0×10^7 Hz

17-13. (a) 1.4 m
(b) 240 Hz

17-15. (a) 4.00 m
(b) 82.8 Hz

17-17. 353 Hz

17-19. 0.502 m

17-21. At least 4 pairs are needed

17-23. 2.7×10^{-8} F

17-25. 12×10^3 Ω

18

Radiated Energy

This entire text has been concerned with the transfer, storage, and control of energy. A major stone in our structure is the idea that energy is conserved. This idea was formally introduced in Chapter 9 and expended in Chapter 10 to the form of Eq. (10-9).

$$\begin{pmatrix}\text{energy added}\\ \text{to system}\end{pmatrix} = \begin{pmatrix}\text{energy lost}\\ \text{by resistors}\end{pmatrix} + \begin{pmatrix}\text{potential energy}\\ \text{stored}\end{pmatrix} + \begin{pmatrix}\text{kinetic energy}\\ \text{stored}\end{pmatrix} \quad (10\text{-}9)$$

Now it can be told: *Eq. (10-9) is not quite correct because in some systems energy can be lost through a process called radiation. We will look briefly at radiation in this last chapter, find out the conditions that promote it, and investigate how this radiated energy is carried away from its source. The ideas about waves in Chapter 17 must be invoked, but they will not do the whole job. The concept of a field (Chapter 8) and particularly the relation between electric and magnetic fields will help you to build a mental framework upon which to hang the pieces of your understanding. In addition, what you learned in Chapters 9 and 10 about the energy of particles is needed to explain some of the behavior of radiated energy.*

18-1 CONSERVATION OF ENERGY REVISITED

Let us look at a simple electrical system—a flashlight. It is an example of a resistive system from which energy is lost in the form of heat generated in the filament of a lamp. If we look closely at the system and do some careful measurements, an interesting fact appears. Meters in the circuit of Fig. 18-1 measure the energy supplied to the lamp in a given time interval. Suppose also that we are able to measure the amount of heat produced in the lamp's

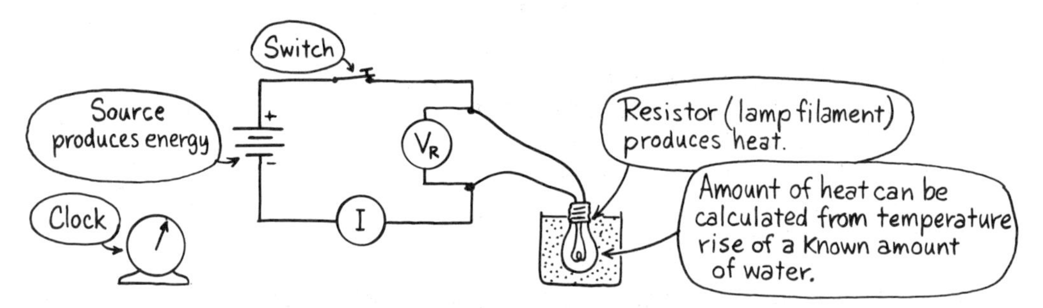

Figure 18-1. Circuit diagram of a flashlight. The amount of heat produced in the bulb is *less* than the energy supplied by the battery.

filament. This is not very hard to do. If the lamp (electrically insulated) is immersed in water, the heat produced raises the water's temperature. The heat produced Q can be calculated using Eq. (3-14), $Q = mc(T_f - T_o)$, where m is the water's mass and $T_f - T_o$ the change in temperature.

Such experiments have been done on all sorts of lamps and the result is always the same. The heat energy produced is always a bit less than the electrical energy put in. What happened? Is energy not conserved? Sure it is. That little bit of leftover energy comes out of the system mostly in the form of light, which is the reason for having the flashlight in the first place. The light is *radiated* energy, and the conservation of energy equation (10-9) must be modified to take this into account:

$$\begin{matrix}\text{energy} \\ \text{added to} \\ \text{system}\end{matrix} = \begin{matrix}\text{energy} \\ \text{lost by} \\ \text{resistors}\end{matrix} + \begin{matrix}\text{potential} \\ \text{energy} \\ \text{stored}\end{matrix} + \begin{matrix}\text{kinetic} \\ \text{energy} \\ \text{stored}\end{matrix} + \begin{matrix}\text{energy} \\ \text{lost by} \\ \text{radiation}\end{matrix} \qquad (18\text{-}1)$$

As we shall soon see, even the heat lost by the resistor might be included as radiant energy. For the sake of our discussion, however, we will consider it separately.

Light energy from the flashlight amounts to only 1 or 2 percent of the total energy lost. The light-producing efficiency of incandescent lamps is small, and you won't be far wrong ignoring the radiation term of Eq. (18-1). Other sorts of objects may radiate energy more efficiently. Objects can be made that are nearly perfect radiators so that the radiation term becomes the most important.

Other examples of radiated energy include all the heat, light, ultraviolet, radio waves, and so forth that come to us from our most important source, the sun. Modern technology depends upon radiated energy in many ways. One example is radio and television signals that are purposely "lost" from transmitter antennas by radiation into surrounding space. Some of that energy is gathered up by the antennas of receivers to be amplified, and we hear or see the result.

As was pointed out at the end of Chapter 17, this kind of energy transfer does not require any substance—mechanical springs and blocks—to travel in. The flashlight shines even better through a vacuum than through space occupied by air or water. Certainly, the sun's energy and that of all the other stars comes to us through nearly empty space. This kind of radiant energy is called *electromagnetic radiation.* It is the subject of the rest of this chapter.

18-2 WHAT CAUSES ELECTROMAGNETIC RADIATION?

Physicists have done a lot of thinking and experimenting on electromagnetic radiation and they have found that it is produced only when electric charge is accelerated. Radio signals for example are produced when charge is made to surge back and forth in a conductor (antenna). If, as in Fig. 18-2a, the antenna is used as a statically charged capacitor, it attracts opposite charges very nicely; but like a stretched spring, it only stores potential energy as long as it remains charged (stretched) and no energy is lost.

If a constant current passes through the antenna (Fig. 18-2b), the system stores energy in the constant magnetic field surrounding the antenna. Energy

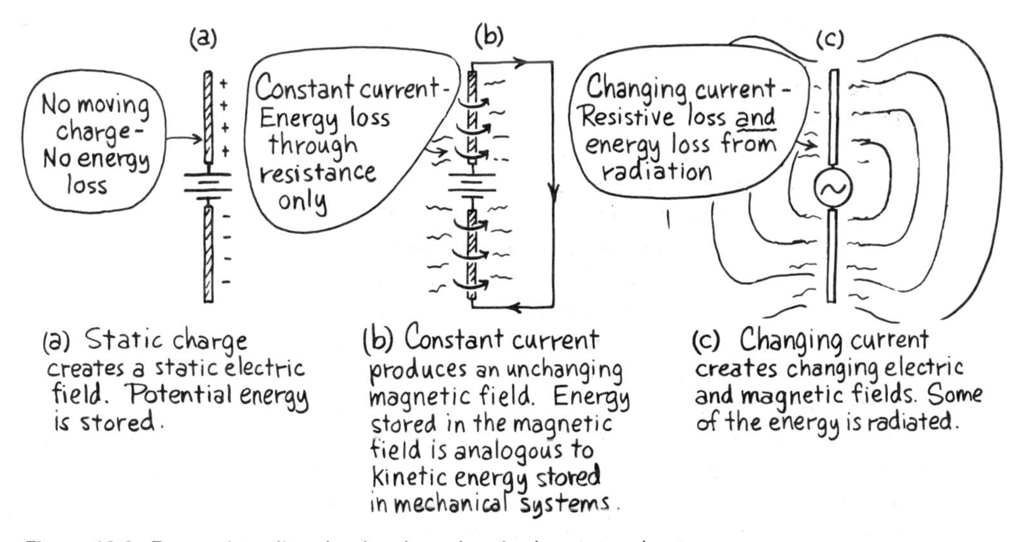

Figure 18-2. Energy is radiated only when electric charge accelerates, and that happens only when current changes.

stored in the magnetic field is analogous to kinetic energy stored in mechanical and fluid systems (Secs. 10-2 and 10-3). The only energy loss is to resistance in the wire.

Experiments show that energy is radiated from such a setup only when the current *changes*. If an ac source is provided, as in Fig. 18-2c, current surges in one direction and then the other. Charge in the conductor is continually accelerating, and the system radiates energy.

The charge does not have to be moving through a wire as in a radio antenna. For example, infrared (the kind of radiation that produces heat when it strikes a surface) is radiated from hot surfaces. Remember (Sec. 3-5) that heat energy is really the kinetic energy of vibrating molecules. Infrared radiation is attributed to the attached charges that are accelerated as they move back and forth with the molecules.

Light is the name for electromagnetic radiation to which our eyes are sensitive. It appears to result from electrons bouncing around within atoms—not from things as large as molecules or even whole atoms.

Whatever the scale or cause, electromagnetic radiation is produced by accelerating charge. A charge can be accelerated in one direction only for so long. In order to sustain the emission, it is necessary to bring the charge back and forth along the same path. Thus sources of radiant energy are necessarily systems that oscillate, and it is natural to talk about the frequency of a radiation source. The difference between one type of electromagnetic radiation and another is attributable to the frequency of its source. Table 18-1 summarizes the frequency ranges of the *electromagnetic spectrum* along with some sources and detectors.

Energy from an oscillating source travels away from the source, a situation reminiscent of Chapter 17. Indeed, as in Sec. 17-5, where the end of a Slinky was shaken, the energy transported away from the source seems to behave like a wave in many respects. The wave propagates or moves in a vacuum at a velocity of 3.00×10^8 m/s. Wave and velocity aspects of electromagnetic radiation are discussed in the next sections.

Different types of electromagnetic radiation may be identified by wavelength as well as by frequency. Wavelength λ and frequency f are related to the velocity c of electromagnetic radiation, which is 3.00×10^8 m/s in a vacuum. The simple relation between them is given by Eq. (17-5), $c = f\lambda$.

A source of radiation normally only converts energy from one form to another. The filament of an incandescent lamp is no more of a light source than anything else until electric charge is forced through it by a source of electrical energy. The flowing electrons jostle other electrons in the filament, causing them to vibrate and emit the energy by radiation. Every object above absolute zero (zero on the Kelvin temperature scale) is continually absorbing and radiating energy. If it absorbs faster than it radiates, we say that the object is heating up. A cooling object is radiating energy faster at that moment than it is absorbing. An object at constant temperature radiates at the same rate that it absorbs energy.

TABLE 18-1 THE ELECTROMAGNETIC SPECTRUM

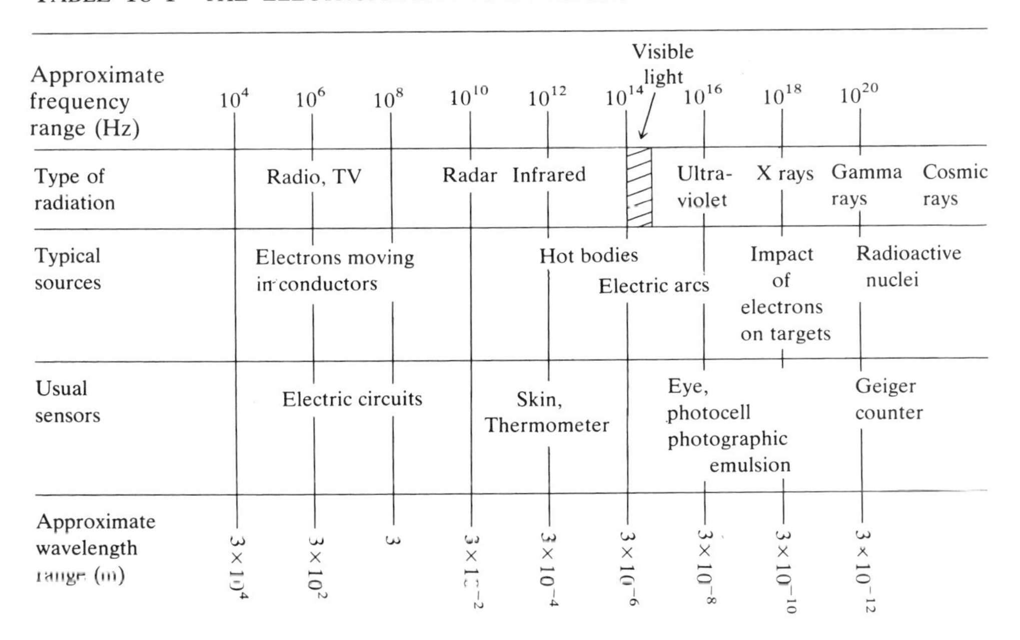

Approximate frequency range (Hz)	10^4, 10^6, 10^8, 10^{10}, 10^{12}, 10^{14}, Visible light, 10^{16}, 10^{18}, 10^{20}
Type of radiation	Radio, TV; Radar; Infrared; Ultra-violet; X rays; Gamma rays; Cosmic rays
Typical sources	Electrons moving in conductors; Hot bodies; Electric arcs; Impact of electrons on targets; Radioactive nuclei
Usual sensors	Electric circuits; Skin, Thermometer; Eye, photocell photographic emulsion; Geiger counter
Approximate wavelength range (m)	3×10^4, 3×10^2, 3, 3×10^{-2}, 3×10^{-4}, 3×10^{-6}, 3×10^{-8}, 3×10^{-10}, 3×10^{-12}

18-3 THE SPEED OF LIGHT

A series of ingenious experiments in combination with some high-powered mathematical modeling has told physicists a great deal about the behavior of radiated energy after it leaves its source. We will look at two aspects of this traveling energy, its speed and its style. In this and succeeding sections when we talk about "light" we really mean all kinds of electromagnetic radiation. Keep that in mind. X rays and radio waves, for example, bend or show interference effects, and they travel through space at the same speed as visible light.

It is a bit disconcerting for students just being introduced to the topic to realize that the speed of light was first successfully measured more than 300 years ago. In 1675, a full century before the American Revolution, a Danish astronomer, Olaus Roemer, measured the time for light from a moon of Jupiter to travel the 186,000,000-mi diameter of the earth's orbit about the sun. He got an answer of 16 min or 960 s, producing a velocity of $186 \times 10^6 \text{ mi}/960 \text{ s} = 194{,}000 \text{ mi/s}$.

About 50 years later (still more than 250 years ago), another and more accurate astronomical method produced a value of 186,300 mi/s. In the ensuing centuries the value has been refined but very little changed. The presently accepted value for the speed of light in free space is about

$$c = 1.86 \times 10^5 \text{ mi/s} = 3.00 \times 10^8 \text{ m/s}$$

Notice that the value given is for a vacuum. This brings up the point that the speed of light is different in different substances. Generally, although not always, the denser a material, the slower light travels in it. It travels fastest in a vacuum (i.e., where no matter is present), but only about $\frac{3}{100}$ percent slower in air. In water the speed is about 75 percent as much (2.3×10^8 m/s). In ordinary glass the speed is down to 65 percent or around 2.0×10^8 m/s. Diamond slows light down to about 42 percent of its maximum speed (1.3×10^8 m/s), and in some natural substances the speed is less than 25 percent of its free space value. It turns out to be rather easy to find the speed of light in a given material once you know it for air (and that value has been with us for over three centuries). The method employs the refraction of light and is discussed in the next section.

18-4 LIGHT SEEMS TO TRAVEL AS A WAVE

The next question to be asked is about the form or style of electromagnetic radiation. There are two logical choices. Light might be delivered in separate little packages or particles of energy like bullets from a machine gun, or it might be some sort of wave phenomenon, as treated in Chapter 17. In fact, as we shall see, electromagnetic radiation shows unmistakable signs of being both things at the same time. If this sounds strange, so be it. In the present state of knowledge about the topic, it is necessary to invoke both wave and particle models to predict behavior under different circumstances.

In this section we will make the case for electromagnetic radiation as waves by studying the *refraction*, *reflection*, and *interference* of light. Having just gone through Chapter 17, it is quite natural to think in terms of waves. It is particularly tempting since it appears that light, produced by an oscillating source, has both a frequency and a wavelength associated with it.

Refraction and Reflection

As light (or other electromagnetic radiation) goes from one medium into another, it changes direction. This bending is called *refraction*. The design of telescopes, microscopes, cameras, eye glasses, and other optical devices depends upon it.

Electromagnetic radiation moves too fast and flows too continuously to really see what is going on that causes a change in direction. The same thing happens, however, to other kinds of transmitted energy. A slower, more easily observable example is what happens to ocean waves as they approach a shoreline.

Water surface waves are quite complicated, but they travel slowly enough so that we can get a good look at them. As shown in Fig. 18-3, surface waves slow down as the water becomes shallower. If waves approach the shore at an angle, the part of a wave to reach shallow water first slows up and the deep

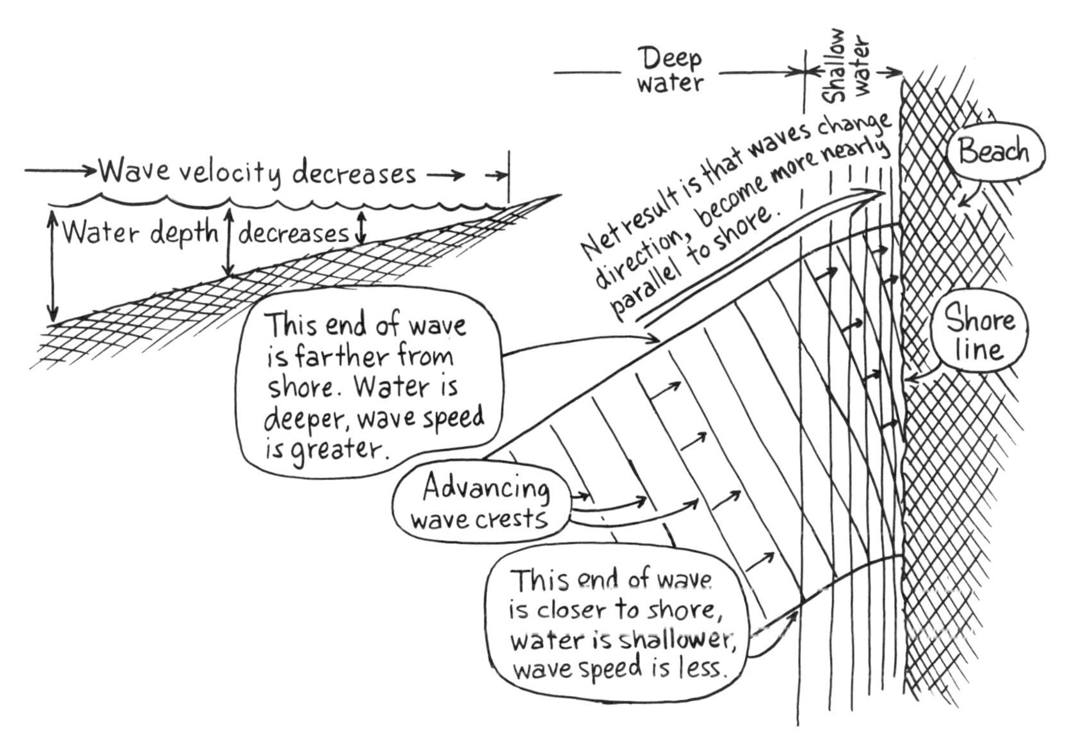

Figure 18-3. Ocean waves change direction (refract) when they approach shallow water at an angle.

water end begins to catch up. The net effect, readily observable in any large body of water, is that the waves change direction, invariably coming in more nearly parallel to the shoreline.

The same thinking seems to work well in predicting what light will do when entering a region of lower velocity. If the velocity change is gradational, as with the ocean wave analogy, exactly the same argument applies. Or the velocity change may be abrupt, as when light enters water or glass. In such a case the analysis is simpler. Figure 18-4 shows a beam of light entering and leaving a flat piece of glass at an angle. The change in direction is simply related to change in speed. The light moving at speed v_1 in air travels distance $s_1 = v_1(t_2 - t_1)$ while distance $s_2 = v_2(t_2 - t_1)$ is covered by the part of the beam moving at speed v_2 in glass.

The ratio of distances is equal to the ratio of speeds.

$$\frac{s_1}{s_2} = \frac{v_1(t_2 - t_1)}{v_2(t_2 - t_1)} = \frac{v_1}{v_2}$$

But these distances s_1 and s_2 cannot be measured directly, so it is more useful

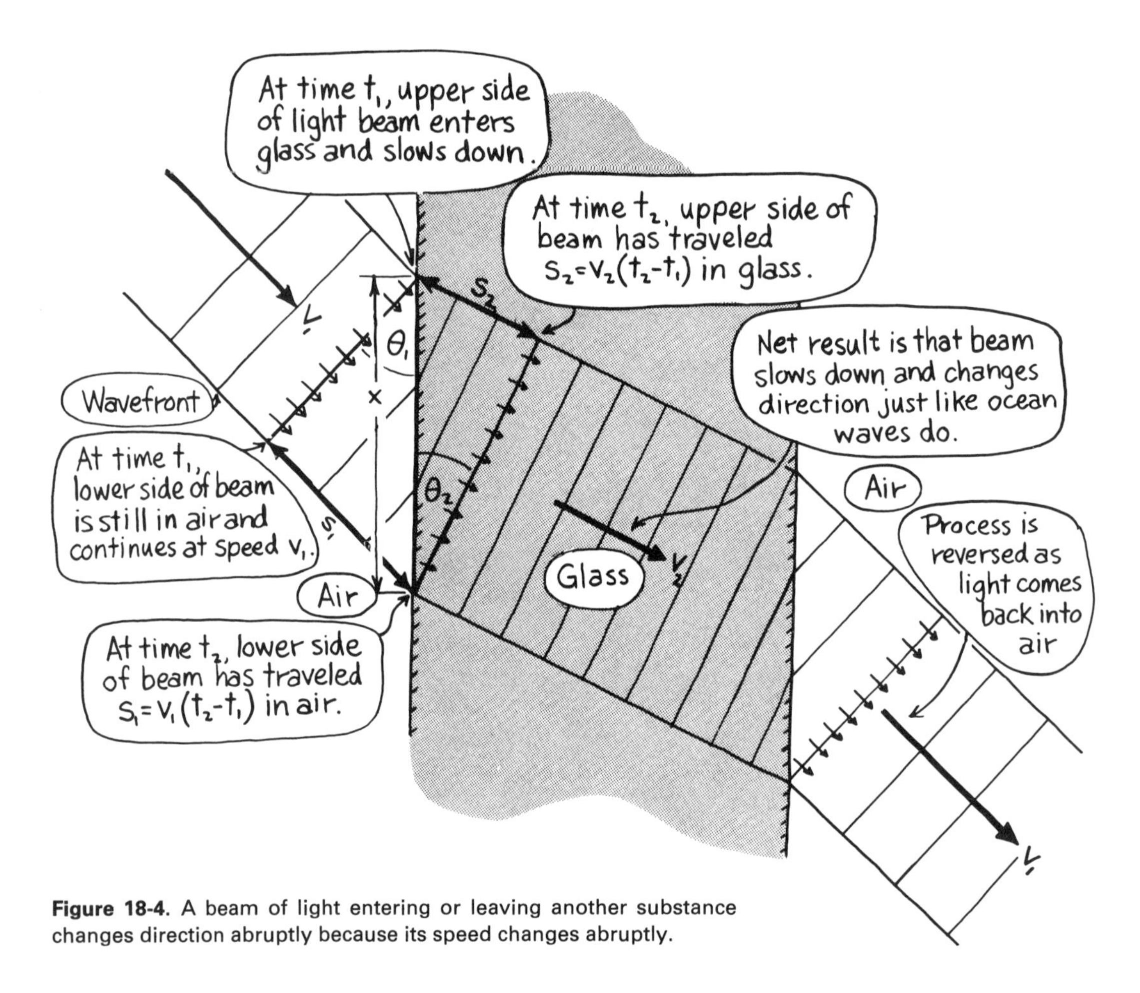

Figure 18-4. A beam of light entering or leaving another substance changes direction abruptly because its speed changes abruptly.

to relate velocity to the angle of incidence θ_1 and the angle of refraction θ_2:

$$s_1 = x \sin \theta_1 \quad \text{and} \quad s_2 = x \sin \theta_2$$

$$\frac{s_1}{s_2} = \frac{x \sin \theta_1}{x \sin \theta_2} = \frac{\sin \theta_1}{\sin \theta_2} = \frac{v_1}{v_2}$$

Thus when light passes from one medium into another we can depend on the fact that

$$\frac{v_1}{v_2} = \frac{\sin \theta_1}{\sin \theta_2} \tag{18-2}$$

Equation (18-2) is called *Snell's law.*

Sometimes it is easier to visualize refraction by looking only at one ray, which is a line perpendicular to the wave front. Figure 18-5 shows incident and refracted rays. A ray behaves just like a tiny flashlight beam. The angle made by a wave front and the surface between the two mediums is the same as the angle made by the ray and a line perpendicular to the surface.

Figure 18-5b also shows a *reflected* ray. The angle of reflection is the same as the angle of incidence. The amount of energy in the reflected and transmitted (refracted) wave depends upon the nature of the two mediums and the angle of incidence. Reflected and transmitted waves were discussed in Sec. 17-6. See if you can sketch a reflected wave front on Fig. 18-5.

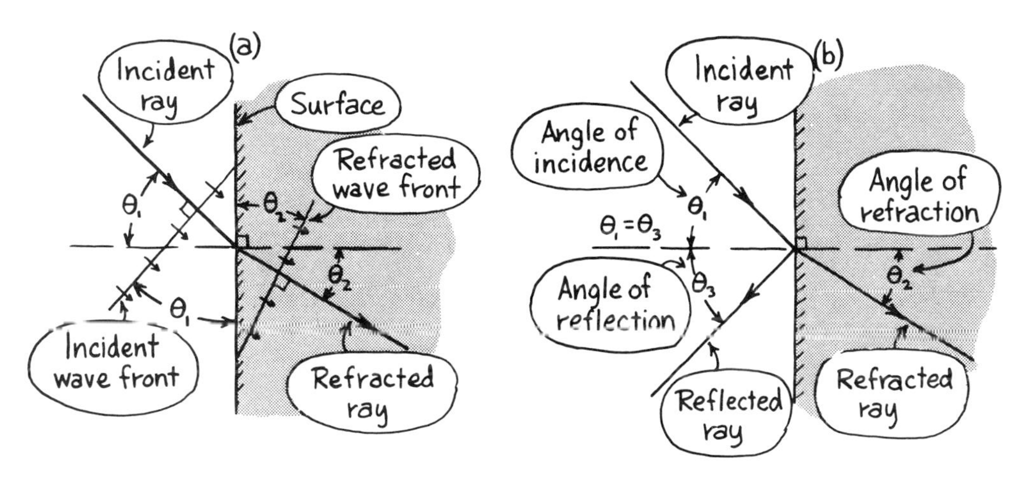

Figure 18-5. Refraction may be easier to see using rays rather than wave fronts. Rays are perpendicular to wave fronts. (a) Incident and refracted rays. (b) Incident and reflected rays. The angle of incidence θ_1 equals angle of reflection θ_3.

The speed of light in a substance can easily be found by measuring angles of incidence and refraction, and applying Eq. (18-2). The ratio of the speed of light c in a vacuum to the speed of light v in a given material is called the *refractive index* n of the material.

$$n = \frac{c}{v} \qquad (18\text{-}3)$$

Refractive index of material (n); Speed of light in vacuum (c); Speed of light in material (v)

The refractive index has been determined accurately for most transparent materials. Some of these values are recorded in Appendix B, Table 9, at the back of the book. Example 18-1 illustrates a practical application.

Example 18-1 A beam of light passes at an angle of incidence θ_1 of 30.0° from air into a sugar solution of unknown concentration. The angle of refraction θ_2 is measured at 20.5°.

(a) What is the refractive index of the solution?
(b) What is the speed of light in the solution?

Solution to (a)

$\begin{Bmatrix} \theta_1 = 30.0° \\ \theta_2 = 20.5° \end{Bmatrix}$

$$n = \frac{c}{v} = \frac{\sin\theta_1}{\sin\theta_2} = \frac{\sin 30.0°}{\sin 20.5°} = \frac{0.500}{0.350} = \boxed{1.43}$$

Solution to (b)

$\{c = 3.00 \times 10^8 \text{ m/s}\}$

$$n = \frac{c}{v} \Rightarrow v = \frac{c}{n} = \frac{3.00 \times 10^8 \text{ m/s}}{1.43} = \boxed{2.10 \times 10^8 \text{ m/s}}$$

Once the refractive index is known, the sugar concentration in the water can be found by consulting tables of refractive index for various sugar concentrations. In this case the solution is 55 percent sugar.

Interference

The close analogy between light and ocean waves is a compelling bit of circumstantial evidence to support the theory that light travels as a wave. If something looks like a duck and swims like a duck, there is a strong inclination to think of it as a duck even though it may turn out to be a decoy.

Interference, as introduced in Sec. 17-7, is another convincing argument for the wave model of electromagnetic radiation. Reflection of a wave in a mechanical spring at a "hard" surface produced a phase reversal. A positive pulse is reflected as a negative one, and vice versa. No phase reversal of the reflected wave occurred at a "soft" surface. Nor was any phase change noted for the transmitted wave. The same is true for electromagnetic radiation. Incident light waves traveling in air undergo a phase reversal when reflected from a "hard" surface, but no reversal occurs when light travelling in glass reflects from a "soft" surface (Fig. 18-6). In optical terms, a "hard" surface is one in which light enters a lower-velocity medium. Since $v = f\lambda$, either frequency f or wavelength λ must decrease if velocity decreases. Experiments show that it is the wavelength that decreases. Refracted waves in both cases, however, undergo no phase reversal.

A number of interesting and useful interference phenomena can be observed and seem to make sense only if light is thought of in terms of waves. One important application involves the use of X rays to investigate the arrangement of atoms in crystalline solids such as metals, ceramic materials, and natural minerals. Crystals feature very orderly latticeworks of atoms, in contrast to other materials such as glass in which atoms form no regular pattern. Figure 18-7 illustrates what happens if a beam of X rays is directed at

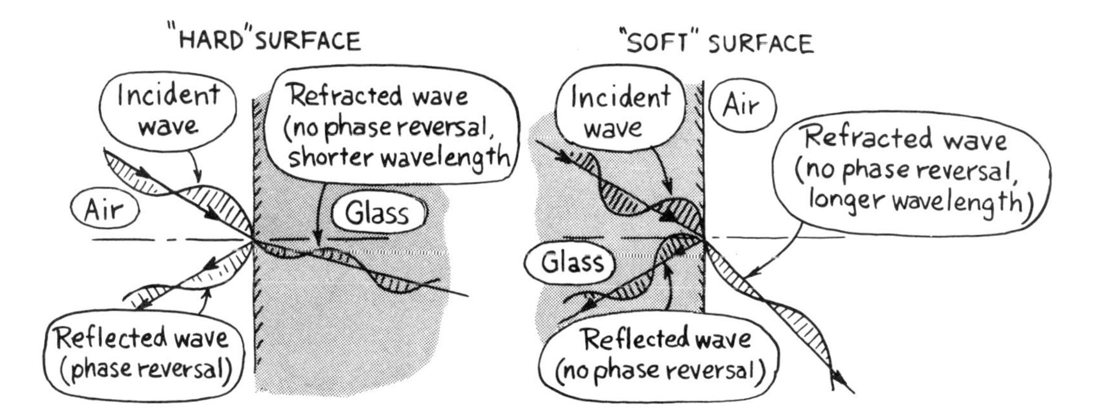

Figure 18-6. Reflected waves undergo phase reversal at a "hard" air-to-glass surface but not at a "soft" glass-to-air surface. Refracted waves change their wavelength but undergo no phase reversal.

the surface of a crystal at an angle θ to the surface. A detector such as a scintillation counter is placed so that it picks up radiation emitted from the surface at the same angle θ. The graph shows how the intensity of the emitted radiation changes as angle θ changes. Angles θ for source and detector are kept equal for the whole experiment.

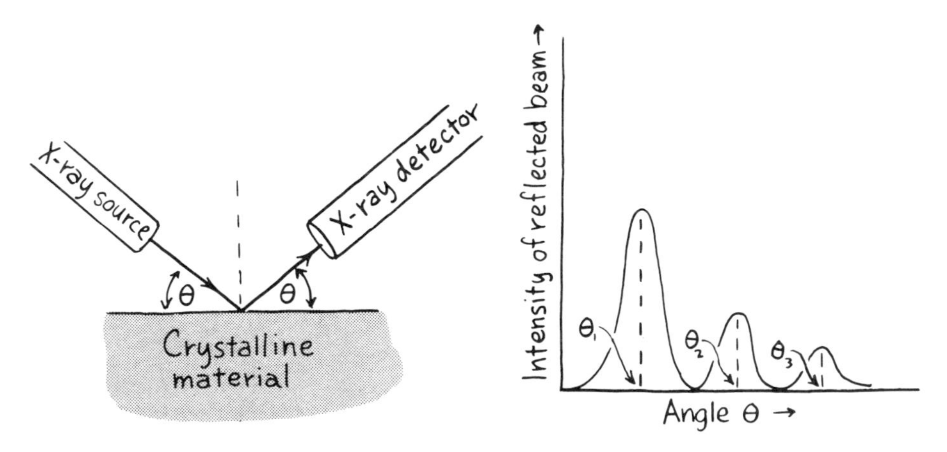

Figure 18-7. A beam of X rays incident on a crystalline material is "reflected" (reemitted) at high intensity only at certain angles. At intermediate angles the reflected beam is of nearly zero intensity.

Such behavior can be explained if we assume that the X-ray energy is transmitted as a wave. Various kinds of evidence indicate that the atoms of many crystals form discrete layers. These layers can be thought of as surfaces from which the X rays reflect at the same angle that they came in. Ray B in

Fig. 18-8, reflecting from a layer deep in the crystal, must travel farther than ray A. If the X rays are waves, A and B will emerge in phase only when the extra distance is one or more whole wavelengths. If the extra distance is not a whole multiple of wavelength λ, each reflected ray will be accompanied by

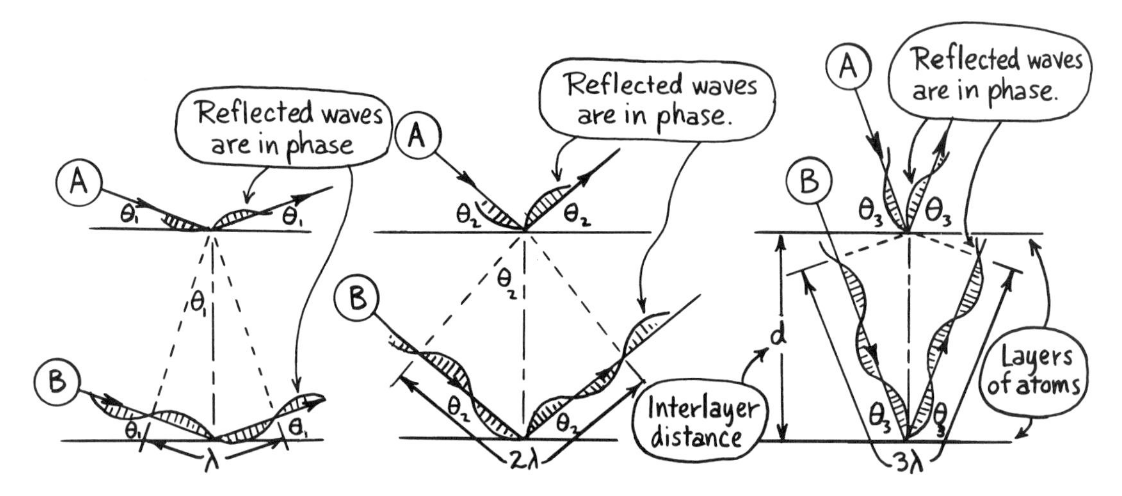

Figure 18-8. Layers of atoms in a crystal effectively act as reflecting surfaces. The peaks of the graph in Fig. 18-7 can be explained if the X rays are thought of as waves. Intensity peaks correspond to constructive interference that happens only at very definite values of θ.

another ray, from some layer deeper in the crystal, that is exactly 180° out of phase. Thus, according to this scheme, reflected intensity should be zero between peaks. That is exactly what is found. Example 18-2 illustrates a practical application.

Example 18-2 The clay mineral halloysite is known to have an interlayer distance d between layers of atoms of 1.01 nm (1.01×10^{-9} m). A sample of halloysite is bombarded with X rays at various angles θ starting near zero. The first intensity peak is detected when $\theta = 4.37°$. What is the wavelength λ of the X rays used?

Solution

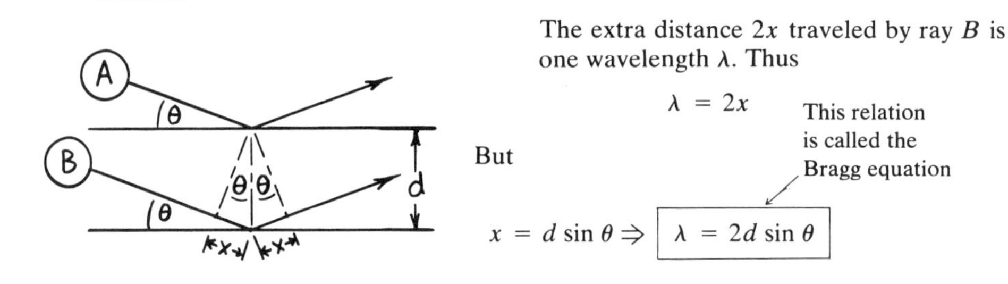

The extra distance $2x$ traveled by ray B is one wavelength λ. Thus

$$\lambda = 2x$$

But

$$x = d \sin\theta \Rightarrow \boxed{\lambda = 2d\sin\theta}$$

This relation is called the Bragg equation

Therefore,

$$\lambda = 2(1.01 \text{ nm})(\sin 4.37°) = 2(1.01 \text{ nm})(0.0762) = \boxed{0.154 \text{ nm}}$$

Once the wavelength of a given X ray is known, the method can be used in reverse to determine the interlayer distance d of unknown materials. This is a very powerful and much used tool in the identification of crystalline substances.

18-5 OPTICAL INSTRUMENTS

The phenomena of reflection and refraction are exploited in a number of useful ways by devices as simple as your bathroom mirror or as complex as a high-resolution microscope. We will take a look in this section at a few relatively simple optical elements that form the basis of more complicated optical instruments.

Plane Mirrors

Everyone knows what a plane (flat) mirror does, but not everyone can explain why. When you look into a plane mirror, you see an accurate reproduction of yourself and your surroundings. Try it and you will note, as shown in Fig. 18-9, that (1) the image is the same size as you are, and (2) the image appears to be as far in back of the mirror as you are in front of it. This well-known relationship can be predicted knowing one simple fact: the angle of reflection is always equal to the angle of incidence. Figure 18-5b showed equal angles of incidence and reflection measured from the normal (perpendicular) to the surface. This relation is referred to as the basic *law of reflection*.

In Fig. 18-10 we direct your attention to the point of your chin. The lines emanating from point A on your chin represent only three of an infinite

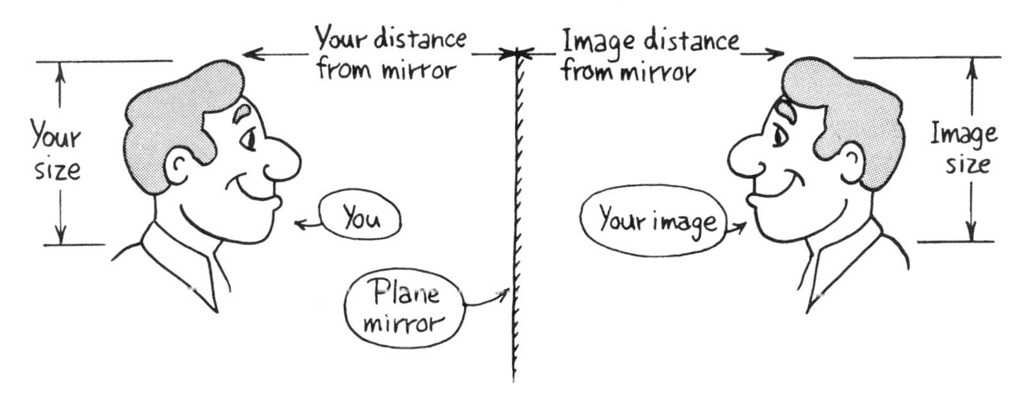

Figure 18-9. Your image in a plane mirror is the same as you are—no better, no worse.

number of rays that go out in all directions. When a ray strikes the mirror, it bounces off according to the basic law of reflection. Each of the three observers (including you) perceives a reflected ray as coming from the same point A'. Each observer or an observer looking from any other angle is fooled into thinking that a chin exists at A'. The chin isn't really there of course. It only looks like it because the light *seems* to be coming from that spot. The resulting illusion is called a *virtual image.* Rays seem to come from but do not actually pass through a virtual image. A real image (to be studied later) is formed by rays that actually pass through it.

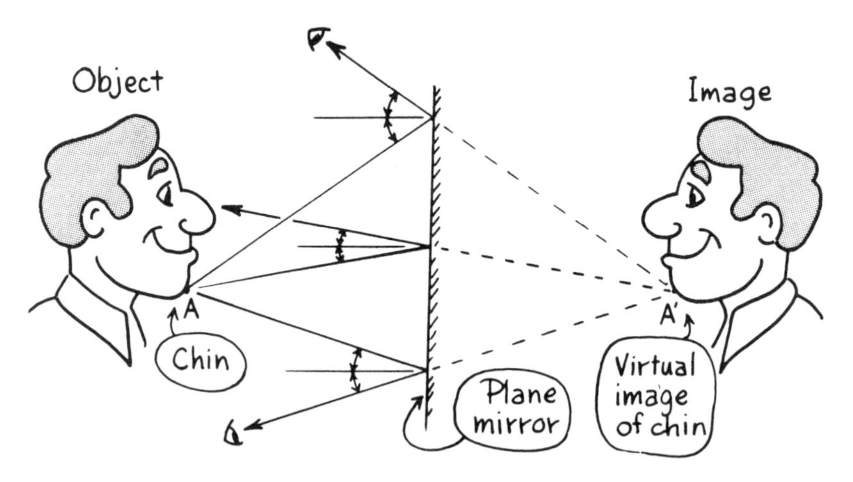

Figure 18-10. Regardless of where an observer is, light from point A always seems to be coming from point A'. A' is a virtual image of A.

It takes only simple geometry to show that the virtual image at A' is as far behind the mirror as the chin at point A is in front. If the same construction is done with some other point (an ear lobe, perhaps) it becomes clear that the image is the same size as the real thing.

Curved Mirrors

If a reflecting surface is not flat, the basic law of reflection still holds. In such a case the normal is perpendicular to the tangent to the curved surface, as shown in Fig. 18-11. Curved mirrors come in two general styles, concave and convex, as illustrated in Fig. 18-12. The diagrams that we will use are necessarily in two dimensions, but don't forget that real life is in three dimensions.

Concave spherical mirrors. Figure 18-13 illustrates some of the features of a concave spherical mirror. An incident light ray close and parallel to the mirror's principal axis reflects and passes through a point called the *principal focus* or *focal point F. Focal length f* is the distance from the mirror to the

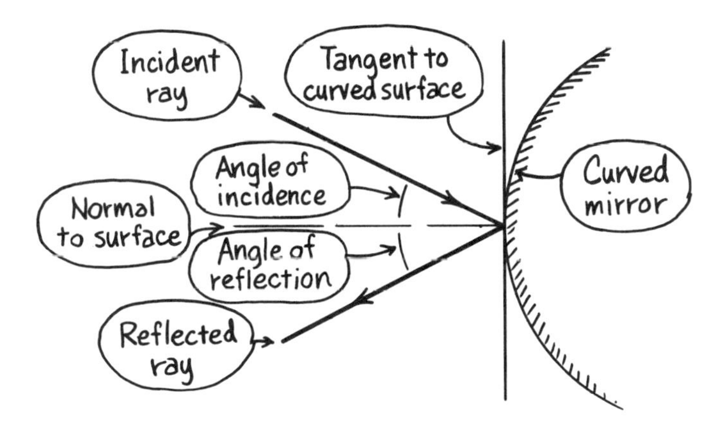

Figure 18-11. Curved reflecting surfaces follow the basic law of reflection.

Figure 18-12. Concave and convex mirrors.

principal focus. It is a simple job of geometry to show that, if the angle of incidence is small, focal length f is half of the radius r. We won't go through the details of that proof.

Figure 18-14 illustrates five situations in which an object (an arrow for simplicity) is placed in front of a concave spherical mirror. The image of the arrow's point is located by tracing at least two rays from that point. Two convenient rays are (1) a ray parallel to the axis that will reflect back through the principal focus F, and (2) a ray heading along a normal toward the center of curvature C that reflects back along the same path. Remember that these rays are only two of an infinite number emanating from the point of the arrow. They are chosen for the diagrams only because they are easy to construct. You can take it on faith or try it yourself to prove that any other ray from the arrow's point will pass through the same image point after it is reflected.

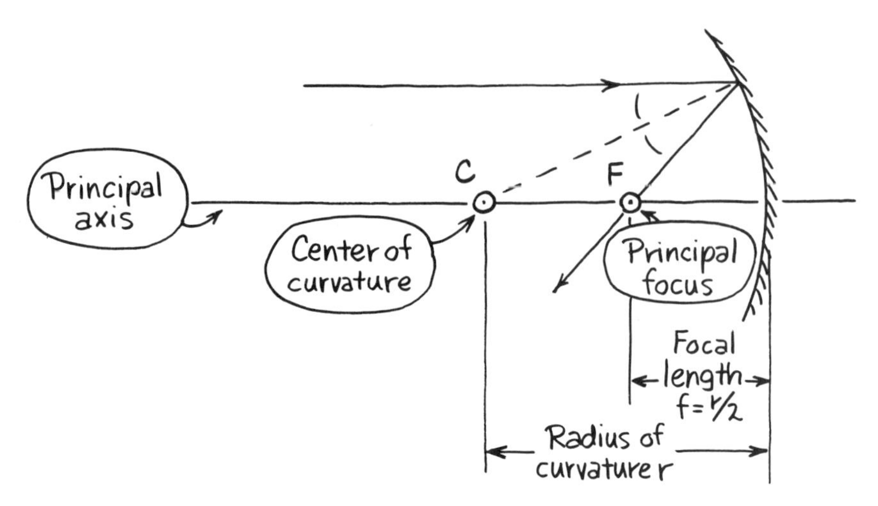

Figure 18-13. Geometric elements of a concave spherical mirror. Focal length f is half of the radius of curvature r.

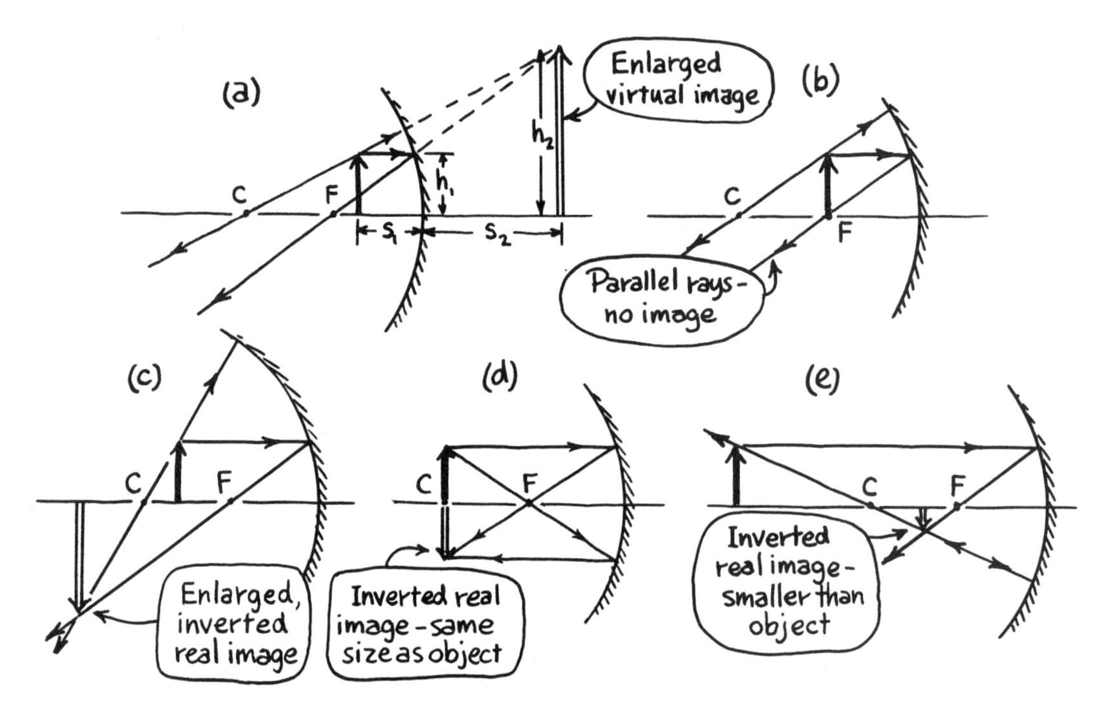

Figure 18-14. The size, position and nature of the image formed by a concave mirror depends on the location of the object. The object distance is s_1 and image distance is s_2. h_1 and h_2 are object and image sizes.

When the object is placed inside the focal point (Fig. 18-14a), reflected rays seem to come from behind the mirror as they did from the plane mirror. The image is virtual, but in this case it is larger than the object and it is farther from the mirror than is the object. If the object is placed at a distance f from the mirror (Fig. 18-14b), reflected rays from any point are parallel. They never do come together and no image is produced. In Figs. 18-14c, d, and e, the object is placed successively farther from the mirror. With the object distance s_1 now greater than focal length f, a different sort of image is produced. Rays from the arrow's tip reflect and come together at a point in front of the mirror. Observers see the rays emanating from this point and perceive an image of the arrow's tip at that location. The image appears to occupy that position in the same way that a virtual image does. But this is different in that it can be projected onto a screen held at that position. A film projector produces this kind of result, called a *real image*. The concave mirror in a large astronomical telescope produces a real image that can be projected onto a photographic plate.

Parabolic reflectors. A spherical concave mirror will not produce a perfectly sharp image. Rays away from the principal axis, although parallel to it, will not pass exactly through the principal focus of a spherical mirror. Concave reflectors for many precise applications take the form of a paraboloid whose cross section is the curve called a *parabola*. A parabolic reflector will reflect any ray parallel to the axis exactly through the principal focus. Figure 18-15a shows that energy from distant sources or from transmitters that produce parallel rays can be be gathered up by the large reflecting surface and concentrated to detectable signal strength. Telescopes used by radio astronomers consist largely of giant parabolic reflectors to concentrate the energy of otherwise undetectable signals from space. The "shotgun" microphone used by broadcasters to pick up sounds of athletes in action uses a parabolic sound reflector.

If a source of radiant energy is placed at the principal focus (Fig. 18-15b), energy is reflected from the parabolic surface as a parallel beam. The result is a

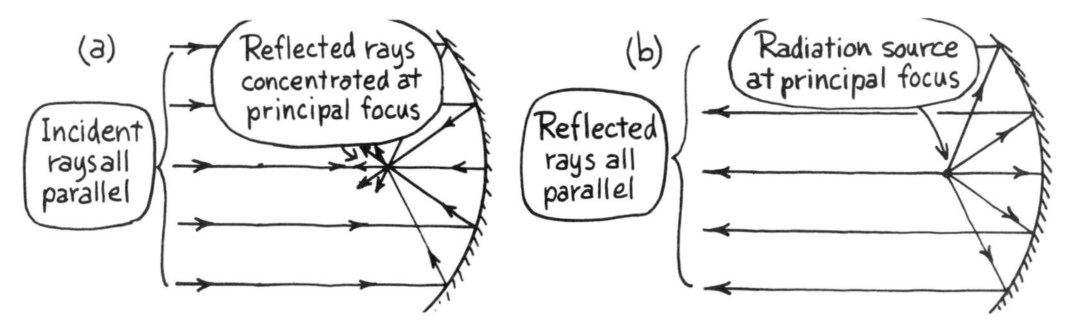

Figure 18-15. (a) A parabolic mirror reflects parallel rays to the principal focus. (b) Rays from the principal focus are reflected by a parabolic mirror as a parallel beam.

spotlight, the transmitting antenna of a microwave relay station, a radar transmitter, or any number of other applications.

Convex spherical mirrors. Convex mirrors are in common use as anti-shoplifting devices in stores. Some rear view mirrors on cars and trucks include a small convex mirror. In both applications they provide an image that, although diminished in size, covers a wider area than does a plane mirror. Figure 18-16 illustrates that a convex mirror forms a virtual image that is right side up and smaller than the object.

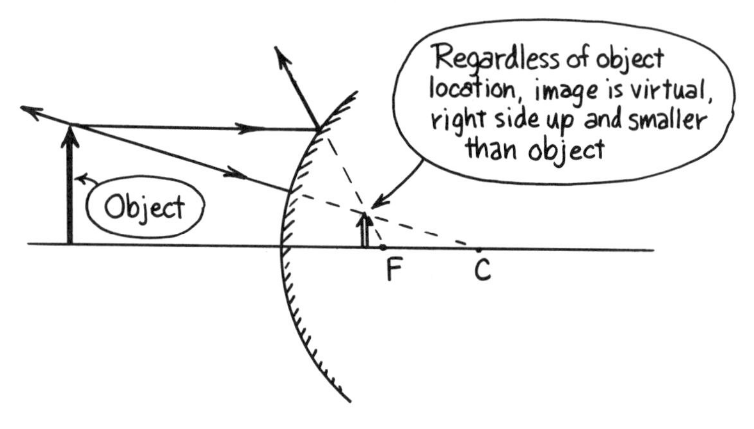

Figure 18-16. A convex spherical mirror can only produce a diminished virtual image.

Lenses

Most practical optical instruments, including cameras, microscopes, eye glasses, and such, rely more on refraction than reflection to form images. If a light ray passes through a transparent object with parallel sides (Fig. 18-17a),

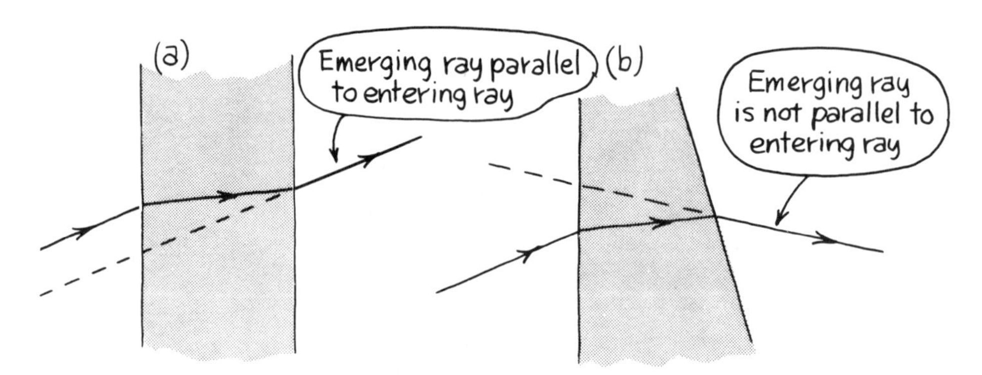

Figure 18-17. A ray changes direction if it passes through an object whose sides are not parallel.

the ray is slightly offset, but it suffers no net change of direction. If the sides are not parallel (Fig. 18-17b), there must be a change of direction in addition to the offset. Lenses are designed to produce this change of direction in an orderly fashion and thus to do much the same work as curved mirrors. Normal manufacturing techniques produce either concave or convex spherical surfaces on pieces of glass. Figures 18-18a, b, and c show some lenses that result in a net convergence (coming together) of light rays. Figures 18-18d, e, and f show some diverging lenses. Converging lenses are thicker in the middle than at the edges; diverging lenses are thicker at the edges.

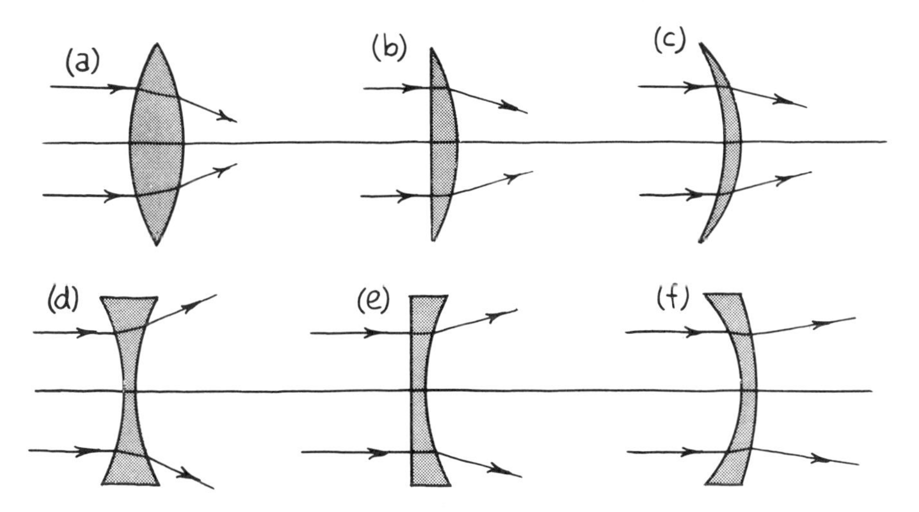

Figure 18-18. Converging lenses (a, b, c) are thickest in the middle and diverging lenses (d, e, f) are thickest at the edges.

The analysis of lenses involves much the same approach that we took with mirrors. As Fig. 18-19 shows, point F, the principal focus or focal point of a lens, is the point at which parallel rays converge or from which they seem to emanate. The focal length f is the distance from the lens to the focal point. One slight difference is that, unlike mirrors, lenses are reversible. Another focal point F' exists at a distance f on the other side of the lens. Ray tracing in lenses is easy to do using a ray parallel to the axis so that it will pass through a focal point, and a ray that passes through the center of the lens. If the lens is thin, the ray through its center will emerge undeflected and not offset enough to matter.

Converging and diverging lenses act much like concave (converging) and convex (diverging) mirrors, respectively. Figures 18-20a, b, c, and d correspond to Fig. 18-14 and show a progression from virtual to real images. The diverging lens of Figs. 18-20e, f, and g can produce only a diminished virtual image, as does the convex mirror of Fig. 18-16.

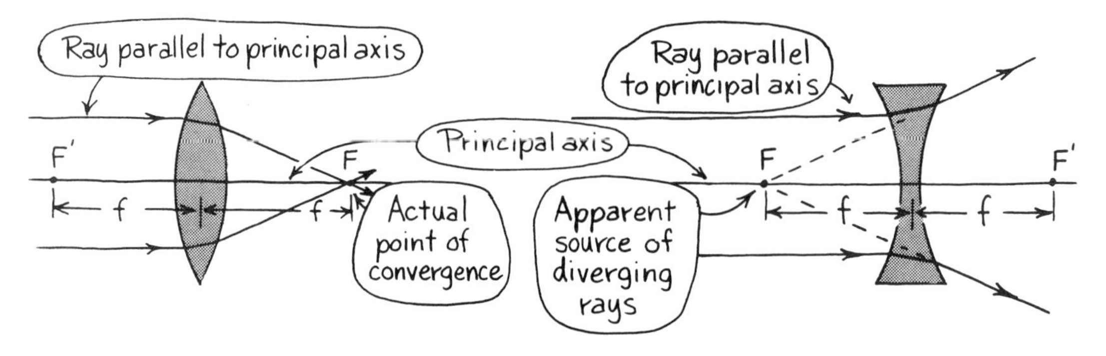

Figure 18-19. The principal focus or focal point (*F* or *F'*) and focal length *f* are defined much the same as for mirrors.

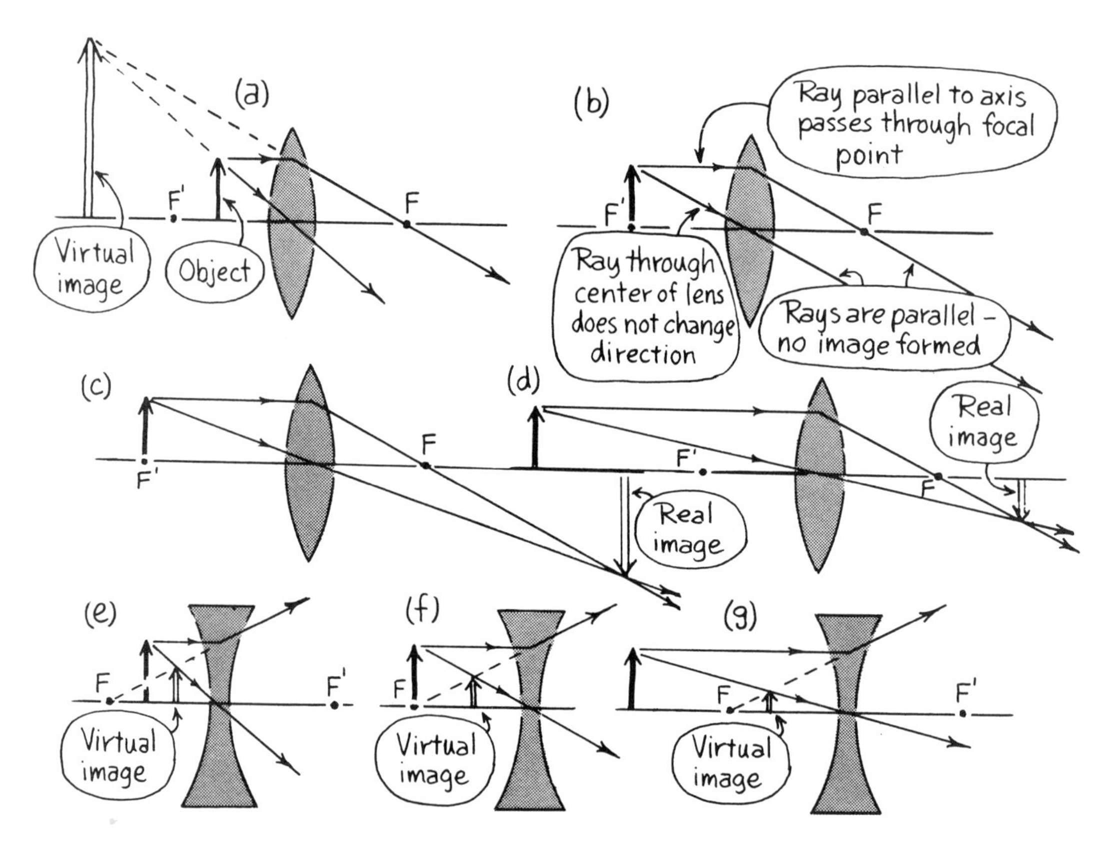

Figure 18-20. Ray diagrams for converging lens (a, b, c, d) and diverging lens (e, f, g).

By a not-too-involved geometric proof, it can be shown that for both mirrors and lenses

$$\frac{1}{f} = \frac{1}{s_1} + \frac{1}{s_2} \tag{18-4}$$

Focal length (f) Object distance (s_1) Image distance (s_2)

And, again for both mirrors and lenses,

$$\frac{h_2}{h_1} = -\frac{s_2}{s_1} \tag{18-5}$$

Image size (h_2) Object size (h_1)

There is an important sign convention to follow in using these equations, which is summarized as follows:

f	s_1	s_2	h
(+) For converging lens or mirror	(+) For real object	(+) For real image	(+) For erect image or object
(−) For diverging lens or mirror	(−) For virtual object	(−) For virtual image	(−) For inverted image or object

The relation of s_1, s_2, h_1, and h_2 to a lens or mirror is illustrated in the following examples.

Example 18-3 A small magnifying glass forms a real image of the house across the street. The image is 7.5 cm from the lens.
(a) What kind of lens is it?
(b) What is its focal length f?

Solution to (a)

If the lens can form a real image it must be **converging**.

Solution to (b)

Rays from far away (across the street is far away compared to 7.5 cm) can be considered parallel. Thus the rays converge at the principal focus F, and the focal length $f = 7.5$ cm.

Example 18-4 The lens of Example 18-3 is used to examine an object 2.0 cm high that is held 5.6 cm from the lens on the side away from the eye.

(a) What kind of image is seen?
(b) Where is the image?
(c) How large is the image?

Solution to (a)

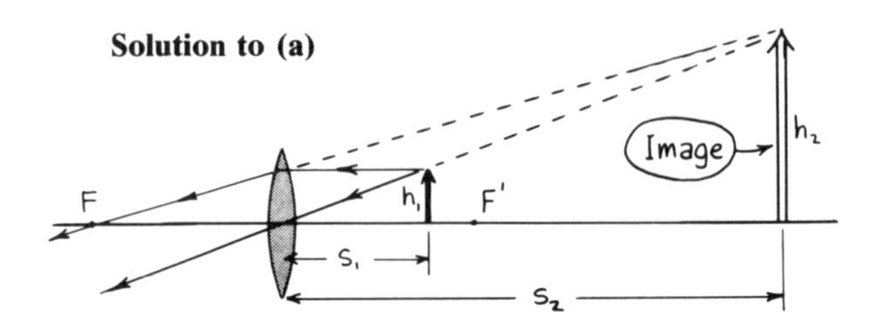

A ray diagram shows the image to be *virtual, enlarged,* and *right side up.*

Solution to (b)

$$\left\{ \begin{array}{l} f = +7.5 \text{ cm} \\ s_1 = +5.6 \text{ cm} \end{array} \right\}$$

$$\frac{1}{f} = \frac{1}{s_1} + \frac{1}{s_2} \Rightarrow \frac{1}{s_2} = \frac{1}{f} - \frac{1}{s_1} = \frac{1}{7.5 \text{ cm}} - \frac{1}{5.6 \text{ cm}} = -0.04524$$

$$s_2 = \frac{1}{-0.04524} = \boxed{-22 \text{ cm}}$$

Negative sign means that image is virtual

Solution to (c)

$$\{h_1 = 2.0 \text{ cm}\}$$

$$\frac{h_2}{h_1} = -\frac{s_2}{s_1} \Rightarrow h_2 = h_1\left(-\frac{s_2}{s_1}\right) = (2.0 \text{ cm})\left(-\frac{-22 \text{ cm}}{5.6 \text{ cm}}\right) = +7.857$$

$$h_2 = \boxed{+7.9 \text{ cm}}$$

Positive sign means that image is erect

Example 18-5 A convex mirror with radius of curvature 3.0 ft is placed high on the wall of a store to provide an overall view of the customers. A shopper 25 ft from the mirror is observed lifting a ladder that is 6.0 ft high.

(a) What kind of image is formed and where is it?
(b) How high is the ladder's image?

Solution to (a)

$$\left\{ \begin{array}{l} s_1 = 25 \text{ ft} \\ r = 3.0 \text{ ft} \\ f = \frac{r}{2} = -1.5 \text{ ft} \end{array} \right\}$$

↑ Diverging mirror

$$\frac{1}{f} = \frac{1}{s_1} + \frac{1}{s_2} \Rightarrow \frac{1}{s_2} = \frac{1}{f} - \frac{1}{s_1} = \frac{1}{-1.5 \text{ ft}} - \frac{1}{25 \text{ ft}}$$

$$\frac{1}{s_2} = -0.707 \Rightarrow s_2 = \boxed{-1.4 \text{ ft}}$$

Virtual image 1.4 ft behind mirror

Solution to (b)

$$\begin{Bmatrix} h_1 = 6.0 \text{ ft} \\ s_2 = -1.4 \text{ ft} \end{Bmatrix}$$

$$\frac{h_2}{h_1} = -\frac{s_2}{s_1} \Rightarrow h_2 = h_1\left(-\frac{s_2}{s_1}\right) = (6.0 \text{ ft})\left(-\frac{-1.4 \text{ ft}}{25 \text{ ft}}\right) = \boxed{+0.34 \text{ ft}}$$

Image is erect

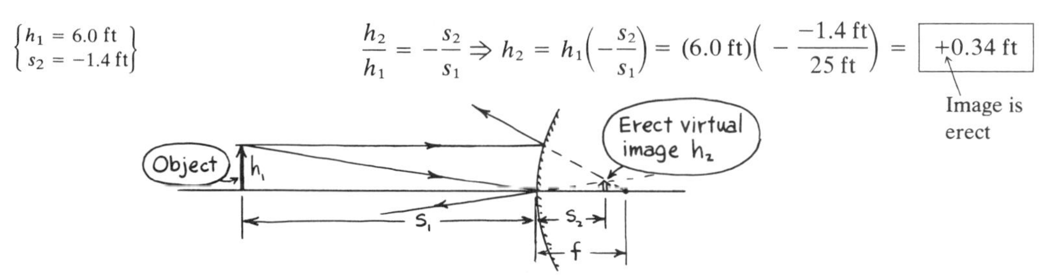

Example 18-6 A concave shaving mirror will focus the sun's rays to a point 27 cm in front of the mirror. In use, the mirror is held 15 cm from the face.

(a) What kind of image is formed and where is it?

(b) How large is the image of a 0.62-cm blemish?

Solution to (a)

Converging mirror

$$\begin{Bmatrix} f = +27 \text{ cm} \\ s_1 = 15 \text{ cm} \end{Bmatrix}$$

$$\frac{1}{f} = \frac{1}{s_1} + \frac{1}{s_2} \Rightarrow \frac{1}{s_2} = \frac{1}{f} - \frac{1}{s_1} = \frac{1}{27 \text{ cm}} - \frac{1}{15 \text{ cm}}$$

$$\frac{1}{s_2} = -0.0296 \Rightarrow s_2 = \boxed{-34 \text{ cm}}$$

Virtual image behind mirror

Solution to (b)

$$\begin{Bmatrix} h_1 = 0.62 \text{ cm} \\ s_2 = -34 \text{ cm} \end{Bmatrix}$$

$$\frac{h_2}{h_1} = -\frac{s_2}{s_1} \Rightarrow h_2 = h_1\left(-\frac{s_2}{s_1}\right) = (0.62 \text{ cm})\left(-\frac{-34 \text{ cm}}{15 \text{ cm}}\right)$$

$$h_2 = \boxed{+1.4 \text{ cm}}$$

Image is erect

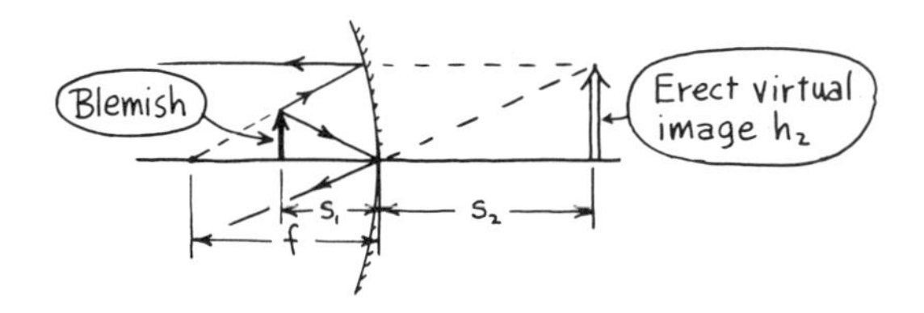

18-6 ELECTROMAGNETIC WAVES

If we are to think of electromagnetic radiation as a wave, it will have to be a very special kind of wave. Unlike all the waves of Chapter 17, this kind travels very nicely through empty space. No physical medium is needed to provide capacitance and inertance for the temporary storage and transfer of energy. But that is alright, because we have seen that electrical systems can store energy in a charged capacitor (Sec. 9-3) and energy in the magnetic field around a current-carrying inductor (Sec. 10-3). Whereas Chapter 17 might well have been titled Mechanical Waves, we are now dealing with electrical and magnetic systems. Thus the term *electromagnetic waves.*

The concept of electromagnetic waves is another idea that is far from new. The electromagnetic theory was published in 1864 by James Clerk Maxwell (England), and it was verified experimentally by Heinrich Hertz (Germany) in 1886. Some knowledge of electric and magnetic fields helps to better understand the structure of electromagnetic waves.

Electric and Magnetic Fields

In Chapter 8 the *magnetic field* was introduced as some sort of colorless, odorless, and tasteless thing that occupied the space between the poles of a magnet. We know it is there because it exerts a force on another magnet in its vicinity. The magnetic field is commonly thought of as an alteration of space in the region of a magnet.

A similar idea is associated with electric charge. Any charged particle *alters space* so that another charged particle experiences a force. The effect can be described or thought of as another kind of field, in this case an *electric field.* In much the same style as for magnetic fields, the electric field is represented by lines showing the direction of a force that would be exerted on a small positive test charge (Fig. 18-21). *Electric field strength E* at a given point is the magnitude and direction of the force exerted per unit positive charge at that point.

$$E = \frac{F}{Q} \tag{18-6}$$

Electric Field Strength E	*Force F*	*Charge Q*
newton/coulomb (N/C)	newton (N)	coulomb (C)

A magnetic field exists between magnets (Fig. 18-22a) that attract or repel one another. Similarly, an electric field exists between charged particles (Fig. 18-22b) that attract or repel one another. When considered as defined, there is no obvious connection between magnetic and electric fields. In fact, if a charged particle is placed beside a magnet (Fig. 18-22c), there is no mutual force as long as everything stands still.

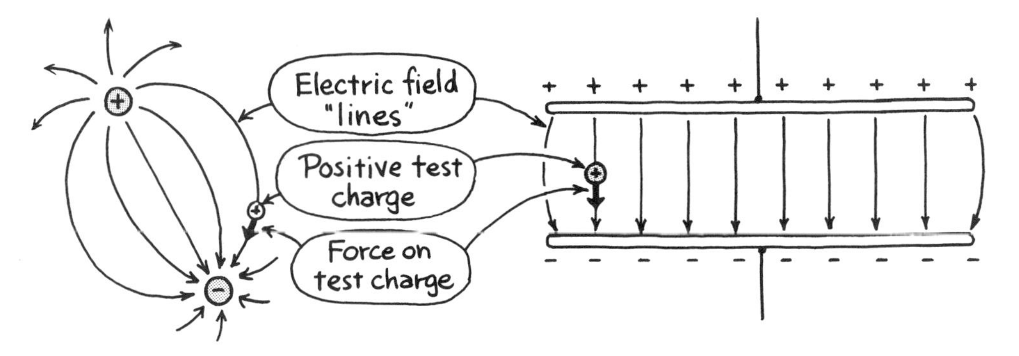

Figure 18-21. Electric field lines by definition point in the direction of force on a positive test charge located in the field. They point radially outward from a positive charge, radially inward toward a negative charge, and straight across from positive to negative plates of a capacitor.

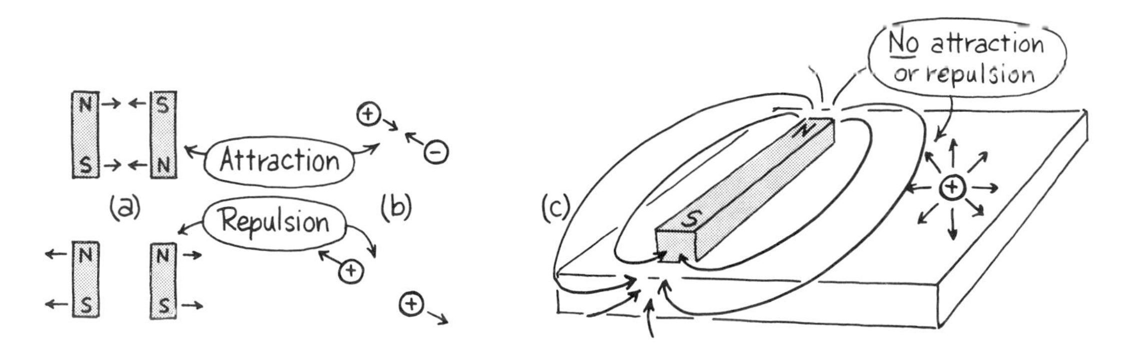

Figure 18-22. (a) Magnets attract or repel each other. (b) Charged particles attract or repel each other. (c) A magnet and a charged particle sitting side by side have no effect on each other.

But suppose that we tilt the table and the charged particle starts to slide toward the magnet. What then? The charge experiences an upward force, as discussed in Sec. 8-1 and detailed in Fig. 18-23a. One way of viewing this result is that the changing electric field in the space around the moving charge produces a magnetic field, and this new magnetic field interacts with the magnet's field to produce the force.

In Fig. 18-23b the permanent magnet has been replaced by an electromagnet whose magnetic field can be changed by varying the current. What happens then? Although the steady unchanging magnetic field did not affect the charge, a force on the charge appears when the magnetic field changes. One useful way of viewing this is that a changing magnetic field produces an electric field that acts on the charged particle.

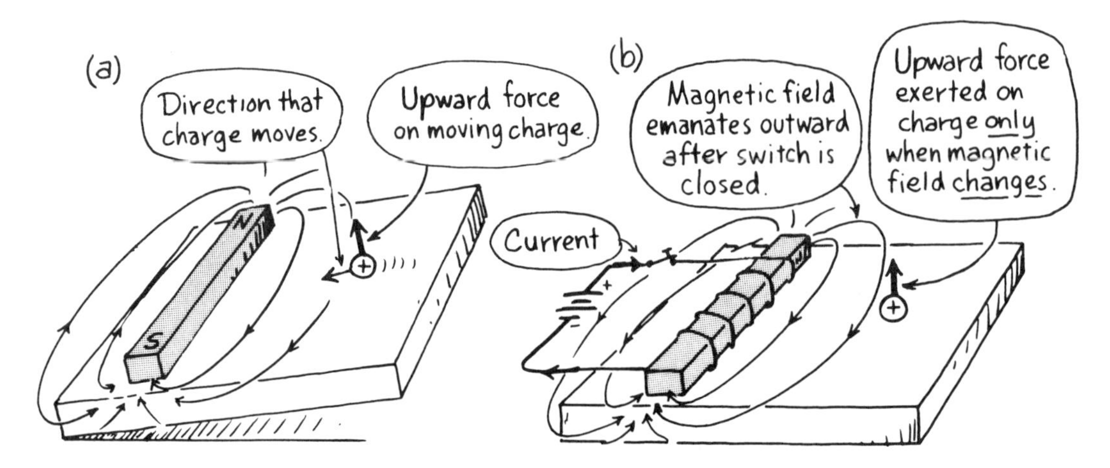

Figure 18-23. The magnet and charged particle exert forces on each other only when (a) the electric field changes or (b) when the magnetic field changes.

Now we are set to attempt an explanation of electromagnetic waves, keeping in mind three points:

1. A changing electric field produces a magnetic field.
2. A changing magnetic field produces an electric field.
3. Electric and magnetic fields store energy.

Propagation of Electromagnetic Waves

Let us return to the antennas of Fig. 18-2 and try to explain how energy escapes. Figure 18-24 interprets the three cases in terms of electric and magnetic fields. In Figs. 18-24a and b, fields exist, but once established they do not change. Situation (a) is like a spring that has been wound up, and situation (b) is like a wheel that has been set spinning. Only one pulse has been generated, essentially a half-cycle, which is not sufficient to produce the wave train that is needed for the continuous flow of energy.

Figure 18-24c represents a materially different situation. The ac source generates changing electric and magnetic fields. A changing electric field in space generates a magnetic field, a changing magnetic field in space generates an electric field, and so it goes. Continuously changing fields propagate outward from the source, carrying energy as they proceed.

The process is analogous to wave propagation in a mechanical medium made up of springs and blocks (Sec. 17-1). A block was responsible for stretching the adjacent spring, and that spring pulled on the next block in line to make it move. There was a continuous change from kinetic energy (motion of the block) to potential energy (stretch of the spring) and back again as a

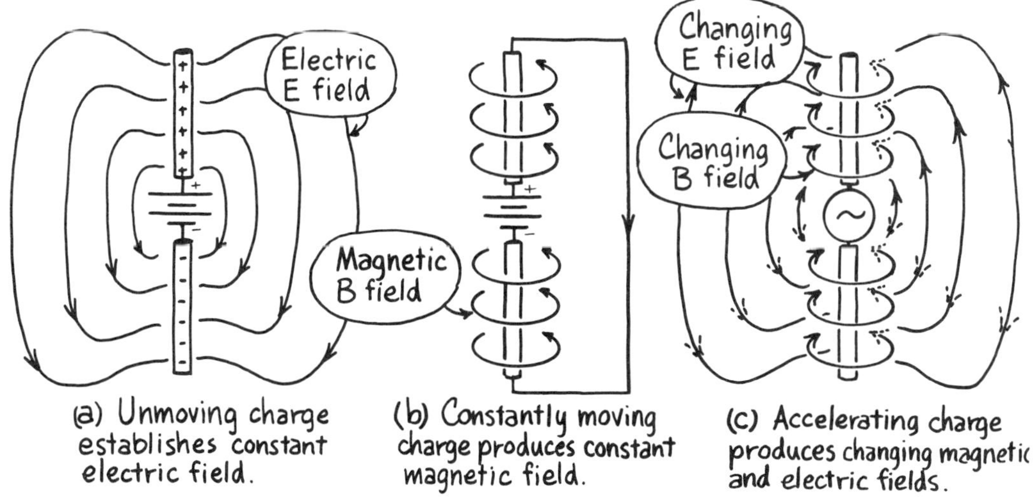

Figure 18-24. Another look at Fig. 18-2 gives us a chance to study how electromagnetic waves are propagated. (a) Unmoving charge establishes constant electric field. (b) Constantly moving charge produces constant magnetic field. (c) Accelerating charge produces changing magnetic and electric fields.

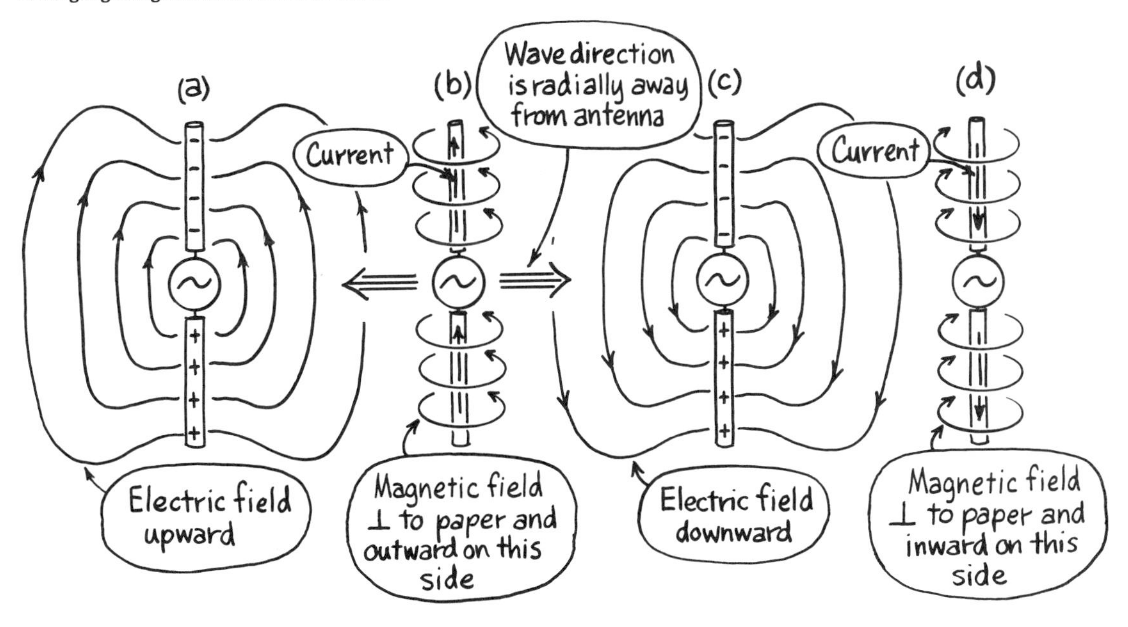

Figure 18-25. As charge surges back and forth in the antenna, the electric field alternates between (a) vertically upward, (b) zero, (c) vertically downward, and (d) zero. Electric field changes parallel to the antenna produce magnetic field changes, as shown, that are either into or out of the paper. The electromagnetic wave propagates radially outward.

wave moved down the line. In the electrical system a changing electric field plays the part of the spring and a changing magnetic field the part of the block. Energy is continually handed down the line, first as energy of the electric field (stored potential energy), and then as energy of the magnetic field (stored "kinetic" energy).

In the setup of Fig. 18-24c, electric charge moves up and down along the antenna. This vertical redistribution of charge produces a vertically changing electric field (Fig. 18-25). Also produced, according to the right-hand rule of Sec. 8-2, is a changing magnetic field that points alternately inward and outward. The changing magnetic field is perpendicular to both the changing electric field and the direction of wave propagation. Electromagnetic energy is carried radially outward from the antenna. Electromagnetic radiation is a transverse wave—actually two simultaneous transverse waves that are mutually perpendicular (Fig. 18-26).

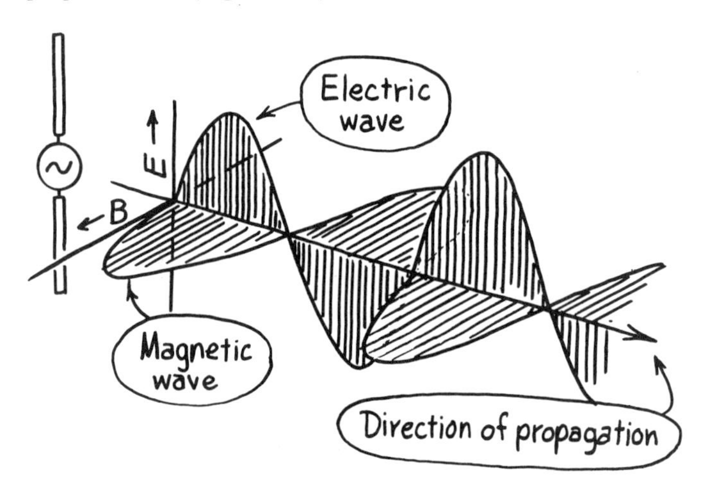

Figure 18-26. An electromagnetic wave consists of mutually perpendicular electric field and magnetic field waves.

18-7 PREDICTING THE SPEED OF LIGHT

If light really does travel in waves, it seems reasonable that the ideas of Sec. 17-4 might be applied to get at its speed. Equation (17-1) is a generalization about the speed of waves.

$$v = \sqrt{\frac{\text{stiffness factor}}{\text{inertance factor}}} = \frac{1}{\sqrt{\begin{pmatrix}\text{inertance}\\ \text{factor}\end{pmatrix} \times \begin{pmatrix}\text{capacitance}\\ \text{factor}\end{pmatrix}}} \qquad (17\text{-}1)$$

The stiffness and inertance factors are proportionality constants that relate force to other factors in a system. The velocity of a longitudinal wave in a rigid rod, for example, is given in Table 17-1 as

$$v = \sqrt{\frac{Y}{\rho}}$$

Young's modulus (Y); Density (ρ)

Young's modulus Y tells how much force it takes to produce a given amount of stretch in the rod. Density ρ tells how much mass is contained in a given volume of the rod, and thus how much force is needed to accelerate a section of rod at a certain rate. Electrical and magnetic systems involve similar constants.

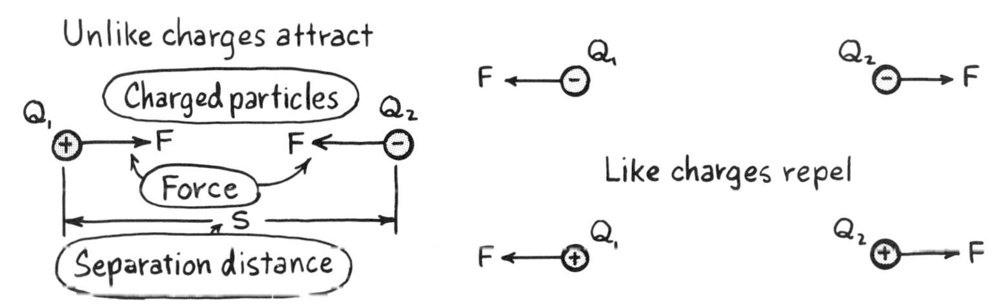

Figure 18-27. The force of attraction (or repulsion) between two charged particles depends directly upon the product of the charges Q_1 and Q_2, and inversely upon the square of the separation distance s.

A well-known relation called *Coulomb's law* gives the force between two separated electrically charged particles (Fig. 18-27). The equation describing this law is

$$F = \frac{1}{4\pi\varepsilon} \times \frac{Q_1 Q_2}{s^2} \tag{18-7}$$

F Force	ε Permittivity	Q_1 | Q_2 Quantity of Charge	s Distance
newton (N)	$\frac{\text{coulomb}^2}{\text{newton} \cdot \text{metre}^2}$ $C^2/(N \cdot m^2)$	coulomb (C)	metre (m)

The constant ε is called the *permittivity* of the substance separating the charges. Its value ε_0 for a vacuum is $\varepsilon_0 = 8.85 \times 10^{-12}\ C^2/(N \cdot m^2)$. Electric fields store potential energy, so we might guess that $1/(4\pi\varepsilon)$ is an electrical stiffness factor analogous to an elastic modulus of a mechanical system, or that permittivity ε is a sort of capacitance factor. Permittivity was first introduced in Sec. 9-2.

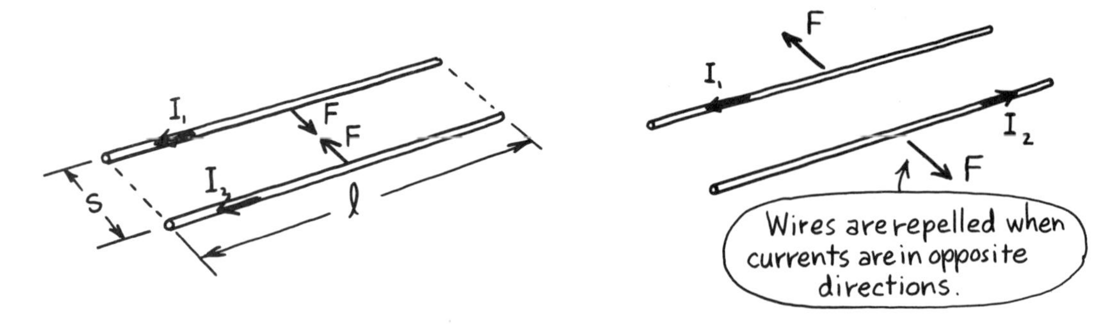

Figure 18-28. Two long parallel wires attract each other if the currents are in the same direction. They repel if the currents are in opposite directions. The magnetic field of one current produces a force on the other current, and vice versa.

A similar kind of equation covers current and magnetic field relations. Two long parallel wires of length ℓ are separated by a distance s and carry currents I_1 and I_2. Experiments show that the wires attract or repel each other depending upon the direction of the currents (Fig. 18-28). This equation is

$$F = \frac{\mu}{2\pi} \times \frac{I_1 I_2 \ell}{s} \qquad (18\text{-}8)$$

F *Force*	μ *Permeability*	I_1 | I_2 *Current*	s *Distance*	ℓ *Length*
newton (N)	$\frac{\text{newton} \cdot \text{metre}}{\text{ampere}^2 \cdot \text{metre}}$ $\text{N} \cdot \text{m}/(\text{A}^2 \cdot \text{m})$	ampere (A)	metre (m)	metre (m)

The constant μ is called the *permeability* of the substance separating the wires. Its value μ_0 for a vacuum is $\mu_0 = 12.57 \times 10^{-7}\ \text{N} \cdot \text{m}/(\text{A}^2 \cdot \text{m})$. Permeability was first discussed in the chapter on magnetism (Sec. 8-2). Magnetic fields store energy analogous to kinetic energy, and so it seems that μ has a chance of being some kind of inertance factor.

An equation for the velocity of light in any medium, first given in Table 18-1, is

$$v = \sqrt{\frac{1}{\mu\varepsilon}} \qquad (18\text{-}9)$$

Using values of μ_0 and ε_0, the velocity of light in a vacuum c is

$$c = \sqrt{\frac{1}{\mu_0 \varepsilon_0}} = \sqrt{\frac{1}{(12.57 \times 10^{-7})(8.85 \times 10^{-12})}} = 3.00 \times 10^8\ \text{m/s}$$

Thus it seems that μ and ε do indeed fill the roles of inertance and capacitance factors. This theoretical calculation of the speed of light, made by Maxwell, is one of his principal triumphs. We will leave it up to you to prove that the units of the calculation are metre/second, the units of velocity.

18-8 ELECTROMAGNETIC RADIATION AS PARTICLES: PHOTOELECTRIC EFFECT

Now that you have been talked into thinking of electromagnetic radiation as waves, we will show you that it isn't necessarily so. Maxwell's ideas (1864) about electromagnetic waves had been experimentally verified by Hertz (1886). Hertz very quickly came upon another phenomenon that serves as one of the most convincing arguments against a simple wave model. In 1887 he was experimenting with electric sparks jumping between charged spheres when he noticed that a spark would jump more readily if the surfaces of the spheres were illuminated with light from another spark. Apparently, the light supplied energy to help dislodge electrons from a surface. This *photoelectric effect* need not be surprising in itself. It is not difficult to imagine electrons resonating to the frequencies of the incident light and soaking up energy from that sinusoidal energy source. This view of things, however, quickly runs into trouble. A couple of curious facts arise from experiments with apparatus like that of Fig. 18-29.

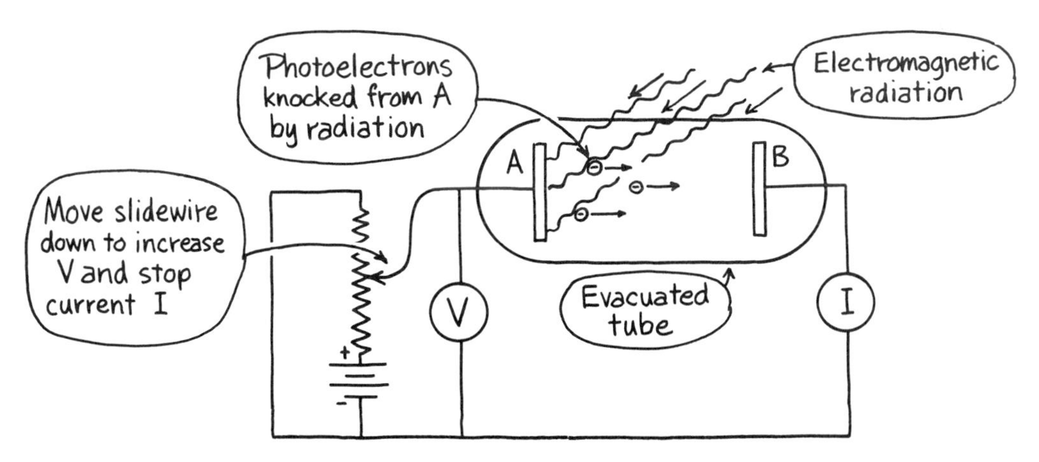

Figure 18-29. An experimental setup to investigate the photoelectric effect. Retarding voltage V can be increased until current I stops completely. This value of V is a measure of the maximum kinetic energy supplied to the electrons by the light.

Light shines through the transparent tube onto a metal surface A from which electrons are dislodged. These *photoelectrons* are emitted with kinetic energies E_K ranging from zero to some maximum value. The photoelectrons in effect climb a potential "hill" on their way to B. If the hill is made high enough, none of the photoelectrons will make it to B and the ammeter will register zero. Voltage V at this point is a measure of the maximum kinetic energy of a photoelectron. Here is an example.

Example 18-7 Light shines from a source onto plate A of the photoelectric apparatus of Fig. 18-29. The photoelectric current is stopped completely ($I = 0$) when the potential difference $V = 0.64$ V. What is the maximum kinetic energy of photoelectrons emitted from plate A?

Solution

$\left\{\begin{array}{l} V = 0.64 \text{ V} \\ Q = 1.6 \times 10^{-19} \text{ C} \end{array}\right\}$

The potential difference between plates A and B is defined as the energy needed per coulomb of charge to move a charged particle from A to B. Zero current means that even the most energetic electron (with charge $e = 1.6 \times 10^{-19}$ C) does not have enough kinetic energy to make the trip.

$$V = \frac{E_K}{Q}$$

$$E_K = VQ = (0.64 \text{ V})(1.6 \times 10^{-19} \text{ C})$$

$$= \boxed{1.0 \times 10^{-19} \text{ J}}$$

If light were strictly a wave phenomenon, we could visualize the beam of light washing over the surface of A like the flame of a blow torch, delivering energy uniformly to electrons at or near the surface. It is easy to check this proposition by a combination of experiment and calculation. Suppose that we use sodium metal for our surface in the simplified setup of Fig. 18-30. If red light of frequency 4.5×10^{14} Hz is used (Fig. 18-30a), there is no effect

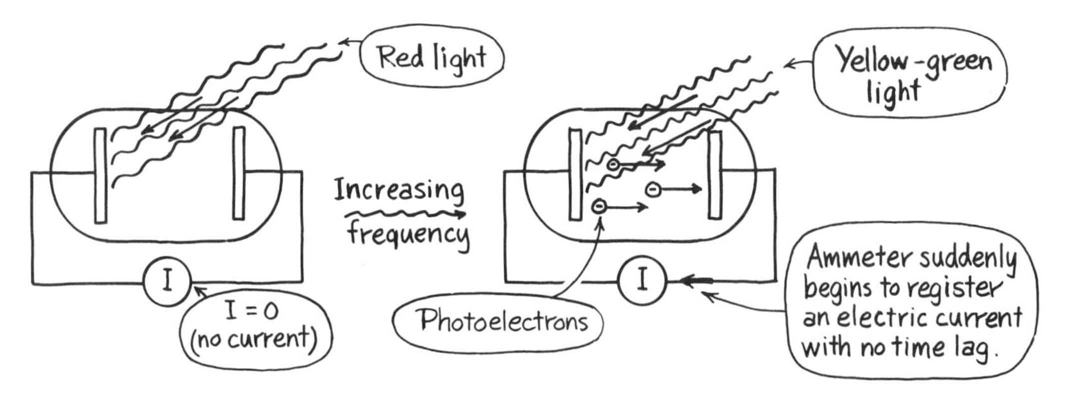

Figure 18-30. (a) Light below threshold frequency does not dislodge electrons no matter how long or how brightly it shines. (b) As soon as the threshold frequency is reached, charge begins to flow. If a brighter source is used, the current increases.

regardless of how long we wait or how bright the red light source is. Electrons in the sodium apparently do not soak up enough energy to help them break free from the surface.

As the frequency of the incident light is slowly increased, we find no change until the light is yellow-green at a frequency of about 5.9×10^{14} Hz. Suddenly at this point the ammeter begins to indicate a current. Current continues to register at frequencies greater than this *threshold frequency*. The threshold frequency is in the ultraviolet for most metals, but it is in the visible range for the alkali metals, which include sodium and potassium.

Another important experimental fact is that photoelectrons are released immediately when light above the threshold frequency hits the surface. Calculations can be made to see if this is reasonable. Suppose that high-frequency visible light (violet) falls on a sodium surface and that the light is just bright enough to produce a detectable photoelectric current. Calculations show that, if all the surface electrons absorb energy at the same rate, it would take nearly a year to supply a given electron with the kinetic energy necessary to free itself from the surface. In reality, it takes less time than can be measured experimentally—less than a nanosecond—to energize a photoelectron.

Another experiment shows that the photoelectric current depends upon illumination of the surface. All else being equal, the more the incident radiation, the more the photoelectric current. The surprising fact is that the maximum energy of an individual photoelectron does not change with illumination. It is found that the kinetic energy of a photoelectron depends only on frequency, as illustrated by the graph of Fig. 18-31. A vital point is that, once the threshold has been reached, the slope of the graph is the same for all photoelectric materials.

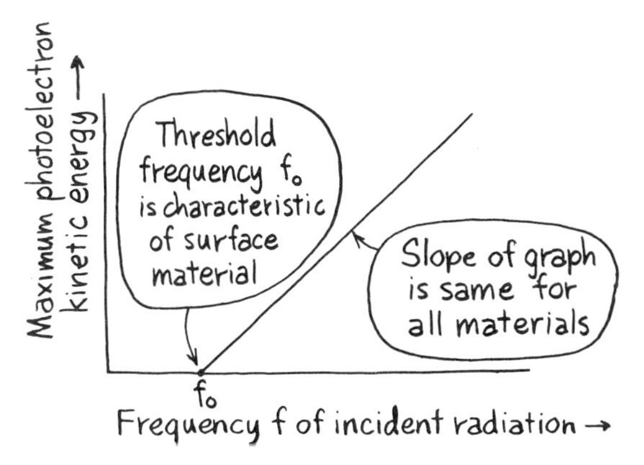

Figure 18-31. The maximum kinetic energy of a photoelectron varies directly with frequency after the threshold frequency f_0 is reached.

The whole puzzle comes together if we can bring ourselves to think of light as discrete bundles of energy. If electromagnetic radiation is more like a hail of bullets than a strong wind, everything will be alright. The bullets affect only those electrons that they hit. The effect on each of those relatively few electrons will be correspondingly greater and more rapid than the diffuse effect of a wind.

The explanation is based largely on an important relation worked out by the German theoretical physicist, Max Planck, in 1900. Planck was working on an apparently different problem—a theoretical explanation of the observed spectrum of perfectly radiating objects, or black-body radiators. Classical mechanics did not work at all, and Planck was able to explain the results by assuming that energy comes in only multiples of a basic amount called a *quantum.* One quantum of energy is the product hf, where f is the frequency and h is an important universal constant later named for Planck.

$$E = hf \tag{18-10}$$

E *Quantum of Energy*	h *Planck's Constant*	f *Frequency*
joule (J)	joule · second (J · s)	hertz (Hz)

The value of Planck's constant h is

$$h = 6.63 \times 10^{-34}\ \text{J} \cdot \text{s}$$

Albert Einstein is credited with using Planck's hypothesis in 1905 to explain the photoelectric effect. Let us take a look at an expanded version of the graph of Fig. 18-31 in light of this idea. Figure 18-32 illustrates Einstein's

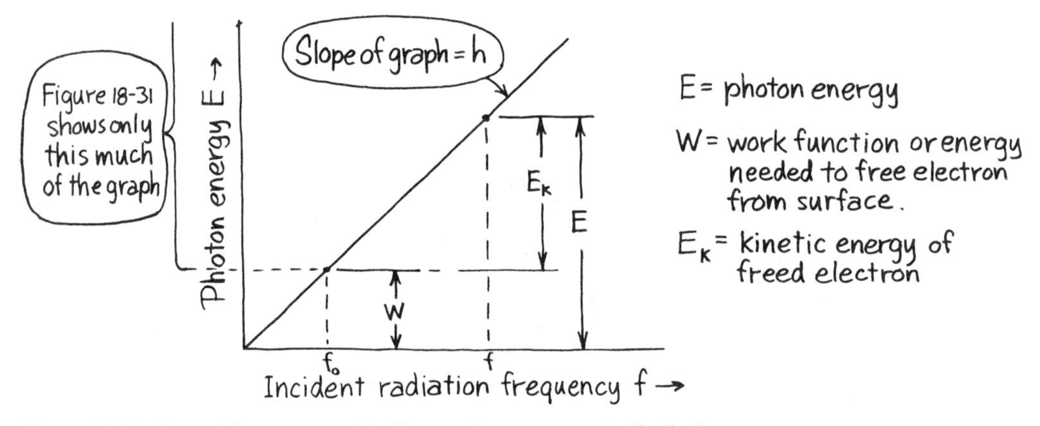

Figure 18-32. Part of the energy added by an electromagnetic "bullet" or photon is used as potential energy to separate an electron from the surface. The rest of it appears as kinetic energy stored in the moving electron.

thought about the distribution of a packet of electromagnetic energy (called a *photon*) as it encounters a surface electron. Some of the energy is needed to raise the electron's potential energy and dislodge it from the surface. The rest of it can go toward accelerating the freed electron. It is as though a certain amount of fuel is available to operate a car (Fig. 18-33). The only catch is that the car is in a hole and some of the fuel must be used to raise it to road level. Only then can the remaining fuel be used to supply kinetic energy to the car.

Figure 18-33. (a) If the car starts in a hole, some of the available energy must be used to raise its potential energy. (b) The rest of the fuel can then go toward increasing the car's kinetic energy.

The part of the photon's energy needed to separate the photoelectron from the surface is called the *work function W* of the surface. The value of W depends upon the material being used. The rest of the photon's energy goes into kinetic energy for the freed photoelectron. Thus, from Fig. 18-32,

$$E = W + E_K \tag{18-11}$$

E Photon Energy	W Work Function	E_K Electron Kinetic Energy
←	joule (J)	→

Einstein theorized that the slope of the graph of Fig. 18-32 was equal to Planck's constant h, and this turned out to be true. Hence $W = hf_0$, $E = hf$, and the photoelectron's kinetic energy $E_K = \frac{1}{2}mv^2$. Combining all this,

$$E = W + E_K$$

or

$$hf = hf_0 + \tfrac{1}{2}mv^2 \qquad (18\text{-}12)$$

h Planck's Constant	f Photon's Frequency	f_0 Threshold Frequency	m Photoelectron's Mass	v Photoelectron's Velocity
joule · second (J · s)	hertz (Hz)	hertz (Hz)	kilogram (kg)	metre/second (m/s)

Thus it turns out that explanation of the photoelectric effect requires electromagnetic energy to be in separate blobs rather than uniformly dispersed as waves. We say that the energy comes in quanta or that energy is quantized. It is tempting to think of this as somehow related to the fact that electric charge comes in discrete packages (the charge of an electron or proton). Be sure to recognize an important difference, however. The energy of a photon (a quantum) is not a universal constant but, as Eq. (18-10) shows, it depends on frequency. A quantum of blue light has more energy than a quantum of red light and less energy than a quantum of ultraviolet. Photons come in different sizes, but all photons of a particular frequency have the same energy. Here are a couple of examples that illustrate photons and the photoelectric effect.

Example 18-8 What is the energy of a photon of green light whose wavelength is 515 nm?

Solution

$\left\{ \begin{aligned} \lambda &= 515\ \text{nm} \\ &= 515 \times 10^{-9}\ \text{m} \\ c &= 3.00 \times 10^{8}\ \text{m/s} \end{aligned} \right\}$

First find the frequency of green light.

$$c = f\lambda \Rightarrow f = \frac{c}{\lambda}$$

Eq. (17-5)

$$f = \frac{3.00 \times 10^{8}\ \text{m/s}}{5.15 \times 10^{-7}\ \text{m}} = 5.82 \times 10^{14}\ \text{Hz}$$

Then use Eq. (18-10) to find the photon energy.

$\{h = 6.63 \times 10^{-34}\ \text{J} \cdot \text{s}\}$

$$E = hf = (6.63 \times 10^{-34}\ \text{J} \cdot \text{s})(5.82 \times 10^{14}\ \text{Hz}) = \boxed{3.86 \times 10^{-19}\ \text{J}}$$

Eq. (18-10)

Example 18-9 The work function for sodium metal is 3.94×10^{-19} J. Blue light of wavelength 450 nm illuminates the sodium surface. What is the kinetic energy of the emitted photoelectrons?

Solution

$$\left\{\begin{aligned} \lambda &= 450\ \text{nm} \\ &= 450 \times 10^{-9}\ \text{m} \\ c &= 3.00 \times 10^{8}\ \text{m/s} \\ h &= 6.63 \times 10^{-34}\ \text{J} \cdot \text{s} \end{aligned}\right\}$$

First calculate the energy of a photon of blue light.

$$f = \frac{c}{\lambda} = \frac{3.00 \times 10^{8}\ \text{m/s}}{4.50 \times 10^{-7}\ \text{m}}$$

$$= 6.67 \times 10^{14}\ \text{Hz}$$

$$E = hf = (6.63 \times 10^{-34}\ \text{J} \cdot \text{s})(6.67 \times 10^{14}\ \text{Hz})$$

$$= 4.42 \times 10^{-19}\ \text{J}$$

$\{W = 3.94 \times 10^{-19}\ \text{J}\}$ Then use Eq. (18-11) to calculate the photoelectron kinetic energy.

$$E = W + E_K$$

$$E_K = E - W = 4.42 \times 10^{-19}\ \text{J} - 3.94 \times 10^{-19}\ \text{J} = \boxed{0.48 \times 10^{-19}\ \text{J}}$$

18-9 PHOTONS HAVE MOMENTUM: THE COMPTON EFFECT

We have not strayed completely from the wave model of radiation. Equation (18-10) defines photon energy in terms of wave frequency. It is a curious mixture of wave and particle ideas, as illustrated by Fig. 18-34. There is nothing so far to prevent us from visualizing photons as little flurries of wave action propagating along with the speed of light. A higher-frequency wave packet involves more wiggles and thus more energy than a low-frequency packet.

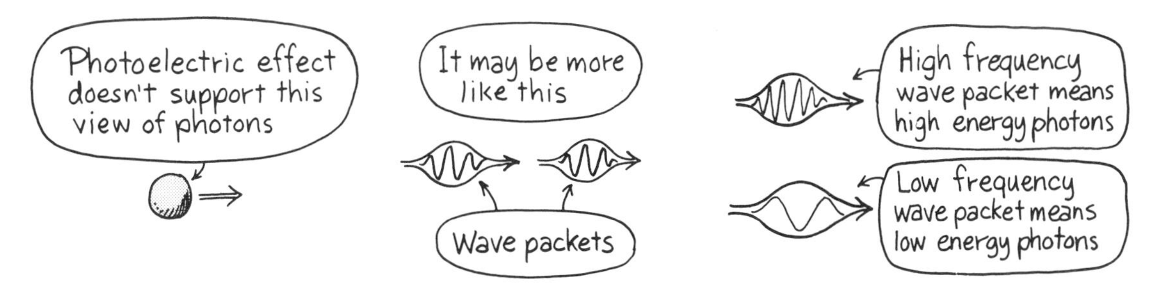

Figure 18-34. The photoelectric effect gives the idea that photons are little concentrated blobs of wave energy.

So it may have seemed until 1922 when an American physicist, Arthur Compton, performed another classic experiment. Compton bombarded electrons in a block of carbon with X rays of a known frequency. He was able to measure the results of collisions between X-ray photons and the electrons. Figure 18-35 shows a simplified summary of what he found. This is highly reminiscent of a couple of billiard balls colliding and rebounding. We won't go through all the details here, but Compton found that the laws of the conservation of energy and momentum predicted the experimental results of the photon collision. The photon and electron can be treated just like colliding billiard balls.

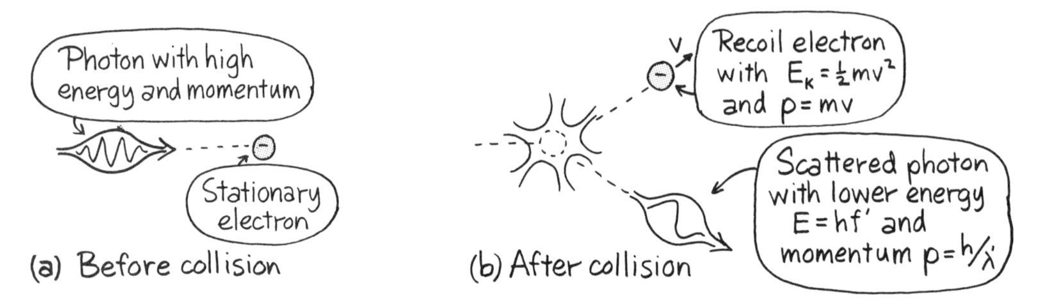

Figure 18-35. The Compton effect. (a) An X-ray photon approaches an electron. (b) After collision the electron moves off in one direction with kinetic energy $\frac{1}{2}mv^2$. The scattered photon moves off in another direction but its frequency is lower. It has lost energy and momentum.

The Compton effect can be summarized mathematically by two equations. First, energy is conserved (see Fig. 18-35):

$$hf = hf' + \tfrac{1}{2}mv^2 \qquad (18\text{-}13)$$

Energy of incident photon (hf); Energy of scattered photon (hf'); Energy of recoil electron ($\frac{1}{2}mv^2$)

But what proved to be astounding is that the photon was shown to possess momentum. Furthermore, momentum was conserved in the collision. The photon's momentum, like any momentum, is the product of mass and velocity. But a photon is not a photon unless it moves at the speed of light c. The photon's mass is defined by Eq. (3-13):

$$E = mc^2 \qquad (3\text{-}13)$$

$$m = \frac{E}{c^2} = \frac{hf}{c^2} \quad \leftarrow \text{Eq. (18-10)}$$

But momentum p is the product of mass and velocity.

$$p = mv = \frac{hf}{c^2}c = \frac{hf}{c} = h\left(\frac{f}{c}\right)$$

But since $c = f\lambda$, $f/c = 1/\lambda$, and

$$p = \frac{h}{\lambda} \qquad (18\text{-}14)$$

p Photon Momentum	h Planck's Constant	λ Photon Wavelength
kilogram $\cdot \frac{\text{metre}}{\text{second}}$ (kg · m/s)	joule · second (J · s)	metre (m)

Here are a couple of examples that illustrate the momentum aspects of electromagnetic radiation.

Example 18-10 Find the momentum of the green light photons in Example 18-8, which have a wavelength of 515 nm.

Solution

$$\left\{\begin{aligned} \lambda &= 515 \text{ nm} \\ &= 515 \times 10^{-9} \text{ m} \\ h &= 6.63 \times 10^{-34} \text{ J} \cdot \text{s} \end{aligned}\right\}$$

$$p = \frac{h}{\lambda} = \frac{6.63 \times 10^{-34} \text{ J} \cdot \text{s}}{5.15 \times 10^{-7} \text{ m}}$$

$$= 1.29 \times 10^{-27} \text{ J} \cdot \text{s/m} \quad \text{or} \quad \boxed{1.29 \times 10^{-27} \text{ kg} \cdot \text{m/s}}$$

Note that the units J · s/m are the same as kg · m/s.

$$\frac{\text{J} \cdot \text{s}}{\text{m}} = \frac{(\text{N} \cdot \text{m}) \cdot \text{s}}{\text{m}} = \text{N} \cdot \text{s} = \left(\frac{\text{kg} \cdot \text{m}}{\text{s}^2}\right) \cdot \text{s} = \text{kg} \cdot \text{m/s}$$

Example 18-11 An X-ray photon of frequency 1.00×10^{18} Hz strikes a stationary electron, giving it a recoil energy of 2.04×10^{-17} J. What is the frequency and momentum of the scattered photon?

Solution

First calculate the frequency f' of the scattered photon.

$$\left\{\begin{aligned} E_K &= 2.04 \times 10^{-17} \text{ J} \\ f &= 1.00 \times 10^{18} \text{ Hz} \\ h &= 6.63 \times 10^{-34} \text{ J} \end{aligned}\right\}$$

$$hf = hf' + E_K$$

$$f' = \frac{hf - E_K}{h} = \frac{(6.63 \times 10^{-34} \text{ J} \cdot \text{s})(1.00 \times 10^{18} \text{ Hz}) - 2.04 \times 10^{-17} \text{ J}}{6.63 \times 10^{-34} \text{ J} \cdot \text{s}}$$

$$= \boxed{9.69 \times 10^{17} \text{ Hz}}$$

$\{c = 3.00 \times 10^8 \text{ m/s}\}$ Then calculate the scattered photon's momentum p'.

$$\lambda' = \frac{c}{f'} = \frac{3.00 \times 10^8 \text{ m/s}}{9.69 \times 10^{17} \text{ Hz}} = 3.09 \times 10^{-10} \text{ m}$$

$$p' = \frac{h}{\lambda'} = \frac{6.63 \times 10^{-34} \text{ J} \cdot \text{s}}{3.09 \times 10^{-10} \text{ m}} = \boxed{2.15 \times 10^{-24} \text{ kg} \cdot \text{m/s}}$$

18-10 PARTICLES AS WAVES

To bring the reasoning full circle, a new proposal was made in 1924 by a French physicist Louis de Broglie. His idea was that, if waves act as particles, what we normally view as particles might well act as waves. He was talking not only about photons but also about electrons, footballs, dogs, cats, planets—anything that moves. De Broglie suggested that the momentum equation (18-4) $p = h/\lambda$ is perfectly valid for particles other than photons.

According to this idea, for example, a 0.15-kg baseball traveling at 40 m/s acts as a wave of wavelength λ that can be found using Eq. (18-14).

$$p = mv = \frac{h}{\lambda}$$

Hence

$$\lambda = \frac{h}{mv} = \frac{6.63 \times 10^{-34}\ \text{J}\cdot\text{s}}{(0.15\ \text{kg})(40\ \text{m/s})} = 1.1 \times 10^{-34}\ \text{m}$$

Wavelength of a fast ball

The result shows that we shouldn't feel surprised not having noticed baseballs doing a sinusoidal wiggle as they approach the plate. The wavelength is so fantastically small that we have no way of detecting it.

If we look at smaller particles, the wavelength ($\lambda = h/mv$) may get large enough to "see" in some way. The mass of an electron for example is 9.11×10^{-31} kg. A 10,000-V potential difference in a TV tube gives an electron a velocity of about 6.0×10^{7} m/s. Using these values,

$$\lambda = \frac{h}{mv} = \frac{6.63 \times 10^{-34}\ \text{J}\cdot\text{s}}{(9.11 \times 10^{-31}\ \text{kg})(6.0 \times 10^{7}\ \text{m/s})} = 1.2 \times 10^{-11}\ \text{m}$$

This is a wavelength on the order of that of X rays, and we have seen in Sec. 18-4 that such waves can be made to show interference effects. This has been done with electron waves, and in fact they have been put to an important practical use in the electron microscope. The size of the smallest particle that can be "seen" by a microscope depends upon the wavelength being used. The very short wavelengths of electrons allow electron microscopes to resolve objects smaller than 1 nm (10^{-9} m).

18-11 HOW CAN WE BELIEVE BOTH THE WAVE AND PARTICLE MODELS?

If you still feel uneasy about the wave–particle nature of things, you are not alone. Here in this final section are some assorted thoughts on the subject that may make you feel better about it.

It Is Partly a Matter of Scale

Our perception of things is greatly influenced by the limitations of our senses. We have already discussed why it is not obvious that large moving objects like baseballs are associated with waves. The wavelengths are too small and the frequencies too high for us to detect. Looking at it the other way, we might ask why the idea of photons has not always been obvious. The answer is simply that most photons are too small to be detected as individuals.

In any kind of radiation sensor, such as a television set, photocell, or an eye, energy must be received at a rate great enough to be distinguished from the background level of thermal agitation of the sensor's molecules. It is calculated that at about 20°C a perfect sensor, the best that can be imagined, must receive approximately 1×10^{-21} J of radiant energy in a time equivalent to the period of the radiation in order not to be masked by thermal noise. The frequency of a photon with 1×10^{-21} J of energy is

$$f = \frac{E}{h} = \frac{1 \times 10^{-21}\ \text{J}}{6.63 \times 10^{-34}\ \text{J} \cdot \text{s}} = 1.5 \times 10^{12}\ \text{Hz}$$

This is a frequency in the infrared region of the spectrum. Individual photons with frequencies above that of infrared waves may be "seen." For frequencies below that of infrared, it is not possible to see an individual photon.

To press the point further, consider a radio–TV transmission wave of frequency $f = 1.0 \times 10^6$ Hz. The energy of a photon is

$$E = hf = (6.63 \times 10^{-34}\ \text{J} \cdot \text{s})(1.0 \times 10^6\ \text{Hz}) = 6.6 \times 10^{-28}\ \text{J}$$

It will take $(1 \times 10^{-21}\ \text{J})/(6.6 \times 10^{-28}\ \text{J/photon}) = 1.5 \times 10^6$ photons (1.5 million photons) to barely be detected under the most favorable conditions. Small wonder that detectors see the flow of photons as a continuous process.

We have been calculating the energies of photons, but it is difficult to get a feeling for how small these energies are. The following example puts it in more understandable terms.

Example 18-12 A speck of dust of mass 1.0×10^{-5} g settles in still air at a velocity of 1.0 cm/s. How many photons of blue light with frequency of 7.0×10^{14} Hz are needed to supply energy equal to the kinetic energy of the dust speck?

Solution

First calculate the dust speck's kinetic energy.

$$\left\{\begin{aligned} m &= 1.0 \times 10^{-5}\ \text{g} \\ &= 1.0 \times 10^{-8}\ \text{kg} \\ v &= 1.0\ \text{cm/s} \\ &= 1.0 \times 10^{-2}\ \text{m/s} \end{aligned}\right\}$$

$$\begin{aligned} E_K &= \tfrac{1}{2}mv^2 = \tfrac{1}{2}(1.0 \times 10^{-8}\ \text{kg})(1.0 \times 10^{-2}\ \text{m/s})^2 \\ &= 5.0 \times 10^{-13}\ \text{J} \end{aligned}$$

Next calculate the energy of a single photon of blue light.

$$\left\{\begin{aligned} h &= 6.63 \times 10^{-34}\ \text{J} \cdot \text{s} \\ f &= 7.0 \times 10^{14}\ \text{Hz} \end{aligned}\right\}$$

$$E = hf = (6.63 \times 10^{-34}\ \text{J} \cdot \text{s})(7.0 \times 10^{14}\ \text{Hz}) = 4.6 \times 10^{-19}\ \text{J}$$

The number of photons needed is

$$\frac{5.0 \times 10^{-13}\ \text{J}}{4.6 \times 10^{-19}\ \text{J/photon}} = \boxed{1.1 \times 10^6\ \text{photons}}$$

← Approximately 1 million photons needed!

We can easily find what sort of radiation will supply photons with energy equal to that of the dust particle.

$$E = hf \Rightarrow f = \frac{E}{h} = \frac{5.0 \times 10^{-13}\ \text{J}}{6.63 \times 10^{-34}\ \text{J} \cdot \text{s}} = 7.5 \times 10^{20}\ \text{Hz}$$

Thus, according to Table 18-1, only gamma or cosmic rays will do the job. Again it is small wonder that individual photons often go unnoticed.

Do not go away with the idea that photons of energy and particles of matter are the same thing. Matter and energy are equivalent and quantitatively related by Eq. (3-13), $E = mc^2$. In fact, matter and energy are interchangeable. One fundamental difference, however, is that photons travel with the speed of light, whereas particles of matter can never travel that fast.

What Is Versus What Might Be

The idea that things coexist as both particles and waves may still seem strange, and so it is a good time to introduce a thought that might well have appeared in Chapter 1 instead of at the end of our text.

Throughout the study of physics we have built conceptual and mathematical models of things. It becomes more apparent in these last chapters that we cannot be sure that our models represent truth. Physicists have come around to the idea that we can only describe how things *seem* to be. Whether or not this is the way things really are, we do not know. The test of a good model is that it allows us to predict the results of an experiment. If the model does not do a good job of this, we change the model, for it seems unlikely that reality will change to accommodate our view of it.

18-12 SUMMARY

In this chapter we have seen that energy can get away from or enter a system in one additional way—as electromagnetic radiation.

1. The principle of conservation of energy has to be expanded to include energy lost by electromagnetic radiation.

$$\begin{pmatrix}\text{energy}\\ \text{added to}\\ \text{system}\end{pmatrix} = \begin{pmatrix}\text{energy}\\ \text{lost by}\\ \text{resistors}\end{pmatrix} + \begin{pmatrix}\text{potential}\\ \text{energy}\\ \text{stored}\end{pmatrix} + \begin{pmatrix}\text{kinetic}\\ \text{energy}\\ \text{stored}\end{pmatrix} + \begin{pmatrix}\text{energy}\\ \text{lost by}\\ \text{radiation}\end{pmatrix} \qquad (18\text{-}1)$$

2. Electromagnetic radiation is produced when electric charge accelerates. To sustain the acceleration, charge must be moved back and forth. Various kinds of electromagnetic radiation appear to differ only in the frequency of the source.

3. The speed of light in a vacuum is a universal constant accurately measured as $c = 3.00 \times 10^8$ m/s $= 1.86 \times 10^5$ mi/s. Light travels slower in any substance such as water, glass, or air. When light passes from one medium into another, the change in velocity makes the light change direction. Directions are related by the Snell's law equation

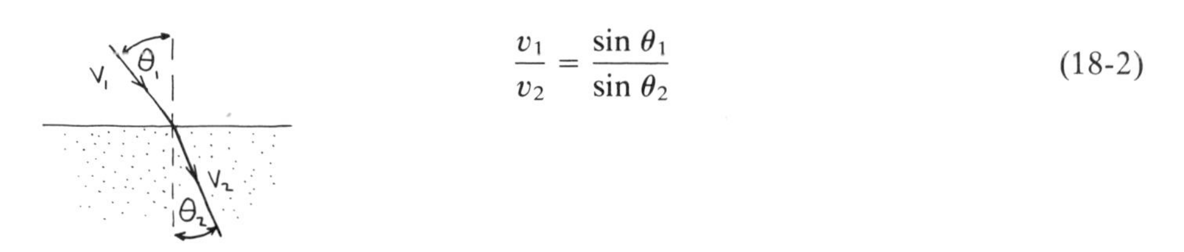

$$\frac{v_1}{v_2} = \frac{\sin \theta_1}{\sin \theta_2} \tag{18-2}$$

The refractive index n of a substance is defined as

$$n = \frac{c}{v} \tag{18-3}$$

Refractive index of substance → n; c ← Speed of light in vacuum; v ← Speed of light in substance

4. Electromagnetic waves reflect from boundaries between adjacent substances. The angle of reflection is the same as the angle of incidence. Reflected waves undergo phase reversal at "hard" surfaces such as air to glass, but not at "soft" surfaces such as glass to air. Refracted waves change their wavelength but undergo no phase reversal.

5. Reflection and refraction of waves are utilized in optical systems that employ mirrors and lenses. Images formed are either real or virtual, depending upon the object's position and the type of mirror (plane, concave, convex) or lens (converging, diverging) used.

6. Electromagnetic radiation behaves in many ways like a wave phenomenon. One of the strongest kinds of evidence is interference effects. X-ray interference effects are put to good use in identifying crystalline materials.

7. Transmission of energy as electromagnetic waves depends upon three facts:

(a) A changing magnetic field produces an electric field.
(b) A changing electric field produces a magnetic field.
(c) Electric and magnetic fields represent stored energy.

Energy is stored alternately by electric and magnetic fields that are perpendicular to each other. The fields propagate at the speed of light.

8. The speed of light in a substance can be calculated using an equation similar to those for wave speeds in Chapter 17.

$$v = \frac{1}{\sqrt{\mu\varepsilon}} \qquad (18\text{-}9)$$

Speed of light in substance → v; permittivity → ε; permeability → μ

The constants μ and ε relate to forces generated by magnetic and electric fields. Their values when substituted into Eq. (18-9) give a value for the speed of light in a substance that is in very close agreement with measured values.

9. The photoelectric effect is one of a number of phenomena that make it necessary to think of electromagnetic radiation as a stream of particles or bundles of energy called photons. Electromagnetic radiation knocks electrons from metal plates in a way that can't be explained by wave theory. The energy of a photon is directly proportional to the frequency of the radiation.

$$E = hf \qquad (18\text{-}10)$$

Quantum of energy → E; Planck's constant $= 6.63 \times 10^{-34}$ J · s → h; Frequency → f

Conservation of energy in the photoelectric effect produces the following relation:

$$hf = hf_0 + \tfrac{1}{2}mv^2 \qquad (18\text{-}12)$$

Photon's energy → hf; Threshold frequency → f_0; Surface work function → hf_0; Photoelectron kinetic energy → $\tfrac{1}{2}mv^2$

10. The Compton effect shows that photons act like mechanical particles. Photons collide with electrons and recoil much like billiard balls.

$$hf = hf' + \tfrac{1}{2}mv^2 \qquad (18\text{-}13)$$

Incident photon energy → hf; Scattered photon energy → hf'; Recoil electron kinetic energy → $\tfrac{1}{2}mv^2$

The momentum of a photon is given by

$$p = \frac{h}{\lambda} \qquad (18\text{-}14)$$

Momentum → p; Photon wavelength → λ

11. Particles of matter act as waves too. Their wavelength λ, using Eq. (18-14), depends upon their momentum $p = mv = h/\lambda$. The frequency of a particle like a baseball is too great to be detected. The frequency of smaller, faster-moving particles such as electrons is easier to detect. The wave nature of moving electrons is exploited in the electron microscope.

12. Our confusion about waves and particles is largely the result of not being able to detect both aspects at the same time. The energy of most photons is unimaginably small, and the frequency of most particles is very large. At best, our ideas of the physical universe are only conceptual or mathematical models of what things *seem* to be—not necessarily what they are.

PROBLEMS

18-1. The speed of light in carbon disulfide is 1.83×10^8 m/s. What is the refractive index of carbon disulfide?

18-2. What is the speed of light in paraffin? (See Appendix B, Table 9, for refractive index.)

18-3. Refraction produces a distorted idea of where things are in a pool of water. If an object appears to be on a line of sight 45° to the normal, what is the angle θ of the actual light path?

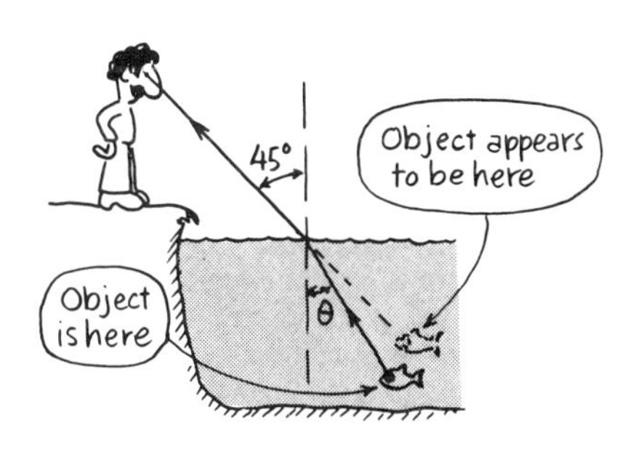

Problem 18-3

18-4. What is the ratio of the speed of light in borosilicate glass to the speed of light in diamond?

18-5. When light goes from a medium of low velocity to a medium of higher velocity, there is some angle θ_1 for which θ_2 is exactly 90° and no light gets out. θ_1 is then called the critical angle. Find the critical angle for borosilicate glass in air.

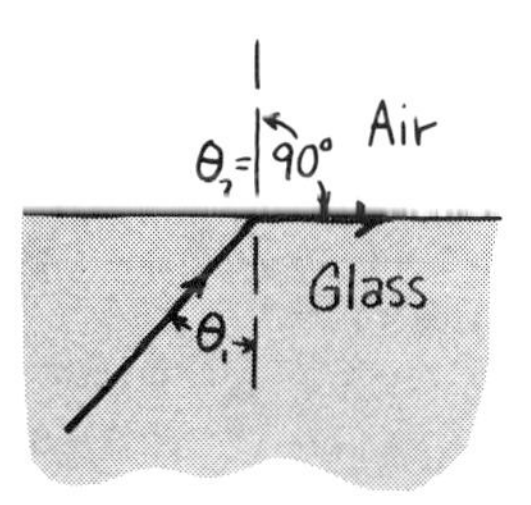

Problem 18-5

18-6. Green light of frequency 5.8×10^{14} Hz enters fresh water. What is its *wavelength* while traveling through the water?

18-7. A beam of X rays of wavelength 0.154 nm is directed at a sample of the mineral microcline. When the X rays are at an angle of 13.9° to the surface of the sample, constructive inteference is detected at the same angle. What is the distance between atomic layers of microcline?

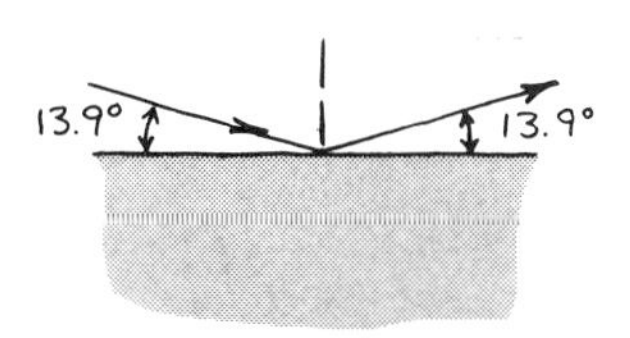

Problem 18-7

18-8. An unknown mineral sample is subjected to X rays of wavelength 2.29×10^{-10} m. Constructive interference is detected when the X-ray beam and detector are at angles of 21.5° to the sample surface. Calculate the interlayer distance and identify the mineral from the accompanying list.

Mineral	*Interlayer Distance (nm)*
Analcite	0.560
Gibbsite	0.483
Quartz	0.335
Talc	0.312

18-9. X rays of wavelength 0.194 nm are directed at the surface of a sample of quartz that has an interlayer distance of 3.35×10^{-10} m. At what angle to the surface would you expect to detect constructive interference?

18-10. What is the energy of a photon of the X rays in Problem 18-7?

18-11. If the concave shaving mirror of Example 18-6 ($f = 0.27$ m) is held 0.76 m from your eye, it will produce a recognizable image of your eye.

(a) Where is the image and what kind is it?

(b) If the pupil of your eye is 3.0 mm in diameter, what is the diameter of its image? Is it upright or inverted?

18-12. Suppose that the mirror of Problem 18-11 is convex instead of concave. The radius of curvature is the same (2×0.27 m) and the eye is 0.76 m from the mirror.

(a) Where is the image of your eye and what kind is it?

(b) If the pupil of your eye is 3.0 mm in diameter, what is the diameter of its image?

18-13. A hand-held magnifying glass used for examining mineral specimens is rated as "10 power." This means that the ratio of image size to object size h_2/h_1 (the magnification) has a value of 10. A mineral sample 0.16 cm long is held 2.0 cm from the lens on the side away from the eye. The image seen is enlarged, erect, and virtual.

(a) How long is the image and where is it?

(b) What is the focal length of the lens?

18-14. A slide projector produces an image 72 in. wide on a screen 23 ft from the lens. The object is a color slide 35 mm wide.

(a) How far is the slide from the projection lens? (*Hint:* The slide is always placed in the projector upside down. h_1 is negative.)

(b) What is the focal length of the projection lens?

18-15. What is the energy of a photon of the X rays in Problem 18-8?

18-16. Cosmic rays with frequencies up to 10^{23} Hz are the highest-frequency electromagnetic radiation yet detected. What is the ratio of the energy of these photons to the energy of the settling dust speck in Example 18-12?

18-17. How many quanta of radiation of 16-m wavelength (short-wave radio) would have the total energy of a single ultraviolet quantum of wavelength 1.54×10^{-8} m?

18-18. The photoelectric work function of carbon is 7.7×10^{-19} J. What threshold frequency of electromagnetic radiation is necessary to dislodge an electron? Where is this in the electromagnetic spectrum?

18-19. What maximum kinetic energy will a photoelectron have that is knocked from a carbon surface by the X rays of Problem 18-8? The photoelectric work function of carbon is 7.7×10^{-19} J.

18-20. Ultraviolet light of wavelength 2.0×10^{-7} m will produce photoelectrons with kinetic energies as high as 2.4×10^{-19} J from a given surface.

(a) What is the work function of the surface?

(b) What is the threshold frequency of the surface?

18-21. The yellow-orange light from a sodium flame has a wavelength of 589 nm. Will such light eject photoelectrons from a sodium surface for which the work function is 3.7×10^{-19} J?

18-22. What is the frequency of an infrared photon whose momentum is 5.1×10^{-30} kg · m/s?

18-23. An X-ray photon of wavelength 0.154 nm strikes an electron and bounces off with wavelength of 0.172 nm. What kinetic energy did the electron gain?

18-24. A photon of frequency 2.0×10^{17} Hz strikes an electron and bounces off with frequency of 1.8×10^{17} Hz. The velocity of the recoil electron is 5.4×10^{6} m/s. Calculate the mass of an electron based on this information.

18-25. A flashlight uses 9.0 W of power. Suppose that 2 percent of this power is radiated as visible light, and let us assume that it is monochromatic (one color) of wavelength 600 nm.

(a) How many photons per second does the flashlight emit?

(b) What is the momentum of one of these photons?

(c) The time rate of change of momentum $\Delta p/\Delta t$ turns out to be *force*. How much force is exerted on a surface illuminated by the flashlight?

18-26. What is the de Broglie wavelength associated with the moving dust speck of Example 18-8?

18-27. What is the de Broglie wavelength of the photoelectron of Problem 18-20? (*Hint:* A photoelectron is *not* a photon.)

ANSWERS TO ODD-NUMBERED PROBLEMS

18-1. 1.64

18-3. 32°

18-5. 41.1°

18-7. 0.321 nm

18-9. 16.8°

18-11. (a) Real image 0.42 m in front of the mirror
(b) 1.7 mm, inverted

18-13. (a) $h_2 = 1.6$ cm; $s_2 = -20$ cm (20 cm behind mirror)
(b) +2.2 cm

18-15. 8.69×10^{-16} J

18-17. 1.0×10^{9} quanta

18-19. 7.9×10^{-18} J

18-21. No, because threshold frequency (5.6×10^{14} Hz) is greater than frequency of sodium light (5.1×10^{14} Hz)

18-23. 1.4×10^{-16} J

18-25. (a) 5.4×10^{17} photons/s
(b) 1.1×10^{-27} kg · m/s
(c) 6.0×10^{-10} N

18-27. 1.0×10^{-9} m

Reference Tables

TABLE A-1 PHYSICAL QUANTITIES AND THEIR SYMBOLS

Symbol	Quantity
A	area
a	acceleration
B	bulk modulus (hydrostatic)
B	magnetic field strength
C	capacitance
c	specific heat
c	speed of light in a vacuum
d	diameter
d	thickness
E	electric field strength
E	electromotive force
E	energy
E_K	kinetic energy
E_P	potential energy
E_R	energy lost through resistor
e	electron charge
F	force
F_C	capacitive force
F_E	excitation force
F_M	inertive force = net force F_{net}
F_m	magnetomotive force (mmf)
F_R	resistive force
f	focal length
f	frequency
f	friction force
f_0	photoelectric threshold frequency
f_r	resonant frequency
f_k	kinetic friction force
f_s	static friction force
G	shear modulus
g	acceleration of gravity
H	magnetic intensity
h	height or depth
h	Planck's constant
I	electric current
$\dot{I}$	rate of change of electric current
I	moment of inertia
K	dielectric constant
k	coefficient of thermal conductivity
k	spring constant
k_r	rotational spring constant
L	inductance
L_f	latent heat of fusion
L_v	latent heat of vaporization
ℓ	length
M	inertance
m	mass
N	normal force
N_R	Reynolds number
n	refractive index
P	power
p	pressure
p_0	atmospheric pressure
p_g	gauge pressure
p	linear momentum
Q	parameter
$\dot{Q}$	parameter rate
$\ddot{Q}$	rate of change of parameter rate
Q	quantity of charge
Q	quantity of heat
Q	quality factor
R	resistance
R_{cr}	critical resistance
R_m	reluctance (magnetic)
r	radius
s	distance
T	period
T	temperature
T_C	Celsius temperature
T_F	Fahrenheit temperature
T_K	kelvin temperature
T_R	Rankine temperature
t	time
t_d	delay time in a transmission line
V	volume
$\dot{V}$	volume flow rate
$\ddot{V}$	rate of change of volume flow rate
V	electrical potential difference
v	velocity
W	work
W	weight
W	photoelectric work function
X_C	capacitive reactance
X_M	inertive reactance
Y	Young's modulus
Z	impedance
α (alpha)	angular acceleration
ε (epsilon)	engineering strain
ε (epsilon)	electrical permittivity
ε_0 (epsilon)	permittivity of a vacuum
η (eta)	efficiency
η (eta)	viscosity
θ (theta)	angle
λ (lambda)	wavelength
μ (mu)	magnetic permeability
μ_0 (mu)	permeability of a vacuum

TABLE A-1—*continued*

μ (mu)	coefficient of friction	τ_{RC} (tau)	time constant of *RC* system
ρ (rho)	density	τ_{RM} (tau)	time constant of *RM* system
ρ (rho)	coefficient of reflection	τ_{RMC} (tau)	time constant of *RMC* system
σ (sigma)	engineering stress	ϕ (phi)	magnetic flux
σ (sigma)	electrical conductivity	ϕ (phi)	phase angle
τ (tau)	torque	ω (omega)	angular velocity

TABLE A-2 UNITS OF MEASUREMENT AND THEIR SYMBOLS

A	ampere (electric current)
A	ampere turn (magnetomotive force)
atm	atmosphere (pressure)
Btu	British thermal unit (heat)
°C	degrees Celsius (temperature)
C	coulomb (electric charge)
cal	calorie (heat)
dyn	dyne (force)
°F	degrees Fahrenheit (temperature)
F	farad (electrical capacitance)
ft	foot (length)
ft · lb	foot · pound (work, energy)
g	gram (mass)
gal	gallon (volume)
gpm	gallons per minute (volume flow rate)
H	henry (electrical inductance)
hr	hour (time)
hp	horsepower (power)
Hz	hertz (frequency)
in.	inch (length)
in. · lb	inch · pound (work, energy)
J	joule (work, energy)
K	kelvin (temperature)
kg	kilogram (mass)
ℓ	litre (volume)
lb	pound (force)
lb · ft	pound · foot (torque)
m	metre (length)
mi	mile (length)
min	minute (time)
N	newton (force)
N · m	newton · metre (torque, energy)
oz	ounce (force)
oz · in.	ounce · inch (torque)
Pa	pascal (pressure)
psi	pounds per square inch (pressure)
°R	degrees Rankine (temperature)
rad	radian (angle)
rms	root mean square

TABLE A-2—*continued*

rpm	revolutions per minute (angular velocity)
s	second (time)
slug	slug (mass)
T	tesla (magnetic field strength)
torr	torr (pressure)
V	volt (potential difference)
Wb	weber (magnetic flux)
W	watt (power)
yd	yard (length)
°	degree (angle)
Ω	ohm (electrical resistance)

TABLE A-3 DECIMAL PREFIXES

Power	*Prefix*	*Symbol*
10^{12}	tera-	T
10^{9}	giga-	G
10^{6}	mega-	M
10^{3}	kilo-	k
10^{-2}	centi-	c
10^{-3}	milli-	m
10^{-6}	micro-	μ(mu)
10^{-9}	nano-	n
10^{-12}	pico-	p

TABLE A-4 SYMBOLS FOR PERIODIC VALUES

Instantaneous	y (lowercase)
rms (root mean square)	Y (uppercase)
Maximum or peak value (amplitude)	Y_m
Average	Y_{av}

TABLE A-5 UNIFIED CONCEPTS

Unified Concept	Mechanical: Translational	Units	Mechanical: Rotational	Units	Fluid	Units	Electrical	Units	Thermal	Units
Force F	Force F	N lb	Torque τ	N · m lb · ft	Pressure p	Pa lb/ft^2	Emf E	V	Temperature T	°C °F
Parameter Q	Distance s	m ft	Angle θ	rad	Volume V	m^3 ft^3	Charge Q	C	Quantity of heat Q	cal Btu
Parameter rate $\dot{Q}$	Velocity $v = \Delta s/\Delta t$	m/s ft/s	Angular velocity $\omega = \Delta\theta/\Delta t$	rad/s	Volume flow rate $\dot{V} = \Delta V/\Delta t$	m^3/s ft^3/s	Electric current $I = \Delta Q/\Delta t$	A	Heat rate $\dot{Q} = \Delta Q/\Delta t$	cal/s Btu/s
Resistance R (Dissipation of energy)	Linear frictional resistance $R = F_R/v$	$\frac{\text{N}}{\text{m/s}}$ $\frac{\text{lb}}{\text{ft/s}}$	Rotational frictional resistance $R = \tau_R/\omega$	$\frac{\text{N} \cdot \text{m}}{\text{rad/s}}$ $\frac{\text{lb} \cdot \text{ft}}{\text{rad/s}}$	Fluid frictional resistance $R = p_R/\dot{V}$	$\frac{\text{Pa}}{\text{m}^3/\text{s}}$ $\frac{\text{lb/ft}^2}{\text{ft}^3/\text{s}}$	Electrical resistance $R = V_R/I$	Ω	Thermal resistance $R = \frac{T_2 - T_1}{\dot{Q}}$	$\frac{°\text{C}}{\text{cal/s}}$ $\frac{°\text{F}}{\text{Btu/s}}$
Capacitance C (Potential energy storage)	Linear spring capacitance $C = s/F_C$	m/N ft/lb	Rotational spring capacitance $C = \theta/\tau_C$	$\frac{\text{rad}}{\text{N} \cdot \text{m}}$ $\frac{\text{rad}}{\text{lb} \cdot \text{ft}}$	Fluid reservoir capacitance $C = V/p_C$	$\frac{\text{m}^3}{\text{Pa}}$ $\frac{\text{ft}^3}{\text{lb/ft}^2}$	Electrical capacitor capacitance $C = Q/V_C$	F	Thermal heat capacity $C = \frac{Q}{T_f - T_o}$	cal/°C Btu/°F
Inertance M (Kinetic energy storage)	Mass $m = F_{net}/a$	kg slug	Moment of inertia $I = \tau_{net}/\alpha$	kg · m^2 slug · ft^2	Fluid inertance $M = p_{net}/\ddot{V}$	$\frac{\text{Pa}}{\text{m}^3/\text{s}^2}$ $\frac{\text{lb/ft}^2}{\text{ft}^3/\text{s}^2}$	Coil inductance $L = V_L/\dot{I}$	H	—	—
Time constant (Resistance and capacitance)	←				$\tau_{RC} = RC$ (s)					→
Time constant (Resistance and inertance)	←				$\tau_{RM} = M/R$ (s)			→	—	—
Resonant frequency (Capacitance and inertance)	←				$f_r = \frac{1}{2\pi\sqrt{MC}}$ (Hz)			→	—	—

TABLE A-6 USEFUL CONVERSION FACTORS*

Length	1 metre (m)	= 39.4 inch (in.)
	1 metre (m)	= 3.28 foot (ft)
	1 inch (in.)	= 2.54 centimetre (cm)
	1 mile (mi)	= 5280 foot (ft)
	1 kilometre (km)	= 0.621 mile (mi)
Angle	1 radian (rad)	= 57.3 degree (°)
	1 revolution (rev)	= 6.28 radian (rad)
Area	1 square metre (m^2)	= 10.8 square foot (ft^2)
	1 square centimetre (cm^2)	= 0.155 square inch ($in.^2$)
Volume	1 cubic metre (m^3)	= 35.3 cubic foot (ft^3)
	1 litre (ℓ)	= 1000 cubic centimetre (cm^3)
	1 cubic foot (ft^3)	= 1728 cubic inch ($in.^3$)
	1 gallon (gal)	= 0.134 cubic foot (ft^3)
	1 gallon (gal)	= 3.79×10^{-3} cubic metre (m^3)
Charge	1 coulomb (C)	= charge of 6.25×10^{18} electrons
Force	1 newton (N)	= 0.225 pound (lb)
	1 newton (N)	= 1.0×10^5 dyne (dyn)
	1 pound (lb)	= 16 ounce (oz)
Mass†	1 kilogram (kg)	= 0.0685 slug (slug)
Pressure	1 atmosphere (atm)	= 1.013×10^5 pascal (Pa)
	1 atmosphere (atm)	= 14.7 pound/square inch (psi)
	1 atmosphere (atm)	= 2.12×10^3 pound/square foot (lb/ft^2)
	1 atmosphere (atm)	= 407 inch of water
	1 atmosphere (atm)	= 760 torr (torr)
Temperature	1 degree Celsius (°C)	= 1.8 degree Fahrenheit (°F)
Energy	1 joule (J)	= 0.738 foot · pound (ft · lb)
	1 joule (J)	= 0.239 calorie (cal)
	1 joule (J)	= 9.49×10^{-4} British thermal unit (Btu)
	1 British thermal unit (Btu)	= 252 calorie (cal)
	1 British thermal unit (Btu)	= 778 foot · pound (ft · lb)
Velocity	60 mile/hour (mi/hr)	= 88 foot/second (ft/s)
	1 kilometre/hour (km/hr)	= 0.621 mile/hour (mi/hr)
	1 metre/second (m/s)	= 2.24 mile/hour (mi/hr)
Angular velocity	1 revolution/minute (rpm)	= 0.1047 radian/second (rad/s)
Volume flow rate	1 gallon/minute (gpm)	= 2.24×10^{-3} cubic foot/second (ft^3/s)
Power	1 watt (W)	= 0.738 foot · pound/second (ft · lb/s)
	1 horsepower (hp)	= 550 foot · pound/second (ft · lb/s)
	1 horsepower (hp)	= 746 watt (W)
	1 watt (W)	= 3.41 British thermal unit/hour (Btu/hr)

*The symbol for each unit is in parenthesis

†*Note*: The *weight* of 1 kilogram is 2.21 pound. The *weight* of 1 slug is 32.2 pound.

TABLE A-7 PHYSICAL CONSTANTS

Quantity	*Symbol*	*Value*
Absolute zero	0 K 0°R	−273°C −460°F
Acceleration of gravity	g	9.81 m/s^2 32.2 ft/s^2
Permeability of a vacuum	μ_0	12.57×10^{-7} T · m/A
Permittivity of a vacuum	ε_0	8.85×10^{-12} C · m/(V · m^2)
Planck's Constant	h	6.63×10^{-34} J · s
Velocity of light in a vacuum	c	3.00×10^8 m/s 1.86×10^5 mi/s
Electron charge	e	1.60×10^{-19} C

Table A-8 TRIGONOMETRIC FUNCTIONS

Angle					Angle				
Degree	Radian	Sine	Cosine	Tangent	Degree	Radian	Sine	Cosine	Tangent
0°	.000	0.000	1.000	0.000					
1°	.018	.018	1.000	.018	46°	0.803	0.719	0.695	1.036
2°	.035	.035	0.999	.035	47°	.820	.731	.682	1.072
3°	.052	.052	.999	.052	48°	.838	.743	.669	1.111
4°	.070	.070	.998	.070	49°	.855	.755	.656	1.150
5°	.087	.087	.996	.088	50°	.873	.766	.643	1.192
6°	.105	.105	.995	.105	51°	.890	.777	.629	1.235
7°	.122	.122	.993	.123	52°	.908	.788	.616	1.280
8°	.140	.139	.990	.141	53°	.925	.799	.602	1.327
9°	.157	.156	.988	.158	54°	.942	.809	.588	1.376
10°	.175	.174	.985	.176	55°	.960	.819	.574	1.428
11°	.192	.191	.982	.194	56°	.977	.829	.559	1.483
12°	.209	.208	.978	.213	57°	.995	.839	.545	1.540
13°	.227	.225	.974	.231	58°	1.012	.848	.530	1.600
14°	.244	.242	.970	.249	59°	1.030	.857	.515	1.664
15°	.262	.259	.966	.268	60°	1.047	.866	.500	1.732
16°	.279	.276	.961	.287	61°	1.065	.875	.485	1.804
17°	.297	.292	.956	.306	62°	1.082	.883	.470	1.881
18°	.314	.309	.951	.325	63°	1.100	.891	.454	1.963
19°	.332	.326	.946	.344	64°	1.117	.899	.438	2.050
20°	.349	.342	.940	.364	65°	1.134	.906	.423	2.145
21°	.367	.358	.934	.384	66°	1.152	.914	.407	2.246
22°	.384	.375	.927	.404	67°	1.169	.921	.391	2.356
23°	.401	.391	.921	.425	68°	1.187	.927	.375	2.475
24°	.419	.407	.914	.445	69°	1.204	.934	.358	2.605
25°	.436	.423	.906	.466	70°	1.222	.940	.342	2.747
26°	.454	.438	.899	.488	71°	1.239	.946	.326	2.904
27°	.471	.454	.891	.510	72°	1.257	.951	.309	3.078
28°	.489	.470	.883	.532	73°	1.274	.956	.292	3.271
29°	.506	.485	.875	.554	74°	1.292	.961	.276	3.487
30°	.524	.500	.866	.577	75°	1.309	.966	.259	3.732
31°	.541	.515	.857	.601	76°	1.326	.970	.242	4.011
32°	.559	.530	.848	.625	77°	1.344	.974	.225	4.331
33°	.576	.545	.839	.649	78°	1.361	.978	.208	4.705
34°	.593	.559	.829	.675	79°	1.379	.982	.191	5.145
35°	.611	.574	.819	.700	80°	1.396	.985	.174	5.671
36°	.628	.588	.890	.727	81°	1.414	.988	.156	6.314
37°	.646	.602	.799	.754	82°	1.431	.990	.139	7.115
38°	.663	.616	.788	.781	83°	1.449	.993	.122	8.144
39°	.681	.629	.777	.810	84°	1.466	.995	.105	9.514
40°	.698	.643	.766	.839	85°	1.484	.996	.087	11.43
41°	.716	.656	.755	.869	86°	1.501	.998	.070	14.30
42°	.733	.669	.743	.900	87°	1.518	.999	.052	19.08
43°	.751	.682	.731	.933	88°	1.536	.999	.035	28.64
44°	.768	.695	.719	.966	89°	1.553	1.000	.018	57.29
45°	.785	.707	.707	1.000	90°	1.571	1.000	.000	∞

Tables of Technical Data

B-1 DENSITIES OF VARIOUS SUBSTANCES
B-2 HEATS OF FUSION (L_f) AND VAPORIZATION (L_v) FOR VARIOUS SUBSTANCES
B-3 SPECIFIC HEATS OF VARIOUS SUBSTANCES AT ORDINARY TEMPERATURES
B-4 COEFFICIENT OF THERMAL CONDUCTIVITY FOR VARIOUS SUBSTANCES
B-5 RESISTIVITY OF VARIOUS SUBSTANCES AT 20°C
B-6 MOMENTS OF INERTIA OF VARIOUS BODIES
B-7 ELASTIC MODULI OF VARIOUS SUBSTANCES
B-8 DIELECTRIC CONSTANT OF VARIOUS SUBSTANCES
B-9 REFRACTIVE INDEX OF VARIOUS SUBSTANCES

TABLE B-1 DENSITIES OF VARIOUS SUBSTANCES

Material	$\rho = \frac{m}{V}$	$\rho = \frac{m}{V}$	$\rho g = \frac{W}{V}$
Solids			
Aluminum	2.70	2.70×10^3	168
Iron (steel)	7.8	7.8×10^3	490
Brass	8.7	8.7×10^3	540
Copper	8.89	8.89×10^3	555
Lead	11.3	11.3×10^3	705
Pine	0.42	0.42×10^3	26
Oak	0.82	0.82×10^3	52
Loose earth	1.2	1.2×10^3	76
Common brick	1.9	1.9×10^3	120
Concrete	2.3	2.3×10^3	145
Ice	0.92	0.92×10^3	57
Liquids			
Mercury	13.6	13.6×10^3	849
Gasoline	0.68	0.68×10^3	42
Ethyl alcohol	0.79	0.79×10^3	49
Lubricating oil	0.90	0.90×10^3	56
Pure water	1.00	1.00×10^3	62.4
Seawater	1.03	1.03×10^3	64.3
Gases at 1 atm of pressure and 0°C			
Hydrogen	0.90×10^{-4}	0.090	5.6×10^{-3}
Air	1.29×10^{-3}	1.29	80.5×10^{-3}
Oxygen	1.43×10^{-3}	1.43	89.2×10^{-3}
Carbon dioxide	1.96×10^{-3}	1.96	122×10^{-3}
Propane	2.02×10^{-3}	2.02	126×10^{-3}

TABLE B-2 HEATS OF FUSION (L_f) AND VAPORIZATION (L_v) FOR VARIOUS SUBSTANCES

Material	Fusion: Liquid ↔ Solid		Vaporization: Liquid ↔ Gas	
	$Q = mL_f$ (Q in Btu, m in lb, L_f in Btu/lb)	$Q = mL_f$ (Q in cal, m in g, L_f in cal/g)	$Q = mL_v$ (Q in Btu, m in lb, L_v in Btu/lb)	$Q = mL_v$ (Q in cal, m in g, L_v in cal/g)
Oxygen	5.9	3.3	92	51
Nitrogen	11.0	6.1	87	48
Mercury	5.1	2.8	128	71
Ethyl alcohol	45	25	367	204
Water	144	80	972	540
Zinc	43	24	855	475

TABLE B-3 SPECIFIC HEATS OF VARIOUS SUBSTANCES AT ORDINARY TEMPERATURES*

Material	$Q = mc(T_f - T_o)$ (Q in cal, m in g, T in °C, c in cal/(g · °C))	$Q = mc(T_f - T_o)$ (Q in Btu, m in lb, T in °F, c in Btu/(lb · °F))
Aluminum	0.217	0.217
Brass	0.094	0.094
Copper	0.093	0.093
Glass	0.199	0.199
Ice	0.55	0.55
Iron (steel)	0.113	0.113
Lead	0.031	0.031
Mercury	0.033	0.033
Silver	0.056	0.056
Water (liquid)	1.000	1.000

*The specific heats of liquids and gases are not constant, but temperature dependent.

TABLE B-4 COEFFICIENT OF THERMAL CONDUCTIVITY FOR VARIOUS SUBSTANCES

Material	$\dot{Q} = kA\dfrac{(T_2 - T_1)}{\ell}$ ($\dot{Q}$: cal/s; A: cm²; T: °C; ℓ: cm) k: $cal \cdot cm/(s \cdot cm^2 \cdot °C)$	$\dot{Q} = kA\dfrac{(T_2 - T_1)}{\ell}$ ($\dot{Q}$: Btu/hr; A: ft²; T: °F; ℓ: in.) k: $Btu \cdot in./(hr \cdot ft^2 \cdot °F)$
Metals		
Aluminum	0.49	1.4×10^3
Brass	0.26	0.75×10^3
Copper	0.92	2.7×10^3
Lead	0.083	0.24×10^3
Silver	0.97	2.8×10^3
Iron (steel)	0.12	0.35×10^3
Zinc	0.26	0.75×10^3
Other solids		
Fire brick	2.5×10^{-3}	7.3
Common brick	1.5×10^{-3}	4.3
Concrete	2.0×10^{-3}	5.8
Glass	2.0×10^{-3}	5.8
Ice	4.0×10^{-3}	12
Rock wool	0.10×10^{-3}	0.29
Various woods	$(0.10 \text{ to } 0.30) \times 10^{-3}$	3.0 to 9.0
Gases		
Air	0.57×10^{-4}	0.17
Helium	3.4×10^{-4}	0.99
Hydrogen	3.3×10^{-4}	0.96
Oxygen	0.56×10^{-4}	0.16

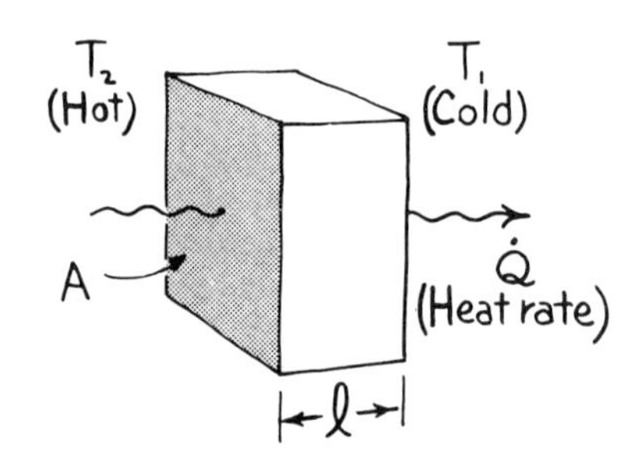

TABLE B-5 RESISTIVITY OF VARIOUS SUBSTANCES AT 20°C

Material (A, ℓ)	$R = \rho \frac{\ell}{A}$ (R: Ω; ℓ: m; A: m²) $\Omega \cdot m^2/m$	$R = \rho \frac{\ell}{A}$ (R: Ω; ℓ: ft; A: Circular mil) $\Omega \cdot$ circular mil/ft
Conductors		
Aluminum	2.6×10^{-8}	16
Copper	1.7×10^{-8}	10.3
Iron (steel)	12×10^{-8}	72
Lead	21×10^{-8}	120
Silver	1.6×10^{-8}	9.6
Semiconductors		
Carbon	3.5×10^{-5}	2.1×10^{4}
Germanium	0.5	3×10^{8}
Insulators		
Plate glass	$\sim 10^{12}$	$\sim 10^{20}$
Rubber	$\sim 10^{12}$	$\sim 10^{20}$

TABLE B-6 MOMENTS OF INERTIA OF VARIOUS BODIES EACH OF MASS m ABOUT THE INDICATED AXIS

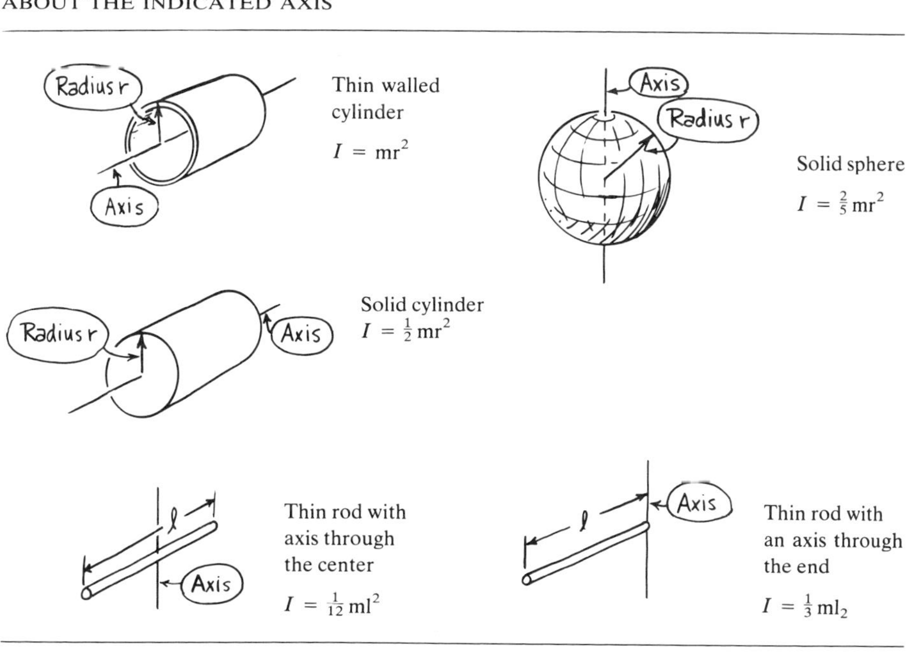

TABLE B-7 ELASTIC MODULI OF VARIOUS SUBSTANCES

Material	Young's Modulus $Y = \frac{F/A_0}{\Delta\ell/\ell_0}$		Shear Modulus $G = \frac{F/A_0}{\tan\theta}$		Bulk Modulus $B = \frac{p}{\Delta V/V_0}$	
	$lb/in.^2$	N/m^2	$lb/in.^2$	N/m^2	$lb/in.^2$	N/m^2
Aluminum	10×10^6	6.9×10^{10}	3.4×10^6	2.3×10^{10}	10×10^6	6.9×10^{10}
Brass	13×10^6	9.0×10^{10}	5.1×10^6	3.6×10^{10}	8.5×10^6	5.9×10^{10}
Copper	16×10^6	11×10^{10}	6.0×10^6	4.1×10^{10}	20×10^6	14×10^{10}
Glass	7.8×10^6	5.4×10^{10}	3.3×10^6	2.3×10^{10}	5.2×10^6	3.6×10^{10}
Iron, cast	13×10^6	9.0×10^{10}	10×10^6	6.9×10^{10}	14×10^6	9.7×10^{10}
Lead	2.3×10^6	1.6×10^{10}	0.80×10^6	0.55×10^{10}	1.1×10^6	0.76×10^{10}
Steel	30×10^6	21×10^{10}	12×10^6	8.3×10^{10}	23×10^6	16×10^{10}
Alcohol, ethyl					1.3×10^5	0.90×10^9
Kerosene					1.9×10^5	1.3×10^9
Mercury					38×10^5	26×10^9
Oil, lubricating					2.5×10^5	1.7×10^9
Water					3.3×10^5	2.3×10^9

TABLE B-8 DIELECTRIC CONSTANT OF VARIOUS SUBSTANCES

$$C = 8.85 \times 10^{-12} K \frac{A}{s}$$

(C: farad; K: dimensionless; A: m^2; s: m)

Material	K
Vacuum	1
Air	1.0006
Glass	5–10
Mica	3–7
Paraffin	2.2
Rubber, neoprene	6.7
Sulfur	4.0

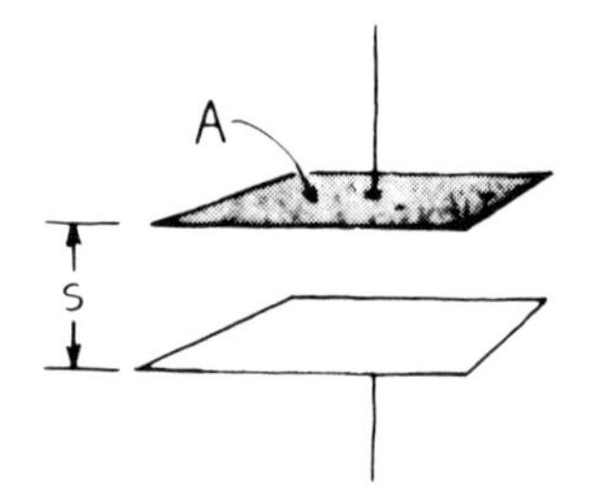

TABLE B-9 REFRACTIVE INDEX OF VARIOUS SUBSTANCES

Material	n
Vacuum	1.0000
Air	1.0003
Water (liquid)	1.33
Ice (solid water)	1.31
Glass	
Borosilicate (Pyrex)	1.52
Flint (soft)	1.6–1.7
Paraffin	1.43
Rock salt	1.53
Diamond	2.42
Fused quartz	1.46
Quartz (crystalline)	1.54
Obsidian (volcanic glass)	1.49

Index

A

D

E

F

M

N

O

X

Y

Z